COLLINS GUIDE TO THE

Pests, Diseases and Disorders

OF GARDEN PLANTS

Stefan T. Buczacki
and
Keith M. Harris

Illustrated by
Brian Hargreaves

COLLINS
St James's Place, London

William Collins Sons & Co Ltd
London · Glasgow · Sydney · Auckland
Toronto · Johannesburg

First published 1981
Reprinted 1982
© S. T. Buczacki, K. M. Harris and B. Hargreaves 1981
ISBN 0 00 219103 2
Filmset by Jolly & Barber Ltd, Rugby
Colour reproduction by Adroit Photo-Litho Ltd, Birmingham
Made and printed in Great Britain by
William Collins Sons and Co Ltd, Glasgow

Contents

CONTENTS

Colour Plates

Acknowledgements

Although we are personally familiar with almost every condition that we describe in this book, we would not claim to be authorities on more than a few of them. Because of this, and because much of the factual information we have required is inaccessible or unpublished, we have had to depend on the advice and opinions of numerous friends and professional colleagues. It is impossible to mention individually all those who have read portions of our script, discussed specific items with us, directed us to obscure literature, provided raw material for the illustrations or helped in other ways. We therefore thank collectively the employees of the Agricultural Research Council, Natural Environment Research Council, British Museum (Natural History), Ministry of Agriculture, Commonwealth Institute of Entomology, Commonwealth Mycological Institute, Royal Horticultural Society, Forestry Commission and of many universities, colleges and agrochemical manufacturers for their time, trouble and patience with our enquiries. While we have welcomed their assistance however, we dissociate them from any errors of fact or advice that may remain and which are entirely our responsibility. We have also relied heavily on the official publications of the above organisations and many of these are mentioned in the bibliography.

We owe a very special debt of gratitude to our artist Brian Hargreaves who has brought his impressive talents to bear upon an astonishing variety of subject matter presented to him in an equally astonishing variety of ways. We hope that his plates will not only aid significantly in identifying problems but also demonstrate the singular aesthetic appeal that can be shown by such ostensibly morbid subjects. To our wives Beverley and Elizabeth and to our families we are of course especially indebted for their unfailing patience and support and for their tolerance not only in finding their lives shared with dead and dying plants but also in making time to tend the gardens that we have neglected!

Preface

This book is a practical guide to the pests, diseases and disorders that commonly affect fruit, vegetable and ornamental plants growing in gardens in Britain and northern Europe. At first sight it may seem a depressingly morbid catalogue of plant failures but we believe that a good understanding of the many different causes of such failures provides the best basis for successfully avoiding them or minimising their effects in the future. Our main aim therefore is to improve plant health and ensure that gardeners derive maximum enjoyment from the plants that they cultivate. It is inevitable that all gardeners will encounter some pests, diseases and disorders sooner or later but we hope that no single gardener will ever have the misfortune of seeing all the afflictions recorded here!

This is a guide book rather than an encyclopaedia, since the subject matter is too vast to be dealt with fully in the space available, but it is in fact more comprehensive than any previous book of this type. Although it is intended primarily for gardeners, it may therefore also be of use to professional horticulturists and to teachers, students and naturalists. Most of the information on pests, diseases and disorders that is readily available to gardeners consists of either a few pages at the end of general gardening books and articles or brief notes in the promotional literature of manufacturers and suppliers of gardening chemicals. Information in the former is often imprecise and in the latter tends to be biased towards the use of chemicals in general and towards a restricted range of active ingredients. We have tried to give accurate and concise information on the symptoms, biology and control of a balanced selection of pests, diseases and disorders. In making this selection we have used published and unpublished records and the advice of many colleagues but we have also been greatly influenced by our own professional experience in Britain. British gardens are therefore our main concern but most of the included pests, diseases and disorders occur throughout northern Europe. We have not attempted to cover southern Europe since the warmer climate there produces quite different conditions, favouring other pests and diseases that are beyond the scope of this book.

How to Use this Book

This book has been specially planned so that gardeners with little or no technical knowledge can quickly identify and treat the pests, diseases and disorders that commonly affect their plants. Problems are therefore approached from the point of view of a gardener dealing with a particular plant or group of plants.

IDENTIFICATION

Correct identification of the cause of trouble is the first and most important step. Careful consideration of the symptoms and circumstances will usually result in a positive diagnosis and, since different plants are affected by different ranges of pests, diseases and disorders, the quickest way of achieving this is to use the plant as a starting point. The first main section of the book (p. 43) therefore consists of an alphabetical listing (A–Z) of the main garden plants that are commonly affected. Under each of these plant entries we indicate the most likely causes of trouble and the reader is then referred from them to the detailed entries in the sections on pests, diseases and disorders, where descriptions of the symptoms can be used to confirm or reject the preliminary diagnosis. Once this diagnosis has been confirmed, information on biology and treatment can be used to determine the best course of action. For example, if rolled leaves are seen on roses, look in the A–Z under **Rose** (p. 114). The entry there will refer to **Rosa** (p. 113), which is the botanical name for the genus to which roses belong, and there you will find that the entry against 'Leaves with leaflets tightly rolled along their length and drooping' leads to 'Leaf-rolling Rose Sawfly' on p. 226. A description of the symptoms is given there, followed by details of the biology of this pest and recommendations for chemical and non-chemical treatment. Similarly, if large, irregular, solid swellings are found on the roots of brassicas, look under **Brassicas** in the A–Z and the entry against 'Roots with irregular swellings not containing caterpillars or holes' will lead to 'Clubroot' on p. 328, where a detailed account of the symptoms, biology and treatment of this disease is given. There are separate entries in the A–Z for 280 different plants or groups of plants. Fruits and vegetables are listed under their common names (apples, pears, parsnips, turnips, etc.) and ornamental plants are listed under the correct botanical name for the genus (*Dahlia*, *Euonymus*, *Rhododendron*, etc.) with cross-references from non-botanical common names, when necessary (e.g., Busy Lizzie – see *Impatiens*; Rubber Plant – see *Ficus*, etc.). It is obviously impracticable to list all garden plants in the A–Z but if a plant is not listed by name, relevant information may be found under one of the group entries (Alpine and Rock Garden Plants; Aquatic Plants; Bedding Plants; Bulbs, Corms, Rhizomes and Tubers; Cacti and Succulents; Climbing Plants; Conifers; Ferns; Glasshouse and House Plants; Hedges; Lawns; Perennials; Seedlings; Shrubs; Trees). Diagnosis of most common conditions will usually be relatively easy but atypical symptoms sometimes develop and symptoms of different troubles are occasionally confusingly similar. Incorrect diagnosis will result in wrong treatment and doubtful cases should therefore be referred to experts for advice (see p. 495).

INFORMATION

Detailed information about each condition is given in the three main sections on Pests (p. 131), Diseases (p. 265) and Disorders (p. 473), which bring together similar groups of problems, so avoiding unnecessary duplication. Within the Pests section these groupings are of closely related organisms (eelworms, slugs and snails, aphids, beetles, flies, mites, birds, mammals, etc.). This arrangement works well for pests, since related species generally damage plants in similar ways, but it does not work so well for fungi, bacteria and other disease organisms, since unrelated species may cause very similar symptoms. The Diseases section is therefore organised more on the basis of symptoms (cankers, rots, etc.) than on close relationships of causative organisms, except for a few types of fungi and for viruses, which are sufficiently distinct to be treated as separate groups. Disorders are simply grouped under the main factors causing them (physiological, mechanical, climatic, nutritional, etc.).

A brief introduction to each of the main sections and to each of the subsidiary groups summarises general information on structure, size, symptoms, biology and, where appropriate, general methods of treatment and these introductions are followed by detailed accounts of individual pests, diseases and disorders. The common English name is given for each pest or disease and is followed by the currently correct scientific name. General distribution, frequency of occurrence and approximate size is indicated, where appropriate, and for the more important entries information is then summarised under sub-headings covering Symptoms, Biology and Treatment. Disorders are dealt with in a similar fashion but they do not have scientific names since they are not caused by living organisms. The arrangement of entries within the main sections varies a little. In the Pests section the pests of fruit and vegetables are dealt with first and then the pests of ornamental plants. There is of course some overlap, since some pest species may affect fruits, vegetables and ornamentals. In the accounts of the major groups of pests (aphids, caterpillars, flies, beetles and mites) the most important species are tabulated in the introduction to indicate the main plants and plant groups that they affect. In the Diseases and Disorders sections the arrangement is different and the entries are presented in alphabetical order of their common names.

Additional information and advice, especially on non-chemical and chemical methods of treatment, is given in the General Introduction (p. 15), which should be read before making detailed use of the rest of the book.

Sources of information in other books and pamphlets are summarised in the Bibliography (p. 493), which also includes details of general and specialist journals and magazines that help to keep gardeners informed of changes such as the appearance of new pests and diseases or the development of new methods of treatment.

Technical terms have been used sparingly and those that have had to be used are defined in the Glossary (p. 496), which is followed by a comprehensive Index (p. 500).

Note:

The advice contained in this book is given in good faith but so many factors affect local conditions that the authors cannot guarantee the success of all of their recommendations, nor can they accept liability for any consequence of their use. They stress that the manufacturer's instructions for all proprietary products should be strictly observed and that expert advice should be sought if any doubt arises.

General Introduction

Pests are those animals that damage cultivated plants; diseases result from infections of plants by certain bacteria, fungi, viruses or mycoplasmas (collectively termed pathogens), while disorders are malfunctions caused by factors such as nutrient deficiencies, drought, water-logging, weather or pollutants that are not living organisms. The many different pests, diseases and disorders that affect cultivated plants are of course only a very small part of the extremely complex system of interactions between all living organisms and their environments and it should be realised that most of the organisms that live on or in plants do no harm and are not pests or pathogens. Indeed many are positively beneficial, good examples being the predaceous and parasitic insects that feed on pests or the root-nodule bacteria and mycorrhizal fungi associated with the roots of some plants.

Deciding whether a pest or disease organism should be classed as important may be very much a matter of opinion, especially in the garden situation. In commercial horticulture and agriculture it is often possible to assess the financial losses caused by a particular pest, disease or disorder and so determine whether it is sufficiently important to justify the expenditure of time and money on investigation and treatment. In gardens it is not so easy, since most gardeners grow plants for their beauty and interest rather than for financial gain. The relative importance attached to a particular pest, disease or disorder must therefore be finally determined by individual gardeners in the light of their own experience and circumstances.

We have selected the pests, diseases and disorders that we consider to be of general importance in Britain and northern Europe. There will not be complete agreement about this selection, even among experts, but we have based it on our own experiences in advising gardeners during the past fifteen years and we have tried to include all conditions that we know are common causes of concern even if they do not always cause appreciable damage. Many uncommon conditions have been left out in order to keep the book to acceptable limits and we stress that it is a guide, not an encyclopaedia. If additional information is needed, it should be sought in the many references listed in the bibliography.

Many (but by no means all) of the problems that arise when plants are cultivated result from disturbances of the natural relationships between the plants and their normal environments. In natural circumstances plants have evolved gradually over millions of years and have therefore become well adapted to soil, seasons, weather and other local environmental factors.

In the wild, most plant species usually grow as individuals dispersed in mixed communities with other species and this dispersion must often limit the impact of potential pests, diseases and disorders. Insects or mites that feed on a particular host plant will tend to spend much time searching for the right plants and will then have less time to spend in feeding and breeding; diseases that spread by direct or close contact will not spread so rapidly, and even disorders, such as nutrient deficiencies, will not have such a marked effect in mixed communities, where different plant species will have different nutrient requirements and may obtain them from different depths in the soil.

In agriculture, forestry, horticulture and gardening these natural circumstances have been abandoned. Plants are often grown with many individuals of the same species in close proximity, or even direct contact, so that the development and spread of pests and diseases is encouraged and the risks of extensive disorders are increased. Additionally, in the course of cultivation, most plants have been greatly changed by selection and breeding to produce higher-yielding, showier, improved forms known as cultivars (varieties) that have often lost much of the natural immunity or tolerance of pests, diseases and disorders that may have existed in their wild ancestors. It is possible to reintroduce and enhance useful levels of tolerance and immunity by further selection and breeding but the considerable research effort that is necessary means that it is generally only undertaken for commercial crops.

Further complications have been caused by international movements of plants by man during the past few centuries. Many plants that have been moved by land, sea or air have inevitably carried pests and diseases into areas where they did not previously exist. Good examples of such pests are the woolly aphid of apples, introduced to Europe from North America about 1787; the Colorado beetle, introduced to Europe from North America about 1922; rhododendron whitefly, probably introduced to Europe from the Himalayas about 1926 and the many species of mealybugs, scale insects and mites that have been introduced into glasshouses from various parts of the tropics and sub-tropics. The most notable recent example of a disease that has spread in this way is that of the virulent strain of the Dutch elm disease fungus introduced to Britain from North America with imported elm logs in the late 1960's with subsequent disastrous effects on our native elms. Other good examples are the bacterial disease fireblight which arrived in Britain from North America in 1957 and has since caused major problems on fruit trees; carnation rust which came to Britain from Europe around 1890 and the very serious chrysanthemum white rust which was brought to Europe with plants imported from Japan in 1963.

It would seem therefore that man's disturbance of nature has opened a Pandora's box of plant ills, but this is far from true. Man's increasing involvement and dependence on cultivated plants, which probably started about 10,000 years ago, has developed into our modern systems of agriculture, forestry, horticulture and gardening and most of the plants that are cultivated in these unnatural conditions are in fact remarkably healthy. Our concern here must be with unhealthy plants, but they are a very small minority of all the plants that are grown every year.

SYMPTOMS

Many pests, diseases and disorders produce characteristic symptoms that make it possible to diagnose the causes of trouble with a fair degree of certainty. These symptoms are described in detail later in the book and the most important or the most typical of each type are illustrated in the colour plates. In all cases we have described the symptoms that are most likely to be noticed by gardeners and we have tried to describe the most typical symptoms. There is obviously bound to be considerable variation in the manifestation of symptoms on different plants, in different seasons and places and on different cultivars, so that the relatively brief descriptions given should not be taken to be absolutely exact for all circumstances. Although many diagnoses are simple and certain, there are also many cases where diagnosis is difficult. If in doubt, seek expert advice, since wrong diagnosis and wrong treatment will waste time and money.

Symptoms are often conspicuous and may give a clear indication of the severity of damage. Good examples of this category are the sudden wilting and collapse of whole plants caused by cabbage root fly on recently transplanted brassicas, by wilt diseases on tomatoes, or by severe frost on potato plants in spring and autumn. In other cases, such as eelworm attacks, virus infections or nutrient deficiencies, symptoms may be less obvious and must be looked for carefully. In all cases it is important to look out for the first signs of symptoms developing since early diagnosis and treatment may make it possible to prevent or minimise damage. Different pests, diseases and disorders affect different parts of plants and symptoms may therefore appear on roots, bulbs, tubers, corms, rhizomes, stems, buds, leaves, flowers, fruits or seeds. In the detailed descriptions reference is made to the main parts affected and, where appropriate, to the time of year when symptoms normally appear. Where pests are involved, brief descriptions of adult and immature stages are given if they are likely to be easily seen on damaged plants but detailed descriptions are not given for those stages, such as adult moths or flies, that are unlikely to be seen. Descriptions of diseases similarly emphasise the gross symptoms produced and do not describe microscopic details of the causative organisms. Sizes of organisms or symptoms where relevant, are indicated by approximate measurements but these give only a rough guide, since there can be considerable variation, even within species.

BIOLOGY

Relevant information about the biology of pests and diseases and about the biological basis of disorders explains how and why plants are damaged and is also most useful in deciding what methods can be used to prevent or limit this damage. Much information has been published about most major pests, diseases and disorders and from the scientific literature we have prepared summaries of the essential facts about overwintering stages and sites, development during the growing season, methods of reproduction and dispersal, host ranges and geographical distribution, and many other topics. There are considerable differences in the general biology of the major groups of pests (eelworms, mites, insects, birds, mammals) and of disease organisms (fungi, bacteria, viruses and mycoplasmas) and these are summarised at the beginning of each relevant part of the book. Similar summaries are also given for important groups of insects (aphids, caterpillars, flies, sawflies, beetles), and for certain groups of fungi (rusts, smuts, downy mildews, powdery mildews).

This arrangement takes advantage of the fact that closely related groups of organisms tend to live, feed and breed in similar ways and it is relatively easy to write about them in general terms. Even so, there is much variation within these restricted groups and most generalisations are subject to some exceptions.

TREATMENT

Specific recommendations for treatment are given in the detailed entries in the pest, disease and disorder sections but there are some general principles that are best summarised here. Our aim in making recommendations for treatment has been to select methods that are safe, practicable and effective when used in gardens. Commercial situations are usually quite distinct, since different criteria apply, and we have not tried to cover them. Firstly, we emphasise that choice of treatment must be personal. Only the individual gardener can decide what monetary, aesthetic or sentimental value is to be placed on his/her plants; what levels of perfection he/she wishes to achieve, and what methods are feasible in his/her circumstances. Some gardeners will prefer to rely entirely on natural controls; others will use varying degrees of non-chemical and chemical control. We have tried to indicate the range of possible effective treatments that are available and have also tried to provide sufficient biological information to indicate when and why treatment may be necessary and how it can best be achieved. Many recommendations made elsewhere in the gardening literature will not be found here and in most cases they will have been omitted because we are not convinced that they are effective and/or safe.

Secondly, we urge the use of non-chemical methods whenever possible. Chemicals are generally a poor substitute for good gardening, are expensive, and may also have harmful side-effects. There are however many cases where chemicals offer the only possibility of prevention and cure and the gardener's choice must then be between the use of chemicals or the risk of serious damage and possible loss of plants. In such cases we recommend chemicals that have been cleared by the Ministry of Agriculture.

Natural Control Factors. Before discussing the many different artificial treatments that can be used to check pests, diseases and disorders, it is worth considering the general background of natural control factors that operate independent of man's intervention and limit many pests and diseases and some disorders. All organisms, including man, tend to increase their populations at rates that would quickly outstrip the resources of their environments if they were not checked and it is the natural control factors that generally ensure that this does not happen. Taking the very simple, but by no means extreme, example of a species of animal that reproduces sexually once a year, the females laying an average of 100 eggs each. If we assume that all of the eggs hatch; that all the young survive to maturity; that individuals do not live more than a year and that the sex ratio averages 50 ♂: 50 ♀; then the .population increase from a single pair of individuals over four years will be

	♂	♀	eggs
1st year	1	1	100
2nd year	50	50	5,000
3rd year	2,500	2,500	250,000
4th year	125,000	125,000	12,500,000

If the population of this species is to remain constant from year to year, only two of every 100 eggs may develop through to sexually mature adults, which means that the mortality must be 98%. If three eggs survive, the population will be increased by 50% and if four win through it will be doubled.

Many pests, especially insects and mites, go through the above process in weeks rather than years and some are able to reproduce asexually, so that the whole population consists of females. Under these conditions truly astronomical increases are possible. To take a familiar example, a single black bean aphid alighting on a broad bean plant in early June could theoretically give rise to a population of 2,000,000,000,000,000 aphids by the end of August, which would be about a million tons of aphids! Similar statistics may be quoted with respect to pathogens: a single honey fungus toadstool may liberate around 10,000,000,000,000 spores in the course of a few days while as many bacterial cells may be present in a single drop of bacterial slime. Fortunately this vast reproductive potential is never fully realised since many natural factors reduce the numbers of individuals surviving to maturity and so keep populations within reasonable limits most of the time. The most important of these factors for pests are climate, quantity and quality of food, and the effects of predators, parasites and diseases. Similarly only a few pathogenic spores survive to alight in the right place and at the right time to infect another plant. Climate often determines whether or not a species can exist at all in a particular area and is therefore most important, especially in northern temperate countries where conditions are quite unsuitable for the survival of many tropical and sub-tropical pests and diseases. Even the native and introduced pests and diseases that are well adapted to survive the winter in a dormant state may suffer heavy mortality in the early spring if a mild period, sufficiently warm to break dormancy, is suddenly followed by cold weather. In addition, the life processes of insects, mites and other invertebrate pests and those of all disease organisms are directly affected by temperature so that rates of movement, growth and reproduction increase when temperatures rise, and fall when they drop.

Humidity is also an important climatic factor, since many pests and disease organisms (especially fungi) are soon desiccated and die or become dormant when humidity is low. Hot, dry periods may therefore check pests, such as aphids, leatherjackets or slugs, and diseases such as downy mildews and many rot-inducing fungi.

Rainfall affects humidity but may also have a direct battering effect on some pests, which are knocked off plants and may not be able to regain them. Rainfall is also important in aiding the dispersal of pathogens; bacterial slime or soil containing fungal spores for instance may be splashed considerable distances by heavy raindrops.

Sunshine has a direct effect on the activity of many adult insects, such as narcissus flies or cabbage white butterflies, so that more eggs are likely to be laid in sunny periods. The effects of seasonal changes of climate are fairly obvious but there can also be local differences of climate within a garden, which may have an important effect on pests and diseases.

The artificial climate maintained in houses and heated glasshouses makes it possible for tropical and subtropical pests and diseases to survive in our cold climate and similar effects may result from the positioning of a hedge or wall, removal of trees, construction of a pond and so on. Food is another important factor affecting pests and diseases. There may be insufficient food available so that individual organisms must compete for it while even if the quantities are adequate, the quality may not be. All pests and disease organisms have certain basic nutritional requirements and although they may be able to survive long adverse periods, they will reproduce most rapidly only when the right combinations of nutrients are available to them in sufficient quantities. Gardens often provide the ideal combination of food plants, both in quantity and quality, and this is one of the reasons why pests and diseases thrive, at least until the food supply starts to give out.

Pest populations may also themselves be checked by various diseases, caused by bacteria, fungi and viruses, and by many different predators and parasites. Epidemic diseases, such as myxomatosis in rabbits' or certain bacterial diseases of caterpillars, have dramatic effects, especially when pest populations are high, but less obvious chronic diseases also take a steady toll. Predators are abundant in most habitats. They hunt or trap prey and each individual predator usually eats many adult or immature stages of the pest. Spiders, birds, fish, frogs and hedgehogs are familiar examples but there are very many more, such as certain capsid bugs that eat three to four thousand fruit tree red spider mites each; anthocorid bugs that can kill fifty red spider mites a day; ladybird larvae that eat up to five hundred aphids each during the three weeks of their development or syrphid fly larvae that can eat as many as twenty large aphids in twenty minutes and may consume about a thousand aphids each during their development. Parasites are usually less conspicuous and more subtle. They generally develop within the bodies of their hosts and one individual host usually provides all the food needed for one or more parasite larvae to complete their development. Many such parasites are the larvae of Hymenoptera, tachinid flies or other groups of insects and, although they are seldom recognised by gardeners, they do play an important part in limiting populations of aphids, caterpillars and other pests. It is however unusual for any parasite or predator to eliminate completely a pest species in natural circumstances since this would result ultimately in the extinction of the parasite or predator species itself and would be a form of biological suicide. What usually happens is that a balance is maintained between the interacting organisms, with parasites and predators reaching peak populations only after pest populations have also reached a peak. Rapid collapse of the pest population is then followed by gradual recovery, which may take two or more years. Because of this cycle, years when a particular pest is unusually abundant are often followed by a number of years when it is at too low a level to cause appreciable damage. Plant diseases are also similarly affected by interactions with other organisms that feed on them or compete with them but the microscopic size of the organisms involved makes it extremely difficult to observe and interpret what is happening in any particular circumstance.

This, briefly, is the basis of the natural control of pests and diseases. Breaking it down into separate factors is of course a gross oversimplification, since most factors interact to varying degrees, producing an exceedingly complex dynamic system which is often glibly referred to as the balance of nature. It is against this complex background that we try to impose our own artificial systems of pest and disease control.

Non-chemical Treatments. The many different types of treatment that have been devised to prevent or limit pests, diseases and disorders can be classified in several ways but the simplest general distinction is between non-chemical and chemical treatments. We use this distinction as we wish to stress the continuing importance of non-chemical treatments, most of which aim at preventing trouble, rather than curing it, although there are some, such as hot water treatment, that can, in the right circumstances, effect better cures than chemicals.

It almost goes without saying that plants are most likely to be healthy if they are grown correctly in the right soil, at the right time and in the right way. Detailed discussion of this aspect of preventive treatment is not possible here, since it would involve virtually the whole subject of gardening, but we would recommend that care be taken to grow plants in situations that suit them. This will not always ensure that they remain free from trouble but it will at least give them the best possible chance of thriving. Plants growing in unfavourable circumstances are less likely to thrive but they too may respond to special care.

Hygiene. The best way to avoid many problems is to prevent their initial establishment in the garden and this can certainly be done for many diseases and some pests. Always obtain new plants from reputable suppliers and scrutinise all new introductions immediately on arrival to ensure that they are free from obvious signs of trouble. If possible, keep new plants in quarantine in a remote part of the garden for a time so that symptoms of incipient trouble may develop and become apparent. Be especially careful when buying in bulbs, corms and tubers and glasshouse and house plants, all of which can easily carry pests and diseases, and ensure that new plantings of bush, cane and tree fruits are healthy, preferably by purchasing certified stocks. Be ruthless with suspect plants and destroy them by burning or by burying them deeply if they are seriously affected by incurable diseases or intractable pests. Do not put diseased plants on compost heaps where pathogens may persist and spread infection when the compost is applied to healthy plants. Do not leave vegetable crops in the ground longer than necessary, since this helps to carry pests and diseases from one crop to the next or from one season to the next and, for the same reason, always burn prunings from fruit and other trees and shrubs as soon as possible. A garden is more likely to be healthy if it is neat and tidy than if it is cluttered with rubbish and debris. Hygiene is particularly important in glasshouses where the protected environment enables many potentially harmful organisms to flourish. Scrupulous clearing away of plant debris and regular cleaning out of water tanks and rain butts is essential therefore. The compost, pots and boxes used to raise seedlings must also be checked; it is not safe to assume that all commercially prepared seedling compost is sterile and small quantities can be sterilised quite easily by 'cooking' them in a domestic oven for about one hour at $150°C$ (gas regulo 2). Pots and boxes should be washed and scrubbed thoroughly after each batch of plants has been raised, and then when all soil has been washed off, rinsed again in a proprietary disinfectant before rinsing in clean water.

Cultivation. Always cultivate ground as thoroughly and frequently as possible since this will reduce populations of soil pests, such as leatherjackets, wireworms and cutworms, and will also keep down weeds, which may harbour pests and diseases. This is particularly important where land is newly broken for the cultivation of vegetables but also applies to the preparation of land for fruit and to the maintenance of beds and borders of ornamentals.

Rotation. The repeated growing of the same or related plants on the same site can cause build-up of soil-borne diseases and pests and may also deplete nutrients. It is possible to avoid some of the more obvious hazards by planning adequate rotations of annually grown plants and this is normal practice when cultivating vegetables. The small size of many modern gardens makes it difficult to rotate plants as often as may be desirable but three groups of plants are particularly important in this respect. If at all possible, potatoes, tomatoes and other solanaceous plants should not follow each other, nor should cabbages, cauliflowers and other brassicas or onions, leeks, shallots and chives. A 3–4 year break is generally sufficient to prevent trouble but once certain pests, such as eelworms, or diseases, such as clubroot or onion white rot have become established, it may be necessary to refrain from growing susceptible plants for much longer. Additionally, with some pest and disease problems it may be wise to avoid growing certain plants in close proximity and such instances are indicated, where appropriate, in the text.

Resistant Plants. Some plants are immune from attack by certain pests and diseases or are quite unaffected by some disorders while others may show varying degrees of susceptibility, which can result from inherited differences in thickness, waxiness or hairiness of the cuticle; chemical composition of the sap and tissues; early maturity; exceptional vigour and many other characters. In theory therefore it may be possible to avoid trouble by growing non-susceptible plants but it is not always easy to give definite advice on this matter, partly because information may not be readily available and partly because resistance or tolerance is seldom absolute and may break down in some circumstances. It is generally easiest to obtain information about commercially grown plants since breeding and selection has developed cultivars that carry genetically based resistance or tolerance to some important pests, diseases and disorders. This is the case with a number of vegetables, such as canker resistant parsnips and root aphid resistant lettuces; some fruits, such as spur blight resistant raspberries and a few ornamentals, such as eelworm resistant chrysanthemums, mildew resistant roses and rust resistant antirrhinums. We have indicated the situations where the use of resistant or tolerant cultivars is feasible and additional information can be found in some seed and nurseryman's catalogues.

Hand-picking. Removal and destruction of pests by hand is an effective method of control if done thoroughly and regularly, and provided that the number of plants to be examined is not too great. Egg batches and young larvae of many pests can often be dealt with simply by crushing them and, although this procedure may seem cruel, it is in fact much quicker and certain than death through slow poisoning by insecticides. House plants and glasshouse plants can easily be scrutinised and aphids, scale insects, mealybugs and other pests located and removed with a soft piece of cloth or a fine brush dipped in soapy water or methylated spirit. Outdoors, search and destroy operations can be mounted against slugs, snails, leatherjackets, chafer grubs, cutworms, tortricid and other caterpillars (including those of many sawflies), egg masses of vapourer moth, lackey moth, cabbage white butterflies and many other species, larvae and pupae of leaf-miners in celery, chrysanthemums, cinerarias and other plants, etc. It may also be possible to check the spread of some diseases by removing and destroying affected leaves or other parts at an early stage but this is usually less effective than the removal of pests.

Barriers. Physical barriers can be used to prevent or limit damage caused by some of the larger pests, especially mammals and birds. Correctly designed fences and gates will exclude deer, rabbits and hares from gardens and permanent cages can be used to protect fruit and vegetables from attack by birds and squirrels. Temporary caging or netting of susceptible plants can also be effective, if used in good time, before the main attack can develop. It is usually quite impossible to erect barriers that will effectively protect plants from insects, mites, eelworms and other small invertebrates but there are a few cases (chrysanthemum eelworm, winter moth, cabbage rootfly) where this can be done and these are dealt with in the appropriate pest entries.

Repellents and Scaring Devices. Many different chemical repellents and various mechanical, visual and psychological scaring devices have been used to prevent pest damage. Some may work for a time but they are mainly used against birds and mammals and these pests quickly become accustomed to most scaring devices and often ignore chemical repellents when food is short.

Traps. Trapping can reduce local populations of mice, rabbits, moles and some other mammals but usually requires special knowledge and experience, combined with patience and persistence. Some other pests (earwigs, codling moth caterpillars, weevils, cockroaches and wasps for example) can be trapped fairly easily but large numbers must be killed before there is any noticeable reduction in populations since trapped individuals are soon replaced by further pests moving in from adjoining areas. Birds should not be trapped since all species are legally protected and although exceptions can be made for some pest species, they are so mobile that local trapping is unlikely to have any real effect.

Biological Control. The subject of biological control has received much publicity in recent years but little has been done to develop this technique for use in gardens. It usually involves the manipulation of parasites, predators or diseases to restrict or eradicate pests and this manipulation may be achieved by the introduction of selected biological control agents from other areas or by encouraging agents that are already present. The idea is an old one and horticultural use dates from at least the thirteenth century, when Chinese farmers put nests of ants into citrus and litchi trees to protect them from pests. In more recent times it has been developed as a sophisticated method for controlling pests, diseases and weeds and has been used successfully against many different kinds of pests in agriculture, horticulture and forestry, especially in the tropics and sub-tropics. The first great success was at the end of the last century when a predaceous Australian ladybird beetle, *Rodolia cardinalis*, was introduced into California to control the cottony-cushion or fluted scale, *Icerya purchasi*, which had become a serious pest in citrus orchards; an equally famous success was the eradication of prickly pear cactus, *Opuntia stricta*, from about 25 million hectares of Australia by the introduction of 2,750 caterpillars of a south American moth, *Cactoblastis cactorum*, in 1925.

Other successes have been less spectacular but equally effective. They have usually involved extensive research, followed by transportation of live parasites, predators or diseases from an area where the pest species is indigenous to an area where it has been accidentally introduced. Sometimes a single introduction is sufficient but in other cases repeated introductions are necessary.

In Britain and northern Europe the most important instances of the deliberate use of this method are the introduction of a parasitic hymenopteran, *Aphelinus mali*, from North America to control woolly aphid (see p. 166), the use of another parasitic hymenopteran, *Encarsia formosa*, to control glasshouse whitefly (see p. 159) and the use of a south American predaceous mite, *Phytoseiulus persimilis* to control glasshouse red spider mites (see p. 248). Both *Encarsia* and *Phytoseiulus* can be used by gardeners and it should be possible to use other parasites, predators and diseases, both in glasshouses and outdoors, but it is diffficult to obtain information and advice and it is especially difficult to obtain supplies of suitable biological control agents. The general principles of biological control can however be applied in any garden simply by encouraging the parasites and predators that are there already, especially the many groups of beneficial insects (ladybirds, lacewings, ground beetles, hoverflies, tachinid flies, ichneumonids, braconids and chalcids for example) and this can best be done by restricting the use of insecticides. Both parasites and predators are easily killed by insecticides, especially those that persist on or in plants for days or weeks, and there is a tendency to use these chemicals when pest populations are at their peak. This is the time when plants have already been damaged and the pest population is about to collapse as a result of the activities of predators, parasites and diseases and other natural control factors. Use of insecticides at this stage can in fact prolong pest infestations and it is much better to try to detect and control pests before this stage has been reached, preferably by using a non-persistent or a selective chemical.

Other Non-chemical Treatments. There are many other non-chemical treatments that can be used in certain circumstances. Sowing dates of carrots and peas can be arranged so that the main periods of pest and pathogen activity are avoided; dormant bulbs and rootstocks of some plants can be treated in regulated hot water baths to kill eelworms or certain pathogens; gladiolus corms can be stored at low temperatures to eliminate thrips, and so on. Details of these and other methods are given at appropriate points in the text.

Chemical Treatments. The regular use of chemicals to control pests and diseases dates from the second half of the nineteenth century and some of the chemicals used then, Bordeaux mixture and nicotine for example, are still in use today. At first the range of materials was restricted but since the 1940's, when the remarkable insecticidal properties of DDT were discovered, a vast array of synthetic chemicals has been developed, primarily for agriculture, horticulture and forestry, and some of these have been made available to gardeners. Now there are chemicals to kill insects (insecticides), mites (acaricides), slugs and snails (molluscicides), nematodes (nematicides), rats and mice (rodenticides), fungi (fungicides), bacteria (bactericides), as well as the many different weed killers (herbicides), growth regulators and fertilisers that are beyond the immediate scope of this book.

Much has been written about the relative advantages and disadvantages of chemical methods of pest and disease control, with extreme claims being made for and against their use. Like most new developments, chemicals have disadvantages, many of which were not immediately apparent when they were first introduced. People have been poisoned, usually by avoidable accidents; birds, fish and other wild animals have been seriously affected, usually through lack of knowledge, and some pests and diseases have become resistant to certain groups of chemicals. On the other hand, it has been possible to control pests and diseases that were previously uncontrollable and, as a direct result of the use of chemicals, the lives of millions of people have been saved, especially by controlling insect vectors of human diseases, and yields of crops have been increased substantially. Public and private debate of the major issues has produced positive improvements in the safe use of chemicals and has also hastened a timely reassessment of the relative importance of chemical and non-chemical treatments. The general conclusion is that best use should be made of all possible methods that will limit pests and diseases to acceptable levels without harmful side effects on man, on wild life, on the general environment, or on plants. Assessment of the many factors that must be considered to achieve this is difficult but excellent guidance is given by the Ministry of Agriculture, Fisheries and Food, especially through the Pesticides Safety Precautions Scheme and through their advisory literature. We have relied heavily on these sources of information.

Pesticides. We use the term pesticides to refer to all chemicals (insecticides, acaricides, fungicides, etc.) that are used to control pests and diseases. It is sometimes used in a narrower sense, referring only to chemicals used to control pests, and may also be used in a wider sense, to include weed killers.

Before any pesticide can be marketed in the United Kingdom, it must be cleared by the Government's Pesticides Safety Precautions Scheme, which is organised by an independent committee of experts. They consider detailed results of tests and experiments carried out during research and development and they then define in detail, for each active ingredient and formulation, the precautions that must be taken to ensure safety in use, the plants on which they may be used, the rates of application and the minimum periods that must elapse between the last application and the harvesting of edible crops. In addition to this scheme, the Ministry of Agriculture, with voluntary cooperation from manufacturers, operates the Agricultural Chemicals Approval Scheme to promote the safe use and easy selection of appropriate chemicals for the control of the main pests and diseases affecting agricultural and horticultural crops. This scheme ensures that the manufacturer's recommendations for safe use are closely scrutinised and all approved products packs are clearly identified by a distinctive capital A, surmounted by a crown, which appears on all pack labels. Until recent years this scheme was extended to cover garden chemicals also but this useful function has now been discontinued.

In 1981, about 55 different chemical pesticides were on sale for garden use, including about 30 different insecticides and about 20 different fungi-. cides, the remainder being various rodenticides, molluscicides and acaricides. We refer to these chemicals by their common chemical names (bromophos, dimethoate, malathion, captan, benomyl, etc.), so avoiding the use of complex chemical names and formulae, and we have summarised essential information about most of them on pp. 33–40. They are sold under about 200 different trade names, with widely used active ingredients, such as malathion, marketed under at least a dozen different trade names. It is difficult to keep up to date with these names, which tend to change frequently, and we have not used them, but current lists of them can be obtained from the British Agrochemicals Association, the Royal Horticultural Society (members only), gardening journals or individual manufacturers. In addition, it is usually possible to find the names of active ingredients on pack labels, though they may be in very small print.

We must also point out that in certain instances a chemical treatment we recommend may be the best available to gardeners but not necessarily the best overall as many products, for reasons of toxicity or marketing factors are available only for commercial use.

Modes of action. Pesticides may act in a number of different ways and these different modes of action affect their possible uses. Most insecticides act by direct external contact with the target organisms, either immediately on application or within a few days. Some insecticides, especially those that kill biting and chewing insects, leave a persistent residue on treated plants, and others are used as persistent poisons in baits (e.g. ant baits) which attract pests to them. Fungicides may act in an eradicant fashion and kill pathogens already existing on the plant or as protectants on healthy tissues when they form a barrier that prevents fungal development and penetration. A few insecticides and fungicides act as fumigants and many of the most recently developed insecticides and fungicides are absorbed into the sap of treated plants and have a systemic action, which is particularly effective against pests that feed on sap or against fungi that permeate plant tissues. Such systemic materials have the added advantages that usually they are required in much smaller quantities, do not suffer from washing off by rain as the externally acting pesticides do and, because they are absorbed into the plant's tissues and moved within, do not require such accuracy in application. Most pesticides can act in more than one way, so that differences in mode of action are not always clear-cut. For example, nicotine has contact and fumigant action; dimethoate has contact and systemic action; HCH has contact, fumigant and slight systemic action, and benomyl has protectant, eradicant and systemic action.

Selectivity. Pesticides also vary considerably in their relative toxicity to different groups of organisms. Some may be highly selective, only killing a restricted group of pests or diseases, while others, said to have broad spectrum activity, can act against many different groups of pests or even against both pests and diseases. Pirimicarb is a good example of a highly selective insecticide that kills aphids very quickly but has little effect on other groups of insects. This is a most useful property where there is a need to check aphids without harming predators and parasites, but is less useful when attempting to control mixed infestations of aphids, whiteflies, caterpillars and mites, which is a common situation on glasshouse and house plants. In such cases a broad-spectrum insecticide, such as HCH, might be more appropriate since it will kill most insects (aphids, capsids, whiteflies, thrips, ants, caterpillars, beetles, flies) and some non-insect pests (woodlice, millepedes). Similarly, among fungicides, oxycarboxin, which mainly controls rusts, and dinocap which mainly controls powdery mildews are highly selective while benomyl is much less selective, being able to control most fungi except Phycomycetes.

These varying degrees of selective action must of course be considered when choosing the most suitable material to control a particular pest or disease, since choice of an ineffective chemical will waste time and money. It should be noted however, that many manufacturers market pesticides in mixtures so that one product may be used to control a number of problems. Caution should be exercised with the 'witches' brews' that claim to control everything and gardeners should not themselves mix different materials together unless the manufacturers state specifically that this may be done.

Persistence. Another way in which pesticides vary is in the length of time for which they can persist on plants or in soil before they break down into inactive residues. During the early development of insecticides, long-term persistence was considered a useful characteristic since a single application of a very persistent pesticide could protect plants from pests or diseases for months or even years. It is now known that such persistence leads to widespread, insidious contamination of the environment, with resultant toxic effects on wild life, mainly by accumulation and concentration of chemicals in food chains, and possible long-term effects on human health. Use of the very persistent organochlorine compounds, especially aldrin, dieldrin and DDT, is now strictly controlled and their use in gardens has been banned since 1971. Most pesticides now available to gardeners are therefore either virtually non-persistent, breaking down after a few days or even a few hours, or moderately persistent, breaking down after a few weeks. This lack of persistence means that more frequent applications may have to be made but it is also a definite advantage when treating edible crops near to harvest time since it reduces the possibility of toxic residues getting into food.

Formulations and methods of application. Pesticides are formulated in various different ways, mainly to make application easier but also to enhance their efficiency. The main formulations for garden use are:

Sprays. The most widely used method of applying pesticides is as water-based sprays. These are prepared by diluting a liquid concentrate or by dissolving or dispersing a powder and the dilute solution is then applied as a fine mist-like spray from a pressurised sprayer. This may be a small hand sprayer, suitable for treating a few glasshouse or house plants; a larger knapsack or other type of hand-operated sprayer, or an expensive motorised sprayer capable of applying many litres of spray to large fruit and ornamental trees. In all cases, the main purpose of spraying is to leave a uniform deposit of chemical over the whole plant by wetting it thoroughly to a point just before excess spray starts to run off the leaves. Most spray formulations contain special additives that improve their wetting and sticking properties but additional wetting agent such as a few drops of washing-up liquid may need to be added to obtain good results on plants, such as brassicas, which have smooth, waxy leaves. When insects are present on plants, they should be thoroughly wetted, although this is not quite so critical when systemic insecticides are used.

Aerosols. These are special formulations of pesticides in small pressurised packs. They are most useful for the occasional treatment of a few house or glasshouse plants and may sometimes be used on a few plants in the open garden but they are generally much too expensive to be used extensively on large collections of plants.

Dusts. These are pesticides that have been formulated in an inert carrier to produce a fine dust that contains the correct concentration of active ingredient. They can be applied by special dusting machines, by puffer packs, or simply by shaking the dust from a suitable container, such as a tin with small holes in the lid, or from a length of old nylon stocking loaded with dust and shaken over plants. Dusts are less versatile than sprays and tend to be less popular, especially since they leave conspicuous deposits on treated foliage and flowers. There are however many cases where they may give better results than sprays and these are indicated later, where appropriate.

Fumigants. Fumigants can generally only be used in confined spaces in glasshouses, conservatories and similar structures or in closed plastic bags or other air-tight containers. Glasshouse fumigants are usually pyrotechnic mixtures, often known as smokes, which release the active ingredients into the air as they burn. Other fumigants may be simply vaporised from hot or cold surfaces and at least one insecticide is formulated in a resin which gradually releases the active ingredient into the air. Only one soil fumigant can be contemplated for garden use; the granular product dazomet which reacts with soil moisture to produce a poisonous gas that must be sealed into the soil with polythene sheet or other means.

Granules. These are usually special formulations of insecticides designed to give slow release of the active ingredient into soil over a period of a month or so to extend the period of protection against soil pests, especially those that attack vegetables.

Seed dressings. These are special formulations of insecticides and fungicides used to coat seeds before sowing. They protect seedlings and young plants from certain pests and diseases and help to get young seedlings established quickly with few losses. They provide a cheap and effective means of giving valuable protection during the first weeks of growth and should be used on seeds of vegetables and ornamentals whenever possible. Sometimes the powder formulations intended for use as sprays can also be used to treat seeds.

Ultra-low-volume sprays. This technique has been used in agriculture and forestry for many years but has only been developed for garden use comparatively recently, mainly through the Turbair system. A range of different pesticides is specially formulated in oil for application from a machine that produces a very fine spray off a rapidly spinning disc, rotated either by a petrol or an electric motor. The sprayers are small and easily portable and might be useful in larger gardens, especially for controlling pests and diseases in small orchards.

Resistance to pesticides. Frequent or continuous exposure of a pest or disease to a particular pesticide can result in the selection of new strains that are resistant to that pesticide, and possibly to related pesticides. This is a case of artificial selection and is similar to the system of selection that has produced many of our modern cultivars of garden plants. The pesticide kills all susceptible individuals but in any population there are likely to be at least some individuals that are less susceptible and are therefore not killed. If there is an inheritable genetic basis for this lack of susceptibility, such as slight differences in structure, behaviour or physiology (e.g. detoxifying mechanisms), it will be passed on to the next generation when the survivors breed. If selection for tolerance continues, through continuing use of the same pesticide, the new strain of the pest or disease may eventually be quite resistant to the pesticide. Resistance to insecticides and acaricides tends to develop most readily in prolific species, such as red spider mites or aphids, which pass through successive generations quickly, so that the whole process of selection is speeded up, and in such cases resistance can appear within a few years of using a new chemical.

Resistance of scale insects to hydrogen cyanide was noted in California before 1916, so it is not a new phenomenon, but with increasing use of chemicals, hundreds of resistant strains of pest species have developed and new strains appear every year. Relatively few of these occur in Britain and the only really important cases at present involve glasshouse red spider mites, glasshouse whitefly, peach-potato aphid and some other aphids, and some flies, especially carrot fly, cabbage root fly and bean seed fly. A similar situation exists with some diseases that have developed tolerance to fungicides, the most important instance in Britain being the tolerance shown by strains of the grey mould fungus, *Botrytis cinerea* to the widely used fungicide benomyl and similar materials.

In all of these cases resistance has almost certainly arisen in commercial situations, where use of chemicals is much greater than in gardens, but once resistant strains have developed, they can easily spread into gardens. Many instances of alleged resistance in gardens result from failure to apply pesticides effectively but in cases where this is definitely not likely to be the cause, it is best to suspect resistance and seek expert advice. Where resistance is known to occur, the choice of pesticides will have to be varied and recommendations are given under the entries for the main pests and diseases concerned.

Phytotoxicity. Pesticides are not only toxic to animals: they may also affect some plants and the injury that results is attributed to the phytotoxicity of the pesticide. The most phytotoxic chemicals are eliminated during the early stages of development of new pesticides since there is no point in achieving good control of a pest or disease if the chemical also kills the plants. Despite this screening, some phytotoxic side-effects still occur with some pesticides, although they are usually restricted to a few relatively susceptible plants. A good example is the severe phytotoxicity of HCH to cucumbers, melons and other cucurbitaceous plants, but most modern pesticides are toxic to some plants and the main hazards are summarised in the following tabulation, which also indicates those pesticides that may taint fruits and vegetables. Information about both phytotoxicity and taint is usually included in the label instructions on pesticide packs and additional guidance may sometimes be obtained from manufacturers.

Benomyl	Do not use on apples (some cultivars) or on strawberry runners.
BHC, See HCH	
Bordeaux mixture	Use cautiously, especially on pears. Consult manufacturer's instructions.
Captan	Do not use on apples (some cultivars) or on soft fruits intended for bottling or freezing.
Dazomet	Do not plant or sow in treated soil for 1–2 months after application.
Diazinon	Do not use on young cucurbits (cucumbers, marrows, melons etc.), young tomatoes or on ferns.
Dichlofluanid	Do not use on strawberries growing under glass or polythene.
Dichlorvos	Do not use on cucurbits (cucumbers, marrows, melons etc.) or on chrysanthemums (some cultivars) and roses. Consult manufacturer's instructions.
Dicofol	Do not use on seedlings and young plants or on young cucurbits (cucumbers, marrows, melons etc.). Consult manufacturer's instructions.
Dimethoate	Use cautiously on cinerarias, flowering cherries, cyclamen, chrysanthemums, begonias, ferns, fuchsias, hydrangeas and poinsettias.
Dinocap	Do not use on chrysanthemums (some cultivars) or on apple 'Golden Delicious' during blossoming.
Formothion	Use cautiously as for dimethoate (see above).
HCH (= BHC)	Do not use on cucurbits (cucumbers, marrows, melons etc.) or on hydrangeas, vines and young tomatoes. Use cautiously on soft fruits and on vegetables, especially potatoes, as it may taint them.
Malathion	Do not use on antirrhinums, crassulas, ferns, gerberas, orchids, petunias, pileas, sweet peas or zinnias and use cautiously on soft fruits and vegetables within a week of harvest as it may taint them.
Quintozene	Do not plant or sow cucurbits (cucumbers, marrows, melons etc.) or tomatoes in treated soil or compost.
Sulphur	Do not use on fruit intended for bottling or freezing and use cautiously on apples, pears, currants and gooseberries. Consult manufacturer's instructions.

| Tar oil | Use only on dormant deciduous woody trees and shrubs. |
| Zineb | Do not use on blackcurrants within a month before harvest. |

Characteristic symptoms of severe phytotoxicity include discoloration and scorching of leaves, often followed by wilting and death. Less obvious symptoms develop in cases of hidden phytotoxicity where exposure to pesticides, notably by the use of seed-dressings or slow-release granules in the soil, can check growth of young plants for a time.

Safety precautions

No garden is ever entirely free from pests and diseases but some will suffer more than others and some plants will be more susceptible than others. If circumstances indicate that chemical control is the only feasible solution, then select and use the correct pesticide with care and follow these basic rules:

Only use chemicals when you really must.

Select suitable materials carefully, preferably well in advance, and ensure that they will be both effective and safe. Check possible phytotoxic effects and observe safe minimum intervals for pre-harvest applications to fruits and vegetables.

Read the pack label carefully before using a pesticide and then apply it at the correct concentration, at the right time, and as thoroughly as possible. Inefficient and late application does more harm than good. Whenever possible, apply pesticides during cool, dry, cloudy, windless weather. During hot, sunny periods, apply pesticides in the late evening, when temperatures and wind speeds are lower.

Do not apply insecticides to flowers that are being visited by hive bees, bumble bees or other pollinating insects. Wait until the flowering period is over or apply insecticides before the flowering period starts.

Store all gardening chemicals safely, well away from children and pets, preferably on a high shelf in a locked shed and always in a shaded position away from direct sunlight and free from frost.

Clean out spraying and other equipment immediately after use and wash out empty containers before putting them in a rubbish bin. Surplus spray and water used to wash sprayers and containers should go into an outside drain, sump or deep hole. Do not contaminate ponds, streams or rivers with insecticides, many of which are highly toxic to fish and other aquatic animals.

Never transfer chemical concentrates or dilute spray solutions into milk bottles, soft drink bottles or other unlabelled containers. Children may be tempted to drink from them and could be poisoned.

Do not apply pesticides with sprayers or other equipment that has been used to apply weed killers. There is always a possibility that traces of weed killer may persist and damage plants.

Pesticides for use in gardens. The following section summarises the characteristic properties and the main uses of the pesticides that are currently available to gardeners. The first tabulation groups them according to their main functions and this is followed by accounts of each chemical, arranged alphabetically under their common chemical names. Trade names are not given (see p. 26).

INSECTICIDES (used against insect pests)

Non-persistent contact insecticides

Bioresmethrin	Pyrethrum
Dichlorvos	Resmethrin
Malathion	Rotenone (derris)
Nicotine	Trichlorphon
Permethrin	

Moderately persistent contact and residual insecticides

Bromophos	Fenitrothion
Carbaryl	HCH (BHC)
Chlorpyrifos	Pirimicarb
Diazinon	Pirimiphos-methyl

Systemic insecticides

Dimethoate	Heptenophos
Formothion	Menazon

Winter washes
Tar oil

Other special uses

Borax	Sodium tetraborate
Mercurous chloride	

ACARICIDES (used against mites)

Dicofol	Lime sulphur

MOLLUSCICIDES (used against slugs and snails)

Methiocarb	Metaldehyde

RODENTICIDES (used against rats and mice)

Coumatetryl	Warfarin
Difenacoum	

BIRD REPELLENTS (used to repel birds)

Ammonium aluminium sulphate	Anthraquinone

WORM KILLERS (used against earthworms)

Chlordane

FUNGICIDES (used against fungi and sometimes against bacteria)

Protectant fungicides

Benomyl	Maneb
Bordeaux mixture	Mercurous chloride
Captan	Quintozene
Carbendazim	Sulphur
Dichlofluanid	Thiophanate-methyl
Dinocap	Thiram
Folpet	Zineb

Systemic fungicides (which may also have eradicant properties)

Benomyl	Thiophanate-methyl
Carbendazim	Triforine
Oxycarboxin	

SOIL PARTIAL STERILANTS (used against all soil inhabiting organisms)

Dazomet	Proprietary phenolic emulsions

Ammonium aluminium sulphate. Formulated as a spray for application to dormant fruit and ornamental trees and bushes to repel bullfinches and other birds. No harmful side-effects but may impart unpleasant metallic taste if it contaminates edible crops, such as winter brassicas.

Anthraquinone. Formulated as a spray for application to dormant fruit and ornamental trees and bushes to repel bullfinches and other birds. No harmful side-effects but may impart unpleasant bitter taste to edible crops.

Benomyl. This is a systemic organic fungicide with both protectant and eradicant properties formulated as a wettable powder for producing sprays, drenches or bulb and corm dips. It is very similar chemically and in its effectiveness to carbendazim and to thiophanate-methyl. It is active against a very wide range of fungi, except phycomycetous types, and is safe to use on almost all vegetable, fruit and ornamental crops. The introduction of this chemical to the commercial market around 1970 revolutionised the control of many plant diseases but a major drawback has been the subsequent build up of strains of a number of fungi with tolerance to it. Because of this and because greater use of the chemical may increase the likelihood of further tolerant strains arising, benomyl should not be used indiscriminately as it once was. The most serious problems with tolerance have arisen in the grey mould fungi (*Botrytis* spp.) and also with apple scab and with these pathogens alternative chemicals such as dichlofluanid should be tried first or used in alternation with benomyl.

BHC. See *HCH.*

Bioresmethrin. A synthetic pyrethroid compound formulated as a spray or aerosol to control aphids, whiteflies, leafhoppers, thrips, some beetles and small caterpillars. Non-persistent, non-phytotoxic and with very low mammalian toxicity. Therefore a very safe insecticide for general use, especially on edible plants.

Borax. Used only in some ant baits. No harmful side-effects.

Bordeaux mixture. A very old and traditional inorganic fungicide with some bactericidal properties, containing a copper compound with hydrated lime or similar ingredients. It is protectant in action but toxic to mammals and has very considerable phytotoxicity; it can for instance cause russetting on fruit and leaf spotting and scorching on a wide range of plants. Bordeaux mixture has been superseded for many purposes by newer compounds but it is still useful as a foliar spray for some diseases, such as certain rusts.

Bromophos. Available only as a special powder formulated for soil application to control cutworms, wireworms, cabbage root fly and chafer larvae attacking vegetables and some ornamental plants. Moderately persistent but can be used up to a week before harvest of edible plants.

Calomel. See Mercurous chloride.

Captan. This is an organic sulphur-based protectant fungicide with low toxicity to mammals and very little phytotoxicity, although a few apple cultivars may be sensitive to it. It is available as a dust and as a wettable powder for producing sprays and is used against a range of diseases, including grey mould on strawberries, apple scab and black spot of roses. It should not be used on soft fruits intended for bottling as it may taint them.

Carbaryl. Available only as a dust for use against caterpillars, beetles, earwigs, wasp nests and woodlice. Moderately persistent but can be used up to a week before harvest of edible plants. It may also be used to control earthworms in lawns. Toxic to bees and fish.

Carbendazim. This systemic organic fungicide is very similar to benomyl and thiophanate-methyl but is only available to gardeners as a mixture with maneb (q.v.).

Chlordane. Used only to control earthworms in lawns or to destroy ant and wasp nests. This is one of the very persistent organochlorine compounds and it is surprising to find that its use in gardens is still permitted. It is toxic to animals and pets should not have access to treated areas for at least two weeks after application. It is best not to use it unless absolutely essential.

Chlorpyrifos. This moderately persistent organophosphorus compound is available only as granules to control soil pests, especially cutworms, wireworms, beetle larvae and fly maggots. It can be used to protect vegetables as well as ornamental plants but must not be used within six weeks of harvesting edible crops.

Coumatetryl. Used only in specially formulated rat and mouse baits.

Dazomet. This is an organic soil partial sterilant which breaks down chemically in soil to release a herbicidal, pesticidal and fungicidal gas. It is of greatest value in gardens where eelworms are persistent vectors of viruses. The chemical has a high toxicity to mammals and is not very easy to apply, the procedure requiring a rotavator and the sealing of the soil with polythene but there are no other soil partial sterilants of comparable effectiveness that can be contemplated for garden use. It is only available in relatively large packs from horticultural wholesalers and gardeners should appreciate that treated land cannot be replanted for several weeks.

Derris. See *Rotenone.*

Diazinon. This is a moderately persistent organophosphorus compound available as sprays for use against aphids, capsids, mealybugs, scale insects, thrips, flies, some other insects and red spider mites or as granules to control soil pests, especially cabbage root fly, carrot fly and other species attacking vegetables. Do not apply within two weeks of harvesting edible crops. It is toxic to bees and fish and may cause phytotoxic damage if applied to ferns or to young cucumbers and tomatoes.

Dichlofluanid. An organic sulphur-based fungicide with protectant action, fairly low toxicity to mammals and low phytotoxicity although it should not be used on strawberries under glass or polythene. It is particularly effective in controlling diseases caused by *Botrytis* sp. and is valuable as an alternative treatment to benomyl. It is available as a wettable powder for producing sprays but three weeks should be allowed between the last spraying and picking edible produce.

Dichlorvos. This non-persistent organophosphorus compound is available only as a specially formulated resin strip for use in glasshouses against aphids, whiteflies, red spider mites and some other pests. It can be used up to a day before harvesting edible crops but should not be used on cucumbers. It is also phytotoxic to roses and to some chrysanthemum cultivars and is toxic to bees.

Dicofol. This useful organochlorine acaricide is only available in a combined pest and disease spray which also contains fungicides and insecticides.

Difenacoum. Available only as a specially formulated rodenticide.

Dimethoate. An important systemic insecticide and acaricide. This organophosphorus compound is available as sprays which are particularly effective against aphids, capsids, leafhoppers, scale insects, mealybugs and other sap-feeding pests and it also controls red spider mites and some other pests. It is moderately persistent and should not be applied to edible crops within a week of harvest. Toxic to fish, birds and other wild animals and to some plants, especially cinerarias, some chrysanthemums, cyclamen, hydrangeas, ferns, begonias, fuchsias and poinsettias.

Dinocap. An organic protectant fungicide with some toxicity to mammals and some phytotoxicity but it is nonetheless very valuable in controlling powdery mildew diseases on a wide range of plants. It is available as a wettable powder for producing sprays, as a dust and also a fumigant smoke. One week should be allowed between the last application and picking edible produce.

Fenitrothion. A moderately persistent organophosphorus compound formulated as sprays to control a wide range of pests, including aphids, capsids, thrips, caterpillars, sawflies, raspberry beetle and gall midges. It should not be applied within two weeks of harvesting edible crops. Toxic to fish, birds and other animals but generally not phytotoxic.

Folpet. An organic sulphur-based protectant fungicide with low toxicity to mammals and low phytotoxicity, available in combination with dinocap as a wettable powder to produce a spray for controlling rose diseases.

Formaldehyde. An organic liquid used as a soil and plant disinfectant for the control of a wide range of pests and diseases but of variable effectiveness. It is commonly available as a 40% aqueous solution known as formalin which emits a pungent and irritating odour when concentrated.

Formothion. A systemic organophosphorus insecticide and acaricide closely related to dimethoate. Available as sprays to control aphids, scale insects, mealybugs and other sap-feeding pests and also effective against red spider mites and some other pests. Phytotoxic to chrysanthemums and probably to the other plants listed under dimethoate and toxic to bees, birds and other animals. Do not apply to edible crops within a week of harvest.

HCH. This is now the official common name for the organochlorine compound that has long been known as BHC, gamma-BHC, gammexane or lindane. It was developed at about the same time as DDT but is not so persistent and has therefore continued in use longer. It is formulated as sprays, aerosols, dusts or smokes, often combined with other insecticides or with fungicides. It is a broad-spectrum insecticide, killing aphids, capsids, leafhoppers, whiteflies, thrips, ants, caterpillars, sawflies, weevils, cabbage root flies, fungus gnats, earwigs, springtails, wireworms, leatherjackets, flea beetles and other insect pests and is also effective against millepedes and woodlice. It is phytotoxic to cucumbers, melons and other cucurbits, to hydrangeas, vines and young tomatoes and may cause musty tainting of currants, gooseberries, raspberries, strawberries and vegetables, especially potatoes, carrots, beetroot, onions and peas. It must therefore be used cautiously on edible crops and, with the appearance of newer insecticides to replace it, its main use will tend to be on ornamentals. It is toxic to bees and to fish and other wild animals.

Heptenophos. A non-persistent systemic organophosphorus insecticide. Available in a spray formulation, combined with permethrin, to control aphids, whiteflies, thrips, capsids, caterpillars and other pests of fruits, vegetables and ornamentals. May be used on edible crops to within a day of harvest. Toxic to bees and fish.

Malathion. This is a very useful non-persistent organophosphorus compound which is used as a spray, dust or aerosol to control many insects, especially aphids, whiteflies, leafhoppers, mealybugs, scale insects, thrips, caterpillars, sawflies, fungus gnats and also gives some control of red spider mites. Edible crops can be used four days after the last application although it is probably best to wait for at least a week if produce is to be preserved by bottling or freezing as this will eliminate any possibility of taint. It should not be used on antirrhinums, crassulas, ferns, gerberas, orchids, petunias, pileas, sweet peas or zinnias as it may be phytotoxic, and it is toxic to birds and fish.

Maneb. An inorganic sulphur and manganese-based protectant fungicide similar to zineb in properties and effectiveness but only available to gardeners in combination with carbendazim.

Menazon. This systemic organophosphorus insecticide is only available in a combined spray formulation with HCH, which restricts its possible uses. It mainly controls aphids but is relatively persistent and should not be applied to edible crops within three weeks of harvest. It is toxic to bees and to other animals.

Mercurous chloride. Generally known as calomel, this inorganic compound has long been used to check cabbage root fly and to control clubroot of brassicas, onion white rot and certain turf diseases. Although it contains mercury, it is not as toxic as other mercury compounds but is a powerful purgative, is highly phytotoxic and should be handled with care. It is available in a dust with low calomel content and also, probably from horticultural wholesalers only, as the pure chemical. There are now safer alternative materials available to gardeners for clubroot, white rot, turf disease and cabbage root fly control.

Metaldehyde. This chemical has been used as the basis of virtually all slug baits since the 1930's and is sold as at least six different brands of slug pellet and as a liquid. It has an anaesthetic effect on slugs but does not kill them outright, so that they can recover, especially in damp weather. It can harm cats and dogs, as well as other animals, and baits should be protected so that access to them is not easy. It should not be applied to edible crops within a week of harvest.

Methiocarb. The main use of this carbamate pesticide is as a slug killer and it is only available as specially formulated slug baits. It is more effective than metaldehyde, especially in moist weather. Do not apply near edible crops within a week of harvest and prevent access of birds, pets and other animals to baits for at least a week.

Nicotine. Pure nicotine is highly toxic but spray concentrates available to gardeners are relatively dilute and therefore not so dangerous. This is a useful non-persistent insecticide effective against aphids, leafhoppers, leaf miners, capsids, thrips, sawflies, small caterpillars and some other pests. It breaks down very rapidly and can be applied to edible crops up to two days before harvest. It shows no marked phytotoxicity and can therefore be used safely on almost any plants. It is toxic to bees and to fish and other animals.

Oxycarboxin. A systemic organic fungicide with fairly low toxicity to mammals and fairly low phytotoxicity. Oxycarboxin was one of the first systemic fungicides to be discovered and it is uniquely effective in controlling rust fungi. It is available as a wettable powder for producing sprays or drenches but is marketed only in large quantities and can probably only be bought from horticultural wholesalers.

Oxydemeton-methyl. A systemic organophosphorus compound, formulated only as an aerosol for aphid control on outdoor plants.

Permethrin. A synthetic pyrethroid compound, formulated as a spray to control aphids, whiteflies, leafhoppers, thrips, beetles and caterpillars. Similar to bioresmethrin and resmethrin but persisting up to three weeks after application. Non-phytotoxic and with very low mammalian toxicity.

Phenolic emulsions. A number of these products are sold either as general garden disinfectants or to control certain soil-borne pests and diseases.

While valuable for disinfecting pots and boxes their effectiveness in other respects is limited but they are discussed, where appropriate, in the text.

Pirimicarb. This carbamate insecticide is highly selective. It kills aphids quickly but does not affect other pests or beneficial insects, such as ladybirds, lacewings, whitefly parasite (*Encarsia*) or red spider mite predator (*Phytoseiulus*). It is therefore especially useful in situations where chemical and biological methods of control are combined into an integrated control system. It is moderately persistent and should not be applied to edible crops within two weeks of harvest but it is not toxic to bees and is virtually non-phytotoxic.

Pirimiphos-methyl. This organophosphorus insecticide and acaricide is available as spray, aerosol or dust for use against aphids, whiteflies, caterpillars, leaf miners, beetles, red spider mites and other pests. The dust formulation is effective against soil pests and can also be used to control woodlice. At least a week should elapse between application and the harvest of edible crops.

Pyrethrum. Natural pyrethrum, which is extracted from flower heads of *Chrysanthemum cinerariaefolium* and some other species, has been used as an insecticide for more than a century. It is formulated as a spray, dust or aerosol to control aphids, whiteflies, leafhoppers, thrips, beetles, smaller caterpillars and some other pests and it breaks down so quickly that it can be used to within a day of harvesting edible crops. It is very safe but often has to be formulated with synergists to enhance its insecticidal activity and will probably be gradually replaced by synthetic pyrethroid compounds, such as bioresmethrin, permethrin and resmethrin.

Quintozene. An organic chlorine-based fungicide that is particularly effective against a wide range of soil inhabiting fungi such as those causing damping off and root and foot rots. It has fairly low toxicity to mammals and low phytotoxicity (except to cucurbits). Quintozene is available to gardeners as a powder in combination with the insecticide carbaryl for use against lawn pests and diseases but as a dust for soil application for control of other diseases it can be bought only in large quantities and in this form is probably only available from horticultural wholesalers.

Resmethrin. This synthetic pyrethroid has been developed in recent years, together with bioresmethrin. It is formulated as a spray or aerosol and controls the same range of pests as natural pyrethrum. It is a very safe insecticide and can be used on edible crops up to a day before harvest.

Rotenone (Derris). Rotenone is the main insecticidal constituent of derris, which is extracted from ground roots of *Derris* plants. It is formulated as sprays or dusts to control aphids, whiteflies, thrips, caterpillars, sawflies, raspberry and flea beetles, red spider mites and some other pests. Residues break down very quickly and it can be applied to edible crops up to a day before harvest. It is a very safe insecticide but is toxic to bees and fish.

Sodium tetraborate. This is only used in a special formulation to control ants.

Sulphur. A chemical element which was an old and traditional remedy to combat a wide range of plant diseases and is a chemical constituent of many modern fungicidal compounds. Elemental sulphur is now only available as a garden product when combined chemically with calcium as lime sulphur or mixed with other fungicides and pesticides as a general panacea. The pure element can be obtained from chemists as flowers of sulphur however and is useful for dusting corms, bulbs and tubers to protect them from rotting organisms during storage and also for the treatment of a few foliar diseases such as some rusts and powdery mildews. It is also used occasionally to control tarsonemid mites. It has low toxicity to mammals and generally low phytotoxicity although some plants said to be 'sulphur-shy' may be damaged by it. Such plants include red and white currants and a number of apple cultivars.

Tar oil. This is used only as a winter wash applied to dormant fruit trees, canes and bushes and to some deciduous ornamental trees and shrubs to kill overwintering eggs of aphids, suckers and winter moths and to check scale insects and mealybugs. It also kills lichens and mosses. The spray can irritate skin, eyes, nose and mouth and can kill non-dormant plant tissues so must be used carefully.

Thiophanate-methyl. This is a systemic organic fungicide very similar to benomyl and is available as a wettable powder and as a liquid concentrate for producing sprays and a root dip for brassicas. Thiophanate-methyl is also available in a very useful formulation as a canker paint for treating wounds and cuts on woody plants.

Thiram. An organic sulphur-based protectant fungicide with low toxicity to mammals and low phytotoxicity. It is effective against a wide range of diseases but is most useful in gardens for controlling a number of foliar diseases of the spotting type, for seed treatment and for soil application against damping off diseases. It is available as a wettable powder for producing sprays and drenches and also in combination with insecticides. Thiram is the fungicide most commonly added to hormone rooting powders to protect cuttings from rotting.

Trichlorphon. This non-persistent organophosphorus insecticide is formulated as a spray to control caterpillars and earwigs. It may also be partially effective against cabbage root fly, carrot fly and onion fly but, unlike other organophosphorus compounds, has little effect on aphids or other sap-feeding pests. It can be used on edible crops up to two days before harvest.

Triforine. A systemic organic fungicide with low toxicity to mammals and low phytotoxicity. It is used primarily to control powdery mildews on ornamentals and is available as a liquid concentrate for producing sprays.

Warfarin. This is an anti-coagulant which is available only in specially formulated mouse and rat baits.

Zineb. An inorganic sulphur and zinc-based protectant fungicide with low toxicity to mammals and low phytotoxicity. It is effective against a wide range of foliar diseases but is most useful against those such as potato blight and downy mildews that are caused by phycomycetous fungi. It is available as a wettable powder for producing sprays. Mancozeb is a somewhat similar material and as it may be more generally available than Zineb, it could be substituted for it in the specific recommendations.

THE LAW AND GARDEN PLANT HEALTH

As with most other aspects of life, the law impinges in a number of ways upon the practice of gardening. It is inappropriate to consider here all the legal restrictions relating to gardens and gardening but as a number bear directly on the subject matter of this book they require explanation. This section of the book relates however only to the regulations prevailing in Britain although similar legislation exists in other countries.

In the general interests of maintaining the health of plants on which peoples' well-being or livelihood depend (and in one way or another this means most plants that gardeners are likely to grow), various Acts of Parliament have been passed. Legislation concerned specifically with plant health dates from the Destructive Insects Act, 1877 which was enacted to prevent the importation into Britain of the Colorado Potato Beetle but it was subsequently extended to other pests and some pathogens and its provisions in their latest form are embodied in the Plant Health Act, 1967. As implied, many of the requirements of these acts relate to the importation of plants and as such are only of direct concern to gardeners when they may wish to bring plants into this country from holidays overseas. No cultivated planting material may be imported without a phytosanitary certificate (in effect, a 'clean bill of health') granted by an inspector in the country of origin. Wild plants (subject to any local or other regulations on their conservation) may however be imported into Britain as a packet not exceeding 2 kg in weight, with passenger's luggage or as unaccompanied baggage provided a licence has been granted in advance by the Plant Health Branch of the Ministry of Agriculture under the Import and Export (Plant Health) (Great Britain) Order 1980. Such a licence, for which no fee is payable, must be surrendered to Customs Officers at the time of landing. A list of the plants imported must be forwarded immediately to the Ministry but it should be pointed out that certain plants, a list of which is issued with the licence, may not be imported under any circumstances. These plants include potatoes, many conifers and sweet chestnuts from all countries, annual and biennial plants from outside Europe and *Prunus* spp. from many parts of Europe. In most cases these prohibitions are to prevent the bringing into Britain of non-indigenous pests and pathogens.

Other provisions have also been made under the Plant Health Act and its predecessors and the principal Orders and their relevance to gardeners are given below.

The various Sale of Diseased Plants Orders, 1927 to 1952, give to the officers of the Ministry of Agriculture power to require anyone offering for sale plants substantially attacked by certain 'insects or pests' to destroy or disinfect them or take other measures to prevent the spread of the problems. A fine not exceeding ten pounds may also be imposed. Although it is exceedingly rare for the Orders to be invoked, gardeners should be aware that they are, in theory at any rate, protected against persons who attempt to sell them plants affected with: fruit tree cankers caused by fungi or bacteria, American gooseberry mildew, silver leaf, blackcurrant mite, woolly aphis, scale insects, brown tail moth, rhododendron bug, apple capsid, potato powdery scab, glasshouse white fly (tomatoes or cucumbers only), clubroot, onion smut and any insect or pest on plants, bulbs, corms, tubers or rhizomes of narcissus, tulip, hyacinth, gladiolus, iris, crocus and potatoes which renders them visibly unfit for planting. In practice the Orders are difficult to enforce as expressions such as 'substantially attacked' or 'visibly unfit for planting' are open to a number of interpretations.

Certain pests and diseases are subject to more specific legislation which renders a number of them 'notifiable', that is, their discovery must by law be reported to officers of the Ministry of Agriculture. Most of these are discussed under the appropriate entries in the main part of the book but the legislation concerned includes the various Colorado Beetle Orders, 1933 to 1950; the Plum Pox (Sharka Disease) Order, 1975; the Potato Cyst Eelworm (Great Britain) Order, 1973; the Red Core Disease of Strawberry Plants Order, 1957 (England and Wales only); the Wart Disease of Potatoes (Great Britain) Orders, 1973 to 1974 and the Fire Blight Disease Orders, 1958 to 1966. Less stringent legislation which does not render the problems notifiable, exists under the Silver Leaf Order, 1923 and the Mediterranean Carnation Leaf-Roller (Great Britain) Order, 1975. Certain similar items of legislation are concerned with the freedom of infestation and disease of some fruit trees and bushes but apply only to Scotland.

Additionally, certain other Acts have a bearing on garden plant health. The first of these relate to the felling of trees and may be a matter of concern when a diseased tree needs to be removed. Under the Forestry Act, 1967 a licence is required for the felling of growing trees (dead trees do not come within the scope of the legislation) but among the exemptions from this are trees in gardens and fruit trees. Such trees may however be subject to a Tree Preservation Order made under the Town and Country Planning Act, 1971 or the Town and Country Planning (Scotland) Act, 1972 and in such circumstances the consent of the local authority imposing the order will be required before felling or lopping may take place. If it can be shown that a tree is dangerous or dead however such work may still be performed despite the existence of a preservation order. The owners and occupiers of land must be notified and have a right to object if a local authority proposes to effect a new tree preservation order.

Mention should also be made of the Protection of Birds Acts, 1954 to 1976 for birds are the only harmful agents described in this book that enjoy legal protection. All species are so protected but certain 'pest' species can be killed in certain ways by 'authorised persons' who will usually be the owners or occupiers of affected land.

Finally, although not really concerned with plant health, it is not inappropriate to remind gardeners that under the provisions of the Conservation of Wild Creatures and Wild Plants Act, 1975 no wild plant may be uprooted without the consent of the owner of the land on which it is growing and some may not be uprooted under any circumstances.

A–Z Key

Abies (Firs)
Leaves shed prematurely

See discussion under
Needle Casts p. 305

Leaves, buds and stems with tufts of white woolly
wax; growth distorted and sometimes dying

Silver Fir Adelgids
p. 183

Stems with cankers and/or dying back of branches
or crown

Cankers and Diebacks
p. 309

Also see **Conifers** and **Trees**

Abutilon
Leaves with fine angular yellow/green variegation

Viruses p. 429

Also see **Glasshouse and House Plants**

Acacia. See **Glasshouse and House Plants** and for
False Acacia see **Robinia**

Acer (Sycamore and Maples)
Leaves with black or less commonly other coloured
spots

Tar Spot p. 393

Leaves of sycamores and field maples with rashes of
small red pimples

Sycamore Gall Mite
P. 254

Bark peels to reveal sooty mass; leaves wilt; tree
unthrifty

Sooty Bark Disease
p. 327

Bark deeply grooved; esp. on hedgerow maples

Winged Cork p. 488

Branches and trunks with brown scales and tufts of
white woolly wax

Horse Chestnut Scale
p. 187

Also see **Trees**

Achimenes. See **Glasshouse and House Plants**

Acidanthera
Corms with sunken lesions; leaves spotted or
browning rapidly

Gladiolus Dry Rot
p. 341 or
Gladiolus Hard Rot
p. 342

Also see **Bulbs etc.**

Aconite. For **Winter Aconite** see **Eranthis**

Acorus (Flag). See **Bulbs etc.**

Adiantum (Maidenhair). See **Ferns**

Aesculus (Horse Chestnut)
Leaves with irregular brown blotches

Leaf Blotch p. 407

Leaves yellowed or with irregular patterns

Viruses p. 434

Branches and trunks with brown scales and tufts of
white woolly wax

Horse Chestnut Scale
p. 187

Also see **Trees**

43

African Marigold. See **Tagetes**

African Violet. See **Saintpaulia**

Agave. See **Cacti and Succulents**

Alder. See **Alnus**

Allium. For ornamental species see **Onion**

Almond. See **Apricot** and **Trees**

Alnus (Alder)

Leaves and/or other parts with powdery white coating	Hazelnut Powdery Mildew p. 299
Leaves blistered and curled; may be yellow coating	Leaf Curls p. 332
Leaves and twigs with tufts of white woolly wax	Alder Psyllid p.157
Leaves with irregular holes and pieces eaten away	Caterpillars p. 194
Branches and stems with numerous small white scales	Willow Scale p. 188
Branches and stems with larger brown scales and white wax egg sacs	Woolly Currant Scale p. 187
Branch or trunk with some decay of wood; fungal brackets may be present	Heart Rot p. 373
Also see **Trees**	

Aloe. See **Cacti and Succulents**

Alpine and Rock Garden Plants

Pests, diseases and disorders seldom cause serious trouble on plants growing in alpine houses and in rock gardens. This is partly due to the general hardiness of this group of plants and to the fact that many of them grow most actively and flower early in the season when temperatures are low and pest and disease organisms are therefore less prevalent. Some pests and diseases are listed under those plant genera that are especially susceptible and the following key can be used to identify general troubles on other plants:

Flowers and leaves torn; plants sometimes uprooted	Birds p. 255
Flowers, leaves and young shoots with irregular holes and pieces eaten away; slime trails on or near plants	Slugs and Snails p. 139
Plants unthrifty, small white insects on roots	Root Mealybugs p. 190 Root Aphids p. 181
Flowers, leaves and young shoots with irregular holes and pieces eaten away, sometimes with silk webbing over damaged parts; no slime trails	Caterpillars p. 194
Leaves, buds and shoots infested by small green, yellow, brown or black wingless and winged insects; foliage sticky, sometimes sooty	Aphids p. 161
Leaves and/or other parts with powdery white coating	Powdery Mildew p. 302
Plants generally unthrifty; roots blackened and/or decayed	Root and Foot Rot p. 361
Also see **Glasshouse and House Plants** and **Cacti and Succulents**	

Alstroemeria
Leaves, stems and young shoots with irregular holes;   Slugs and Snails
  slime trails present on or near plants   p. 139
Roots eaten by white caterpillars in soil   Swift Moths p. 201

Althaea (Hollyhocks)
Leaves with small, ragged holes, mainly on younger
  leaves   Capsid Bugs p. 151
Leaves, buds and flowers with larger holes   Caterpillars p. 194
Leaves yellowing and drying, numerous small mites   Glasshouse Red
  on undersides, sometimes with silk webbing   Spider Mites
                                                           p. 247
Leaves and stem with raised orange-brown pustules   Rust p. 278
Stem and/or root with large irregular swelling   Crown Gall p. 330

Alyssum
Leaves of young seedlings with numerous small
  round holes and pits   Flea Beetles p. 238
Leaves and/or other parts with powdery white coating   Powdery Mildew
                                                           p. 302
Roots with irregular nodular swellings   Clubroot p. 328

Amaryllis
Leaves red-brown and becoming shrivelled from tip   Narcissus Leaf
  downwards   Scorch p. 412
Leaves mottled   Hippeastrum Mosaic
                                                           Virus p. 434

For indoor Amaryllis (Hippeastrum) see **Glasshouse
  and House Plants**

Amelanchier
Flowers and/or leaves shrivelled; shoots may exude   Fireblight p. 316
  slime and die back
Also see **Trees**

Androsace. See **Alpine and Rock Garden Plants**

Anemone
Flowers and/or buds with fluffy grey mould growth   Grey Mould p. 346
Flower buds killed or producing distorted flowers   Scabious Bud
                                                           Eelworm p. 136
Plants unthrifty; roots with short dark longitudinal   Root-Lesion
  slits   Eelworms p. 139
Plants unthrifty; white, centipede-like creatures in
  soil   Symphylids p. 144
Leaves and/or other parts with powdery white coating   Powdery Mildew
                                                           p. 302
Leaves mottled; flowers of dark-coloured types may
  be streaked with pale colour   Viruses p. 434
Leaves with small, cup-like bodies on undersides   Plum Rust p. 283
Leaves with pale yellowish spots; off-white mould
  growth beneath   Downy Mildew p. 295
Leaves and/or stem with dark streaks and blisters   Smut p. 288
Leaves of young seedlings with numerous small
  round holes and pits   Flea Beetles p. 238

Leaves, stems, flowers and buds with irregular large
holes · Caterpillars p. 194

Leaves with irregular pieces eaten away; slime trails
present · Slugs and Snails
p. 139

Leaves, stems, flowers and buds sticky, sometimes
sooty, small insects present · Aphids p. 161

Roots eaten by white caterpillars in soil · Swift Moths p. 201

Roots eaten by dirty grey-brown caterpillars in soil · Cutworms p. 200

Annuals. See **Bedding Plants**

Anthurium. See **Glasshouse and House Plants**

Antirrhinum (Snapdragons)

Flowers and shoot tips distorted, buds sometimes
killed · Strawberry Mite
p. 251

Flowers with holes eaten in bases · Bumble Bees p. 233

Flowers, leaves, stems and buds sticky, sometimes
sooty, small insects present · Aphids p. 161

Leaves (undersides) and/or stem with chocolate
brown pustules · Rust p. 269

Leaves with dark spots · Leaf Spots p. 393

Leaves with pale spots and holes · Shot Hole p. 394

Leaves and/or other parts with powdery white coating · Powdery Mildew
p. 302

Aphelandra. See **Glasshouse and House Plants**

Apple

During their long period of growth apple trees may be affected by many different
pests, diseases and disorders. Only the most important of these are noted here. For
more detailed information consult specialist publications (see Bibliography
p. 493).

The following key to symptoms has been divided into sections for easier
reference and, since the fruit is the end-product of cultivation, damage to mature
fruit is dealt with first.

FRUIT

Fruit in store with large pieces eaten away, often with
small tooth marks showing · Mice p. 261

Fruit with rotten cores tunnelled by caterpillars, often
with holes to exterior · Codling Moth
p. 197

Fruit with superficial galleries in flesh, often under
leaves attached to surface of fruit by silk webbing · Fruit Tortrix Moths
p. 198

Fruit with small pieces pecked away · Birds p. 255

Fruit misshapen and with curved, discoloured scars in
skin; fruitlets with rotten cores, mostly falling
prematurely · Apple Sawfly p. 224

Fruitlets irregularly misshapen after earlier infestation by aphids · Rosy Apple Aphid
p. 165

Fruit misshapen, with superficial bumps and discoloured patches on skin · Capsid Bugs p. 151

Fruit with soft brown rot and usually off-white cottony pustules; associated twigs may die back · Brown Rot p. 350

Fruit with round, dark, saucer-like depressions; usually only in store — Bitter Rot p. 349

Fruit leathery and cracked; often on windfalls and low hanging fruit — Leathery Rots p. 354

Fruit rotted and with grey powdery mould growth; usually in store — Grey Mould Rot p. 352

Fruit with brown sunken area around calyx — Dry Eye Rot p. 353

Fruit otherwise rotted; especially in store — Minor Rots p. 354

Fruit split or hollow — Splitting p. 488

Fruit scorched following very hot weather — High Temperature Injury p. 484

Fruit with corky patches both on and within flesh — Boron Deficiency p. 479

Fruit with dark spots or pits both on and within flesh — Calcium Deficiency p. 476

Fruit with very dark green more or less rounded corky patches on surface only — Scab p. 335

Fruit with dark green mouldy patches on surface; easily wiped off — Sooty Blotch p. 420

Fruit abnormally small and red — Nitrogen Deficiency p. 474

FLOWERS

Flowers fail to open and petals turn brown, small insect larvae sometimes present inside closed flowers — Apple Blossom Weevil p. 241

Flowers and flower buds discoloured, failing to develop normally, often sticky and with small flattened insects present — Apple Sucker p. 156

Flowers and flower buds with holes; small looper-type caterpillars often present — Winter Moths p. 196

Flowers and then spur leaves wilt — Brown Rot Blossom Wilt p. 350

Flowers and/or buds die suddenly — Frost p. 482 but also see Bud Rot p. 345

LEAVES

Leaves with small, irregular, tattered holes, especially at tips of shoots — Capsid Bugs p. 151

Leaves with larger, irregular holes and pieces eaten away — Caterpillars p. 194 Leaf Weevils p. 241 Chafers p. 235

Leaves, shoots and buds infested by small insects; foliage sticky and sooty and sometimes discoloured and distorted — Apple Aphids p. 165

Leaves bronzing and drying in summer, numerous small mites on undersides — Fruit Tree Red Spider Mite p. 249

Leaves with fine light flecking, moult skins of insects sometimes on undersides — Leafhoppers p. 154

Leaves and young shoots with frothy masses of spit-like substance — Froghoppers p. 153

Leaves with irregular yellowish streaks and patterns in early summer — Apple Mosaic Virus p. 436

Leaves scorched at margins in spring — Frost p. 482

Leaves and/or other parts with powdery white coating — Powdery Mildew p. 297

Leaves and/or other parts with sticky black coating — Sooty Mould p. 303

Leaves with silvery sheen; branches may die back — Silver Leaf p. 324

Leaves with irregular dark green blotches — Scab p. 335

Leaves more or less bleached; veins remain dark green — Iron Deficiency p. 478

Leaves with irregular marbling; may drop prematurely — Magnesium Deficiency p. 477

BUDS

Buds pecked in winter and spring, causing lack of blossom and leaves — Birds p. 255

BRANCHES, TWIGS AND SHOOTS

Branches and twigs with tufts of white woolly wax, often associated with hard, irregular woody swellings — Woolly Aphid p. 166

Branches tunnelled by large caterpillars, causing wilting of shoots and leaves — Leopard Moth p. 199

Branches die back in absence of stem lesions; leaves silvery — Silver Leaf p. 324

Branches die back; stem base and roots decayed; toadstools and bootlace-like strands may be present — Honey Fungus p. 375

Branches with canker lesions; may die back beyond lesions — Canker p. 310

Branches die back; masses of small, usually pink pustules present — Coral Spot p. 314

Branches with parasitic leathery plant growing from them — Mistletoe p. 427

Branches and/or stem with hard irregular swelling — Crown Gall p. 330

Branches with masses of proliferating shoots (witches' brooms) (not Britain) — Proliferation Mycoplasma p. 436

Branches (esp. of cv. 'Lord Lambourne') pliable and rubbery — Rubbery Wood Mycoplasma p. 436

Shoots exude slime and die back; flowers and/or leaves shrivelled — Fireblight p. 316

TRUNKS AND STEMS

Trunks and stems of young trees with bark stripped near and just above ground level — Rabbits p. 259 Voles p. 262 Hares p. 260

Trunks with small round holes, often with fine wood dust emerging; affected trees may die — Shot-hole Borer p. 244

ROOTS

Roots and stem base decayed; toadstools and bootlace-like strands may be present and branches die back — Honey Fungus p. 375

GENERAL DEBILITY

Trees generally unthrifty	Latent Viruses p. 436
	Specific Replant Disease p. 428
	Deficiencies of Nitrogen p. 474
	Phosphorus p. 475
	and Potassium p. 475
Also see **Trees**	

Apricot, Almond, Nectarine and Peach

Fruits with holes eaten in ripening flesh	Birds p. 255 Wasps p. 232
Fruit with soft brown rot and usually off-white cottony pustules; associated twigs may die back	Brown Rot p. 350
Fruit otherwise rotted; especially in store	Minor Rots p. 354
Fruit with brown-green, more or less rounded corky patches on surface (not apricot)	Prunus Scab p. 337
Flowers and/or buds die suddenly	Frost p. 482
Flowers and then spur leaves wilt	Brown Rot Blossom Wilt p.350
Leaves with irregular yellowish streaks and patterns	Viruses p. 437
Leaves scorched in early spring	Frost p. 482
Leaves and/or other parts with powdery white coating	Powdery Mildew p. 300
Leaves and/or other parts with sticky black coating	Sooty Mould p. 303
Leaves with silvery sheen; branches may die back	Silver Leaf p. 324
Leaves with rounded green-brown blotches (not apricot)	Prunus Scab p. 337
Leaves blistered, crumpled and reddish in early summer	Peach Leaf Curl p. 332
Leaves with more or less regular holes; no gummy branch lesions	Peach Shot Hole p. 414
Leaves with brown or black pustules beneath; small yellow spots above	Plum Rust p. 283
Leaves more or less bleached; veins remain dark green	Iron Deficiency p. 478
Leaves and shoots infested by small insects, often sticky and sooty	Aphids p. 161
Leaves discoloured, bronzing and drying, with numerous small mites on undersides	Glasshouse Red Spider Mites p. 247 Fruit Tree Red Spider Mite p. 249
Leaves with large irregular holes and pieces eaten away	Caterpillars p. 194
Branches and stems with brown, waxy scales, leaves often sticky and sooty	Scale Insects p. 185
Branches die back in absence of stem lesions; leaves silvery	Silver Leaf p. 324
Branches and/or stems with many gummy lesions; may die back and leaves have irregular holes	Bacterial Canker p. 320
Branches and/or stem with hard irregular swelling	Crown Gall p. 330

Trees generally unthrifty

Specific Replant
  Disease p. 428
Deficiencies of
  Nitrogen p. 474
  Phosphorus p. 475
  and Potassium
  p. 475

Also see **Trees**

Aquatic Plants
Aquatic plants are seldom affected by pests, diseases or disorders although water-lilies are sometimes damaged by insects or affected by leaf spots (see Nymphaea). Insecticides should not be used on aquatic plants growing in ponds and streams since they kill fish and other aquatic animals.

Aquilegia
Leaves and/or other parts with powdery white coating

Powdery Mildew
  p. 302

Leaves spotted

Leaf Spots p. 394

Leaves and shoots infested by small insects, often sticky and sooty

Aphids p. 161

Arabis
Leaves with yellowish spots or irregular patterns

Viruses p. 437

Araucaria (Monkey-puzzle). See **Conifers** and **Trees**

Arbor Vitae. See **Thuja**

Arbutus (Strawberry Tree)
Fruits eaten

Birds p. 255

Leaves with small irregular spots

Leaf Spots p. 394

Also see **Trees**

Arctotis
Leaves, shoots and flowers infested by small insects; leaves often sticky and sooty

Aphids p. 161

Armeria
Leaves with brown pustules

Rust p. 269

Artemisia
Leaves with pale brown pustules

Tansy Rust p. 286

Leaves and shoots with frothy masses of spit-like substance

Froghoppers p. 153

Leaves and shoots infested by small insects, sometimes sticky and sooty

Aphids p. 161

Roots infested by small white insects; plants unthrifty

Root Aphids p. 181

Arum Lily. See **Zantedeschia**

Aruncus
Leaves extensively eaten, often reduced to a skeleton   Spiraea Sawfly
of veins; caterpillars may be present   p. 228

Asclepias
Leaves with yellowish mottling or other irregular
patterns   Viruses p. 438

Ash. See **Fraxinus**

Asparagus (Vegetable)
Leaves and stems eaten, grey/black larvae and/or   Asparagus Beetle
yellow/black beetles may be present   p. 239
Shoots eaten, slime trails present on or near plants   Slugs p. 139
Shoots with masses of brown powdery pustules   Rust p. 270
Shoots blacken and die suddenly in spring   Frost p. 482
Shoots with fluffy grey mould growth   Grey Mould Rot
                                                          p. 352
Plants die back with decay at base and/or on roots   Root and Foot Rot
                                                          p. 361
Plants feeble and stunted; roots decayed with matted   Violet Root Rot
dark strands   p. 360

Asparagus (Ornamental). See **Glasshouse and
House Plants**

Aspidistra. See **Glasshouse and House Plants**

Asplenium. See **Ferns**

Aster (China). See **Callistephus**

Aster (Michaelmas daisies)
Flowers malformed, often killed, leaving rosettes of
green calyces; young shoots distorted, lateral buds   Strawberry Mite
killed   p. 251
Leaves and shoot tips infested by small insects, often
sticky and occasionally sooty   Aphids p. 161
Leaves with irregular holes   Caterpillars p. 194
Leaves and shoots with white froth covering wingless
insects   Froghoppers p. 153
Leaves discoloured, growth stunted   Chrysanthemum
                                                          Eelworm p. 134
Leaves and/or other parts with powdery white coating   Powdery Mildew
                                                          p. 302
Leaves wilt; stem interior with brown or black streaks   Wilt p. 306
Shoots eaten near soil, slime trails present on or near
plants   Slugs p. 139

Aubergine (Egg Plant)
Leaves discoloured, bronzing and drying, numerous   Glasshouse Red
small mites on undersides   Spider Mites
                                                          p. 247
Leaves and/or other parts variously crumpled and   Nicotiana Viruses
distorted   p. 455

Aubrieta. See **Bedding Plants**

Auricula. See **Primula**

Autumn Crocus. See **Colchicum**

Azalea. See **Rhododendron**

Barberry. See **Berberis**

Bay Laurel. See **Laurus**

Beans (Dwarf or French, Runner and Broad)

Seeds fail to germinate and/or seedlings grow poorly and may die	Mice p. 261
	Slugs p. 139
	Bean Seed Fly p. 219
	Millepedes p. 143
Plants stunted and deformed, sometimes collapsing	Stem Eelworm p. 132
	Pea Cyst Eelworm p. 138
Pods of broad beans with triangular pieces pecked out and beans removed	Jays p. 258
Pods rotted and with fluffy grey mould growth	Grey Mould Rot p. 352
Pods with rounded greasy spots (runner and French beans only)	Halo Blight p. 395
Pods with rounded reddish spots; reddish veins on undersides of leaves (runner and French beans only)	Anthracnose p. 395
Seeds with rounded dark brown lesions within; plant unthrifty	Manganese Deficiency p. 479
Leaves discoloured, yellowing, bronzing and dying, many small mites on undersides	Glasshouse Red Spider Mites p. 247
Leaves and shoots infested by small black insects, sometimes sticky and sooty	Black Bean Aphid p. 171
Leaves with small semi-circular notches eaten out of edges, producing a scalloped effect	Pea and Bean Weevils p. 242
Leaves with reddish veins beneath; pods with rounded reddish spots (runner and French beans only)	Anthracnose p. 395
Leaves with dry lesions surrounded by marked yellow halo; pods may have greasy spots (runner and French beans only)	Halo Blight p. 395
Leaves with yellowish patches on uppersides and purple-white mould growth beneath (broad beans only)	Pea Downy Mildew p. 295
Leaves with chocolate brown spots (broad beans only)	Chocolate Spot p. 395
Leaves and other parts with pale brown pustules beneath (broad beans only)	Rust p. 270
Leaves crumpled and/or with irregular yellowish patterns	Viruses p. 438
Leaves wilt (oldest first); stem with dark streaks within, extending well above ground level	Wilt p. 306

Leaves wilt; stem rotted at base; no dark streaks within stem well above ground level	Root and Foot Rot p. 361
Leaves with scorched margins; oldest affected first	Potassium Deficiency p. 475
Leaves turn blue-green or yellowish and may wilt	Copper Deficiency p. 480
Stems streaked; leaves with reddish veins beneath (runner and French beans only)	Anthracnose p. 395
Stems and roots rotted and blackened at base	Root and Foot Rot p. 361
Stems rotted at base; white cottony mould present	Sclerotinia Disease p. 369
Stem with long dark streaks within; leaves wilt	Wilt p. 306
Roots blackened and decayed	Root and Foot Rot p. 361
Roots with small nodular swellings	Root Nodules p. 428

Bean Tree. See **Catalpa**

Bedding Plants
Ornamental annuals and biennials that are grown for summer bedding need special care if they are to do well during their short period of growth. They are often raised in warmth and, if not adequately hardened-off, the soft, succulent growth will be susceptible to rotting diseases. Handle fragile young plants gently when planting out and always try to keep a few plants in reserve to fill gaps caused by early losses. During their initial growth in frames and glasshouses young plants may be affected by many of the pests, diseases and disorders of **Glasshouse and House Plants** (see p. 80 for key) and of **Seedlings** (see p. 116). In addition, transplants or direct-sown plants growing outdoors may be damaged by soil pests, especially **Wireworms** (p. 236), **Leatherjackets** (p. 215), **Chafers** (p. 235), **Cutworms** (p. 200) and **Swift Moths** (p. 201) and soil-borne pathogens such as those causing **Root and Foot Rots** (p. 361), all of which may cause sudden wilting, collapse and death. Among diseases frequently found affecting the leaves of some genera are **Downy Mildews** (p. 291), **Powdery Mildews** (p. 296), **Grey Mould** (p. 346) and certain **Leaf Spots** (p. 391) and **Stem and Leaf Rots** (p. 379). **Slime Moulds** (p. 424) may occasionally infest bedding plants and the commonest disorder is probably **Magnesium Deficiency** (p. 477) which gives rise to pale interveinal areas on the leaves. Other troubles are noted under the main plant genera.

Beech. See **Fagus**

Beet (Spinach and Beetroot)

Leaves and other above ground parts with yellow-brown pustules	Rust p. 270
Leaves pale and toughened with grey-purple mould on undersides	Downy Mildew p. 292
Leaves with small, more or less circular pale spots	Leaf Spots p. 396
Leaves with brown spots	Blackleg p. 363
Leaves toughened, crumpled and/or with abnormal colouration	Viruses p. 439
Leaves with inter-veinal yellowing and sometimes dead patches also; edges may roll upwards; older leaves affected first	Manganese Deficiency p. 479

Leaves infested by small black insects   Black Bean Aphid
                                                              p. 171
Leaves with conspicuous brown blotch mines   Beet Leaf Miner
                                                              Fly p. 217
Roots with irregular holes and cavities   Cutworms p. 200
                                                              Swift Moths p. 201
Roots decayed with mottled dark strands; plants often   Violet Root Rot
  feeble   p. 360
Roots with scab-like patches of corky tissue   Common Scab p. 336
Roots with shiny black areas and red-brown decay   Blackleg and Dry
  beneath   Rot p. 363
Roots with rough surface patches and dark patches or   Boron Deficiency
  rings within flesh   p. 479
Roots with irregular swellings   Crown Gall p. 330
Seedlings collapse with blackened and shrivelled stem
  base   Blackleg p. 363
Plants run to flower prematurely   Bolting p. 487

Beetroot. See **Beet**

Begonia
Leaves with irregular mottling; plants may be stunted   Viruses p. 439
Leaves and/or other parts with powdery white coating   Powdery Mildew
                                                              p. 302
Leaves wilted; stem interior with brown or black
  streaks   Wilt p. 306
Leaves with rough warty outgrowths on undersides   Oedema p. 488
Leaves infested by small white winged insects; plants   Glasshouse
  sometimes sticky and sooty   Whitefly p. 158
Leaves infested by green or yellow insects, sometimes
  sticky and sooty   Aphids p. 161
Leaves with yellow to dark-brown or black areas be-   Chrysanthemum
  tween veins   Eelworm p. 134
Leaves distorted, puckered, slightly thickened and
  with margins curled downwards; sometimes with   Tarsonemid Mites
  rusty colour; flowers malformed   p. 251
Leaves and stems with colonies of soft-bodied insects   Glasshouse
  covered with white wax powder and filaments   Mealybugs p. 189
Leaves with discoloured areas, yellowing and silver-
  ing; small narrow-bodied insects present   Thrips p. 191
Leaves with brown scales on undersides; foliage
  sticky and sooty   Scale Insects p. 185
Plants unthrifty; roots with numerous irregular hard   Root-Knot
  lumps   Eelworms p. 136
Roots, tubers and rhizomes eaten; small C-shaped
  legless larvae present   Vine Weevil p. 242
Also see **Bulbs etc. Bedding Plants** and
  Glasshouse and House Plants

Berberis (Barberry)
Leaves with irregular mottling and/or crumpling   Viruses p. 470
Leaves with orange spots on upper surface and pus-
  tules beneath   Rust p. 271
Branches with masses of proliferating shoots (witches'
  brooms) (rare in Britain)   Rust p. 271

Betula (Birches)

Leaves and buds infested by small insects, foliage often sticky and sooty	Birch Aphids p. 180
Leaves with irregular holes and pieces eaten away	Caterpillars p. 194
	Weevils p. 241
	Sawflies p. 224
Leaves with masses of yellow pustules on undersides	Rust p. 271
Leaves with powdery white coating	Powdery Mildew p. 302
Leaves with irregular yellowish streaks and patterns	Viruses p. 439
Branches with masses of proliferating shoots like large birds' nests	Witches' Brooms p. 334
Stem and/or branches with dry, crumbly rot; fungal brackets may be present	Birch Polypore p. 373
Also see **Trees**	

Biennials. See **Bedding Plants**

Bilberry. See **Vaccinium**

Birch. See **Betula**

Blackberry, Dewberry

Fruit rotted and with fluffy grey mould growth	Grey Mould p. 352
Fruits malformed and maggoty	Raspberry Beetle p. 237
Flowers killed suddenly	Frost p. 482
Leaves and young shoots with masses of spit-like substance	Froghoppers p. 153
Leaves with light mottling, insect moult skins often on undersides	Leafhoppers p. 154
Leaves with small tattered holes	Capsid Bugs p. 151
Leaves and/or other parts with powdery white coating	Powdery Mildew p. 302
Leaves with orange-yellow and/or black pustules on undersides	Rusts p. 271
Leaves with greyish circular purple-bordered spots and no pustules beneath	Purple Blotch p. 396
Stems with orange, slit-like lesions	Common Rust p. 271
Stems with greyish elliptical purple-bordered spots	Cane Spot or Purple Blotch p. 396
Stems with hard, irregular swellings	Crown Gall p. 330
Roots with hard, irregular swellings	Crown Gall p. 330

Black Currants. See **Currants**

Bluebell. See **Endymion**

Box. See **Buxus**

Brassicas (Broccoli, Brussels Sprouts, Cabbages, Cauliflowers, Chinese Cabbages, Kale and Kohlrabi)

Leaves and/or other parts with powdery white coating	Powdery Mildew p. 297
Leaves with brownish angular patches and greyish mould beneath	Downy Mildew p. 292

Leaves variously spotted	Leaf Spots p. 397
Leaves and/or other parts with small, often concentric clusters of creamy droplets	White Blister p. 304
Leaf blades narrow; leaves appearing whip-like; esp. cauliflower	Molybdenum Deficiency p. 481
Leaves with blue-green colouration, often marginal scorching and plant generally unthrifty. On older leaves first	Potassium Deficiency p. 475
Leaves small, pale but with yellow, red or purplish tints, on older leaves first; plant generally unthrifty but no obvious root damage	Nitrogen Deficiency p. 474
Leaves stunted and brittle and, when cut, brownish patches in cauliflower curds and stem	Boron Deficiency p. 479
Leaves within cabbage heads and Brussels sprout buttons with small dark patches (internal browning)	Calcium Deficiency p. 476
Leaves with inter-veinal yellowing; usually on older leaves first	Manganese Deficiency p. 479
Leaves with inter-veinal yellowing giving pronounced marbled effect	Magnesium Deficiency p. 477
Leaves and/or stem with rough warty outgrowths	Oedema p. 488
Leaves variously affected with irregular yellowish or dark green patterns, crinkling and/or crumpling	Viruses p. 440
Leaves with dark margins; plants wilt, remain stunted and may decay (unusual in Britain)	Black Rot p. 350
Leaves wilting, plants collapsing and dying, usually in dry weather after transplanting; white maggots feeding on roots	Cabbage Root Fly p. 218
Leaves infested by small white-winged insects and oval yellow scales, especially on undersides; leaves sticky and sooty	Cabbage Whitefly p. 159
Leaves infested by dense colonies of grey-green insects covered in white powdery wax	Cabbage Aphid p. 172
Leaves pecked and torn, often reduced to a skeleton of veins	Wood Pigeon p. 256
Leaves of young plants with many small circular holes and pits	Flea Beetles p. 238
Leaves with irregular holes and pieces eaten away; caterpillars present, often eating into hearts	Cabbage Caterpillars p. 201
Leaves with fine white flecking, small elongate insects present	Thrips p. 191
Leaves with irregular holes eaten out, slime trails present on or near plants	Slugs and Snails p. 139
Heads with soft and slimy decay; usually no mould growth	Bacterial Soft Rot p. 348
Heads with soft rot and fluffy grey mould growth	Grey Mould Rot p. 352
Curds with dark internal patches	Downy Mildew p. 292 or Boron Deficiency p. 479 Former tends to give black patches and the latter brown

Curds fail to form	Blindness p. 487
Stem with dark internal patches (esp. cauliflower)	Boron Deficiency p. 479
Stem with rough warty outgrowths	Oedema p. 488 or Herbicide Damage p. 491
Stem with elongate sunken areas at ground level; sometimes completely girdled	Canker and Blackleg p. 364
Stems eaten at soil level, plants collapsing and dying	Cutworms p. 200
Stems and roots with rounded irregular hollow outgrowths (galls) containing legless white larvae	Turnip Gall Weevil p. 242
Roots with irregular swellings not containing larvae or holes	Clubroot p. 328
Seedlings die, usually in patches. Roots and/or stem base rotted	Damping-off p. 362
Seedlings unthrifty and with tough wiry stems	Wirestem p. 362
Seedlings die with ashen spots on cotyledons	Blackleg p. 364
Seedlings with yellow specks on upper leaf and cotyledon surfaces and off-white mould beneath	Downy Mildew p. 292
Plants run to flower	Bolting p. 487

Broad Beans. See **Beans**

Broccoli. See **Brassicas**

Broom. See **Cytisus** and **Sarothamnus**

Brussels Sprouts. See **Brassicas**

Buddleia

Leaves narrowed, crumpled or otherwise distorted	Viruses p. 441
Leaves with small, tattered holes, especially on young shoots	Capsid Bugs p. 151
Leaves distorted, growth of young shoots checked, buds sometimes killed	Chrysanthemum Eelworms p. 134

Bulbocodium

Leaves and corms with blister-like swellings and dark brown powdery contents	Colchicum Smut p. 289
See **Bulbs etc.**	

Bulbs, Corms, Rhizomes and Tubers

Bulbs, corms, rhizomes and tubers should always be carefully examined before planting, to detect any signs of pests and diseases which might prove difficult to eradicate after planting. The most important symptoms at this stage are:

Bulbs feel soft when squeezed and show circles of brown discoloured tissues when cut across with a knife	Stem Eelworm p. 132
Bulbs feel soft when squeezed, rotting cavity in centre, often with fly maggot present	Large Narcissus Fly p. 221
Bulbs, corms, rhizomes or tubers otherwise soft and slimy	Bacterial Soft Rot p. 348
Bulbs, corms, rhizomes or tubers with decay and blue-green mould growth	Blue Mould Rot p. 339

Bulbs with red markings around necks and showing
    internal red discolouration of scales when cut across   Tarsonemid Mites
    with knife   p. 251
Bulbs, corms, rhizomes or tubers with wounds in-
    fested by small, shining white mites with long hairs   Bulb Mite p. 254

The main symptoms appearing after planting are:

Leaves weak, distorted and discoloured, sometimes
    with small yellow bumps; flowers weak and
    distorted   Stem Eelworm p. 132
Leaves thin and grass-like; flowers weak and distorted   Large Narcissus Fly
     p. 221

Bulbs, corms, rhizomes or tubers with irregular holes
    and pieces eaten away below and above ground;
    slime trails present   Slugs p. 139
(Millepedes (p. 143) are often associated with slug
    damage)
Bulbs, corms, rhizomes or tubers with small round
    holes; long, narrow, yellow insect larvae present   Wireworms p. 236
Bulbs and bases of stems eaten, stems sometimes sev-
    ered at ground level; caterpillars in soil   Cutworms p. 200
Bulbs, corms, rhizomes or tubers with base and roots   Violet Root Rot
    rotted with matted dark strands   p. 360
Corms and rhizomes eaten; small C-shaped, legless
    white larvae present   Vine Weevil p. 242

Plants fail to flower   Blindness p. 487

Also see separate entries under main plant genera

Busy Lizzie. See **Impatiens**

Buxus (Box)
    Leaves at tips of shoots forming tight clusters, like
      miniature cabbages   Box Sucker p. 157
    Leaves with dark brown pustules above and below   Rust p. 272
    Leaves otherwise spotted   Leaf Spots p. 397
    Branches and twigs die back in absence of root damage   Dieback p. 313
    Also see **Trees** and **Hedges**

Cabbage. See **Brassicas**

Cacti and Succulents
    These plants are generally trouble-free, providing they are well grown in good
    conditions. They can survive long periods of drought but may be damaged by
    over-watering, which encourages rotting of soft tissues (see p. 379). The main
    pests are those that commonly affect **Glasshouse and House Plants** (p. 80) and
    the commonest disorder is the rough warty condition of stems and/or leaves
    brought about by Oedema (p. 488).

Calceolaria
    Leaves irregularly mottled   Viruses p. 441

Plants grown indoors in houses and glasshouses are likely to be attacked by pests of
    Glasshouse and House Plants (see p. 80) and those grown outdoors may be

affected by the main groups of pests attacking **Alpine Plants** (p. 44) and **Bedding Plants** (p. 53).

Calendula

Leaves and/or other parts with powdery white coating	Powdery Mildew p. 302
Leaves with circular, gradually darkening spots	Smut p. 288
Stems with irregular swellings	Crown Gall p. 330

Also see **Bedding Plants**

Callistephus (China aster)

Leaves wilt; stem interior with brown or black streaks	Wilt p. 306
Leaves with irregular mottling and yellowing; flowers malformed	Viruses p. 441

Also see **Bedding Plants**

Calluna (Heather)

Stem and branches with irregular, small, usually greyish growths	Lichen p. 425
Stem and branches overgrown with leafless, twining, reddish shoots	Dodder p. 427
Stem and branches die back; leaves sparse and yellowed; some death of roots, especially older ones close to stem (common)	Phytophthora Root Death p. 377
Stem and branches die back; tiny toadstools may occur on old dead roots (uncommon)	Dieback p. 317
Shoots grazed	Rabbits p. 259

Camellia

Leaves and/or flowers and buds scorched or killed suddenly	Frost p. 482
Leaves more or less bleached; veins remain dark green	Iron Deficiency p. 478
Leaves and/or shoots with rough warty outgrowths	Oedema p. 488
Leaves variously spotted	Leaf Spots p. 398
Leaves with irregular mottling	Viruses p. 441
Leaves with flat yellow or brown scales on undersides, often sticky and sooty on upper surfaces	Scale Insects p. 185
Flowers pecked and torn	Birds p. 255
Stems infested by wingless insects covered in white wax powder and filaments; leaves often sticky and sooty	Mealybugs p. 189
Roots eaten by C-shaped, white legless grubs	Vine Weevil p. 242

Campanula

Leaves with yellowish spots and off-white mould beneath	Downy Mildew p. 295
Leaves and shoots with irregular holes, slime trails present on or near plants	Slugs and Snails p. 139
Leaves and/or other parts with orange or red pustules	Coltsfoot Rust p. 275
Leaves otherwise spotted	Leaf Spots p. 398
Leaves with irregular yellowish mosaic and distortion	Viruses p. 441
Stem rotted at base; white cottony mould present	Sclerotinia Disease p. 369
Stems and shoots with small masses of froth covering wingless insects	Froghoppers p. 153

Capsicum (Pepper)
Fruit with dark, toughened areas at blossom end (blossom end rot) — Calcium Deficiency p. 476
Leaves and/or stem with rough warty outgrowths — Oedema p. 488
Leaves and/or other parts irregularly mottled and distorted — Nicotiana Viruses p. 455
Leaves discoloured, bronzing and drying, numerous small mites on undersides — Glasshouse Red Spider Mites p. 247
Leaves and shoots infested by small insects; leaves sometimes sticky and sooty — Aphids p. 161

Caraway
Stem rotted at base; white cottony mould present — Sclerotinia Disease p. 369

Carnation. See **Dianthus**

Carpinus (Hornbeam)
Leaves with powdery white coating — Hazelnut Powdery Mildew p. 299
Leaves with large irregular holes and pieces eaten away — Caterpillars p. 194
Leaves with small yellow or green insects on undersides; foliage sticky and sooty — Leafhoppers p. 154
Branches with masses of proliferating shoots — Witches' Broom p. 334
Also see **Trees**

Carrot
Leaves and/or other parts with powdery white coating — Powdery Mildew p. 302
Leaves spotted on upper surface; white mould beneath — Downy Mildew p. 295
Leaves prematurely yellowed and then blackened and killed; roots stunted — Leaf Blight p. 380
Leaves reddened and with fine yellowish mottling; plants stunted — Viruses p. 441
Leaves infested by small green insects, foliage sometimes distorted and discoloured — Willow-Carrot Aphid p. 174
Root tunnelled by maggots, often with secondary rotting; foliage discoloured, growth stunted — Carrot Fly p. 217
Roots with larger holes eaten in them, grey-brown or white caterpillars in soil — Cutworms p. 200 / Swift Moth p. 201
Roots with elongate crater-like spots (cavity spot) — Calcium Deficiency p. 476
Roots split — Splitting p. 488
Roots with soft and slimy rot — Bacterial Soft Rot p. 348
Roots (in store) with white cottony mould growth — Sclerotinia Disease p. 369
Roots decayed and with matted dark strands to which soil adheres — Violet Root Rot p. 360
Roots (in store) with mealy sometimes grey-green rot — Black Rot p. 351
Roots (in store or overwintering in ground) with black sunken lesions on crown — Liquorice Rot p. 352
Seedlings die with black patches on stem, leaves and/or cotyledons — Black Rot p. 351

Caryopteris
Leaves with small ragged holes, especially at tips of
shoots — Capsid Bugs p. 151

Castanea (Sweet Chestnut)
Leaves with powdery white coating — Oak Powdery Mildew
p. 299

Leaves yellow and wilt in absence of root damage — Wilt p. 306
Stem and/or branches with lesions and dieback — Cankers p. 326
Branches die back, leaves sparse and yellowed; tissue — Phytophthora Root
at stem base may be blackened and decayed — Death p. 377
Stem with some decay of wood; fungal brackets may — Oak Heart Rot
be present — p. 376
Stem with foul smelling fluxes; branches may die back — Slime Flux p. 325
Also see **Trees**

Catalpa (Indian bean-tree)
Leaves turn yellow and wilt in absence of root damage — Wilt p. 306
Also see **Trees**

Cattleya. See **Orchids**

Cauliflower. See **Brassicas**

Ceanothus
Leaves and/or flowers scorched or killed — Frost p. 482
Leaves more or less bleached; veins remain dark green — Iron Deficiency
p. 478

Stems with brown or white scales, leaves sticky and
sooty — Scale Insects p. 185
Also see **Trees**

Cedar (Cedrus). See **Conifers** and **Trees**. For **Western Red Cedar** see **Thuja**

Cedrus (Cedar, deodar). See **Conifers** and **Trees**

Celeriac. See **Celery**

Celery, Celeriac
Leaves irregularly mottled or yellowed and crumpled — Viruses p. 442
Leaves variously spotted — Leaf Spot p. 399
Leaves yellowed; leaf stalks discoloured and split; no — Boron Deficiency
root damage — p. 479
Leaves with pale yellow spots and white mould — Downy Mildew
growth beneath — p. 295
Leaves with yellow–brown blotch mines — Celery Fly p. 217
Stalks with irregular pieces eaten out, slime trails
present — Slugs p. 139
Plants collapsing, stems eaten through at soil level — Cutworms p. 200
Plants collapsing, roots and bases of stems tunnelled
by maggots — Carrot Fly p. 217
Leaf stalks with soft and slimy decay — Bacterial Soft Rot
p. 348

Leaf stalks with soft rot at base and white cottony mould growth — Sclerotinia Disease p. 369

Leaf stalks rotted and with patches of fluffy grey mould growth — Grey Mould Rot p. 352

Leaf stalks split but not discoloured — Splitting p. 488

Leaf stalks split and with brownish patches; leaves yellowed — Boron Deficiency p. 479

Leaf stalks and centre of crown blackened (black heart) — Calcium Deficiency p. 476

Roots and stem base rotted and/or with black lesions below soil level — Root Rot p. 361

Roots decayed with matted dark strands — Violet Root Rot p. 360

Seedlings die in patches — Damping-off p. 362

Plants run to flower — Bolting p. 487

Centaurea

Leaves and/or other parts with powdery white coating — Powdery Mildew p. 302

Leaves with pale yellowish patches and whitish mould growth beneath — Lettuce Downy Mildew p. 293

Leaves and/or other parts with brownish pustules — Rust p. 273

Cercis (Judas tree)

Branches die back; masses of small, usually pink pustules present — Coral Spot p. 314

Also see **Trees**

Chaenomeles (Japanese Quince)

Leaves with irregular yellowish streaks and patterns and/or crumpled — Viruses p. 470

Leaves more or less bleached; veins remain dark green — Iron Deficiency p. 478

Flowers pecked and torn — Birds p. 255

Shoots exude slime and die back; flowers and/or leaves may shrivel — Fireblight p. 316

Chamaecyparis (False Cypress)

Leaves yellowed and sparse; shoots may die back; roots close to stem base may die and patches of dead bark develop on stem — Phytophthora Root Death p. 377

Also see **Conifers** and **Trees**

For true cypresses see **Cupressus**

Cheiranthus (Wallflower)

Flowers of dark coloured types with pale streaks; leaves may be malformed — Viruses p. 442

Leaves with brownish angular patches and greyish mould growth beneath — Downy Mildew p. 292

Leaves wilted; stem interior with brown or black streaks — Wilt p. 306

Leaves and/or other parts with small, often concentric clusters of creamy droplets — White Blister p. 304

Leaves and shoots eaten, slime trails present on and near plants — Slugs p. 139

Leaves of young plants with numerous small round
  holes and pits   Flea Beetles p. 238
Leaves with irregular holes   Caterpillars p. 194
Plants collapsing, roots eaten by maggots   Cabbage Root Fly
                                                       p. 218

Stem with irregular swellings   Crown Gall p. 330
Roots with irregular swellings not containing larvae
  or holes   Clubroot p. 328

Cherry (fruiting and flowering ornamental)
Fruit eaten when ripe   Birds p. 255
Fruit undersized; ripen late and unevenly; taste bitter   Little Cherry
                                                       Mycoplasma p. 443
Fruit shrivelled and with dark pustules   Bitter Rot p. 349
Fruit with soft brown rot and usually off-white cot-
  tony pustules   Brown Rot p. 350
Fruit with dark green, velvety spots   Scab p. 336
Fruit split   Splitting p. 488
Flower buds eaten in winter, branches with bare
  lengths devoid of flowers in spring   Bullfinch p. 255
Flowers and/or leaves scorched or killed suddenly   Frost p. 482
Leaves scorched and/or killed suddenly in spring   Frost p. 482
Leaves with dark green, velvety spots   Scab p. 336
Leaves curled and reddened in early summer   Leaf Curl p. 332
Leaves at tips of shoots tightly curled and infested by
  colonies of black insects; foliage sticky and sooty   Cherry Blackfly
                                                       p. 169
Leaves with irregular holes and pieces eaten away   Caterpillars p. 194
Leaves wilt in absence of root damage or other stress   Wilt p. 306
Leaves with silvery sheen; branches may die back   Silver Leaf p. 324
Leaves and/or other parts with sticky black coating Sooty Mould p. 303
Leaves and/or other parts with powdery white coating   Plum Powdery
                                                       Mildew p. 300
Leaves more or less bleached; veins remain dark green   Iron Deficiency
                                                       p. 478
Leaves with yellow-brown blotches; later wither and
  remain hanging on tree over winter   Leaf Scorch p. 399
Leaves with more or less regular holes   Peach Shot Hole
                                                       p. 414
Leaves with irregular patterns or variously
  malformed   Viruses p. 442
Leaves wither and tip of terminal shoot droops   Wither Tip p. 350
Branches die back in absence of stem lesions; leaves
  silvery   Silver Leaf p. 324
Branches die back; root and stem base decayed; toad-
  stools and bootlace-like strands may be present   Honey Fungus p. 375
Branches and/or stem oozing gum; may be some die-
  back and also spotting and holes in leaves   Bacterial Canker
                                                       p. 320
Branches otherwise dying back   Diebacks p. 320
Branches with masses of proliferating shoots   Witches' Brooms
                                                       p. 334
Stem with some decay of wood; fungal brackets may
  be present   Plum Heart Rot
                                                       p. 378
Stem with decay at base; toadstools and/or bootlace-
  like strands may be present and branches die back Honey Fungus p. 375
Stem with hard irregular swelling   Crown Gall p. 330

Trees generally unthrifty

Specific Replant
Disease p. 428 and
Deficiencies of
Nitrogen p. 474
Phosphorus p. 475 and
Potassium p. 475

Also see **Trees**

Chervil
Leaves with pale yellowish spots; white mould growth
beneath

Downy Mildew
p. 295

Chestnut. See **Castanea.** For **Horse Chestnut** see
Aesculus

Chicory
Leaves with irregular or more or less circular ragged
holes

Lettuce Ring Spot
p. 409

Leaves and crown blackened and stunted

Calcium Deficiency
p. 476

Plants wilted; stem rotted at base; white cottony
mould present

Sclerotinia Disease
p. 369

China Aster. See **Callistephus**

Chinese Cabbage. See **Brassicas**

Chionodoxa
Flowers with brown-black powdery masses in centre

Scilla Anther Smut
p. 290

Also see **Bulbs etc.**

Chives
See **Onion**; many of the problems on onions affect
chives also

Chlorophytum (Spider plant)
Also see **Glasshouse and House Plants**
Leaves with brown, dead tips – a common problem of
unknown origin but see

Dry Air p. 482

Christmas Rose. See **Helleborus**

Christmas Tree. See **Picea**

Chrysanthemum (Including glasshouse Chrysanthemums and Marguerite)
Many different pests, diseases and disorders affect chrysanthemums, especially
those that are grown for exhibition, either outdoors or indoors, and those that are
grown as decorative glasshouse and house plants. Much published information is
available on these popular plants and in Britain the National Chrysanthemum
Society publishes a booklet dealing with the main pests and diseases.
Flowers with ragged holes eaten in petals   Earwigs p. 148
Flowers with large pieces eaten out of petals   Caterpillars p. 194
Flowers discoloured, with fine light flecking; many
    small, narrow-bodied insects present   Thrips p. 191
Flowers malformed, flower buds sometimes killed   Capsid Bugs p. 151

Flowers stunted, ragged and deformed	Aspermy Virus p. 443
Flowers green coloured	Green Flower Disease p. 443
Flowers and/or buds rotted and with fluffy grey mould growth	Grey Mould p. 346
Flowers with small brown spots or translucent lesions; blooms shrivel (Britain only)	Petal Blight p. 345
Leaves with brown pustules beneath; pale yellowish spots above	Rust p. 274
Leaves with dirty-white pustules beneath; pale yellowish spots above	White Rust p. 274
Leaves and/or other parts with powdery white coating	Powdery Mildew p. 303
Leaves variously spotted with dark more or less circular lesions	Leaf Blotch p. 400
Leaves wilt; stem interior with brown or black streaks	Wilt p. 306
Leaves with irregular yellowish markings and/or crumpled	Soil-borne Viruses p. 444
Leaves with brown or black areas between veins, older leaves dying	Chrysanthemum Eelworm p. 134
Leaves discoloured, bronzing and dying, many small mites on undersides	Glasshouse Red Spider Mites p. 247
Leaves with sinuous white mines	Chrysanthemum Leaf Miner p. 222
Leaves with irregular holes and pieces eaten away	Caterpillars p. 194
Leaves with small ragged holes, especially at tips of shoots, buds sometimes killed	Capsid Bugs p. 151
Leaves with small white winged insects on undersides, foliage sticky and sooty	Glasshouse Whitefly p. 158
Leaves, stems, buds and flowers infested by small winged and wingless green, yellow, brown or red insects; foliage sticky and sooty	Aphids p. 161
Leaves, buds and stems with small protruding growths (galls); shoots distorted	Chrysanthemum Gall Midge p. 220
Shoots eaten when young, leaves with holes, slime trails present on or near plants	Slugs p. 139
Stem keels over at point of decay lesion; fluffy grey mould present	Stem Rot p. 380
Stem rotten, usually at base; cottony white mould present	Sclerotinia Disease p. 369
Stem with irregular swellings	Crown Gall p. 330
Roots, stems and shoot tunnelled by maggots	Chrysanthemum Stool Miner p. 223
Plants entirely midget	Stunt Viroid p. 444
Plants generally unthrifty; roots blackened and/or decayed	Root and Foot Rot p. 361
Plants or cuttings generally unthrifty with decay of parts of stem and/or leaves	Ray Blight p. 345

Also see **Bedding Plants** and **Perennials**

Cineraria (*Senecio cruentus*)

Leaves and/or other parts with powdery white coating	Powdery Mildew p. 302
Leaves with pale yellowish spots and whitish mould growth beneath	Lettuce Downy Mildew p. 293

Leaves and/or other parts with blisters and small orange and white pustules or cup-like bodies	Rust p. 275
Leaves variously spotted	Leaf Spots p. 400
Leaves, flowers, buds and stems infested by colonies of winged and wingless yellow, green or pink insects; foliage sticky and sooty	Aphids p. 161
Leaves with sinuous white mines	Chrysanthemum Leaf Miner p. 222
Leaves with small white winged insects on undersides; foliage sticky and sooty	Glasshouse Whitefly p. 158

Also see **Glasshouse and House Plants**

Citrus. See **Glasshouse and House Plants**

Clarkia

Leaves with pale yellowish spots and off-white mould growth beneath	Downy Mildew p. 295

Clematis

Leaves and/or other parts with powdery white coating	Powdery Mildew p. 302
Leaves variously mottled or malformed	Viruses p. 470
Leaves wilt; shoots and entire plant may die back	Dieback p. 313
Leaves sticky and sooty, leaves and stems infested by colonies of small wingless and winged insects	Aphids p. 161
Leaves sticky and sooty, stems and undersides of leaves infested by brown or yellow scales	Scale Insects p. 185
Leaves with small notches eaten out of edges	Weevils p. 241
Leaves with small ragged holes, especially at tips of shoots, buds sometimes killed	Capsid Bugs p. 151
Flowers with ragged holes eaten in petals	Earwigs p. 148
Shoots eaten at or just above soil level, leaves with irregular holes, slime trails present	Slugs and Snails p. 139

Also see **Climbing Plants**

Climbing Plants

Temperatures are often slightly higher near walls, especially if these are part of a heated building, and plants climbing on them may be affected by some of the pests and diseases that attack **Glasshouse and House Plants** (p. 80). They may also be susceptible to drought if they are protected from prevailing winds and rain. Young plants are often attacked by **Slugs and Snails** (p. 139), **Earwigs** (p. 148) and **Woodlice** (p. 145) that thrive near walls, especially if there are many crevices and some accumulations of organic debris to shelter them. Some climbing plants such as Clematis, Honeysuckle and Ivy may themselves cause trouble if they are allowed to grow over other plants.

Clivia

Leaves with red-brown spots	Leaf Spot p. 400

Cobnut. See **Corylus**

Codiaeum (Croton). See **Glasshouse and House Plants**

Colchicum (Autumn Crocus)

Leaves, corms and/or other parts with dark, blister-like swellings	Smut p. 289
Corms with dryish rot and whitish mould growth; plants stunted or fail to emerge	Tulip Grey Bulb Rot p. 344

Also see **Bulbs etc.**

Coleus. See **Glasshouse and House Plants**

Comfrey. See **Symphytum**

Conifers

Most conifers are relatively free from pests and diseases. This may be partly due to the comparative toughness of their leaves and to the production of resin, which may repel pests and seal wounds. Young trees may need protection from **Deer** (p. 263), **Squirrels** (p. 260), **Rabbits** (p. 259), and **Voles** (p. 262), all of which can cause serious damage by stripping bark from stems. Older trees are mainly susceptible to attack by **Adelgids** (p. 182), a group of pests that is restricted to conifers, by some **Aphids** (p. 161) and by **Caterpillars** (p. 194) of moths and **Sawflies** (p. 224). The premature shedding of leaves occurs fairly commonly and is discussed under **Needle Casts** (p. 305) while the masses of proliferating shoots known as **Witches' Brooms** (p. 334) are also seen sometimes. Trunks of mature trees are often affected by **Resin Bleeding** (p. 313) or by a number of rotting diseases such as **Honey Fungus** (p. 375), **Brown Cubical Rot** (p. 373), **White Pocket Rot** (p. 378) and **Red Rot** (p. 378) while one of the commonest causes of death of cypresses in gardens is **Phytophthora Root Death** (p. 377). **Mistletoe** (p. 427) grows sometimes on conifers in parts of Europe while the growth of toadstools under trees is often an indication of the beneficial presence of **Mycorrhiza** (p. 429) on the roots.

Also see separate entries for the main plant genera and for **Trees**

Conophytum. See **Cacti and Succulents**

Convallaria (Lily-of-the-Valley)

Leaves with orange cup-like structures beneath (rare in Britain)	Rust p. 275
Leaves variously spotted	Leaf Spots p. 400
Leaves with soft brown region at base with mould growth	Paeony Grey Mould Blight p. 385
Rhizomes eaten by large white caterpillars	Swift Moths p. 201

Coreopsis. See **Bedding Plants** and **Perennials**

Corms. See **Bulbs etc.**

Cornflower. See **Centaurea**

Cornus (Dogwood)

Leaves with various irregular pale streaks and mottles or distortions	Viruses p. 470

Corylus (hazel, cobnut, filbert)

Nuts with neat small holes drilled in shells	Nut Weevil p. 244
Nuts cracked open and eaten	Squirrels p. 260
Leaves with powdery white coating	Powdery Mildew p. 299

Leaves with irregular yellowish streaks and mottles	Viruses p. 444
Brown-purple scaly and leafless plants growing beneath trees.	Toothwort p. 426
Leaves with irregular holes and pieces eaten away	Caterpillars p. 194
Buds abnormally enlarged	Nut Gall Mite p. 253

Also see **Trees**

Cosmos. See **Bedding Plants**

Cotinus

Leaves wilt; stem interior with brown or black streaks	Wilt p. 306

Cotoneaster

Fruits eaten	Birds p. 255
Leaves sticky and sooty, stems and leaves infested by brown or grey scales	Scale Insects p. 185
Leaves sticky and sooty, shoots infested by small wingless and winged insects	Aphids p. 161
Leaves with irregular holes, foliage sometimes drawn together with silk webbing	Caterpillars p. 194 esp. Hawthorn Webber p. 209
Leaves with various irregular streaks and mottles or distortions	Viruses p. 429
Shoots exude slime and die back; flowers and/or leaves shrivelled	Fireblight p. 316
Stems with tufts of white wax wool covering small dark wingless insects	Woolly Aphid p. 166

Courgette. See **Cucumber**

Crassula

Stem decayed	Stem Rot p. 379
Roots eaten by C-shaped white legless larvae	Vine Weevil p. 242

Also see **Cacti and Succulents**

Crataegus (Hawthorn)

Leaves with various irregular streaks and mottles or distortion	Viruses p. 470
Leaves and/or other parts with powdery white coating	Powdery Mildew p. 302
Leaves variously spotted	Leaf Spots p. 401
Leaves silver; branches may die back	Silver Leaf p. 324
Leaves with irregular holes and pieces eaten away	Caterpillars p. 194
Shoot tips stunted, with rosettes of small leaves	Hawthorn Button Top Midge p. 221
Shoots with vivid red/orange colouration (rare in Britain)	Rust p. 277
Shoots exude slime and die back; flowers and/or leaves shrivelled	Fireblight p. 316
Branches and/or stem with hard irregular swelling	Crown Gall p. 330
Branches and/or stem with canker lesions; may die back beyond lesion	Apple Canker p. 310
Branches with parasitic leathery plant growing from them	Mistletoe p. 427

Also see **Trees** and **Hedges**

Cress
Seedlings die in patches   Damping-off p. 362

Crinum
Leaves turn yellow and shrivel; young shoots and/or
    other parts with gradually spreading brown   Narcissus Leaf
    blotches   Scorch p. 412
Also see **Bulbs etc.**

Crocosmia. See **Bulbs etc.**

Crocus. For **Autumn Crocus** see **Colchicum**
Note: It is important to check both corm and leaf
    symptoms on growing plants as several problems
    with very similar symptoms can affect crocuses

Flowers torn, often completely spoiled   Birds p. 255
Shoots eaten off, often carried away   Squirrels p. 260
                                                  Mice p. 261
Leaves yellowed; sheaths decayed at ground level and   Gladiolus Dry Rot
    masses of tiny black bodies present   p. 341
Leaves with brown-purple spots bearing tiny black   Gladiolus Hard Rot
    bodies   p. 342
Leaves with small, rounded red-brown spots and later   Gladiolus Scab
    shrivelling   p. 342
Leaves turn yellow, rot at base and are easily pulled   Hyacinth Black
    away from corm   Slime p. 342
Leaves emerge feebly; corms when dug up have dry   Tulip Grey Bulb Rot
    rot and adhering soil   p. 344
Leaves become yellow and die; corms with rot spread-   Narcissus Basal Rot
    ing upwards from base   p. 343
Corms with small sunken lesions, visible with scales   Gladiolus Dry Rot
    removed   p. 341 or
                                                       Gladiolus Hard Rot
                                                       p. 342

Corms with rounded pale yellow spots   Gladiolus Scab
                                                   p. 342
Corms (in store) with firm, dry rot and blue-green   Blue Mould Rot
    mould growth   p. 339
Corms with rot spreading upwards from base and   Narcissus Basal Rot
    with rings of dark sunken spots at top   p. 343
Corms with dark pustules on membranous sheaths
    (Holland only)   Rust p. 275
Also see **Bulbs etc.**

Croton (Codiaeum). See **Glasshouse and House-
plants**

Cucumber (also Courgette, Marrow, Melon, Pumpkin, Squash and Zucchini)
These plants are most susceptible to pests when grown under glass in glasshouses
or frames and are generally less likely to be attacked when grown in the open
garden.
Fruit with small greyish spots oozing liquid; later a
    dark green mould   Gummosis p. 352
Fruit not formed or warty   Viruses p. 444

Fruit with sunken lesions	Anthracnose p. 401
Leaves and young shoots infested by colonies of green, yellow, pink or black winged and wingless insects; growth sometimes distorted, foliage sticky and sooty	Aphids p. 161
Leaves with small white winged insects on undersides; foliage sticky and sooty	Glasshouse Whitefly p. 158
Leaves discoloured, yellowing, bronzing and dying; many small mites on undersides, sometimes with fine silk webbing	Glasshouse Red Spider Mites p. 247
Leaves and/or other parts with powdery white coating	Powdery Mildew p. 298
Leaves wilt; stem interior with black or brown streaks	Wilt p. 306
Leaves crumpled and/or with irregular yellowish mottling	Viruses p. 444
Leaves and/or other parts thickened, twisted or otherwise distorted	Hormone Herbicide Injury p. 491
Leaves with very pale green, often translucent spots	Anthracnose p. 401
Leaves with brown, angular leaf spots exuding drops of liquid	Bacterial Leaf Spot p. 401
Leaves shrivelled and/or with greyish spots and rounded holes	Leaf Blotch p. 402
Leaves with pale brown and dry margins	High Temperature Injury p. 484
Stem with patches of soft rot and fluffy grey mould growth	Stem and Fruit Rot p. 381
Stem and/or other parts thickened, twisted or otherwise distorted	Hormone Herbicide Injury p. 491
Stem and/or other parts with patches of soft rot, bearing small black bodies	Stem Rot p. 381
Stem rotted at base; cottony white mould growth present	Sclerotinia Disease p. 369
Stem with soft irregular swelling	Crown Gall p. 330
Stem and/or other parts with soft and slimy rot; no mould growth	Bacterial Soft Rot p. 348
Roots decayed; sometimes with tough unaffected core; plants unthrifty	Black Root Rot or other Root and Foot Rot p. 361
Plants unthrifty, roots with irregular solid swellings	Root Knot Eelworm p. 136
Plants unthrifty; no stem lesions but roots decayed	Black Root Rot or other Root and Foot Rot p. 361
Plants wilt in absence of stem or root lesions	Drought p. 482
Plants more or less covered in slimy, cushion-like bodies of varying colour	Slime Mould p. 424

Cupressus (Cypress)

Leaves yellowed and sparse; shoots may die back; roots close to stem may die and patches of dead bark develop on stem	Phytophthora Root Death p. 377
Branches with resinous canker lesions; may be dieback beyond lesion (Monterey cypress only)	Canker p. 315

Also see **Conifers** and **Trees**

Currants (Black, Red and White)

Fruit sparse	Reversion Disease p. 445
Fruits eaten as they ripen	Birds p. 255
Leaves sticky and sometimes sooty, often with red or yellow blisters and other distortions; small green or yellow wingless and winged insects present on undersides of leaves and on young shoots	Aphids p. 161
Leaves with small tattered holes, especially on younger leaves at tips of shoots	Capsid Bugs p. 151
Leaves at tips of shoots twisted, failing to develop normally	Black Currant Leaf Midge p. 214
Leaves scorched suddenly in spring	Frost p. 482
Leaves silvery; branches may die back	Silver Leaf p. 424
Leaves with small, irregular brownish spots or blotches	Leaf Spot p. 402
Leaves and/or other parts with deep orange or red pustules	Gooseberry Cluster-Cup Rust p. 277
Leaves with small yellow pustules and bristly structures on undersides	White Pine Blister Rust p. 286
Leaves and/or other parts with powdery white or felted brown coating	American Gooseberry Mildew p. 298
Leaves with inter-veinal yellowing	Manganese Deficiency p. 479
Leaves with pale yellow banding along veins	Gooseberry Vein Banding Virus p. 449
Buds abnormally enlarged in spring, dying later	Black Currant Gall Mite p. 252
Stems and branches infested by brown, waxy scales, sometimes with conspicuous tufts of white wax wool	Scale Insects p. 185
Branches die back; leaves with silver sheen	Silver Leaf p. 324
Branches die back; may be toadstools and bootlace-like strands at stem base; root decayed close to stem	Honey Fungus p. 375
Branches die back; masses of small, usually pink pustules present	Coral Spot p. 314
Branches die back in absence of lesions or other damage	Grey Mould Dieback p. 317
Branches die back; small, hoof-like fungal growth at stem base	Collar Rot p. 374
Branches and/or stem with hard, irregular swelling	Crown Gall p. 330

Cyclamen

Hardy outdoor cyclamen are not seriously affected by pests, diseases or disorders, although they may suffer from some of the conditions noted under **Alpine and Rock Garden Plants.** For conditions affecting pot cyclamen also see **Glasshouse and House Plants** and **Bulbs etc.**

Flowers and/or buds shrivelled and with fluffy grey mould	Grey Mould p. 346
Leaves and/or other parts with powdery white coating	Powdery Mildew p. 302
Leaves wilt; stem interior with black or brown streaks	Wilt p. 306
Leaves variously spotted	Leaf Spots p. 403
Leaves irregularly striped with pale colour	Viruses p. 445

Corm with soft and slimy rot — Bacterial Soft Rot
p. 348

Corm with firm, dry rot and blue-green mould growth — Blue Mould Rot
p. 339

Plants unthrifty; roots decayed — Root and Foot Rot
p. 361

Cymbidium. See **Orchids**

Cypress. See **Chamaecyparis** and **Cupressus**

Cypripedium. See **Orchids**

Cytisus (Broom)

Plants die back in absence of root damage — Broom Dieback
p. 313

Buds develop into large irregular growths (galls) — Broom Gall Mite
p. 254

Daffodil. See **Narcissus**

Dahlia

Flowers and/or buds shrivelled and with fluffy grey mould growth — Grey Mould
p. 346

Flowers with tattered petals — Earwigs p. 148

Flowers with light flecking on petals, many small long-bodied winged and wingless insects present — Thrips p. 191

Leaves with numerous pale spots — Smut p. 289

Leaves variously crumpled or with irregular, pale yellowish patterns — Viruses p. 446

Leaves wilt; stem interior with brown or black streaks — Wilt p. 306

Leaves with irregular holes and pieces eaten away; no slime trails present on or near plants — Caterpillars p. 194

Leaves with irregular holes and pieces eaten away; slime trails present on or near plants — Slugs and Snails
p. 139

Leaves with small tattered holes, especially at tips of shoots; buds sometimes killed and flowers distorted — Capsid Bugs p. 151

Leaves discoloured, yellowing, bronzing and dying; many small mites on undersides of leaves and sometimes on dry shoots, with fine silk webbing — Glasshouse Red Spider Mites p. 247

Leaves, shoots and flowers infested by colonies of black, green or yellow wingless and winged insects; foliage sticky and sometimes sooty — Aphids p. 161

Stem with soft, irregular swelling — Crown Gall p. 330

Stem base rotted; cottony white mould growth present — Sclerotinia Disease
p. 369

Damson. See **Plum**

Daphne

Leaves variously spotted — Leaf Spot p. 403

Leaves variously mottled and/or crumpled — Viruses p. 446

Leaves and young shoots infested by colonies of small winged and wingless insects; leaves sticky — Aphids p. 161

Datura
Leaves and/or other parts variously mottled and/or   Nicotiana Viruses
   crumpled   p. 455
Also see **Glasshouse and House Plants**

Day Lily. See **Hemerocallis**

Delphinium (including Larkspur)
Leaves and/or other parts with powdery white coating   Powdery Mildew
    p. 302
Leaves with bituminous black blotches in absence of
   stem rot   Black Blotch p. 403
Leaves and flowers with irregular holes and pieces   Caterpillars p. 194
   eaten away   Earwigs p. 148
Leaves and young shoots eaten; slime trails present on   Slugs and Snails
   or near plants   p. 139
Leaves, shoots and flowers infested by yellow, green
   or pink wingless and winged insects; foliage sticky,
   sometimes sooty   Aphids p. 161
Leaves variously crumpled or with irregular yel-
   lowish patterns   Viruses p. 446
Stem base rotted; cottony white mould growth   Sclerotinia Disease
   present   p. 369
Stem base and/or roots rotted   Root and Foot Rot
    p. 361
Stem or crown variously rotted   Stem Rots and
    Blights p. 381
Roots eaten by large white caterpillars in soil   Swift Moths p. 201

Dendrobium. See **Orchids**

Deodar. See **Cedrus**

Dewberry. See **Blackberry**

Dianthus (Carnations, pinks and sweet williams)
Flowers distorted; anthers filled with dark powdery   Carnation Anther Smut
   mass   p. 288
Flowers of dark coloured types with pale streaks   Carnation Vein
    Mottle Virus p. 447
Flower buds with brownish decay and internal mould
   growth   Bud Rot p. 345
Flowers, flower buds and young shoots eaten and   Carnation Tortrix
   spun together with fine silk webbing; caterpillars   Moth p. 207
   usually present
Flower petals with light flecking, sometimes wither-
   ing, small narrow-bodied insects present   Thrips p. 191
Flowers, buds, leaves and stems infested by colonies
   of yellow, green or dark-coloured wingless and
   winged insects; foliage sticky and sooty   Aphids p. 161
Leaves with small yellow-brown pustules   Rust p. 273
    (Carnation),
    p. 286 (Sweet
    William)

Leaves and/or other parts with powdery white coating	Powdery Mildew p. 302
Leaves distorted and/or with irregular yellowish mottling	Viruses p. 447
Leaves wilt; stem interior with brown or black streaks	Wilt p. 306
Leaves variously spotted	Leaf Spots p. 403
Leaves and stems tunnelled by fly maggots producing extensive white mines; plants may wilt and die	Carnation Fly p. 223
Leaves discoloured, yellowing, bronzing and dying; many small mites present, sometimes with fine silk webbing	Glasshouse Red Spider Mites p. 247
Stems pitted or grooved	Carnation Etched Ring Virus p. 447
Stems with mass of proliferating, leafy shoots	Leafy Gall p. 331
Leaves variously spotted	Leaf Rot p. 380
Roots rotted; plants unthrifty	Root and Foot Rot p. 361
Plants unthrifty, roots infested by small white wingless insects	Root Aphids p. 181

Also see **Glasshouse and House Plants**

Digitalis (Foxglove)

Leaves and/or other parts with powdery white coating	Powdery Mildew p. 302
Leaves with pale yellowish spots and whitish mould growth beneath	Downy Mildew p. 295
Leaves otherwise spotted	Leaf Spots p. 404

Dill

Leaves with brown or black pustules (not Britain)	Rust p. 276

Dogwood. See **Cornus**

Doronicum

Leaves and/or other parts with powdery white coating	Powdery Mildew p. 302

Douglas Fir. See **Pseudotsuga**

Dracaena

Leaves variously spotted	Leaf Spots p. 404

Also see **Glasshouse and House Plants**

Echeveria

Leaves with orange, cup-like dimples	Houseleek Rust p. 279

Also see **Cacti and Succulents**

Egg Plant. See **Aubergine**

Elaeagnus

Branches die back; masses of small, usually pink pustules present	Coral Spot p. 314

Also see **Trees**

Elderberry. See **Sambucus**

Elm. See **Ulmus**

Endive
Leaves and/or stems with black or brown pustules · Centaurea Rust
p. 273
Leaves with irregular or more or less circular ragged · Lettuce Ring Spot
holes · p. 409

Endymion (Bluebell)
Leaves and/or stems with dirty yellowish spots bear-
ing brown pustules · Rust p. 272
Also see **Bulbs etc.**

Epiphyllum. See **Cacti and Succulents** and
Glasshouse and House Plants

Eranthis (Winter Aconite)
Flowers and flower buds pecked · Birds p. 255
Flowers, flower buds, leaves and stems infested by
small insects; foliage sticky and sooty · Aphids p. 161

Erica (Heath, Heather)
Leaves and young shoots grazed · Rabbits p. 259
Stem and branches with irregular, small, usually
greyish growths · Lichen p. 425
Stem and branches die back; leaves sparse and yel-
lowed; some death of roots, especially older ones · Phytophthora Root
close to stem · Death p. 377

Erythronium
Leaves with yellowish or brown spots and pustules · Rust p. 276
Also see **Bulbs etc.**

Escallonia
Leaves with various irregular pale streaks and mottles
or distortions · Viruses p. 470

Eucalyptus
Leaves and/or shoots scorched or killed · Frost p. 482
Leaves and young shoots distorted, small flat insects · Blue-Gum Sucker
present; foliage sticky and sooty · p. 158
Also see **Trees**

Euonymus
Leaves and/or other parts with powdery white coating · Powdery Mildew
p. 302
Leaves with yellowish mosaic or mottle patterns · Viruses p. 447
Stem with hard, irregular swelling · Crown Gall p. 330
Leaves, shoots and young stems infested by small · Black Bean Aphid
black insects; foliage sticky and sooty · p. 171
Stems with colonies of white scales · Euonymus Scale p. 188
Also see **Trees**

Euphorbia (including **Poinsettia**). See **Glasshouse
and House Plants**

Fagus (Beech)

Leaves scorched in early spring	Frost p. 482
Leaves with powdery white coating	Oak Powdery Mildew p. 299
Leaves variously spotted	Leaf Spots p. 396
Leaves yellowed and sparse; shoots may die back; roots close to stem may die and patches of dead bark develop	Phytophthora Root Death p. 377 but also see Bark Disease p. 311 and Iron Deficiency p. 478
Leaves with irregular holes and pieces eaten away	Caterpillars p. 194 Weevils p. 241
Leaves mined, browning and curling	Weevils p. 241
Leaves with tufts of white wax wool on undersides covering yellow-green insects; foliage sticky and sooty	Beech Aphid p. 180
Leaves with small yellow or green active winged and wingless insects on undersides	Leafhoppers p. 154
Branches die back; masses of usually pink pustules present	Coral Spot p. 314
Branches and/or stem with canker lesions; branches beyond lesion may die back	Canker p. 312
Stems and trunks with conspicuous areas of white wax wool covering minute insects	Beech Scale p. 187
Stem or branches with large areas of dead bark which may drop off; may be white woolly growth associated. Leaves become yellow and trees are generally unthrifty	Bark Disease p. 311
Stem with canker lesions; branches beyond lesion may die back	Canker p. 312
Stem with decay of various types; fungal brackets often present	Heart Rots p. 372 and cross-refer to other rots
Roots close to stem die; patches of dead bark on stem; shoots may die back and leaves become yellowed and sparse	Phytophthora Root Death p. 377

Also see **Trees**

Fatsia. See **Glasshouse and House Plants**

Ferns

Ferns growing outdoors are generally free from serious troubles but ferns growing in glasshouses and houses may be affected by some of the pests, diseases and disorders of **Glasshouse and House Plants** (p. 80). Variously-coloured, but usually brown pustules on the fronds (not to be confused with the fern spore masses which are also brown but usually regularly arranged) occur fairly commonly and are the symptoms of **Rust** (p. 277).

Fern (Asparagus). See **Glasshouse and House Plants**

Ficus (Rubber Plant). See **Glasshouse and House Plants**

Fig

Fruit rotted and with fluffy grey mould growth	Grey Mould Rot p. 346
Leaves scorched in spring	Frost p. 482
Leaves with irregular yellow-green blotches	Viruses p. 447
Branches die back; masses of small, usually pink pustules present	Coral Spot p. 314
Branches with target-like canker lesions; may be some dieback	Canker p. 315
Also see **Trees**	

Filbert. See **Corylus**

Fir. For true firs see **Abies**; for Douglas Fir see **Pseudotsuga**

Flag. See **Acorus**

Forget-Me-Not. See **Myosotis**

Forsythia

Flowers and flower buds pecked	Birds p. 255
Leaves with small ragged holes, especially at tips of shoots	Capsid Bugs p. 151
Leaves with yellowish banding along veins; may be some distortion	Viruses p. 448
Leaves variously spotted in absence of shoot dieback	Leaf Spots p. 405
Leaves and young shoots wilt; leaves may have small brownish spots and buds and flowers become blackened	Lilac Blight p. 318
Shoots with firm irregular swellings	Gall p. 331

Foxglove. See **Digitalis**

Fraxinus (Ash)

Leaves with powdery white coating	Hazelnut Powdery Mildew p. 299
Leaves with irregular yellowish streaks and patterns	Viruses p. 488
Branches die back in crown; trees take on 'stag's-head' appearance	Dieback p. 311
Branches and/or stem with canker lesions; may be some dieback	Canker or Bacterial Canker p. 311
Stem with some decay of wood; fungal fruiting bodies may be present and branches die back	Heart Rot p. 372
Stems encrusted by colonies of white scales; foliage sticky and sometimes very sooty	Willow Scale p. 188
Trees overgrown with evergreen climbing plant	Ivy p. 427
Also see **Trees**	

Freesia

Flowers fail to open fully; leaves with yellowish mottling	Bean Yellow Mosaic Virus p. 448
Flowers and/or buds shrivel; fluffy grey mould growth present	Grey Mould p. 346
Flowers, flower buds, stems and leaves infested by small insects; foliage sticky, sometimes sooty	Aphids p. 161
Flowers and leaves with large irregular holes; slime trails present on or near plants	Slugs p. 139
Flowers and leaves with large irregular holes; no slime trails present	Caterpillars p. 194
Leaves discoloured; many small mites present on undersides	Glasshouse Red Spider Mites p. 247
Leaves with rows of tiny white flecks	Freesia Streak Virus p. 448
Leaves with gradually enlarging brown or other coloured spots	Iris Leaf Spot p. 407 but see also Gladiolus Core Rot p. 340 and Gladiolus Scab p. 342
Leaves with brown-purple spots bearing tiny black bodies	Gladiolus Hard Rot p. 342
Leaves turn yellow and are decayed at ground level where masses of tiny black bodies are present on them	Gladiolus Dry Rot p. 341
Corms with small sunken lesions, visible when scales removed	Gladiolus Dry Rot p. 341 or Gladiolus Hard Rot p. 342
Corms with rounded pale yellow spots	Gladiolus Scab p. 342
Corms (in store) with firm, dry rot and blue-green mould growth	Blue Mould Rot p. 339

Also see **Bulbs** and **Glasshouse and House Plants**

Fritillaria (Fritillary)

Leaves with brown or blackish leaf pustules (rare in Britain)	Lily Rust p. 280
Leaves turn yellow and plants topple over; bulbs with dry rot	Hyacinth Black Slime p. 342

Also see **Bulbs etc.**

Fritillary (Fritillaria). See **Fritillaria**

Fuchsia

Leaves with small winged insects and yellow scales on undersides; foliage sticky and sooty	Glasshouse Whitefly p. 158
Leaves and stems infested by green, yellow or pink wingless and winged insects; foliage sticky and sooty	Aphids p. 161
Leaves discoloured, yellowing, bronzing and dying; many small mites on undersides of leaves	Glasshouse Red Spider Mites p. 247
Leaves with small tattered holes, especially at tips of young shoots	Capsid Bugs p. 151

Leaves with pale yellow patches and orange pustules beneath (rare in Britain) — Rust p. 276

Leaves wilt; stem interior with brown or black streaks — Wilt p. 306

Also see **Glasshouse and House Plants**

Gaillardia

Leaves with dark brown spots; may shrivel later (rare in Britain) — Smut p. 289

Leaves with pale yellowish spots and whitish mould beneath — Lettuce Downy Mildew p. 293

Galanthus (including Snowdrop)

Flowers pecked and torn — Birds p. 255

Leaves turn yellow and shrivel; young shoots and/or other parts with gradually spreading brown blotches — Narcissus Leaf Scorch p. 412

Leaves turn yellow and decay at ground level where masses of tiny black bodies are present on them — Gladiolus Dry Rot p. 341

Leaves become brown and rotten; plants stunted and fail to flower; bulbs turn soft and pulpy — Grey Mould Blight p. 381

Bulbs with many small sunken lesions, visible when scales removed — Gladiolus Dry Rot p. 341

Also see **Bulbs etc.**

Galtonia

Leaves with fine mottling of white and green — Hyacinth Viruses p. 450

Also see **Bulbs etc.**

Gardenia

Stem with sunken or swollen lesions close to ground level — Canker p. 316

Garlic. See **Onion**

Genista. See **Cytisus**

Geranium

Leaves with brown, black or orange pustules on undersides — Rust p. 276

Also see **Pelargonium**

Gerbera

Flowers and leaves with large irregular holes; slime trails on or near plants — Slugs p. 139

Leaves, flower buds and shoots infested by small yellow, green, pink or dark-coloured wingless and winged insects; foliage sticky and sooty — Aphids p. 161

Leaves with small white winged insects on undersides; foliage sticky and sooty — Glasshouse Whitefly p. 158

Leaves with large, rounded brown spots — Leaf Spots p. 405

Stem and/or roots rotted; plants unthrifty — Root and Foot Rot p. 361

Geum

Leaves with pale yellowish spots and off-white mould beneath   Downy Mildew p. 295

Gladiolus

Flowers with disrupted colour pattern   Viruses p. 448

Flowers and leaves discoloured, with light flecking, later withering and dying; many small narrow-bodied insects present, especially under leaf bases   Gladiolus Thrips p. 192

Leaves, flowers, buds and shoots infested by small black, green or yellow wingless and winged insects; foliage sticky and sooty   Aphids p. 161

Leaves and flowers with irregular holes; slime trails present on or near plants   Slugs and Snails p. 139

Leaves and flowers with irregular holes; no slime trails present   Caterpillars p. 194

Leaves variously streaked or mottled with pale markings   Viruses p. 448 But also see Corm Rot and Yellows p. 341

Leaves with dark brown blisters which burst to release powdery contents   Smut p. 289

Leaves with gradually enlarging brown spots   Iris Leaf Spot p. 407 but also see Core Rot p. 340, Hard Rot p. 342 and Scab and Neck Rot p. 342

Leaves turn yellow and decay at ground level where masses of tiny black bodies are present on them   Dry Rot p. 341

Corms (in store) with firm, dry rot and blue–green mould growth   Blue Mould Rot p. 339

Corms with central soft rot to the extent that the core may drop out   Core Rot p. 340

Corms with dark brown, concentrically ridged surface lesions   Corm Rot and Yellows p. 341 but also see Dry Rot p. 341 and Hard Rot p. 342

Corms with rounded pale yellow spots   Scab and Neck Rot p. 342

Roots eaten by white caterpillars in soil   Swift Moths p. 201

Also see **Bulbs etc.**

Glasshouse and House Plants

The special conditions in which glasshouse and house plants are grown favour many pests and some diseases. Some of the pests are native species that thrive when they are protected from extremes of climate and from the predators and parasites that attack them outdoors and some are species that have been accidentally introduced from tropical and sub-tropical countries, usually without the parasites and predators that check them in their original environment. High temperatures and humidities may also encourage the development and spread of diseases and, in addition, pot-grown plants are especially liable to drought, waterlogging, and other disorders. During winter, plants may be chilled or frosted if temperatures fall too low and in summer they may be scorched by strong sun if not adequately shaded.

Good hygiene and careful routine inspection of plants can prevent much trouble and, if chemicals must be used, it is generally easier to treat glasshouse and house plants than it is to treat outdoor plants. Fumigation is effective in glasshouses, providing they are reasonably air-tight, and biological control of some pests may also be feasible. The following key can be used to identify the main causes of trouble:

Plants infested by small, wingless and winged, green, yellow, brown or black insects, often in dense colonies on leaves, young shoots, stems, buds and flowers; plants sticky, often sooty, growth checked and sometimes distorted — Aphids p. 161

Plants infested by small, white, winged insects, especially on undersides of young leaves; plants sticky, sometimes sooty — Glasshouse Whitefly p. 158

Plants infested by soft-bodied, wingless insects covered in white wax powder and filaments, often forming clusters in leaf axils and on stems; plants sticky, sometimes sooty — Glasshouse Mealybugs p. 189

Plants infested by brown, yellow or white, flat or raised scales, especially on undersides of leaves and also on stems; foliage sticky, sometimes sooty — Scale Insects p. 185

Plants infested by small winged and wingless yellow-green insects feeding on undersides of leaves and causing coarse light mottling on upper surfaces; empty moult skins often remain attached to leaves — Glasshouse Leafhoppers p. 154

Plants infested by small green or red-brown mites, especially on undersides of leaves; fine light flecking on upper surfaces later turns yellow and leaves may dry out and die; fine silk webbing may cover affected parts — Glasshouse Red Spider Mites p. 247

Plants infested by narrow-bodied yellow, brown or black wingless and winged insects, especially on leaves and in flower buds and flowers; leaves and flowers with fine white flecking and other discolourations — Thrips p. 191

Plants with irregular holes and pieces eaten out of leaves and flowers; stems and buds sometimes damaged; slime trails present on or near plants — Slugs and Snails p. 139

Plants with irregular holes and pieces eaten out of leaves, buds and flowers; tips of shoots sometimes drawn together with fine silk webbing; no slime trails present — Caterpillars p. 194

Plants with young growth distorted; leaves sometimes thickened and curled, buds killed and flowers disfigured; very small brown mites and white eggs present in buds and growing points — Tarsonemid Mites p. 251

Plants with roots, corms or tubers eaten by C-shaped legless grubs, especially in pots — Vine Weevil p. 242

Plants with roots developing irregular hard swelling, sometimes affecting most of root system; plants growing poorly — Root Knot Eelworms p. 136

Plants with many small dark flies running over surface of soil or potting medium; white, legless larvae with dark heads sometimes present among roots and at bases of stems — Sciarid Flies p. 215

Plants with minute, wingless insects jumping from surface of soil or potting compost, especially when plants are grown in pots — Springtails p. 146

Flowers and/or buds fail to form or drop prematurely — See under Drought p. 482

Flowers and/or buds rotten with fluffy grey mould growth present — Grey Mould p. 346

Leaves generally dull; later wilted, toughened and turning brown — Drought p. 482

Leaves yellowed; sometimes with dry angular blotches and generally stunted; roots may become rotten — Waterlogging p. 485 but see also Root and Foot Rot p. 361 and Light p. 484

Leaves and/or other parts with powdery white coating — Powdery Mildew p. 302

Leaves and/or other parts rotten and bearing fluffy grey mould growth — Grey Mould (Follow guidelines on p. 346)

Leaves variously spotted or malformed — See Leaf Spots p. 391 and discussion under Localised Pollutants p. 491

Also check Plates 21–22 giving symptoms of — Mineral Nutrient Deficiency

Roots rotten; plants generally unthrifty — Root and Foot Rot p. 361

Also see **Cockroaches** (p. 149), **Crickets** (p. 147), **Woodlice** (p. 145), and **Symphylids** (p. 144), which are sometimes troublesome in glasshouses and see separate entries for important plants **(Begonia, Chrysanthemum, Senecio (Cineraria), Dianthus, Freesia, Fuchsia, Primula, Rhododendron (Azalea), Saintpaulia).**

Gleditschia
Leaves with various yellowish mosaic-like patterns — Viruses p. 449

Globe Artichoke
Flower heads shrivelled and with fluffy grey mould growth — Grey Mould p. 346

Leaves with pale yellowish spots and whitish mould growth beneath — Lettuce Downy Mildew p. 293

Globe Flower. See **Trollius**

Gloxinia. See **Sinningia**

Godetia
Leaves of seedlings and young plants with small holes and pits — Flea Beetles p. 238

Leaves wilt; stem interior with brown or black streaks — Wilt p. 306

Golden Rod. See **Solidago**

Gooseberry

Fruit rotted and with fluffy grey mould growth	Grey Mould Rot p. 352
Fruit scorched following hot weather	High Temperature Injury p. 484
Flowers die suddenly; leaves may be scorched	Frost p. 482
Buds pecked out during winter and early spring	Birds p. 255
Leaves with irregular holes and pieces eaten away, sometimes stripped to veins	Gooseberry Sawfly p. 225 or Magpie Moth p. 198
Leaves and shoot tips infested by small grey-green insects; leaves curled	Gooseberry Aphid p. 170
Leaves with small tattered holes, especially at tips of shoots	Capsid Bugs p. 151
Leaves and/or other parts with orange or red pustules	Cluster-Cup Rust p. 277
Leaves with small yellow pustules and bristly structures on undersides	White Pine Blister Rust p. 286
Leaves and/or other parts with powdery white or felted brown coating	American Mildew p. 298
Leaves with pale yellow banding along veins	Vein Banding Virus p. 449
Leaves scorched suddenly in spring	Frost p. 482
Leaves silvery; branches may die back	Silver Leaf p. 324
Leaves variously spotted	Currant Leaf Spot p. 402
Leaves with inter-veinal yellowing	Manganese Deficiency p. 479
Branches die back; leaves with silvery sheen	Silver Leaf p. 324
Branches die back in absence of lesions or other damage	Grey Mould Dieback p. 317
Branches die back; masses of small, usually pink pustules present	Coral Spot p. 314
Branches and/or stem with hard, irregular swelling	Crown Gall p. 330
Branches die back; small hoof-like fungal growth at stem base	Currant Collar Rot p. 374
Branches die back; may be toadstools and bootlace-like strands at stem base; root decayed close to stem	Honey Fungus p. 375

Gorse. See **Ulex**

Grape Hyacinth. See **Muscari**

Grape-Vine

Fruits partially eaten as they ripen; skin punctured and contents removed	Wasps p. 232
Fruit rotted and with fluffy grey mould growth present	Grey Mould Rot p. 352
Fruit with pale pustules and some rotting	Bitter Rot p. 349
Fruit scorched after hot weather or when close to glass in glasshouse	High Temperature Injury p. 484
Flowers killed suddenly; leaves may be scorched	Frost p. 482
Leaves and/or other parts with powdery white coating	Powdery Mildew p. 302
Leaves with pale yellowish spots on upper surfaces and off-white mould beneath	Downy Mildew p. 293

Leaves with irregular grey spots and holes	Anthracnose p. 405
Leaves and shoots wilt in absence of lesions or other damage	Drought p. 482
Leaves scorched in spring	Frost p. 482
Leaves and/or other parts with sticky black coating	Sooty Mould p. 303
Leaves turn yellow and drop prematurely; roots covered with cobweb-like mould	White Root Rot p. 370
Leaves with rough, warty outgrowths on undersides	Oedema p. 488
Leaves variously distorted, mottled or abnormally disposed	Viruses p. 449
Leaves discoloured, yellowing, bronzing and dying, sometimes falling prematurely; many small mites on undersides	Glasshouse Red Spider Mites p. 247
Leaves showing red discolourations, followed by rapid collapse and death of plants; small yellow insects on roots (extremely rare in Britain)	Vine Phylloxera p. 184
Leaves and stems infested by yellow and brown waxy scales; foliage sticky and sooty	Scale Insects p. 185
Stems with clusters of soft-bodied wingless insects covered in white wax powder and filaments; foliage sticky and sooty	Glasshouse Mealybugs p. 189
Shoots with rough, warty outgrowths	Oedema p. 488
Shoots wilt and die; elongate tumours on stems (not Britain)	Dead-Arm Disease p. 317
Stem with hard, irregular swelling	Crown Gall p. 330
Stem with toadstools and/or bootlace-like strands at base; roots close to stem may be decayed and branches die back	Honey Fungus p. 375
Roots covered with cobweb-like mould	White Root Rot p. 370
Roots decayed close to stem; toadstools and/or bootlace-like strands may be present	Honey Fungus p. 375

Gypsophila

Stem rotted at base; cottony white mould growth present	Sclerotinia Disease p. 369

Hawthorn. See **Crataegus**

Hazel Nuts. See **Corylus**

Heaths and Heathers. See **Erica** and **Calluna**

Hebe. See **Veronica**

Hedera (Ivy)

Leaves with various distortions and/or yellowish mottled patterns	Viruses p. 470
Leaves spotted	Leaf Spots p. 408

Also see **Climbing Plants** and **Glasshouse and House Plants**

Hedges

Most hedging plants are sufficiently robust to be little affected by pests, although they may be regularly infested by **Caterpillars** (p. 194), **Aphids** (p. 161), **Leafhoppers** (p. 154) and other groups of pests. Some diseases may be more prevalent in hedges, since close contact of roots and other parts may favour rapid transmission. Diseases commonly affected by this type of transmission, which results in death of patches of the hedge while the remainder stays healthy, are **Honey Fungus** (p. 375) and **Phytophthora Root Death** (p. 377). Leakage of pollutants, such as gas from an underground source or the dumping of road salt on streets, can give rise to similar effects (see p. 489).

Also see separate entries for the main plant genera.

Helenium

Flowers green and malformed	Viruses p. 450
Leaves and shoots distorted; leaf blades narrowed and curled	Stem Eelworm p. 132
Leaves drawn together at tips of shoots with fine silk webbing protecting small caterpillars	Tortrix Moths p. 207
Shoots and leaves eaten when young; slime trails present on or near plants	Slugs and Snails p. 139
Leaves wilt; stem interior with brown or black streaks	Wilt p. 306

Helianthus (Sunflower)

Flowers shrivelled and with fluffy grey mould growth present	Grey Mould p. 346
Leaves with brown, black or other coloured pustules on undersides (not Britain)	Rust p. 278
Leaves at base of plants with irregular holes and pieces eaten away, sometimes reduced to a skeleton of veins; slime trails present on or near plants	Slugs and Snails p. 139
Stem rotted at base; cottony white mould growth present	Sclerotinia Disease p. 369

Helichrysum

Leaves with pale yellowish spots and whitish mould growth beneath	Lettuce Downy Mildew p. 293

Helleborus (including Christmas rose)

Flowers and leaves infested by pale green insects; leaves sticky, sometimes sooty	Aphids p. 161
Leaves with pale yellowish spots on upper surfaces and off-white mould growth beneath	Downy Mildew p. 295
Leaves with large, more or less circular, concentrically marked brown blotches	Leaf Spot p. 406

Hemerocallis (Day Lily)

Leaves with gradually enlarging brown spots	Iris Leaf Spot p. 407

Hemlock. See Tsuga

Herbs. See entries for main plant genera

Heuchera

Stem with masses of short or thickened shoots, especially near ground level	Leafy Gall p. 331

Hibiscus. See **Glasshouse and House Plants**

Hippeastrum. See **Glasshouse and House Plants**

Holly. See **Ilex**

Hollyhock. See **Althaea**

Honeysuckle. See **Lonicera**

Hornbeam. See **Carpinus**

Horse Chestnut. See **Aesculus**

Horse Radish

Leaves with pale spots, usually with brown margins	Pale Leaf Spot p. 407
Leaves and/or other parts with small, often concentric clusters of creamy droplets	White Blister p. 304
Leaves with large, yellow blotches or yellowish mosaic	Virus p. 450
Roots with irregular swellings not containing caterpillars or holes	Clubroot p. 328

Houseleek. See **Sempervivum**

House Plants. See **Glasshouse and House Plants**

Hosta

Leaves and young shoots eaten away at tips	Rabbits p. 259
Leaves and young shoots eaten, leaves with irregular holes; slime trails present on or near plants	Slugs and Snails p. 139

Hoya

Leaves with various yellowish mosaic-like patterns	Asclepias Viruses p. 438

Hyacinthus

Leaves with fine mottling of white and green	Virus p. 450
Leaves die back from tips and shrivel; masses of fluffy grey mould growth present	Fire p. 382
Leaves turn yellow; plants topple over after flowering; leaf bases rotted and readily pulled away from bulb	Black Slime p. 342
Bulb (in store) with firm, dry rot and blue-green mould growth	Blue Mould Rot p. 339
Bulbs rot in ground and plants fail to emerge or are stunted with brownish blotches; bulbs with bright yellow slime when cut	Yellows p. 343
Bulbs rot in ground and plants fail to emerge; bulbs with dry rot and soil adhering to whitish mould	Tulip Grey Bulb Rot p. 344
Bulbs with soft and slimy rot	Bacterial Soft Rot p. 348
Roots blackened and rotted; decay not usually extending into bulb	Root and Foot Rot p. 361

Plants grown in water culture covered with pale yellowish slimy mass — Slime Mould p. 424
Also see **Bulbs etc.**

Hydrangea
Flowers shrivelled and with fluffy grey mould growth — Grey Mould p. 346
Flowers stunted and remain green (note that hydrangea flowers are normally green before they open fully) — Virus p. 451
Leaves with many small ragged holes, especially at tips of shoots; buds sometimes killed — Capsid Bugs p. 151
Leaves twisted and distorted, especially at tips of shoots; stems scarred, growth stunted — Stem Eelworm p. 132
Leaves and/or other parts with powdery white coating — Powdery Mildew p. 302

Leaves with yellowish mottled and/or irregular patterns — Virus p. 451
Leaves variously spotted — Leaf Spots p. 407
Leaves and/or stems with patches of rot bearing fluffy grey mould growth — Stem and Leaf Rot p. 382
Leaves more or less bleached; veins remain dark green — Iron Deficiency p. 478

Roots and/or stem base blackened and rotten — Root and Foot Rot p. 361

Also see **Glasshouse and House Plants** for conditions affecting pot hydrangeas

Hypericum (St John's Wort)
Leaves with yellow spots on upper surfaces and orange pustules beneath — Rust p. 279

Ilex (Holly)
Leaves with brown and yellow blotch mines — Holly Leaf Miner p. 223

Leaves with brown or yellow scales on undersides; foliage sticky and sooty — Scale Insects p. 185
Leaves with greyish spots — Leaf Spots p. 407
Also see **Trees**

Impatiens (including Busy Lizzie)
Flowers and/or buds shrivelled and with fluffy grey mould growth present — Grey Mould p. 346
Leaves with pale brown blotches — Leaf Spots p. 407
Also see **Glasshouse and House Plants**

Ipomoea (Morning Glory). See **Climbing Plants**

Iris
Flowers of dark coloured types with streaks of paler colour — Viruses p. 451
Leaves with small brown pustules — Rust p. 279
Leaves spotted with brown or grey — Leaf Spot p. 407
Leaves yellowed or mottled; may be general stunting and disruption of flower colour — Viruses p. 451

Leaves with yellow but darkening streaks; terminal parts turn red-brown and shrivel (bulbous irises only) Ink Disease p. 343

Leaves with irregular holes and pieces eaten; slime trails present on or near plants Slugs and Snails p. 139

Leaves eaten, sometimes extensively; no slime trails present Caterpillars p. 194 / Iris Sawfly p. 227

Bulbs or rhizomes with soft and slimy rot Bacterial Soft Rot p. 348

Bulbs with black patches or streaks; may be extensive decay Ink Disease p. 343

Bulbs with dry rot and soil adhering to whitish mould when dug up; plants emerge feebly Tulip Grey Bulb Rot p. 344

Bulbs with brown rot and thin white strands of mould between scales White Root Rot p. 370

Also see **Bulbs etc.**

Ivy. See **Hedera**

Ixia

Corms with dry rot and with soil adhering to whitish mould when dug up; plants emerge feebly Tulip Grey Bulb Rot p. 344

Corms with brown rot and thin white strands of mould between outer tissues White Root Rot p. 370

Also see **Bulbs etc.**

Japanese Quince. See **Chaenomeles**

Jasmine. See **Jasminum**

Jasminum (Jasmine)

Leaves with yellow variegation Viruses p. 451

Leaves and young shoots infested by small insects; foliage sticky and sooty Aphids p. 161

Jerusalem Artichoke

Stem base and/or tubers rotted and with cottony white mould growth present Sclerotinia Disease p. 369

Tubers holed and tunnelled, often extensively Slugs p. 139 / Swift Moths p. 201

Judas Tree. See **Cercis**

Juglans (Walnut)

Nuts variously rotted Nut Rots p. 361

Leaves with blister-like growths (galls) Gall Mites p. 252

Leaves with brown or yellowish blotches; may be dark blotches on green nuts also Leaf Blotch p. 423

Leaves scorched in early spring Frost p. 482

Branches die back; masses of small, usually pink pustules present Coral Spot p. 314

Branches and/or stem with hard, irregular swelling Crown Gall p. 330

Branches with masses of proliferating shoots (witches' brooms) Viruses p. 451

Stem with foul smelling fluxes; branches may die back Slime Flux p. 325

Stem with some decay of wood; fungal brackets may
be present and branches die back

Ash Heart Rot p. 372
or Elm Heart Rot
p. 374

Also see **Trees**

Juniper. See **Juniperus**

Juniperus (Juniper)
Leaves and shoots drawn together by small caterpillars spinning fine silk webbing; foliage browns and dies

Juniper Webber
Moth p. 209

Leaves discoloured, yellowing and browning; small mites present, producing some silk webbing

Conifer Spinning
Mite p. 249

Leaves and stems with small, round, grey-white scales, sometimes forming extensive encrustations

Scale Insects
p. 185

Shoots with orange-brown elongated swellings

Mountain Ash Rust
p. 281 or
Hawthorn Rust p. 277

Also see **Conifers** and **Trees**

Kalanchoe
Stem rotted

Cactus and Succulent
Stem Rot p. 379

Also see **Cacti and Succulents**

Kale. See **Brassicas**

Kerria
Leaves yellowed or with irregular patterns
Twigs die back

Viruses p. 470
Twig Blight p. 317

Kohl-Rabi. See **Brassicas**

Laburnum
Leaves yellowed or with irregular patterns
Leaves with powdery white coating

Viruses p. 470
Powdery Mildew
p. 302

Leaves with pale yellow spots and whitish mould growth beneath

Downy Mildew
p. 295

Leaves variously spotted; usually in grey or brown
Leaves with brown blotch mines

Leaf Spots p. 408
Laburnum Leaf
Miner p. 210

Leaves with regular, almost semicircular pieces cut away

Leaf Cutter Bees
p. 233

Branches die back; fungal fruit bodies may be present

Dieback p. 317 or
Silver Leaf p. 324

Also see **Trees**

Lachenalia
Leaves with fine mottling of white and green

See Hyacinth Viruses
p. 450

Leaves with yellow but darkening streaks; terminal parts turn red-brown and shrivel

Iris Ink Disease
p. 343

Bulb softened and with rich brown rot spreading from basal ring

Narcissus Basal Rot
p. 343

Bulbs with black patches or streaks; may be extensive
decay
Also see **Bulbs etc.**

Iris Ink Disease
p. 343

Larch. See **Larix**

Larix (Larch)
Leaves scorched in early spring; shoots may be killed
Leaves with tiny orange and white cups

Frost p. 482
Birch Rust p. 271
Poplar Rust p. 283
Willow Rust p. 287
Leaves turn brown or pale-coloured and drop
prematurely
Leaves and shoots with tufts of white wax wool cover-
ing small dark insects
Leaves eaten, especially at tips of shoots
Branches die back; canker lesions may be present and
trees appear tufted
Stem with some decay of wood; fungal brackets may
be present

See discussion under
Needle Casts p. 305

Adelgids p. 182

Sawflies p. 224
Canker and Dieback
p. 318
Brown Cubical Rot
p. 373 or
Oak Heart Rot
p. 376

Also see **Conifers** and **Trees**

Larkspur. See **Delphinium**

Lathyrus (Sweet Pea)
Flowers pecked and torn
Flowers and leaves discoloured with light flecking and
more extensive silvering; narrow bodied wingless
and winged insects present
Flowers, leaves and shoots infested by small winged
and wingless green or yellow insects; plants some-
times sticky and sooty
Flowers of dark coloured types with paler streaks
Leaves with irregular holes; young shoots eaten; slime
trails present on or near plants
Leaves discoloured, yellowing and drying out; many
small mites on undersides
Leaves and/or other parts with powdery white coating

Birds p. 255

Thrips p. 191

Aphids p. 161
Viruses p. 452
Slugs and Snails
p. 139
Glasshouse Red Spider
Mites p. 247
Powdery Mildew
p. 302 but also see
White Mould p. 388
Leaves with pale yellowish spots and purple-white
mould growth beneath
Leaves wilt; stem interior with brown or black streaks
Leaves with various mottle or mosaic patterns or
otherwise distorted
Stem with soft irregular swelling
Stem with masses of short or thickened shoots, es-
pecially near ground level
Stem with decay at ground level; may extend to roots

Pea Downy Mildew
p. 294
Wilt p. 306

Viruses p. 452
Crown Gall p. 330

Leafy Gall p. 331
Root and Foot Rot
p. 361
Stem with decay, usually at ground level and with
cottony white mould growth
Roots with many small, hard swellings

Sclerotinia Disease
p. 369
Root Nodules p. 428

Laurel, Bay. See **Laurus**

Laurel, Cherry. See **Prunus**

Laurel, Portugal. See **Prunus**

Laurus (Bay laurel)
Leaves with edges curled and thickened; small insects present, usually covered with white wax wool — Bay Sucker p. 157
Leaves with flat yellow or brown scales on undersides; foliage sticky and sooty — Scale Insects p. 185
Also see **Trees**

Lavandula (Lavender)
Flowers and/or buds shrivelled and with fluffy grey mould growth — Grey Mould p. 346
Leaves with small, brown circular spots — Leaf Spot p. 409
Shoots wilt suddenly and may die back — Shab Disease p. 318
Shoots and stems with small masses of white froth covering wingless insects; growth sometimes distorted — Froghoppers p. 153

Lavatera (Tree-mallow)
Leaves and stem with raised, orange-brown pustules a few mm in diameter — Hollyhock Rust p. 278
Leaves and/or other parts with brown, ovoid spots, 5 cm or more in length — Leaf Spot p. 408

Lavender. See **Lavandula**

Lawns
The establishment and maintenance of fine lawns is a specialised aspect of gardening which is dealt with in detail in many gardening books. Good control of pests and diseases is essential if consistently good results are to be achieved and this usually involves good management combined with some chemical treatments. The following key can be used to identify the main causes of trouble:
Seed removed from newly-sown areas — Birds p. 255
Soil scratched and newly-sown seed disturbed — Cats p. 264, Birds p. 255
Grass yellowing, browning and dying, often in patches — Leatherjackets p. 215, Chafers p. 235, Dogs p. 264, Lawn Turf Rots p. 365
Grass with gelatinous green bodies among leaves — Algae p. 426
Grass with white or greyish, often branched growths — Lichen p. 425
Grass smothered with variously coloured, slimy, often sponge-like bodies — Slime Moulds p. 424
Leaves with powdery white coatings — Powdery Mildews p. 302
Leaves variously spotted — Leaf Spots p. 406
Mounds of loose soil thrown up on surface of lawns to about 10 cm high or higher — Moles p. 262
Mounds of fine soil thrown up on surface to about 10 mm high — Burrowing Bees p. 233 or Ants p. 230
Small casts of soil extruded from holes on surface — Earthworms p. 142

Leek

Leeks have many problems in common with **Onion** (q.v.) but a few, such as Rust and White Tip, are much commoner

Also see **Leek Moth** p. 204

Lettuce

Plants wilt and die; stem severed at soil level	Cutworms p. 200
Plants unthrifty; roots attacked by thin yellow-brown larvae	Wireworms p. 236
Plants unthrifty; roots infested by small white insects	Root Aphids p. 181
Leaves with pale yellowish or brown, angular spots and whitish mould growth beneath	Downy Mildew p. 293
Leaves shrivelled at margins	Marginal Leaf Spot p. 409
Leaves with irregular or more or less circular ragged holes	Ring Spot p. 409
Leaves otherwise spotted	Leaf Spots p. 409
Leaves with pale inter-veinal areas giving marbled effect	Beet Western Yellows Virus p. 452
Leaves irregularly blistered and distorted and with pale yellow bands along veins	Big Vein Disease p. 452 or Lettuce Mosaic Virus p. 453
Leaves pecked and torn	Birds p. 255
Leaves infested by small green or yellow insects	Aphids p. 161
Leaves with irregular holes; slime trails present; seedlings may be killed	Slugs and Snails p. 139
Leaves with irregular holes; no slime trails present	Caterpillars p. 194
Heads with soft and slimy rot	Bacterial Soft Rot p. 348
Heads rotted and with fluffy grey mould growth present; stems may rot completely	Grey Mould p. 352 but also see Sclerotinia Disease p. 369 and Root and Foot Rot p. 361
Plants run to flower	Bolting p. 487

Ligustrum (Privet)

Leaves turn yellow and wilt in absence of root damage	Wilt p. 306
Leaves yellow or with pale yellowish spots and patterns (visible even on 'golden' cultivars)	Viruses p. 453
Leaves with brown spots	Leaf Spots p. 417
Leaves with disfiguring brown blotch mines	Lilac Leaf Miner p. 210
Leaves with white flecking and silvery sheen; small, narrow-bodied insects present	Privet Thrips p. 193
Leaves rolled tightly and discoloured, especially at tips of shoots; leaves fall prematurely	Privet Aphid p. 179
Stems with encrustations of small white scales	Willow Scale p. 188
Branches or stem with irregular hard swellings	Crown Gall p. 330
Roots covered in cobweb-like mould growth; leaves may turn yellow and drop prematurely	White Root Rot p. 370

Also see **Hedges**

Lilac. See **Syringa**

Lilium (Lily)

Plants infested by red beetles and grubs; holes eaten in leaves, stems, buds, flowers and seed-capsules	Lily Beetle p. 240
Flowers distorted; leaves irregularly mottled; plants unthrifty	Viruses p. 453 esp. Tulip Breaking Virus p. 454
Leaves with mottling or distortion	Viruses p. 453
Leaves with rusty brown pustules (rare in Britain	Rust p. 280
Leaves with brown spots which often bear mould growth (not Britain)	Leaf Spot p. 410
Leaves with red-brown, often elliptical spots; may wither and hang down (common)	Lily Disease p. 383
Leaves and young shoots grazed	Rabbits p. 259
Leaves, young shoots and other parts infested by small insects; leaves sticky, sometimes sooty	Aphids p. 161
Bulbs dug up and eaten	Squirrels p. 260
Bulbs tunnelled; young shoots and leaves with irregular holes; slime trails present	Slugs and Snails p. 139
Bulbs with dry rot and with soil adhering to whitish mould when dug up; plants emerge feebly	Tulip Grey Bulb Rot p. 344
Bulbs (in store) with firm, dry rot and blue-green mould growth	Blue Mould Rot p. 339
Roots blackened and rotted; decay not usually extending into bulb	Root and Foot Rot p. 361
Also see **Bulbs etc.**	

Lily. See **Lilium**

Lily-Of-The-Valley. See **Convallaria**

Lime. See **Tilia**

Lithops. See **Cacti and Succulents**

Lobelia

Leaves mottled and/or distorted	Viruses p. 454
Also see **Bedding Plants**	

Loganberry. See **Raspberry**

Lonicera (Honeysuckle)

Leaves and young shoots infested by dense colonies of blue-green insects; growth distorted, foliage very sticky and sooty	Honeysuckle Aphid p. 178
Leaves with powdery white coating	Powdery Mildew p. 302
Leaves with tiny yellow-white cup-like bodies on undersides	Rust p. 278
Leaves variously spotted	Leaf Spots p. 406
Also see **Climbing Plants**	

Lunaria

Leaves and/or other parts with small, often concentric
clusters of creamy droplets   White Blister p. 304
Roots with irregular swellings   Clubroot p. 328

Lupin. See **Lupinus**

Lupinus (Lupin)

Leaves and/or other parts with powdery white coating   Powdery Mildew
                                                                                     p. 302
Leaves with dark green bands along veins or various
irregular yellowish patterns   Viruses p. 454
Leaves with purplish or grey-brown spots   Leaf Spot p. 410
Stem with irregular swellings   Crown Gall p. 330
Stem rotted at base; white cottony mould growth   Sclerotinia Disease
present   p. 369
Roots blackened and decayed; plant unthrifty   Root and Foot Rot
                                                                                     p. 361
Roots with small nodular swellings   Root Nodules p. 428

Lychnis

Leaves and young shoots infested by small insects;
growth distorted   Aphids p. 161
Leaves, shoots and stems with small masses of white
froth covering small wingless insects   Froghoppers p. 153

Magnolia

Buds and flowers pecked or torn   Birds p. 255 or Grey
                                                                                     Squirrel p. 260
Flowers or buds shrivel and die suddenly   Frost p. 482
Flowers and/or buds with powdery grey mould
growth   Grey Mould p. 346
Leaves with tattered holes   Capsid Bugs p. 151
Leaves with irregular pale spots   Leaf Spot p. 410
Leaves with irregular yellowish patterns   Viruses p. 454
Branches die back; masses of small, usually pink pus-
tules present   Coral Spot p. 314

Mahonia

Leaves with pale brown pustules on undersides   Rust p. 280
Leaves with tiny orange cups on undersides   Berberis Rusts
                                                                                     p. 271

Maidenhair Fern. See **Ferns**

Maize. See **Sweet Corn**

Mallow (Tree). See **Lavatera**

Malus. See **Apple.** Many of the problems affecting
apples also occur on ornamental **Malus**

Malva (Mallow). See **Althaea**

Mammillaria. See **Cacti and Succulents**

Maple. See **Acer**

Marguerite. See **Chrysanthemum**

Marigold. See **Calendula** and **Tagetes**

Marrow. See **Cucumber**

Matthiola (Stock)
Flowers of dark coloured types with paler streaks and
distortion; leaves may be mottled or distorted   Viruses p. 454
Leaves and/or other parts with small, often concentric
clusters of creamy droplets   White Blister p. 304
Roots with irregular swellings   Clubroot p. 328
Leaves of young plants with small round holes and
pits   Flea Beetles p. 238
Leaves of older plants with irregular holes and pieces
eaten away   Caterpillars p. 194
Leaves and young shoots infested by small wingless
and winged insects; foliage sticky and sooty   Aphids p. 161
Roots eaten by maggots; plants wilt   Cabbage Root Fly
                                                     p. 218

Medlar
Fruit with soft brown rot and usually off-white cot-
tony pustules; associated twigs may die back   Brown Rot p. 350
Leaves with powdery white coating   Powdery Mildew
                                                     p. 297
Leaves with irregular yellowish streaks and patterns
and/or some distortion   Viruses p. 470
Leaves with large, dark brown blotches   Leaf Blotch p. 411
Also see **Trees**

Melon. See **Cucumber**

Mesembryanthemum
Leaves with pale yellowish spots and white mould   Downy Mildew
growth beneath   p. 295

Metasequoia. See **Conifers** and **Trees**

Michaelmas Daisy. See **Aster**

Mint
Plants distorted and bearing masses of small orange
pustules   Rust p. 280

Monkey-Puzzle Tree (Araucaria). See **Conifers**
and **Trees**

Monstera. See **Glasshouse and House Plants**

Montbretia
Leaves turn yellow and decay at ground level where   Gladiolus Dry Rot
masses of tiny black bodies are present on them.   p. 341
Leaves with yellow but darkening streaks; terminal   Iris Ink Disease
parts turn red-brown and shrivel   p. 343
Also see **Bulbs etc.**

Morning Glory. See **Ipomoea**

Mountain Ash. See **Sorbus**

Mulberry
    Leaves with irregular yellowish streaks and patterns
      and/or some distortion — Viruses p. 470
    Leaves variously spotted — Leaf Spot or
      Bacterial Leaf Spot
      p. 411

    Leaves with powdery white coating — Powdery Mildew
      p. 302

    Branches die back; masses of small, usually pink pus-
      tules present — Coral Spot p. 314
    Branches with small canker lesions; may die back — Canker and Dieback
      beyond lesion — p. 319
    Also see **Trees**

Mullein. See **Verbascum**

Muscari (Grape Hyacinth)
    Flowers with brown-black powdery masses in centre — Scilla Anther Smut
      p. 290

    Leaves and/or stems with dirty yellowish spots bear-
      ing brown pustules — Bluebell Rust p. 272
    Bulbs with soft and slimy rot — Bacterial Soft Rot
      p. 348

    Also see **Bulbs etc.**

Mushrooms (Cultivated)
    Stalks and caps tunnelled by maggots — Sciarid Flies p. 215
    Stalks and caps with irregular pieces eaten away; slime
      trails present — Slugs p. 139

Myosotis (Forget-me-not)
    Plants infested by small green or pink wingless and
      winged insects; growth distorted, foliage sticky and
      sooty — Aphids p. 161
    Flowers and/or buds with fluffy grey mould growth — Grey Mould p. 346
    Leaves and/or other parts with powdery white coating — Powdery Mildew
      p. 302

    Leaves with pale yellow spots and white mould — Downy Mildew
      growth beneath — p. 295
    Leaves with small off-white spots; no mould growth
      even in damp conditions — Smut p. 290

Myrtle. See **Myrtus**

Myrtus (Myrtle)
    Leaves variously spotted — Leaf Spot p. 411

Narcissus

Bulbs soft at planting time, with internal brown discolouration and decay; growth poor and distorted	Stem Eelworm p. 132
Bulbs soft at planting time, with internal cavity tunnelled by maggots; leaves thin and weak and flowers fail to develop	Large Narcissus Fly p. 221
Bulbs in soil with holes eaten in from the outside, often extensively tunnelled	Slugs p. 139
Bulb softened and with rich brown rot spreading from basal ring	Basal Rot p. 343
Bulbs (in store) with firm, dry rot and blue-green mould growth	Blue Mould Rot p. 339
Bulbs with dry rot and with soil adhering to whitish mould when dug up; plants emerge feebly	Tulip Grey Bulb Rot p. 344
Bulbs with brown rot and strands of thin white mould between scales	White Root Rot p. 370
Flowers with pale brown spots which spread rapidly until blooms destroyed	Fire p. 383
Leaves variously streaked or distorted, plants often unthrifty	Viruses p. 455
Leaves with red-brown scorch, spreading downwards from tip; gradually shrivel and die; flower stalks and flowers may also have brown discolouration	Leaf Scorch p. 412
Leaves turn brown at base; upper parts yellow and wilted and soon covered with fluffy grey mould	Smoulder p. 384
Leaves with grey or yellow spots on which white mould grows	White Mould p. 384
Roots blackened and rotted; decay not usually extending into bulb	Root and Foot Rot p. 361

Also see **Bulbs etc.**

Nasturtium. See **Tropaeolum**

Nectarine. See **Apricot**

Nemesia. See **Bedding Plants**

Nerine. See **Bulbs etc.**

Nerium (Oleander)

Shoots and/or other parts with soft, spongy growth, canker lesions or other distortion	Gall p. 331

Also see **Glasshouse and House Plants**

Nicotiana (including Tobacco)

Leaves with irregular, yellowish streaks and patterns and/or various forms of distortion; plants generally unthrifty	Viruses p. 455
Leaves, especially of tobacco seedlings, with yellowish patches and dense blue-white mould beneath (rare in Britain)	Tobacco Blue Mould p. 295

Also see **Bedding Plants**

Nymphaea (Water lily)

Leaves with red-brown concentrically marked spots	Leaf Spot p. 412
Leaves with extensive irregular holes eaten by beetles and larvae on upper surfaces	Water-Lily Beetle p. 239
Buds infested by small green or brown insects; growth distorted	Water Lily Aphid p. 179

Also see **Aquatic Plants**

Oak. See **Quercus**

Odontoglossum. See **Orchids**

Oleander. See **Nerium**

Onion, Shallot, Chives, Garlic, Ornamental Allium, Leek

Leaves turn yellow and die from tip downwards; various other leaf symptoms may occur also either simultaneously or later and include several types of mould growth. It is difficult to distinguish between the various different causes	See discussion under Leaf Rot p. 384
Leaves with dull streaks or blisters of dark powder; plants often distorted	Smut p. 290
Leaves and or other parts with red-orange pustules (esp. on chives and leeks)	Rust p. 279
Leaves with short yellow streaks at base and variously distorted	Viruses p. 456
Leaves swollen and distorted; bulbs soft, tending to crack at base	Stem Eelworm p. 132
Leaves wilting and yellowing; bulbs and stems tunnelled by maggots; young plants often killed	Onion Fly p. 219
Leaves with fine light flecking and more extensive silver sheen; small narrow-bodied insects present	Onion Thrips p. 192
Bulbs covered with white cottony mould in which are tiny black bodies (esp. on salad onions)	White Rot p. 367
Bulbs (in store) soften from neck downwards; fluffy grey mould growth present	Neck Rot p. 356
Bulbs soften, roots and leaves shrivel	Shanking p. 385
Bulbs with soft and slimy rot	Bacterial Soft Rot p. 348
Bulbs with masses of small black bodies beneath outer scales; inner tissues may be stained yellow	Smudge p. 413
Plants run to flower	Bolting p. 487

Opuntia. See **Cacti and Succulents**

Orchids

Flowers with brownish streaks and spots	Viruses p. 456
Leaves variously spotted	Leaf Spots p. 413
Leaves with irregular streaks and patterns and often distortion of entire plant	Viruses p. 456

Also see **Glasshouse and House Plants**

Ornithogalum. See **Hyacinthus** and **Bulbs etc.**

Paeonia (Paeony)

Leaves with irregular yellowish rings or patterns	Viruses p. 457
Leaves with grey-brown, reddish-bordered spots	Blotch p. 413
Leaves with soft brown region at base; stem base brown and decayed with fluffy grey mould growth	Grey Mould Blight p. 385
Leaves wilt; stem interior with brown or black streaks	Wilt p. 306
Leaves and young shoots distorted and discoloured and growth checked in early spring	Chrysanthemum Eelworm p. 134
Roots eaten by white caterpillars in soil	Swift Moths p. 201

Palms

Leaves with dark warty scabs bearing yellow masses of spores	False Smut p. 413

Also see **Glasshouse and House Plants**

Pansy. See **Viola**

Papaver (Poppy)

Flowers and/or buds shrivelled and with fluffy grey mould growth	Grey Mould p. 346
Leaves and/or other parts with powdery white coating	Powdery Mildew p. 302
Leaves with pale yellowish spots and whitish mould growth beneath	Downy Mildew p. 295
Leaves, stems and flower buds infested by colonies of small black insects	Aphids p. 161

Paphiopedilum. See **Orchids**

Parsley

Leaves and/or other parts with powdery white coating	Powdery Mildew p. 302
Leaves with pale yellowish spots and white mould growth beneath	Umbellifer Downy Mildew p. 295
Leaves with small angular brown spots (common)	Leaf Spot p. 413
Leaves with brown or black pustules (uncommon)	Dill Rust p. 276
Leaves turning yellow and red; growth poor; fly maggots feeding on roots	Carrot Fly p. 217
Leaves with yellow blotch mines	Celery Fly p. 217
Plants stunted with yellowish or reddened leaves	Viruses p. 457

Parsnip

Leaves and/or other parts with powdery white coating	Powdery Mildew p. 302
Leaves with pale yellowish spots and white mould growth beneath	Umbellifer Downy Mildew p. 295
Leaves with pale, brown-bordered spots	Leaf Spot p. 413 but also see Canker p. 356
Leaves with net-like veinal yellowing	Viruses p. 457
Leaf bases swollen and splitting, crowns rotting	Stem Eelworm p. 132
Leaves with yellow blotch mines	Celery Fly p. 217
Roots tunnelled by white maggots	Carrot Fly p. 217
Roots with soft and slimy rot	Bacterial Soft Rot p. 348

Roots (especially in store) with brown rot and cottony
white mould growth; mainly around crown — Sclerotinia Disease
p. 369

Roots rotted and with matted dark strands to which
soil adheres — Violet Root Rot
p. 360

Roots split but not usually decayed — Splitting p. 488

Roots with other forms of decay or lesion, especially
around crown — Canker p. 356

Parthenocissus (Virginia Creeper)

Leaves with pale yellowish spots and whitish mould
growth beneath — Grapevine Downy
Mildew p. 293

Leaves with greyish, dark-bordered spots — Leaf Spot p. 414

Also see **Climbing Plants**

Passiflora (Passion Flower)

Leaves with irregular yellowish mosaics or patterns,
crumpling and/or other distortion — Viruses p. 457

Also see **Climbing Plants**

Passion Flower. See Passiflora

Paulownia

Branches with masses of proliferating shoots (witches'
brooms) — Viruses p. 458

Also see **Trees**

Pea

Seeds removed before they germinate — Birds p. 255
Mice p. 261

Seeds with dark areas on cotyledons when pulled
apart — Manganese Deficiency
p. 479

Seedlings with pieces eaten away as they emerge
through soil, often failing to establish — Slugs p. 139
Millepedes p. 143
Bean Seed Fly p. 219

Seedlings pecked and pulled out of soil — Birds p. 255

Plants growing poorly, yellowing and dying; small
yellow or brown cysts on roots — Pea Cyst Eelworm
p. 138

Pods with caterpillars inside, eating holes in peas — Pea Moth p. 203

Pods and leaves with light flecking and silvery sheen
on surfaces, small narrow-bodied insects present — Pea Thrips p. 192

Pods with triangular pieces pecked away and peas
removed — Birds p. 255

Pods, flower stalks and leaves with brown-yellow,
sunken spots — Leaf and Pod Spot
p. 414

Pods rotted, usually at flower end and with fluffy grey
mould growth — Grey Mould p. 352

Leaves and/or other parts with powdery white coating — Powdery Mildew
p. 299

Leaves with pale yellowish patches and purple-white
mould growth beneath — Downy Mildew
p. 294

Leaves wilt, stem interior with brown or black streaks
extending well above ground level — Wilt p. 306

Leaves with brown or black pustules on undersides
(rare in Britain) — Rust p. 281 Also see
Broad Bean Rust
p. 272

Leaves variously mottled and/or distorted	Viruses p. 458
Leaves with blue-green tint; may later turn yellow and wilt	Copper Deficiency p. 480
Leaves with small semi-circular notches eaten out of edges, producing a scalloped effect	Pea and Bean Weevils p. 242
Leaves and shoots infested by colonies of green, pink or yellow winged and wingless insects; foliage sticky and growth sometimes checked	Aphids p. 161
Stem rotted at base with cottony white mould growth present	Sclerotinia Disease p. 369
Stem with brown or blackish lesions at base; roots blackened or rotten; plants unthrifty	Root and Foot Rot p. 361
Roots with small nodular swellings	Root Nodules p. 428
Seedlings die; often in groups	Damping-off p. 362

Peach. See **Apricot**

Pear

Fruits pecked as they ripen; (peck-holes often extended by wasps)	Birds p. 255
Fruits distorted, with irregular bumps and scars	Capsid Bugs p. 151
Fruit with soft brown rot and usually off-white cottony pustules; associated twigs may die back	Brown Rot p. 350
Fruit with round, dark, saucer-like depressions; usually only in store	Bitter Rot p. 349
Fruit leathery and cracked; often on windfalls and low-hanging fruit	Leathery Rots p. 354
Fruit otherwise rotted, especially in store	Minor Rots p. 354
Fruit scorched following very hot weather	High Temperature Injury p. 484
Fruit with very dark brown more or less rounded corky patches	Scab p. 337
Fruit with dark green mouldy patches on surface; easily wiped off	Sooty Blotch p. 420
Fruit dimpled; sometimes wholly misshapen	Stony Pit Virus p. 458
Fruitlets enlarged and distorted, falling prematurely; small yellow or white maggots in cavities	Pear Midge p. 213
Flowers and/or buds die suddenly	Frost p. 482 but also see Apple Bud Rot p. 345
Flowers and then spur leaves wilt	Brown Rot Blossom Wilt p. 350 but also see Bacterial Blossom Blight p. 347
Flowers and flower buds discoloured, failing to develop normally; small flattened insects present, foliage sticky and sooty	Pear Sucker p. 156
Leaves and shoots infested by small black or brown insects, often in dense colonies; foliage sticky and sooty	Aphids p. 161
Leaves with rashes of blister-like pustules, at first yellow but later turning dark brown	Pear Leaf Blister Mite p. 253
Leaves with irregular holes and pieces eaten away	Caterpillars p. 194
Leaves with surface tissues eaten away, exposing fine veins; shining black larvae present	Pear and Cherry Slugworm p. 225

Leaves with irregular yellowish streaks and patterns	Viruses p. 458
Leaves scorched at margins in spring	Frost p. 482
Leaves and/or other parts with powdery white coating	Apple Powdery Mildew p. 297
Leaves and/or other parts with sticky black coating	Sooty Mould p. 303
Leaves with irregular dark brown blotches	Scab p. 337
Leaves with small brown blisters	Leaf Curl p. 332
Leaves with numerous grey, dark-bordered spots	Leaf Fleck p. 414
Leaves with red spots on uppersides and rough brown structures beneath (rare in Britain)	Rust p. 281
Branches die back; stem base and roots decayed; toadstools and bootlace-like strands may be present	Honey Fungus p. 375
Branches with canker lesions; may die back beyond lesions	Canker p. 310
Branches die back; masses of small, usually pink pustules present	Coral Spot p. 314
Branches and/or stem with hard irregular swelling	Crown Gall p. 330
Shoots exude slime and die back; flowers and/or leaves shrivelled	Fireblight p. 316 but also see Bacterial Blossom Blight p. 347
Branches otherwise die back	See discussion under Decline Mycoplasma p. 459
Stems become blistered and bark cracks away when 1–2 years old	Blister Canker Virus p. 459
Stems with some decay of wood; fungal brackets may be present	Elm Heart Rot p. 374
Stem base and roots decayed; toadstools and bootlace-like strands may be present and branches die back	Honey Fungus p. 375
Trees generally unthrifty	See discussion under Decline Mycoplasma p. 459, also Specific Replant Disease p. 428 and Deficiencies of Nitrogen p. 474, Phosphorus p. 475 and Potassium p. 475

Also see **Apple** and **Trees**

Pelargonium

Leaves with yellowish rings, streaks or other patterns	Viruses p. 459
Leaves with pale spots on uppersides and more or less rounded brown pustules beneath	Rust p. 282
Leaves otherwise spotted	Leaf Spots p. 415
Leaves and/or stems with rough, warty outgrowths	Oedema p. 488
Stem, especially of cuttings, pinched and shrivelled just above ground level	Blackleg p. 385

Also see **Bedding Plants** and **Glasshouse and House Plants**

Penstemon

Leaves and young shoots distorted and scarred; growth checked	Chrysanthemum Eelworm p. 134

Peperomia
Leaves and/or stems with rough warty outgrowths   Oedema p. 488
Also see **Glasshouse and House Plants**

Peppers. See **Capsicum**

Perennials
Ornamental plants grown as herbaceous perennials, usually in mixed borders and beds, remain on the same site for a number of years and may therefore suffer from a gradual build-up of pests and diseases as well as loss of vigour resulting from depletion of nutrients and competition between plants. Sustained good performance of these plants can often be encouraged by regular maintenance and renewal. Clear debris in autumn and winter; divide, reduce, mulch and support plants in spring and cultivate, manure and replant borders and beds when necessary, which may be every five years or so on light infertile soils but can be much longer on fertile soils. The following key can be used to identify the main causes of trouble and additional entries are given for some of the more important plant genera.

Leaves, buds, shoots, stems or flowers infested by small wingless and winged green, yellow, brown or pink insects; young growths sometimes checked and distorted; foliage sticky, sometimes sooty   Aphids p. 161

Leaves and young shoots distorted, sometimes discoloured; growth poor   Eelworms p. 132

Leaves and/or other parts with powdery white coating   Powdery Mildew p. 296

Leaves with various yellowish streaks and patterns   Viruses p. 429

Leaves variously spotted   Spots p. 391

Leaves and shoots with small masses of froth covering pink or white wingless insects; growth sometimes distorted   Froghoppers p. 153

Leaves with irregular holes and pieces eaten away; no slime trails present   Caterpillars p. 194

Leaves with irregular holes and pieces eaten away; young shoots eaten; flowers and buds sometimes attacked; slime trails on or near plants   Slugs and Snails p. 139

Leaves with small tattered holes, especially near tips of shoots; buds sometimes killed and flowers distorted   Capsid Bugs p. 151

Stem with irregular swellings   Crown Gall p. 330

Roots blackened and decayed; plants unthrifty   Root and Foot Rot p. 361

Roots and other underground parts eaten by white caterpillars living in soil   Swift Moths p. 201

Roots and other underground parts eaten by C-shaped white beetle larvae living in soil   Chafers p. 235

Roots and other underground parts eaten by grey-brown caterpillars living in soil; stems sometimes severed at soil level   Cutworms p. 200

Roots and other underground parts eaten by thin, long, wiry white or yellow beetle larvae living in soil; corms, tubers and rhizomes often with small cylindrical holes and tunnels   Wireworms p. 236

Periwinkle. See **Vinca**

Petunia

Leaves with irregular, yellowish streaks and patterns
and/or various forms of distortion; plants generally
unthrifty

Nicotiana Viruses
p. 455 also discussion
under Leafy Gall
p. 331

Also see **Bedding Plants**

Philadelphus

Leaves with irregular yellowish streaks and patterns
and/or distortion

Viruses p. 470

Leaves and young shoots infested by small black insects; growth distorted and checked

Black Bean Aphid
p. 171

Phillyrea

Leaves with small white, winged insects and/or flat
round scales on undersides; upper surfaces sticky
and sooty

Phillyrea Whitefly
p. 161

Philodendron. See **Glasshouse and House
Plants**

Phlox (annual and perennial)

Leaves and/or other parts with powdery white coating

Powdery Mildew
p. 302

Leaves with pale, more or less circular, dark-bordered
spots

Leaf Spot p. 415

Leaves and young shoots of perennial phlox deformed; leaves reduced to narrow frills along mid-
ribs

Stem Eelworm p. 132

Stem with irregular swelling

Crown Gall p. 330

Stem and/or other parts generally distorted

See discussion under
Leafy Gall p. 331

Also see **Perennials** and **Bedding Plants**

Phyllitis. See **Ferns**

Picea (Spruce)

Leaves with yellow bands in summer; white or yellow
pustules beneath in autumn

Rhododendron Rust
p. 284 or
Needle Rust p. 285

Leaves discolour and drop prematurely

See discussion under
Needle Casts p. 305

Leaves discoloured, sometimes falling prematurely;
small green insects present

Green Spruce Aphid
p. 181

Leaves discoloured, yellowing and bronzing, sometimes falling prematurely; small mites and silk
webbing present

Conifer Spinning
Mite p. 249

Leaves and stems with tufts of white wax wool

Adelgids p. 182

Buds at tips of shoots enlarged and modified to form
galls that resemble miniature pineapples

Spruce Gall
Adelgids p. 182

Stem with resinous lesions; branches die back

Pine Dieback p. 319

Stem with elongate fluting or strips of dead bark

Canker p. 326

Stem with some decay of wood; fungal brackets may
be present

Brown Cubical Rot
p. 373

Also see **Conifers** and **Trees**

Pine. See **Pinus**

Pink. See **Dianthus**

Pinus (Pine)

Leaves discolour and drop prematurely	See discussion under Needle Casts p. 305 also Snow Blight p. 325
Leaves drop prematurely but leaving tuft at branch tips	Dieback p. 319
Leaves eaten, especially at tips of shoots, caterpillars usually present	Caterpillars p. 194 or Pine Sawflies p. 228
Leaves, shoots and stems with tufts of white wax covering small dark insects	Adelgids p. 182
Shoots turning brown at tips and dying; caterpillars tunnelling into growing points	Pine Shoot Moth p. 211
Branches with yellowish or white and orange swellings or blisters; dieback and death may follow (five-needled pines only)	Blister Rust p. 286
Stem with pronounced 'S'-shaped kink (two-needled pines only)	Twisting Rust p. 282
Stems or branches with large black resinous cankers bearing orange pustules; branches may die back (Scots Pine only)	Resin-Top Rust p. 282
Stem with some decay of wood, fungal brackets may be present	Brown Cubical Rot p. 373 or Oak Heart Rot p. 376

Also see **Conifers** and **Trees**

Plane. See **Platanus**

Platanus (Plane)

Leaves with large, angular brown blotches	Anthracnose p. 415
Leaves yellowed and sparse; shoots may die back; roots close to stem may die back and patches of dead bark develop	Phytophthora Root Death p. 377
Stem with some decay of wood; fungal brackets may be present	Ash Heart Rot p. 372

Also see **Trees**

Pleione. See **Glasshouse and House Plants** and **Orchids**

Plum and Damson

Fruit with small holes pecked from the outside as it ripens	Birds p. 255
Fruit with larger holes, often extending under the skin at edges	Wasps p. 232
Fruits and fruitlets tunnelled, often with external holes exuding wet brown frass; fruitlets fall prematurely	Plum Sawfly p. 224
Fruits tunnelled by caterpillars, causing rotting and premature ripening	Plum Fruit Moth p. 198
Fruit with soft brown rot and usually off-white cottony pustules	Brown Rot p. 350

Fruit with green-brown, velvety spots	Scab p. 337
Fruit split	Splitting p. 488
Fruit with numerous, small dark spots	Fly Speck p. 405
Fruit with dark green mouldy patches on surface; easily wiped off	Sooty Blotch p. 420
Fruit twisted with pocket-like hollow on one side	Pocket Plums p. 332
Fruit (red-coloured types) with dark bands and lines; (yellow and dark coloured types) with grooves and pits on surface	Plum Pox Virus p. 460
Flowers and/or leaves scorched or killed suddenly.	Frost p. 482
Flowers and then spur leaves wilt	Brown Rot Blossom Wilt p. 350
Leaves at tips of shoots tightly curled and distorted; colonies of small insects present inside curled leaves	Leaf-curling Plum Aphid p. 168
Leaves with dense colonies of green and yellow insects on undersides; upper surfaces often very sticky and sooty	Mealy Plum Aphid p. 168
Leaves with irregular holes and pieces eaten away	Caterpillars p. 194
Leaves at tips of shoots with small tattered holes; fruits sometimes misshapen and discoloured	Capsid Bugs p. 151
Leaves on foliar spur or terminal shoot wither	See Brown Rot Spur Blight and Wither Tip p. 350
Leaves and/or other parts with powdery white coating	Powdery Mildew p. 300
Leaves with brown or black pustules beneath; small yellow spots above	Plum Rust p. 283
Leaves with silver sheen; branches may die back	Silver Leaf p. 324
Leaves with irregular patterns or variously malformed	Viruses p. 460
Leaves with orange or red blotches (not Britain)	Leaf Blotch p. 415
Leaves and/or other parts with sticky black coating	Sooty Mould p. 303
Leaves with brown leaf spots and/or holes	Bacterial Canker p. 320
Leaves small and narrow with irregular margins and thickened blades; fruit sparse	Prune Dwarf Virus p. 461
Branches die back in absence of stem lesions; leaves silvery	Silver Leaf p. 324
Branches die back; root and stem base decayed; toadstools and bootlace-like strands may be present	Honey Fungus p. 375
Branches and/or stem oozing gum; may be some dieback and also spotting and holes in leaves	Bacterial Canker p. 320
Branches with masses of proliferating shoots	Witches' Brooms p. 334
Stem with some decay of wood; fungal brackets may be present	Plum Heart Rot p. 378
Stem with hard, irregular swelling	Crown Gall p. 330
Tree generally unthrifty	See Specific Replant Disease p. 428 and Deficiencies of Nitrogen p. 474, Phosphorus p. 475 and Potassium p. 475

Also see **Trees**

Poinsettia (Euphorbia). See **Glasshouse and House Plants**

Polyanthus. See **Primula**

Polygonatum (Solomon's Seal)
Leaves with large holes and pieces eaten away, sometimes reduced to a skeleton of veins; blue-grey insect larvae present — Solomon's Seal Sawfly p. 227
Also see **Bulbs etc.**

Poplar. See **Populus**

Poppy. See **Papaver**

Populus (Poplar)

Leaves with surface tissues eaten away by metallic blue, red or green beetles and black-spotted larvae	Poplar Leaf Beetles p. 240
Leaves with diffuse, irregular yellowish spots	Viruses p. 461
Leaves with numerous small yellowish or black pustules beneath	Rusts p. 283
Leaves variously spotted	Leaf spots p. 416 or Leaf and Shoot Blight p. 322
Leaves blistered; may be a powdery yellow coating	Leaf Curl p. 332
Leaves with silvery sheen; branches may die back	Silver Leaf p. 324
Leaf stalks of Lombardy poplars with conspicuous pouch-like growths (galls) enclosing colonies of small insects	Lettuce Root Aphid p. 173
Shoots die back; leaf spots may be present	Leaf and Shoot Blight or Twig Dieback p. 322
Branches die back; leaves with silvery sheen	Silver Leaf p. 324
Branches and young shoots crack and ooze creamy slime; large branches and stems with canker lesions	Bacterial Canker p. 321
Branches with parasitic leathery plant growing from them (esp. on hybrid black poplars)	Mistletoe p. 427
Stems with flattened or sunken lesions; branches may die back	Canker and Dieback p. 321
Stems or branches with hard irregular swelling	Crown Gall p. 330
Stem with foul smelling fluxes; branches may die back	Slime Flux p. 325 but also see Bacterial Canker p. 321
Brown-purple scaly and leafless plants growing beneath trees	Toothwort p. 426

Also see **Trees**

Potato

Leaves discoloured, growth poor, plants often collapsing and dying; small white, yellow or brown cysts on roots	Potato Cyst Eelworm p. 137
Leaves infested by small green, yellow or pink wingless and winged insects; foliage sticky, sometimes discoloured	Aphids p. 161

Leaves with light flecking, sometimes browning; small winged and wingless insects and cast skins on undersides	Potato Leafhoppers p. 155
Leaves with small ragged holes, especially at tips of shoots	Capsid Bugs p. 151
Leaves eaten by black and yellow striped beetles and/or by pink-red larvae (rare in Britain)	Colorado Beetle p. 239
Leaves with black, soft patches bearing powdery white mould growth	Blight p. 385
Leaves with dark brown, concentrically zoned spots (uncommon in Britain)	Early Blight p. 387
Leaves browned or scorched at margins; plant unthrifty	Potassium Deficiency p. 475
Leaves small and pale (symptom spreading from lower to upper leaves); plant unthrifty; few tubers	Nitrogen Deficiency p. 474
Leaves rolled; shoots leggy; many very small tubers	Calcium Deficiency p. 476
Leaves rolled (lower first) stiff and brittle; may rattle if shaken	Leaf Roll Virus p. 462
Leaves rolled (upper first) and wilt; stem base black and slimy; roots rotten	Black Leg p. 368
Leaves with pronounced inter-veinal yellowing (lower first)	Manganese Deficiency p. 479
Leaves with dark streaks or spots along veins; leaves turn yellow and hang by thread	Virus Y p. 462
Leaves roughened or puckered; yellow green mottling; few tubers	Virus Y p. 462
Leaves with indefinite yellowish mottling or patterns	Viruses p. 461
Stem base black and slimy; tubers may be rotten and leaves rolled	Black Leg p. 368
Stem base rotten and with white cottony mould growth	Sclerotinia Disease p. 369
Stem with rough, brown and cracked lesions on underground parts	Blight p. 385
Tubers with small round holes and tunnels; thin yellow-brown larvae present in tubers and/or soil	Wireworms p. 236
Tubers with larger tunnels and holes, often extensively damaged	Slugs p. 139
Tubers with large pieces eaten out of their surface	Chafers p. 235, Cutworms p. 200, Swift Moths p. 201
(Note: Millepedes (p. 143) may often be present in damaged tubers but are usually of only secondary importance)	
Tubers crack and wrinkle; tissues dry out and have mealy consistency	Potato Tuber Eelworm p. 134
Tubers turn green	Light p. 484
Tubers split or with internal hollows but not usually decayed and no mould growth present	Mechanical Injury p. 489 or Splitting p. 488
Tubers soft and rubbery when lifted	Drought p. 482
Tubers with small black scurfy patches	Black Scurf p. 368
Tubers with silvery sheen; usually only apparent in store	Silver Scurf p. 416
Tubers with small brownish pimples	Skin Spot p. 416
Tubers soft and rotten; stem base black and slimy	Black Leg p. 368

Tubers (in store) with soft and slimy rot — Bacterial Soft Rot p. 348

Tubers with scabby spots of corky tissue — Common Scab p. 336 but also see Powdery Scab p. 337

Tubers with rough and warty outgrowths — Wart p. 333

Tubers (in store) with soft, brown rot and fluffy grey mould growth — Grey Mould p. 352

Tubers swollen or cracked and with a ring of creamy yellow ooze seen when cut open – later becoming soft and slimy (not Britain) — Bacterial Ring Spot p. 357

Tubers shrunken or wrinkled and with cracks of cavities lined with fluffy white, blue or pink mould (very common) — Dry Rot p. 357

Tubers with small thumb-print-like lesions from which rot extends — Gangrene p. 358

Tubers with rubbery consistency when cut; tissues turn pink within half an hour — Pink Rot p. 359

Tubers with small dark lesions which ooze liquid when touched — Watery Wound Rot p. 359

Tubers with indefinite dark patches within — often caused by Frost p. 482 but there are several other causes including Viruses p. 461

Tubers with more or less definite dark lines or crescentic patterns within but no decay — Viruses p. 461

Plants with masses of proliferating shoots — Viruses p. 461

Primula (including Auricula and Polyanthus)

Flowers and flower buds pecked and torn — Birds p. 255

Leaves with irregular holes, especially near soil surface; slime trails present — Slugs p. 139

Leaves discoloured, yellowing and later bronzing, drying out and dying; small mites present — Glasshouse Red Spider Mites p. 247 or Bryobia Mites p. 250

Leaves, buds, stems and flowers infested by small green, yellow or pink wingless and winged insects; plants sticky and sooty — Aphids p. 161

Leaves with pale yellowish spots and whitish mould growth beneath — Downy Mildew p. 295

Leaves with yellowish mosaic pattern; plant often stunted — Viruses p. 464

Leaves variously spotted — Leaf Spots p. 417

Leaves with tiny cup-like bodies or brownish pustules on undersides — Rust p. 284

Stem base and/or roots rotted; plants unthrifty — Root and Foot Rot p. 361

Also see **Bedding Plants** and **Glasshouse and House Plants**

Privet. See **Ligustrum**

Prunus. See separate entries for **Almond, Cherry** and **Plum.** Ornamental species and cultivars are generally affected by the same pests, diseases and disorders as those affecting their fruiting counterparts.
   Also see **Trees**

Pseudotsuga (Douglas Fir)

Needles discolour and drop prematurely	See discussion under Needle Casts p. 305
Leaves and stems with tufts of white wax wool covering small dark insects; foliage sooty	Adelgids p. 182
Stem with decay of wood which breaks into cubical fragments; fungal brackets may be present	Brown Cubical Rot p. 373

Also see **Conifers** and **Trees**

Pumpkin. See **Cucumber**

Pyracantha

Fruit and leaves with small brownish spots	Scab p. 338
Leaves and stems with tufts of white wax wool	Woolly Aphids p. 166
Leaves eaten and sometimes drawn together with silk webbing	Caterpillars p. 194
Leaves and stems infested by brown, waxy scales; foliage sticky and sooty	Scale Insects p. 185
Shoots die back; masses of small, usually pink pustules present	Coral Spot p. 314
Shoots exude slime and die back; flowers and/or leaves shrivelled	Fireblight p. 316

Pyrus. See **Pear;** many of the problems affecting pears can occur on ornamental *Pyrus* species also

Quercus (Oak)

Leaves with powdery white coating	Powdery Mildew p. 299
Leaves with irregular yellowish streaks and patterns	Viruses p. 464
Leaves with small yellow spots which enlarge and turn brown	Leaf Spot p. 412
Leaves with pink-purple blisters	Leaf Curl p. 332
Leaves with small orange pustules	Rust p. 281
Leaves with irregular holes and pieces eaten away	Caterpillars p. 194, Chafers p. 235
Leaves with yellow and brown mottling on uppersides, small yellow insects on undersides; leaves may fall prematurely	Oak Phylloxera p. 185
Leaves, buds, stems, acorns or roots with unusual growths (galls) of varied forms	Gall Wasps p. 229
Branches die back with or without canker lesions	Cankers and Diebacks p. 319
Branches with dormant buds burst into leaf late in season	Disturbed Growth p. 487
Stem with some decay of wood but confined to base; fungal brackets may be present	Butt Rot p. 376
Stem with decay of wood which extends well above ground level; fungal brackets may be present	Heart Rot p. 376
Stem with skin-like fungal fruit body on bark and decay of wood within	Pipe Rot p. 377

Also see **Trees**

Quince. For **Japanese Quince** see **Chaenomeles**

Flowers and/or buds die suddenly	Frost p. 482
Fruit with soft brown rot and usually off-white cottony pustules; associated twigs may die back	Brown Rot p. 350
Fruit with round, dark, saucer-like depressions; usually only in store	Bitter Rot p. 349
Leaves and/or other parts with powdery white coating	Apple Powdery Mildew p. 297
Leaves with irregular red, but later darker spots	Leaf Blight p. 417
Leaves with large, dark brown blotches	Leaf Blotch p. 418
Leaves with silvery sheen; branches die back	Silver Leaf p. 324
Branch and/or stem with hard irregular swelling	Crown Gall p. 330
Branches die back; leaves with silvery sheen	Silver Leaf p. 324

Also see **Trees**

Radish

Radishes have many problems in common with their close relatives, **Brassicas** (q.v.) although they also commonly have rough, corky patches on the roots caused by	Common Scab p. 336

Ranunculus

Leaves and/or other parts with powdery white coating	Powdery Mildew p. 302

Raspberry and **Loganberry**

Fruits malformed and infested by small white beetle larvae	Raspberry Beetle p. 237
Fruit rotted and with powdery grey mould growth	Grey Mould p. 352
Fruit crumbly	See discussion under Viruses p. 464
Flowers killed suddenly	Frost p. 482
Leaves and/or other parts with powdery white coating	Strawberry Powdery Mildew p. 301
Leaves with yellowish mosaics or mottles; plants may be unthrifty	Mosaic Viruses p. 464
Leaves with yellow pustules on uppersides or orange and/or black pustules beneath	Rust p. 284
Leaves more or less bleached; veins remain dark green	Iron Deficiency p. 478
Leaves greasy and curl downwards; canes stunted (cultivar 'Lloyd George' only)	Bushy Dwarf Disease p. 465
Leaves with small yellow spots or pale colouring of veins; plants stunted	Arabis Mosaic Virus p. 465
Leaves wilt; canes with blue striping on one side	Wilt p. 306
Leaves, canes and/or other parts with purple grey-centred spots	Cane and Leaf Spot p. 418
Leaves and shoots infested by small green or yellow wingless and winged insects; foliage sticky and sooty	Aphids p. 161
Leaves with light flecking, sometimes with more extensive discolouration; small yellow insects and cast skins on undersides	Leafhoppers p. 154
Leaves with irregular holes and pieces eaten away, sometimes reduced to a skeleton of veins	Caterpillars p. 194 or Sawflies p. 224
Leaves with small tattered holes, especially near tips of shoots	Capsid Bugs p. 151

Shoots shrivel and die; pink-red caterpillars tunnelling in shoots and buds	Raspberry Moth p. 199
Shoots and leaves with white froth covering small wingless insects	Froghoppers p. 153
Canes with small pink or orange-red larvae under discoloured areas of bark, especially near cracks and wounds	Raspberry Cane Midge p. 214
Canes with dark, cracked patches just above soil level	Cane Blight p. 322
Canes with dark purple patches in summer; silvery sheen in winter	Spur Blight p. 323
Canes with hard, irregular swellings	Crown Gall p. 330
Canes die back; roots and stem base decayed; toadstools and boot-lace-like strands may be present	Honey Fungus p. 375

Red Currants. See **Currants**

Rhizomes. See **Bulbs etc.**

Rhododendron (including **Azalea**)

Flowers and/or buds shrivelled and with fluffy grey mould growth	Grey Mould p. 346
Flower buds covered with numerous tiny black pinhead-like structures	Bud Blast p. 347
Flowers become limp and slimy and hang on bushes, often with a powdery white coating	Petal Blight p. 347
Leaves mottled yellow on uppersides with rusty brown deposits on undersides; insects present on undersides during summer	Rhododendron Bug p. 150
Leaves with small white winged insects and yellow scales on undersides, especially at tips of shoots; foliage sticky and sooty	Rhododendron Whitefly p. 160 or Azalea Whitefly p. 160
Leaves with small irregular notches eaten out of edges, especially on lower leaves	Vine Weevil p. 242
Leaves and buds infested by winged insects with conspicuous red stripes on green wing cases	Rhododendron Leafhopper p. 154
Leaves of pot azaleas with brown blotch mines; leaves may fall prematurely	Azalea Leaf Miner p. 211
Leaves more or less bleached; veins remain dark green	Iron Deficiency p. 478
Leaves wilt in absence of obvious damage or stress factors	Wilt p. 306
Leaves with brown or orange pustules on undersides	Rust p. 284
Leaves variously spotted	Leaf Spots p. 418
Leaves, buds and/or flowers with small, irregular swellings (esp. on glasshouse azaleas)	Gall p. 334
Shoots with irregular small, usually greyish growths	Lichen p. 425
Branches die back; leaves sparse and yellowed; tissues at stem base may be blackened and decayed	Phytophthora Root Death p. 377
Stem with large, irregular, woody swellings	Crown Gall p. 330

Rhubarb

Leaves with pale yellowish spots and whitish mould growth beneath	Downy Mildew p. 295
Leaves with pale, brown-bordered spots	Horseradish Pale Leaf Spot p. 407
Leaves with mosaic patterns or irregular yellowish streaks	Viruses p. 465

Leaves wilt; crown with brown or black streaks within	Wilt p. 306
Leaves discoloured and distorted, crowns soft and rotten	Stem Eelworm p. 132
Crown with soft, chocolate brown rot and internal cavities; terminal bud dies	Crown Rot p. 387
Roots and crown base rotted; toadstools and bootlace-like strands may be present	Honey Fungus p. 375
Roots eaten by white caterpillars in soil	Swift Moths p. 201

Rhus (Sumach)

| Buds and/or leaves shrivelled or scorched suddenly | Frost p. 482 |
| Leaves wilt in absence of obvious damage or stress factors | Wilt p. 306 |

Also see **Trees**

Ribes. For ornamental *Ribes* see **Currants**

Robinia (False acacia)

Leaves with various irregular yellowish patterns and/or distortion	Viruses p. 466
Leaves wilt in absence of obvious damage or stress factors	Wilt p. 306
Branches with parasitic, leathery plant growing from them	Mistletoe p. 427
Branches with masses of proliferating shoots (witches' brooms)	Viruses p. 466

Also see **Trees**

Rosa (Roses)

Flowers, flower buds and leaves with irregular holes and pieces eaten away	Caterpillars p. 194 Chafers p. 235
Flowers with fine light flecking; narrow-bodied insects present	Thrips p. 191
Flowers and/or buds shrivelled and with fluffy grey mould growth	Grey Mould p. 346
Leaves, buds, shoots and stems infested by green, pink or brown wingless and winged insects; foliage sticky, sometimes sooty	Aphids p. 161
Leaves with fine white mottling, sometimes causing extensive discolouration; small green or yellow insects and their cast skins present on undersides	Leafhoppers p. 154
Leaves with small tattered holes, especially at tips of shoots	Capsid Bugs p. 151
Leaves with regular, almost semicircular pieces removed from edges	Leaf-Cutter Bee p. 233
Leaves with surface tissues eaten away, exposing veins; green or yellow larvae present	Rose Slug Sawfly p. 226
Leaves with leaflets tightly rolled along their length and drooping	Leaf-Rolling Rose Sawfly p. 226
Leaves discoloured, yellowing, bronzing and sometimes dying; small mites present on undersides	Glasshouse Red Spider Mites p. 247
Leaves and/or other parts with powdery white coating	Powdery Mildew p. 300
Leaves with dark, blackish spots or blotches but if mould present on undersides also see	Black Spot p. 419 Downy Mildew p. 295

Leaves and/or other parts with bright orange pustules	Rust p. 285
Leaves with small yellow or black pustules on undersides	Rust p. 285
Leaves with various irregular yellowish streaks and patterns	Viruses p. 466
Leaves variously spotted	Leaf Spots p. 420
Leaves more or less bleached; veins remain dark green	Iron Deficiency p. 478
Leaves wilt in absence of obvious damage or stress factors	Wilt p. 306
Leaves with silvery sheen; branches may die back	Silver Leaf p. 324
Shoots with hard, irregular swelling (*Note:* Such swellings associated with graft unions or on heavily and regularly pruned stems may well be normal)	Crown Gall p. 330
Shoots with various types of canker lesions and/or dieback	Cankers p. 323
Shoots arise in dense groups around scion bud of maiden plants	See Viruses p. 466
Plants generally unthrifty	See Specific Replant Diseases p. 428 and deficiencies of Nitrogen p. 474, Phosphorus p. 475 and Potassium p. 475

Rose. See **Rosa**

Rowan. See **Sorbus**

Rubber Plant (Ficus). See **Glasshouse and House Plants**

Rudbeckia. See **Perennials**

Saintpaulia (African Violet)

Flowers distorted, growth poor; small mites present in buds and growing points	Tarsonemid Mites p. 251
Leaves and/or stems rotted; sometimes mould growth present	Stem and Leaf Rots p. 387

Also see **Glasshouse and House Plants**

Note: Correct diagnosis of the varied symptoms shown by Saintpaulias can prove difficult. If in doubt, seek expert advice.

Salix (Willow)

Leaves with irregular holes and pieces eaten away, sometimes reduced to a skeleton of veins	Caterpillars p. 194 Sawflies p. 224 or Willow Leaf Beetles p. 240
Leaves with conspicuous hard raised yellow or bright red bean-like growths (galls)	Willow Bean-Gall Sawfly p. 228
Leaves discoloured, yellowing, bronzing, drying out and often falling prematurely; small mites, eggs and silk webbing present	Red Spider Mites p. 246

Leaves and shoots infested by small yellow or green wingless and winged insects; foliage sticky and sooty — Aphids p. 161

Leaves with orange or yellowish pustules on undersides — Rusts p. 287

Leaves with black, bitumen-like spots — See Acer Tar Spot p. 393

Leaves with silvery sheen; branches may die back — Silver Leaf p. 324

Leaves with small red-brown spots; small lesions on twigs; entire tree dies back (esp. weeping willow) — Anthracnose p. 327

Stems and branches infested by dense colonies of grey-brown wingless insects; foliage usually very sticky and sooty and attractive to wasps — Large Willow Aphid p. 180

Stems infested by small white scales; foliage sticky and sooty — Willow Scale p. 188

Branches die back; trees take on stag's head form; sticky liquid oozes from bark cracks (*Salix alba* only) — Watermark Disease p. 327 but also see Slime Flux p. 325

Branches with hard irregular swellings — Crown Gall p. 330

Branches with parasitic leathery plant growing from them (esp. on *Salix fragilis*) — Mistletoe p. 427

Stem and/or branches with some decay of wood; fungal brackets may be present — Heart Rot p. 379

Trees with brown or purple scaly leafless plant growing beneath — Toothworts p. 426

Also see **Trees**

Salsify

Flower heads filled with black powdery mass (rare in Britain) — Smut p. 290

Leaves and/or other parts with powdery white coating — Powdery Mildew p. 302

Leaves and/or other parts with small, often concentric clusters of creamy droplets — White Blister p. 305

Salvia

Leaves with small tattered holes, especially at tips of shoots — Capsid Bugs p. 151

Also see **Glasshouse and House Plants** and **Bedding Plants**

Sambucus (Elderberry)

Leaves with various yellowish streaks and patterns and/or distortion — Viruses p. 466

Stems infested by black wingless and winged insects — Elder Aphid p. 178

Sansevieria

Leaves with more or less rounded sunken, red-brown spots — Leaf Spot p. 420

Also see **Glasshouse and House Plants**

Sarothamnus (Broom)

Shoots blacken and shrivel; leaves may drop prematurely — Dieback p. 313

Savoy. See **Brassicas**

Saxifraga (Saxifrage)

Leaves with dark red-brown pustules   Rust p. 285
Also see **Cacti and Succulents**

Saxifrage. See Saxifraga

Scabiosa (Scabious)

Flowers and leaves distorted, buds killed   Scabious Bud
                                             Eelworm p. 136
Leaves and/or other parts with powdery white coating   Powdery Mildew
                                             p. 302

Scabious. See Scabiosa

Schizanthus

Leaves and/or other parts with powdery white coating   Powdery Mildew
                                             p. 302

Also see **Glasshouse and House Plants**

Schizostylis. See Bulbs etc.

Scilla (Squill)

Flowers with brown-black powdery masses in centre   Anther Smut p. 290
Leaves and/or stems with dirty yellowish spots bear-
  ing brown pustules   Bluebell Rust p. 272
Leaves turn yellow; plants topple over after flowering;   Hyacinth Black
  leaf bases rotten and readily pulled away from bulb   Slime p. 342
Bulbs with decay and green-blue mould growth   Blue Mould Rot
                                             p. 339

Also see **Bulbs etc.**

Scorzonera

Flower heads filled with black powdery mass (rare in
  Britain)   Salsify Smut p. 290
Leaves and/or other parts with powdery white coating   Powdery Mildew
                                             p. 302

Leaves and/or other parts with small, often concentric
  clusters of creamy droplets   White Blister p. 305

Sea Kale

See **Brassicas**, although the following symptoms
are especially common on Sea Kale:
Roots with dark ring within   Black Rot p. 350
Roots rotted and with matted dark strands   Violet Root Rot
                                             p. 360

Sedum. See Cacti and Succulents

Seedlings

Young seedlings are especially susceptible to pests and diseases and may also suffer
from disorders, especially those caused by under- or over-watering and by low
temperatures during germination and establishment. Diagnosis of the causes of
trouble may be difficult since seedlings quickly die and rot away, leaving no
symptoms, but the following key can be used to identify the main causes of trouble.

Seedlings pulled out of soil; pecked and torn   Birds p. 255

Seedlings scratched out of soil	Cats p. 264
Seedlings collapse and die; stems eaten into at soil level, roots damaged	Cutworms p. 200 Leatherjackets p. 215 Wireworms p. 236
Seedlings collapse and die, usually in patches, and with blackening and decay of the stem base and/or fluffy grey mould growth present	Damping-off p. 362
Seedlings with leaves yellowed, sometimes with dry angular blotches, and generally stunted; roots may rot	Waterlogging p. 485
Seedlings with leaves variously discoloured – check Plates 21–22 giving symptoms of Mineral Nutrient Deficiency	
Seedlings with leaves and growing points eaten away	Slugs p. 139, Woodlice p. 145, Crickets p. 147, Cockroaches p. 149
Seedlings, especially of brassicas, with small round pits and holes; jumping beetles present	Flea Beetles p. 238
Seedlings with leaves minutely pitted; minute, wingless jumping insects present	Springtails p. 146

Sempervivum (Houseleek)

Leaves with orange, cup-like dimples	Rust p. 279

Also see **Cacti and Succulents**

Senecio. For *Senecio cruentus* see **Cinerarea** and for succulent forms see **Cacti and Succulents**

Leaves and/or other parts with powdery white coating	Powdery Mildew p. 302

Also see **Bedding Plants**

Sequoia, Sequoiadendron (Redwood). See **Conifers** and **Trees**

Shallot. See **Onion**

Shrubs

Ornamental shrubs are generally robust, once established, but may be affected by some of the pests, diseases and disorders noted under **Trees**. Also see separate entries for the main plant genera.

Sidalcea

Leaves and/or stem with raised orange-brown pustules	Hollyhock Rust p. 278

Sinningia (Gloxinia). See **Glasshouse and House Plants**

Skimmia

Leaves more or less bleached; veins remain dark green	Iron Deficiency p. 478
Leaves scorched suddenly in spring; blossom may be killed	Frost p. 482

Snapdragon. See **Antirrhinum**

Snowdrop. See **Galanthus**

Solanum
> Ornamental *Solanum* species may be expected to have many problems in common with **Potato** *(S. tuberosum)* q.v.
>
> Leaves with irregular, yellowish streaks and patterns and/or various forms of distortion; plants generally unthrifty — Nicotiana Viruses p. 455

Solidago (Golden rod)
> Leaves and/or other parts with powdery white coating — Powdery Mildew p. 302
>
> Also see **Perennials**

Solomon's Seal. See **Polygonatum**

Sorbus (including Mountain Ash)
> Berries eaten — Birds p. 255
>
> Leaves with yellow or brown blister-like growths — Pear Leaf Blister Mite p. 253
>
> Leaves with irregular pieces eaten away — Caterpillars p. 194
>
> Leaves with orange-yellow pustules on undersides (mountain ash only) — Rust p. 281
>
> Shoots exude slime and die back; flowers and/or leaves shrivelled — Fireblight p. 316
>
> Also see **Trees**

Sparmannia. See **Trees**

Spider Plant. See **Chlorophytum**

Spinach (including Spinach Beet)
> Leaves with brown blotch mines — Beet Leaf Miner p. 217
>
> Leaves with small black wingless and winged insects — Black Bean Aphid p. 171
>
> Leaves with pale yellowish spots and grey-purple mould growth beneath — Downy Mildew p. 292
>
> Leaves with small rounded white spots — Leaf Spot p. 421
>
> Leaves yellowed and curled or otherwise distorted; symptoms on inner leaves first; entire plant may later wilt and die — Viruses p. 467
>
> Leaves yellowed between veins; symptoms on outer leaves first which may curl upwards at edges — Manganese Deficiency p. 479
>
> Plants run to flower — Bolting p. 487

Spiraea
> Leaves with irregular yellowish streaks and patterns — Viruses p. 429

Spruce. See **Picea**

Squash. See **Cucumber**

Squill. See **Scilla**

St John's Wort. See **Hypericum**

Stapelia

Leaves with yellowish mottling or other irregular patterns	See Asclepias Viruses p. 438

Also see **Glasshouse and House Plants**

Stephanotis

Leaves with yellowish mottling or other irregular patterns	See Asclepias Viruses p. 438

Also see **Glasshouse and House Plants**

Sternbergia

Leaves with red-brown scorch, spreading downwards from tip; gradually shrivel and die; flower stalks and flowers may also have brown discolouration	Narcissus Leaf Scorch p. 412

Also see **Bulbs etc.**

Stock. See **Matthiola**

Stranvaesia

Shoots exude slime and die back; flowers and/or leaves shrivelled	Fireblight p. 316

Also see **Trees**

Strawberry

Fruits pecked and eaten	Birds p. 255
Fruits with large pieces eaten out; slime trails present on and near plants	Slugs and Snails p. 139
Fruits with small cavities, seeds eaten, black beetles present under trusses and straw	Strawberry Beetles p. 237
Fruit with soft rot and fluffy grey mould growth	Grey Mould p. 352
Fruit leathery, brown and shrivelled; fruit stalks shrivelled	Leathery Fruit Rot p. 354
Fruits with brown sunken patches bearing many seeds esp. on cultivar 'Paxton'	Hard Rot p. 359
Flowers killed suddenly; young leaves may be scorched	Frost p. 482
Flowers deformed and with green colouration	Green Petal Disease p. 467
Leaves and/or other parts with powdery white coating; leaves curl upwards at margins	Powdery Mildew p. 301
Leaves and/or other parts variously spotted	Leaf Blotch, Leaf Scorch or Leaf Spot p. 421
Leaves more or less bleached; veins remain dark green	Iron Deficiency p. 478
Leaves with yellow margins in late summer; plants stunted	Yellow Edge Disease p. 467
Leaves with various yellowish patterns or streaks, distortion and/or crumpling	Viruses p. 467
Leaves wilt, (outer first); stem interior with brown or black streaks	Wilt p. 306
Leaves wilt (inner first); crown interior brown and may be rotten	Crown Rot p. 369

Leaves and shoots infested by small green or yellow wingless and winged insects; foliage sticky and sooty	Aphids p. 161
Leaves with small white winged insects and yellow scales on undersides; foliage sticky and sooty	Honeysuckle Whitefly p. 161
Leaves discoloured, yellowing, bronzing and dying; small mites and silk webbing present	Glasshouse Red Spider Mite p. 247
Leaves and shoots distorted; small light brown mites present in buds and growing points	Strawberry Mites p. 251
Leaves crumpled and distorted; white marks alongside veins; growth stunted	Eelworms p. 132
Stems eaten through at ground level; plants wilt and die	Cutworms p. 200
Stem with masses of short or thickened shoots, especially near ground level	Leafy Gall p. 331
Plants wilt and die; roots eaten by small white legless larvae in soil	Vine Weevil p. 242
Plants stunted; older leaves turn brown and hang down; root with black regions	Root Rot p. 361
Plants stunted; central leaves small; outer, brown and stiff; central core of roots (other than in summer) appears red	Red Core p. 370
Plants otherwise unthrifty	Specific Replant Disease p. 428

Strawberry Tree. See **Arbutus**

Streptocarpus. See **Glasshouse and House Plants**

Succulents. See **Cacti and Succulents**

Sunflower. See **Helianthus**

Sumach. See **Rhus**

Swede, Turnip

Leaves and/or other parts with powdery white coating	Powdery Mildew p. 297
Leaves with pale yellowish patches and greyish mould beneath	Crucifer Downy Mildew p. 292
Leaves with yellowish inter-veinal mottling, crinkling and stunting	Turnip Mosaic Virus p. 440
Leaves with tiny more or less circular dirty white spots with purplish border	White Spot p. 398
Leaves and/or other parts with small, often concentric clusters of creamy droplets	White Blister p. 304
Leaves with dark margins; plants wilt, remain stunted and may decay (unusual in Britain)	Black Rot p. 350
Leaves of seedlings with small circular holes and pits	Flea Beetles p. 238
Stems eaten through at ground level; plants wilt and die	Cutworms p. 200
Roots with holes eaten out by white caterpillars in soil	Swift Moths p. 201
Roots tunnelled by white maggots	Cabbage Root Fly p. 218
Roots with rough surface patches and dark patches or rings within flesh	Boron Deficiency p. 479

Roots with pale, elongated lesions on exposed parts; eventually forming deep cracks — Dry Rot p. 364 but also see Splitting p. 488

Roots decayed with matted dark strands — Violet Root Rot p. 360

Roots with soft and slimy rot — Bacterial Soft Rot p. 348

Roots with irregular fairly soft swellings and often some decay; plants may be stunted — Clubroot p. 328

Roots with hard irregular swellings; no decay; plants otherwise normal — Hybridisation Nodules p. 428

Roots (in store) with soft rot and grey fluffy mould growth — Grey Mould p. 352

Roots with scab-like patches of corky tissue — Common Scab p. 336

Sweet Corn (and ornamental maize)
Seeds removed from soil after sowing — Birds p. 255
Leaves with yellow streaks, later fraying; fly maggots tunnelling in shoots, killing growing points and checking growth — Frit Fly p. 216
Cobs malformed and containing black powder — Smut p. 290
Cobs fail to develop; whitish stripes on leaves — Boron Deficiency p. 479

Sweet Pea. See **Lathyrus**

Sweet William. See **Dianthus**

Sycamore. See **Acer**

Symphytum (Comfrey)
Leaves with yellowish or yellow-brown pustules on undersides — Rust p. 275

Syringa (Lilac)
Leaves with powdery white coating — Hazelnut Powdery Mildew p. 299

Leaves with yellowish patterns or streaks and/or malformation — Viruses p. 468
Leaves variously spotted; some dieback of shoots may occur — Leaf Spots p. 410 or Blight p. 318
Leaves with brown blotch mines — Lilac Leaf Miner p. 210
Leaves with semi-circular pieces cut out of edges — Leaf Cutter Bee p. 233
Stems infested by small white scales; foliage sticky and sooty — Willow Scale p. 188
Also see **Trees**

Tagetes (African marigold). See **Bedding Plants**

Tanacetum (Tansy)
Leaves and/or other parts with pale brown pustules — Rust p. 286

Tansy. See **Tanacetum**

Taxus (Yew)
Leaves eaten away, leaving bare sections on shoots, often with bark removed — Weevils p. 241
Shoot tips enlarged to form growths (galls) like miniature artichokes — Yew Gall Midge p. 220
Shoots distorted, new growth weak and spindly; buds infested by small mites — Yew Gall Mite p. 254
Stems with hemispherical brown scales; leaves often sticky and sooty — Scale Insects p. 185
Branches die back; leaves sparse and yellowed; tissues at stem base may be blackened and decayed — Phytophthora Root Death p. 377
Also see **Conifers** and **Trees**

Thuja/Thujopis (Western Red Cedar, Arbor Vitae)
Needles turn brown and die; twigs may die back; esp. on young plants — Needle Scorch p. 422
Also see **Conifers** and **Trees**

Tiger Flower. See **Tigridia**

Tigridia (Tiger Flower). See **Bulbs etc.**

Tilia (Lime)
Leaves and buds infested by small green or yellow wingless and winged insects; foliage often very sticky and sooty — Aphids p. 161
Leaves with bright red growths (galls) projecting from upper surfaces — Nail Gall Mite p. 254
Leaves with irregular holes and pieces eaten away — Caterpillars p. 194
Leaves discoloured, yellowing, bronzing and dying; small mites and silk webbing present; trunks sometimes covered in glistening sheets of webbing during late summer and autumn — Red Spider Mites p. 246
Leaves and/or other parts with sticky black coating — Sooty Mould p. 303
Leaves variously spotted — Leaf Spots p. 410
Leaves wilt in absence of obvious damage or stress factors — Wilt p. 306
Branches die back; masses of small, usually pink pustules present — Coral Spot p. 314
Branches with parasitic leathery plant growing from them — Mistletoe p. 427
Branches die back; leaves sparse and yellowed; tissues and stem base may be blackened and decayed — Phytophthora Root Death p. 377
Stems with sunken lesions; branches may die back (not Britain) — Canker and Dieback p. 319
Stem and/or branches with cracks which ooze reddish yellow gum — Bleeding Canker p. 312
Also see **Trees**

Tobacco. See **Nicotiana**

Tomato
Fruit split; no rotting or mould growth — Splitting p. 488
Fruit with small white spots; esp. late in season — Ghost Spotting p. 422

Fruit with soft and slimy rot (usually in store); no mould growth — Bacterial Soft Rot p. 348

Fruit with tough dryish rot and sometimes white mould growth; may be dark blotches on leaves and stems — Blight p. 385 but also see Buck-Eye Rot p. 360 and Stem and Fruit Rot p. 389

Fruit with tough, dark coloured patch at blossom end ('blossom end rot') — Calcium Deficiency p. 476

Fruit with hard dark green or sometimes yellow patches when remainder ripens — High Temperature p. 484

Fruit ripen with irregular patches of ripe and unripe tissue — Tobacco Mosaic Virus p. 468 Potassium Deficiency p. 475 or Drought p. 482

Fruit hollow — Drought p. 482, Potassium Deficiency p. 475 or Hormone Herbicide Injury p. 491

Flowers and/or buds shrivelled and with fluffy grey mould growth — Grey Mould p. 346

Flowers fail to set fruit satisfactorily — Viruses p. 468 or Drought p. 482

Leaves and shoots infested by green or pink-brown wingless and winged insects; leaves and fruits sticky and sooty — Aphids p. 161

Leaves with white winged insects and yellow scales on undersides; foliage and fruits sticky and sooty — Glasshouse Whitefly p. 158

Leaves with irregular white spots; yellow insects and cast skins on undersides — Glasshouse Leafhoppers p. 154

Leaves discoloured, yellowing, bronzing and dying; small mites on undersides — Glasshouse Red Spider Mite p. 247

Leaves discoloured, plants wilting and dying; small white, yellow or brown cysts on roots — Potato Cyst Eelworm p. 137

Leaves wilting, growth poor; roots with hard irregular growths (galls) — Root-Knot Eelworm p. 136

Leaves with sinuous white mines — Leaf Miner p. 222

Leaves with dark blotches which may bear white mould growth; fruit may also be rotted (usually outdoor plants only) — Blight p. 385

Leaves with dark, usually concentrically zoned blotches (unusual in Britain) — Early Blight p. 387

Leaves with pale yellow patches on upper surface with grey-brown mould beneath (usually glasshouse plants only) — Leaf Mould p. 388

Leaves and/or other parts, thickened, twisted or otherwise distorted — Hormone Herbicide Injury p. 491 but also see Viruses p. 468

Leaves (undersides) and/or stems with rough warty outgrowths — Oedema p. 488

Leaves with numerous circular or angular spots (unusual in Britain) — Leaf Spot p. 422

Leaves with yellowish mottling and/or distortion — Viruses p. 468

Leaves wilt in absence of obvious damage or stress factors; no dark streaking in stem	Viruses p. 468 or Drought p. 482
Leaves wilt; stem interior with brown or black streaks well above soil level	Wilt p. 306 or Bacterial Canker p. 388
Leaves wilt; stem undamaged except possibly at base; roots blackened or decayed	Root and Foot Rots p. 361
Leaves with irregular silvery sheen	Silvering p. 488
Leaves curl in absence of obvious damage or stress factors plant otherwise normal – this is quite normal especially on outdoor tomatoes and no action is necessary	
Leaves turn blue-green, then yellow and wilt; plants unthrifty	Copper Deficiency p. 480
Leaves with marked interveinal yellowing on older leaves first	Magnesium Deficiency p. 477
(Note however that this symptom can arise commonly on tomatoes quite normally as leaves begin to age and decline and/or are deeply shaded)	
Stem with brown streaks; may split open to reveal cavities	Bacterial Canker p. 388, Pith Necrosis p. 389, Sclerotinia Disease p. 369 or Stem and Fruit Rot p. 389
Stem with patches of rot and grey mould growth esp. around leaf scars and pruning cuts	Grey Mould Stem Rot p. 388
Stem with large irregular swelling	Crown Gall p. 330
Roots blackened and/or decayed; plants unthrifty	Root and Foot Rots p. 361
Roots with rough and warty appearance	Potato Powdery Scab p. 337

Tradescantia. See **Glasshouse and House Plants**

Trees

Young trees may be seriously damaged during their early growth by **Rabbits** (p. 259), **Hares** (p. 260), **Voles** (p. 262) and some other mammals that strip bark from the stems just above soil level. Mechanical barriers, such as spiral card or plastic tree protectors or wire netting should be used to prevent damage in areas where these pests are known to be present. Growth of young trees may also be checked by **Aphids** (p. 161) and **Caterpillars** (p. 194) feeding on buds, leaves and young shoots. The most important symptoms are:

Stems of young plants with bark stripped away at or above soil level	Voles p. 262, Rabbits p. 259, Hares p. 260, Deer p. 263, Squirrels p. 260
Leaves with irregular holes and pieces eaten away, sometimes reduced to a skeleton of veins	Caterpillars p. 194, Sawflies p. 224, Chafers p. 235
Leaves, buds and young shoots infested by small wingless and winged insects, especially on undersides of leaves; foliage sticky and sooty	Aphids p. 161

Leaves, buds or other parts with abnormal growths (galls) of varied form	Gall Midges p. 213, Gall Mites p. 252, Gall Wasps p. 229
Leaves with powdery white coating	Powdery Mildew p. 296
Leaves with sticky black coating	Sooty Mould p. 303
Leaves wilt in absence of obvious damage or stress factors	Wilt p. 306
Foliage of large part of tree, esp. facing prevailing wind, browned	Wind Scorching p. 486
Branches die back; masses of small, usually pink pustules present	Coral Spot p. 314
Branches with masses of proliferating shoots like large birds' nests	Witches' Brooms p. 334
Branches die back; leaves sparse and yellowing; tissues and stem base may be blackened and decayed	Phytophthora Root Death p. 377
Branches die back; roots and stem base decayed; toadstools and bootlace-like strands may be present	Honey Fungus p. 375
Branches with leathery parasitic plants growing from them	Mistletoe p. 427
Branches with some gross abnormality, e.g. twisted; flattened; or deeply fluted	Disturbed Growths p. 487
Branches with bark stripped, buds eaten and/or fruits damaged	Squirrels p. 260
Stems and/or branches with hard, irregular swelling	Crown Gall p. 330
Stems shattered, often for total height of tree	Lightning p. 485
Stems and/or branches broken off	Wind p. 486
Stems with foul-smelling liquid oozing from bark	Slime Flux p. 325 but also see Bleeding Canker p. 312
Stems with some decay of wood; fungal brackets may be present.	Ash Heart Rot p. 372, Beech Heart Rot p. 372, Elm Heart Rot p. 374, Brown Oak p. 374, Oak Heart Rot p. 376, Willow Heart Rot p. 379
Trees with green or greyish growth on bark	Lichen p. 425, Algae p. 426
Trees overgrown by climbing plants	Ivy p. 427, Honeysuckle p. 427, Clematis p. 426
Trees with brown or purple scaly leafless plants growing beneath	Toothworts p. 426
Trees with toadstools or fungal brackets growing from or beneath them	Tree and Shrub Rots p. 371, Mycorrhiza p. 429
Trees generally unthrifty	See discussion under Mechanical Injury p. 489

Also see **Conifers** and individual entries for the main tree genera

Trillium. See **Bulbs etc.**

Tropaeolum (Nasturtium)

Leaves with reddish spots or blotches with yellowish halo — Leaf Spot p. 423

Leaves with various irregular mottles and/or distortions; flowers may have pale streaks — Viruses p. 470

Leaves with irregular holes and pieces eaten away — Cabbage Caterpillars p. 201

Leaves and stems infested by small black insects — Black Bean Aphid p. 171

Stems with masses of short or thickened shoots; esp. near ground level — Leafy Gall p. 331

Tsuga (Hemlock)

Needles turn brown and drop prematurely — See discussion under Needle Casts p. 305

Also see **Conifers** and **Trees**

Tubers

Tubers with soft and slimy rot — Bacterial Soft Rot p. 348

Tubers rotted and with matted dark strands — Violet Root Rot p. 360

Tubers with decay and blue-green mould growth — Blue Mould Rot p. 339

Also see individual entries under main plant genera, especially under **Potato**, and see **Bulbs etc.**

Tulipa (Tulip)

Flowers, leaves and stems distorted — Stem Eelworm p. 132

Flowers of dark coloured types with pale streaks — Breaking Virus p. 470

Buds and leaves turn yellow or reddish and shrivel — Shanking p. 390

Leaves variously streaked and/or distorted — Viruses p. 470

Leaves with regular patterns of holes in early spring; slime trails may be present — Slugs p. 139

Shoots, buds and/or other parts withered and with fluffy grey mould growth — Fire Disease p. 390

Bulbs in store eaten — Mice p. 261

Bulbs in store infested by brown or yellow wingless insects, especially on shoots — Tulip Bulb Aphid p. 178

Bulbs dug up after planting — Squirrels p. 260

Bulbs (in store) with firm, dry rot and blue-green mould growth — Blue Mould Rot p. 339

Bulbs with brown rot and strands of thin white mould between scales — White Root Rot p. 370

Bulbs, when cut through, with yellow spots in conducting tissues; produce stunted plants with silvery leaf streaks — Yellow Pock p. 391

Bulbs with dry rot and with soil adhering to whitish mould when dug up; plants emerge feebly — Tulip Grey Bulb Rot p. 344

Roots blackened and rotted; decay not usually extending into bulb — Root and Foot Rot p. 361

Also see **Bulbs etc.**

Turf. See **Lawns**

Turnip. See **Swede**

Ulmus (Elm)

Leaves wilt, turn yellow and die; entire tree becomes unthrifty and dies back; no root damage but many galleries visible when bark peeled from trunk — Dutch Elm Disease p. 307

Leaves with green or brown blisters — Leaf Curl p. 332

Leaves with various irregular yellowish streaks or mottles — Viruses p. 471

Leaves variously spotted — Leaf Spots p. 404

Leaves with rashes of small blister-like growths (galls) — Gall Mites p. 252

Bark of dead trees tunnelled by larvae making characteristic radiating galleries — Elm Bark Beetles p. 244

Branches die back; leaves wilt; turn yellow and die; entire tree becomes unthrifty; no root damage but many galleries visible when bark peeled from trunk — Dutch Elm Disease p. 307

Branches die back; masses of small, usually pink pustules present — Coral Spot p. 314

Branches with deeply grooved bark — Winged Cork p. 488

Stem rotten or hollowed extensively; fungal brackets may be present; decay often associated with wounds — Heart Rot p. 374 or Ash Heart Rot p. 372

Stem with some decay of wood, usually close to ground level; fungal brackets may be present — Butt Rot p. 374

Stem with foul-smelling fluxes; branches may die back — Slime Flux p. 325

Stem and sometimes entire crown overgrown by evergreen climbing plant — Ivy p. 427

Stem with large irregular swelling — Crown Gall p. 330

Also see **Trees**

Vaccinium (including Bilberry)

Leaves with irregular yellow streaks or patterns and/or other distortion — Viruses p. 471

Valeriana (Valerian)

Leaves and/or other parts with powdery white coating — Powdery Mildew p. 302

Leaves and/or other parts with tiny yellow/white cuplike structures or brownish pustules — Rust p. 286

Vallota. See **Bulbs etc.**

Vanda. See **Orchids**

Verbascum (Mullein)

Leaves and/or other parts with powdery white coating — Powdery Mildew p. 302

Verbena. See **Bedding Plants**

Veronica (including Hebe)

Leaves and/or other parts with pale yellowish spots on upper surfaces and off-white mould growth beneath — Downy Mildew p. 295

Leaves variously spotted; shoots may die back and plants become unthrifty	Leaf Spot p. 406

Viburnum

Flowers and/or buds shrivelled and with fluffy grey mould growth	Grey Mould p. 346
Flowers and/or buds killed suddenly; leaves may be scorched	Frost p. 482
Leaves variously spotted	Leaf Spots p. 423
Leaves crumpled and tightly curled	Viburnum Aphid p. 178
Leaves of *Viburnum tinus* sticky and sooty; white-fringed black scales on undersides	Viburnum Whitefly p. 160
Also see **Trees**	

Vinca (Periwinkle)

Leaves with brownish pustules on undersides; plants may be distorted and fail to flower	Rust p. 282

Vine. See **Grape Vine** for both fruiting and ornamental types, and see **Parthenocissus** for **Virginia Creeper**

Viola (Viola, Violet, Pansy)

Leaves and/or other parts with powdery white coating	Powdery Mildew p. 302
Leaves with pale yellowish spots on uppersides and whitish mould growth beneath	Downy Mildew p. 295
Leaves and leaf stalks with dark blisters containing blackish powder	Smut p. 291
Leaves and/or other parts with green blister bearing orange or brown pustules	Rust p. 286
Leaves with various irregular yellowish mottles, crumpling and/or other distortions	Viruses p. 471
Leaves variously spotted	Leaf Spots p. 423
Leaves with edges rolled and thickened; white or orange larvae present inside	Violet Leaf Midge p. 220
Leaves discoloured, yellowing, bronzing and dying; small mites present on undersides, sometimes with silk webbing	Glasshouse Red Spider Mite p. 247
Leaves, buds and shoots infested by green, yellow or pink wingless and winged insects; foliage sticky and sooty	Aphids p. 161
Leaves with irregular holes and pieces eaten away; slime trails present on or near plants	Slugs and Snails p. 139
Stem dry and brittle immediately above soil level	Stem Rot p. 391
Stem base and/or roots blackened and decayed	Root and Foot Rots p. 361

Violet. See **Viola** and for **African Violet** see **Saintpaulia**

Virginia Creeper. See **Parthenocissus**

Vitis (Vines). See **Grape Vine** for both fruiting and ornamental types

Wallflower. See **Cheiranthus**

Walnut. See **Juglans**

Watercress

Leaves with pale yellowish spots on uppersides and off-white mould growth beneath	Downy Mildew p. 292
Leaves with yellowish mottling or distortion	Viruses p. 471
Leaves with pale yellowish circular spots bearing tiny brown bodies but no mould growth	Leaf Spot p. 424
Plants stunted and yellowed; roots swollen, stunted and often bent, crook-like	Crook Root p. 334

Water-Lily. See **Nymphaea**

Willow. See **Salix**

White Currants. See **Currants**

Wistaria

Flowers pecked and torn	Birds p. 255
Leaves with yellowing of veins or yellowish mottling	Viruses p. 471
Leaves variously spotted	Leaf Spot p. 424
Branches die back; masses of small, usually pink pustules present	Coral Spot p. 314

Also see **Climbing Plants**

Yew. See **Taxus**

Yucca

Leaves with ovoid, dark brownish spots	Leaf Spot p. 424

Also see **Cacti and Succulents** and **Glasshouse and House Plants**

Zantedeschia (Arum Lily)

Leaves and/or other parts with grey-black, thumb-print like spots	Leaf Spot p. 424
Leaves irregularly mottled or with yellowish streaks and patterns and distortion	Begonia Viruses p. 439
Corms and/or other parts with soft and slimy rot	Bacterial Soft Rot p. 348
Roots and/or stem base blackened and decayed	Root and Foot Rots p. 361

Also see **Bulbs etc.** and **Glasshouse and House Plants**

Zinnia

Flowers and/or buds shrivelled and with fluffy grey mould	Grey Mould p. 346

Also see **Bedding Plants**

Zucchini. See **Cucumber**

PLATE 1   Eelworms, Slugs, Millepedes, Woodlice

Stem Eelworm *Ditylenchus dipsaci*   132
1. Bloated onion plant with swollen, distorted leaves.
2. Live eelworms in water, as seen through a × 10 magnifier.
3. Infested narcissus plant showing damaged flower and weak growth.
4. Narcissus leaves with characteristic raised bumps ('spickels').
5. Infested narcissus bulb cut across to show brown rings of dead tissue.

Chrysanthemum Eelworm *Aphelenchoides ritzemabosi*   134
6. Chrysanthemum plant showing discolouration of upper leaves and collapse of lower leaves.

Root-knot Eelworms *Meloidogyne* spp.   136
7. Root system of infested plant showing characteristic hard irregular swellings (galls).

Yellow Potato Cyst Eelworm *Globodera rostochiensis*   137
8. Cysts on potato roots, as seen through a × 10 magnifier.

Slugs   139
9. Potato tuber showing tunnels eaten out by subterranean slugs.
10. Garden slug.
11. Slug eggs in soil.

Millepedes   143
12. Black millepede.
13. Spotted millepedes.

Woodlice   145
14. Adult woodlouse on rotting wood.

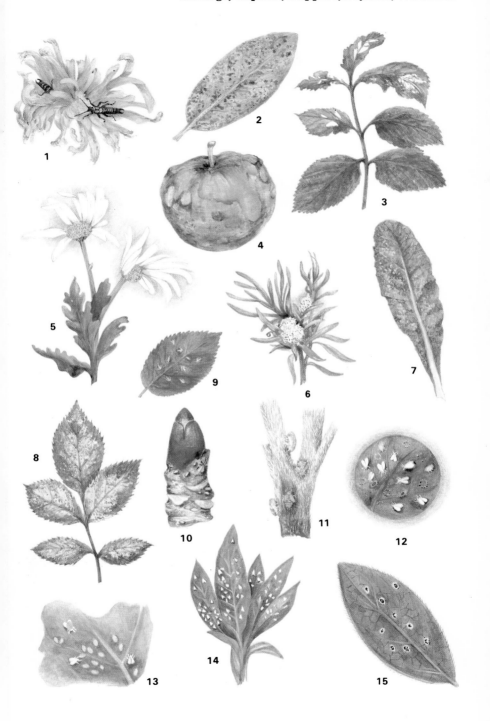

PLATE 3 **Aphids**

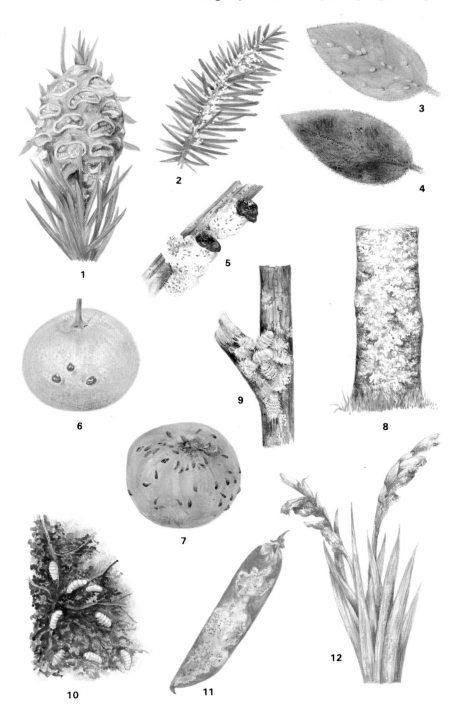

1

2

3

4

5

6

7

8

9

10

11

12

PLATE 5 **Caterpillars**

Winter Moths 196
1. Hornbeam leaves showing typical injury caused by caterpillars of the winter moth, *Operophtera brumata*, and related species feeding on buds and young growths in spring.
2. Caterpillar of the mottled umber moth, *Erannis defoliaria*, feeding on hornbeam leaf.

Codling Moth *Cydia pomonella* 197
3. Apple fruit damaged by codling moth caterpillar tunnelling into core.

Leopard Moth *Zeuzera pyrina* 199
4. Caterpillar tunnelling in woody branch of apple tree.

Cutworms 200
5. Caterpillars of the turnip moth, *Agrotis segetum*, in soil.

Swift Moths 201
6. Caterpillar of the garden swift moth, *Hepialus lupulinus*, in soil.

Cabbage Caterpillars 201
7. Caterpillar of the large white butterfly, *Pieris brassicae*, feeding on cabbage leaf.
8. Caterpillar of the small white butterfly, *Pieris rapae*.

Pea Moth *Cydia nigricana* 203
9. Caterpillars feeding on peas inside a pea pod.

Silver Y Moth *Autographa gamma* 203
10. Caterpillar feeding on leaf of glasshouse chrysanthemum.

Angle Shades Moth *Phlogophora meticulosa* 205
11. Caterpillar feeding on leaf of glasshouse plant.

Carnation Tortrix Moth *Cacoecimorpha pronubana* 207
12. Caterpillar on damaged leaf surrounded by silk webbing which it spins to draw leaves together into a protective covering.

Vapourer Moth *Orgyia antiqua* 205
13. Caterpillar feeding on leaves of dahlia.

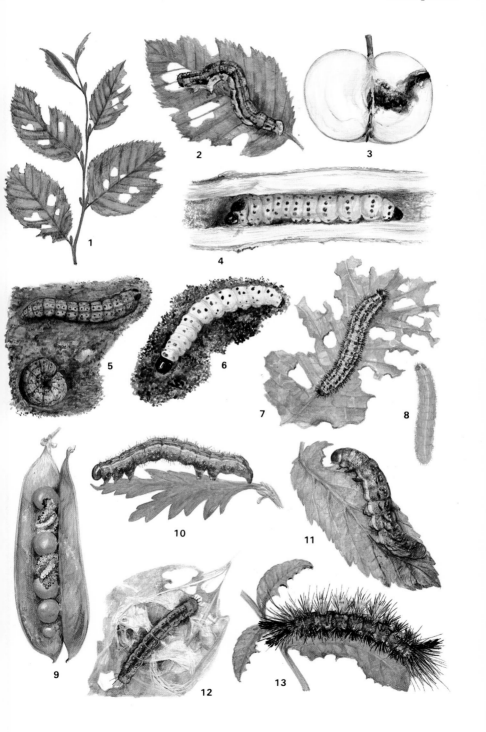

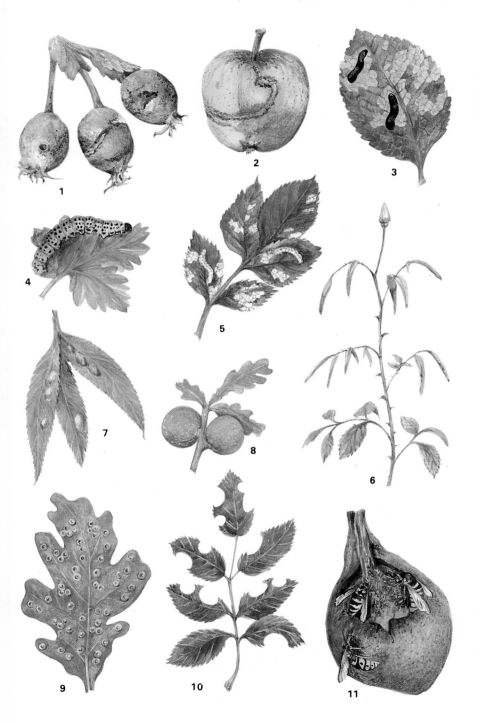

PLATE 8 **Beetles**

PLATE 9   **Mites, Birds**

Glasshouse Red Spider Mite *Tetranychus urticae*   247
1. Typical symptoms of early stages of attack on leaves of *Impatiens*.
2. Adult mites and eggs on underside of leaf, as seen through a ×10 magnifier.

Fruit Tree Red Spider Mite *Panonychus ulmi*   249
3. Overwintering eggs on branch of apple tree.

Conifer Spinning Mite *Oligonychus ununguis*   249
4. Typical discolouration of leaves of *Picea* resulting from mite infestation.

Strawberry Mite *Tarsonemus pallidus*   251
5. Pot cyclamen buds showing characteristic discolouration caused by mites feeding in the buds. Flowers developing from damaged buds are discoloured and distorted.

Bulb Scale Mite *Steneotarsonemus laticeps*   252
6. Leaves of narcissus showing typical scarring caused by mites feeding on young leaves.

Black Currant Gall Mite *Cecidophyopsis ribis*   252
7. Enlarged buds of black currant ('big bud') caused by mite infestation. This symptom is generally most obvious in early spring.

Pear Leaf Blister Mite *Eriophyes pyri*   253
8. Pear leaf showing typical blistering and discolouration caused by mites feeding in young tissues.

Nail Gall Mite *Eriophyes tiliae*   254
9. Nail galls projecting from upper surface of lime leaf.

Bird Damage   255
10. Pear buds attacked by bullfinches in late winter.
11. Cabbage plant attacked by wood pigeons.
12. Ripening apple pecked by blackbird.

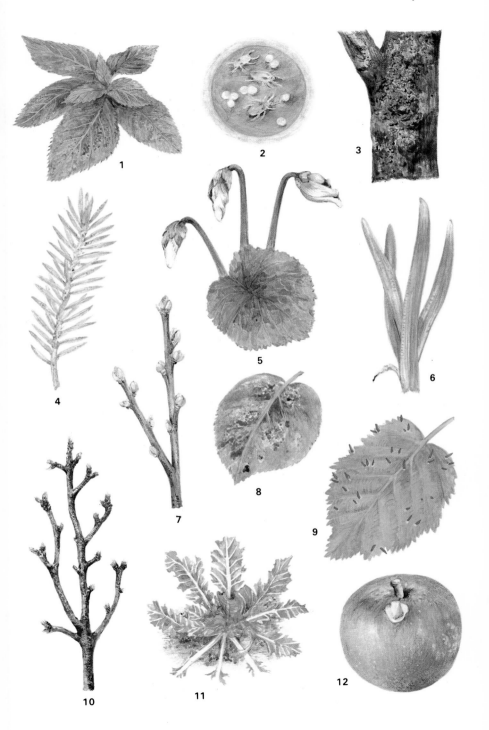

1

2

3

4

5

6

7

8

9

10

11

12

PLATE 10   **Rusts, Smuts**

Blackberry Common Rust *Phragmidium violaceum*   271
 1. Aeciospore-producing lesion on stem.
 2. Aeciospore and urediniospore (yellow) and teliospore (black) -producing pustules on underside of leaf.

Blackberry Stem Rust *Kuehneola uredinis*   271
 3. Aeciospore-producing pustules on upperside of leaf.

Carnation Rust *Uromyces dianthi*   273
 4. Plant bearing masses of urediniospore-producing pustules.

Chrysanthemum Rust *Puccinia chrysanthemi*   274
 5. Urediniospore-producing pustules on underside of leaf.

Chrysanthemum White Rust *Puccinia horiana*   274
 6. Plant bearing typical white pustules brought about by the germination *in situ* of teliospores to produce basidiospores.

Hollyhock Rust *Puccinia malvacearum*   278
 7. Teliospore-producing pustules on underside of leaf.

Leek Rust *Puccinia allii*   279
 8. Slit-like urediniospore-producing lesions on leaf.

Mint Rust *Puccinia menthae*   280
 9. Plant distorted by systemic infection and bearing two or three types of spore-producing pustule.

Pelargonium Rust *Puccinia pelargonii-zonalis*   282
 10. Urediniospore-producing pustules on underside of leaf.

Plum Rust *Tranzschelia discolor*   283
 11. Upperside of plum leaf with characteristic yellow spots indicating positions of pustules beneath.
 12. Powdery urediniospore-producing pustules on underside of plum leaf.
 13. Cluster cups on underside of anemone leaf; the alternate host.

Rose Rust *Phragmidium tuberculatum*   285
 14. Masses of aeciospores on briar shoot.
 15. Yellow urediniospore and black teliospore-producing pustules on underside of rose leaf.

Mahonia Rust *Cumminsiella mirabilissima*   280
 16. Upperside of leaf with dark spots and red colouration; often, but not invariably associated with the disease.
 17. Underside of leaf with powdery urediniospore and teliospore-producing pustules.

Sweet Corn Smut *Ustilago maydis*   290
 18. Cob distorted by systemic infection and bursting to reveal masses of black teliospores.

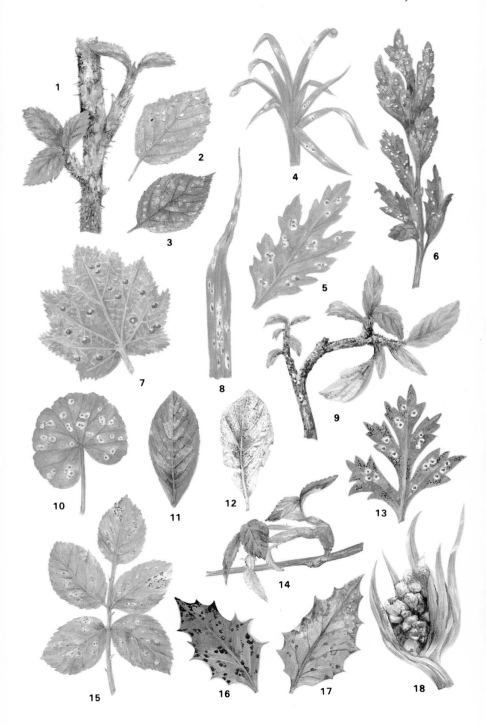

PLATE 11 # Downy Mildews, Powdery Mildews, Sooty Mould, White Blister, Needle Cast, Wilts

PLATE 12   Cankers, Diebacks, Stem and Twig Blights, Galls, Witches' Brooms, Leaf Curls

Apple Canker *Nectria galligena* f.sp. *mali*   310
1. Four to five-year-old lesion on young stem.

Plum and Cherry Bacterial Canker *Pseudomonas mors-prunorum*   320
2. Canker symptom on cherry branch with copious exudation of gum.
3. 'Shot-hole' symptoms on cherry leaves in early summer.

Coral Spot *Nectria cinnabarina*   314
4. Conidia-producing pustules on old pea stick.
5. Clusters of perithecia on dying gooseberry twig.

Fireblight *Erwinia amylovora*   316
6. Scorched appearance of foliage on hawthorn hedge.

Clubroot *Plasmodiophora brassicae*   328
7. Swollen lateral roots on swede.
8. 'Finger and toe' symptoms on young cauliflower root.
9. 'Club' symptom on cabbage.

Potato wart *Synchytrium endobioticum*   333
10. Knobbly gall on tuber.

Crown Gall *Agrobacterium radiobacter* var. *tumefaciens*   330
11. Elm trunk with a type of gall common on many trees and believed to be caused by the crown gall organism.
12. Gall at base of stem on young apple tree.

Witches' broom *Taphrina turgida*   334
13. Bird's nest type brooms on silver birch; probably the commonest host plant for witches' brooms.

Peach Leaf Curl *Taphrina deformans*   332
14. Appearance of affected foliage in early summer.
15. Single affected shoot in summer with white coating imparted by the layer of ascospores formed on the tissues.

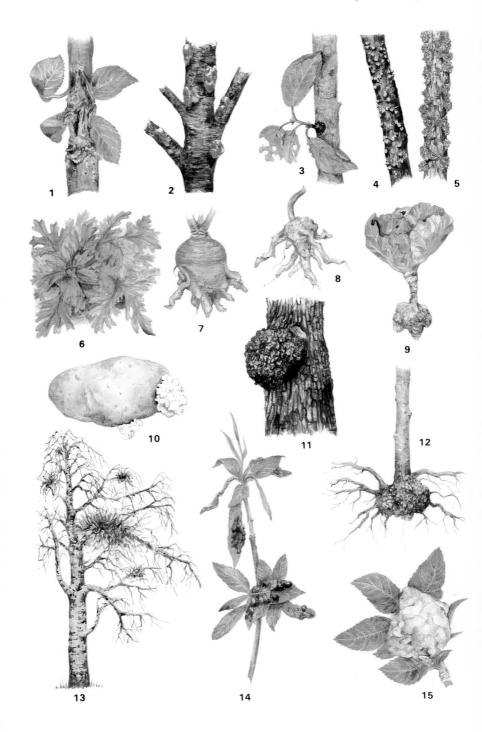

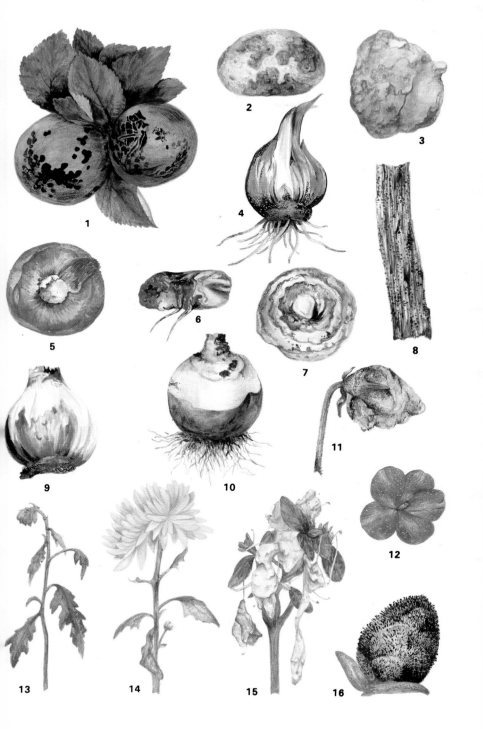

PLATE 14   **Fruit and Vegetable Rots**

Bacterial Soft Rot *Erwinia carotovora*   348
  1. Soft, slimy decay spreading down from neck into carrot root.
  2. Slimy outer leaves of Brussels sprout buttons.

Violet Root Rot *Helicobasidium purpureum*   360
  3. Carrot root enmeshed by deep purple mycelium.

Brown Rot *Sclerotinia fructigena*   350
  4. Decay spreading from diseased to healthy apple fruit.
  5. Small canker bearing conidial pustules on affected apple shoot.

Blue Mould Rot *Penicillium* sp.   354
  6. Affected apple fruit in store.

Bitter Rot *Pezicula malicorticis*   349
  7. Depressed lesion developed on apple in store.

Onion Neck Rot *Botrytis allii*   356
  8. Greyish mould growth around the softened neck in store.

Onion White Rot *Sclerotium cepivorum*   367
  9. Decayed roots, softening at bulb base, and small black sclerotia over
     bulb surface of salad onion.

Potato Dry Rot *Fusarium* sp.   357
  10. Shrinkage and concentric wrinkling of tuber with fluffy mould growth.

Potato Gangrene *Phoma* sp.   358
  11. 'Thumb-print' lesions on tuber.

Potato Pink Rot *Phytophthora erythroseptica*   359
  12. Section through affected tuber as seen about half an hour after cutting;
      the pink later darkens to black.

Grey Mould *Botrytis cinerea*   352
  13. Raspberries.
  14. Grapes.
  15. Broad bean, with decay beginning characteristically at one end of the
      pod.

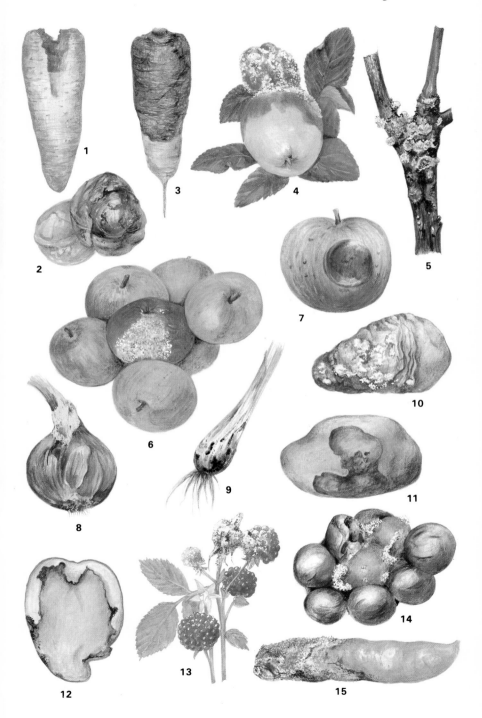

PLATE 16 **Spots**

PLATE 17 **Spots**

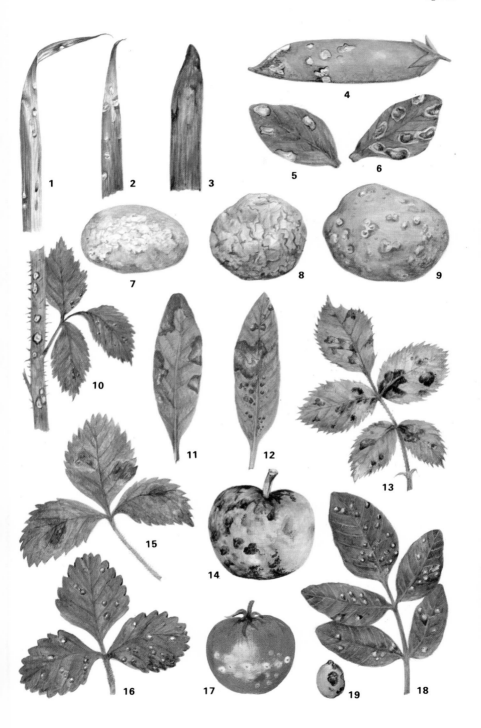

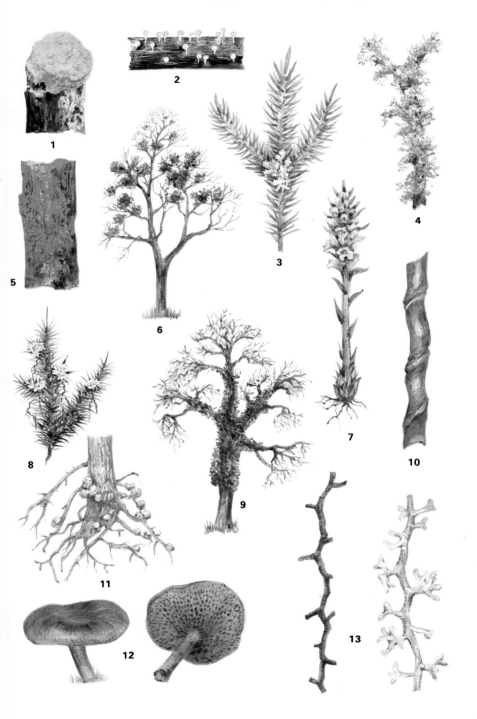

PLATE 19 **Viruses and Virus-like Organisms**

Refer to pp. 429–32 for general types of symptom.

Mosaics and Mottles
1. Abutilon mosaic, p. 434.
2. Cucumber mosaic, p. 444.
3. Raspberry mosaic, p. 464.

Line Patterns and Ringspots
4. Apple mosaic, p. 436.
· 5. Rose mosaic, p. 466.
6. Chaenomeles (apple mosaic virus), p. 470.

Vein Yellowing
7. Forsythia (arabis mosaic virus), p. 448.

Vein Banding and Yellowing
8. Cauliflower mosaic, p. 440.
9. Narcissus yellow stripe, p. 455.

· Necrotic Spots
10. Cabbage black ring spot (turnip mosaic virus), p. 440.
11. Apple star-crack, p. 435.

Enations
12. Pea enation, p. 458.

Leaf Roll
13. Potato leaf roll, p. 462.

PLATE 20 **Viruses and Virus-like Organisms**

Refer to pp. 429–32 for general types of symptom.

Flower-breaking
1. Wallflower (turnip mosaic virus), p. 442.
2. Tulip, p. 470.

Stunting
3. Rubus stunt (raspberry), p. 439.
4. Apple rubbery wood, p. 436.

Leaf Crinkling
5. Lettuce big vein, p. 452.

Leaf Narrowing
6. Tomato mosaic (fern-leaf) (tobacco mosaic virus), p. 468.

Fruit Discolouration
7. Plum pox; symptoms on cultivar 'Victoria', p. 460.

Fruit Malformation
8. Pear stony pit, p. 458.

Tuber Discolouration
9. Potato spraing; tuber in section (tobacco rattle virus), p. 462.

Yellowing
10. Strawberry yellow edge; normal and affected leaf, p. 467.

Grassiness
11. Narcissus, p. 455.

Green Flowers
12. Chrysanthemum, p. 443.
13. Strawberry, p. 467.

PLATE 21   Mineral Nutrient Deficiencies

PLATE 22   **Mineral Nutrient Deficiencies**

Magnesium   477
  1. Potato.
  2. Swede.
  3. Lettuce.
  4. Sweet corn.

Manganese   479
  5. Potato.
  6. Beetroot.
  7. Blackcurrant.
  8. Pea, split in half; marsh spot.

Boron   479
  9. Swede.
  10. Cauliflower.
  11. Parsnip.

Molybdenum   481
  12. Cauliflower.

Calcium   476
  13. Tomato; blossom-end rot.
  14. Apple; internal bitter-pit.
  15. Lettuce.
  16. Brussels sprout, split in half to show internal browning.

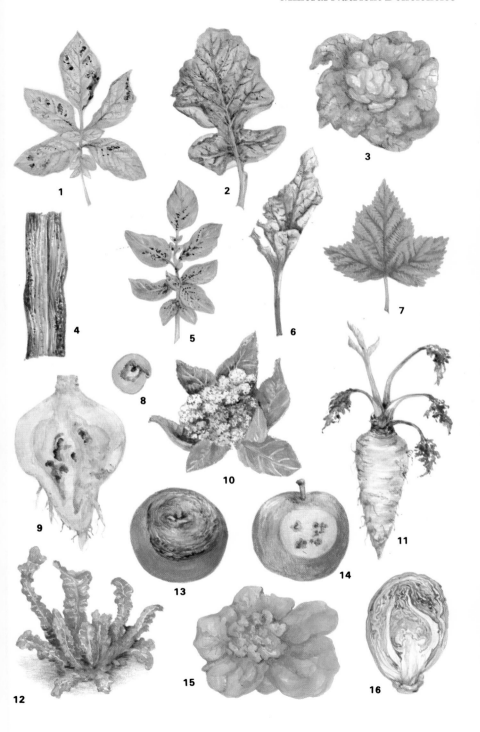

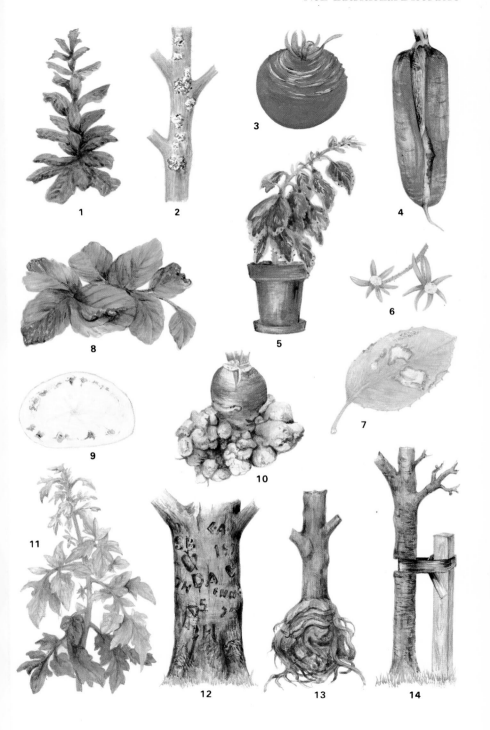

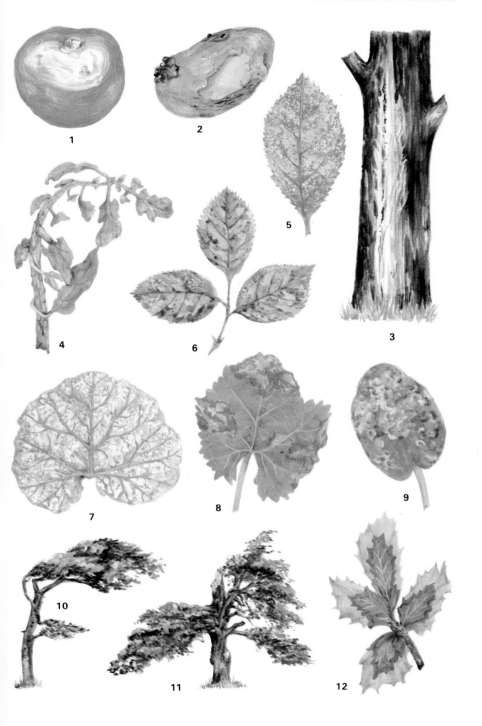

Pests

The pests that are dealt with in this section are either vertebrate or invertebrate animals. Of the vertebrates, only a few species of birds and mammals are important pests in gardens. Invertebrate animals are generally much more numerous than vertebrates, both in numbers of individuals and in numbers of species, and it is therefore not surprising to find that most pests are invertebrates. Of these, the majority are insects which, with a world total of about a million species described, and possibly as many yet to be recognised and described, far outnumber all other groups of animals. Fortunately only a small proportion of all insects are pests but the list of pest species is still a very long one. In addition to insects, a few smaller groups of invertebrates, especially eelworms, slugs and snails, millepedes, woodlice and mites, contain some pest species.

Pests mainly affect garden plants by feeding directly on living plant tissues, either by biting, chewing and ingesting them or by piercing them to feed on sap. In most cases only the immature stages feed on plants but sometimes the adults do also. Symptoms vary considerably and all parts of plants may be affected by some pest or other. In a few instances whole plants may be eaten or killed but damage is usually restricted to some particular part (leaf, stem, bud, flower, fruit etc.). Damage done to flowers, to fruits and to vegetables at or near harvest is of obvious importance, since it has a direct effect on both the quantity and the quality of garden produce, but less conspicuous damage to leaves, stems and roots may impair growth and may also make plants more susceptible to invasion by pathogens. Some pests may also produce special effects, such as the development of galls, which are abnormal growths of plant tissues, or of leaf-mines.

Pests also have important indirect effects on plants, especially by transmitting viruses and some other organisms that cause diseases, and also by fouling plants with excretions. The sticky honeydew that is excreted by aphids and other sap-feeding insects is a particular nuisance since it encourages the growth of sooty moulds on plants and this, in addition to making plants look unsightly, must also reduce photosynthetic activity by cutting off light from leaves.

Because of the great range in size, from gall mites that are less than 0.25 mm long to fallow deer that stand about a metre high, and because of similar great variation in the biology of different groups of pests, their impact on plants and also the methods used to prevent or minimise damage differ substantially from group to group. In the following pages basic information about the most important garden pests is summarised. Eelworms, slugs and snails, earthworms, millepedes and woodlice are dealt

131

with first, followed by the main groups of insects, which take up most of this section, then mites, birds and finally mammals.

EELWORMS
(Plate 1)

Also known as nematodes (Nematoda). Microscopic, usually 1–2 mm long, about 0.1 mm wide, just visible against a dark background. Seen through a microscope they look like miniature translucent eels (Pl. 1). Head end is relatively blunt, with characteristic minute internal mouth-spear in many species. Tail tapers. Female cyst eelworms swell to form spherical brown cysts about 0.5 mm diameter (Pl. 1) and root-knot eelworms swell to pear shape. Males are generally smaller than females. World-wide distribution with many different species living in soil, water, animals (some as important parasites) and in living and dead plant tissues. Plant parasitic species feed internally or externally on plant tissues by puncturing cells with the mouth-spear and extracting fluid contents. This causes discolouration, distortion and death of affected plants. Some species also transmit virus diseases (see p. 429). Females lay minute eggs in plant tissues or in soil. Larvae usually hatch within a few days and feed, grow and moult to adults in 2–3 weeks. Populations increase rapidly in favourable conditions and a single infested plant may contain many millions of active eelworms. Normal development only proceeds in fluid environments but non-feeding dormant larvae and dormant eggs in cysts can survive desiccation for many years. Plant parasitic eelworms can move a few metres a year through soil but are mostly spread in infested plants, plant debris and soil carried by man, animals, wind or water.

STEM EELWORM *Ditylenchus dipsaci*   **Pl. 1**
Also known as stem and bulb eelworm and some of its biological races are referred to as narcissus eelworm, tulip eelworm, onion eelworm and phlox eelworm. This species has been recorded from more than 400 different species of wild and cultivated plants. World-wide distribution.
Symptoms. Diagnosis of stem eelworm attack on symptoms alone is not always easy since similar symptoms may be produced by other pests and by some diseases and disorders. Doubtful cases should be referred to specialists, if possible.
Fruit. Only strawberries are seriously damaged. Plants growing slowly in spring are stunted and leaves are crinkled, crumpled and brittle. Leaf and flower stalks are thickened, often with brown cores. Ripening fruits develop light coloured soft patches. Leaf eelworms (p. 134), tarsonemid mites (p. 251) and virus diseases (p. 429) cause similar symptoms.
Vegetables. Onions: plants swell at base in spring and early summer. Leaves are swollen and distorted, producing the condition known as onion bloat (Pl. 1). Bulbs crack and rot, plants die. Similar symptoms develop on shallots, chives, garlic and leeks. Parsnips: crowns rot, dry and split; leaf bases swell and split. Symptoms may be confused by presence of other pests and diseases. Similar symptoms develop on carrots in some areas. Beans: severe attacks on broad beans stunt and malform plants and distort

pods. Light infestations produce red to black patches at bases of stems causing collapse of plants at soil level. French and runner beans are also occasionally attacked. Stems swell, blister and brown and severely damaged plants are stunted, with leaves tightly bunched together. Rhubarb: crowns of young plants rot at ground level in spring. Stalks swell at base, split and rot. A bacterium, *Erwinia rhapontici*, is sometimes said to be associated with this condition (see p. 387). Also occasionally attacks peas, potatoes, lettuce, spinach and swedes, producing various similar symptoms.

Ornamentals. Narcissus, tulips, hyacinths, scillas and snowdrops are often seriously damaged. Infested bulbs feel soft, especially at necks, and show discoloured brown rings of dead tissue when cut across with a knife (Pl. 1). Plants grown from infested bulbs are malformed, with distorted leaves and flowers (Pl. 1) and may not grow at all. Exact symptoms vary considerably. Daffodil leaves develop characteristic small green or yellow bumps ('spickels') (Pl. 1). Tulip leaves tend to split longitudinally, flower stalks bend and petals stay green. Perennial phlox show progressive narrowing of upper leaves which are reduced to a mid-rib with a narrow, frilly leaf blade. Similar symptoms develop on oenotheras. Aubrietias campanulas, gypsophilas, heleniums, heucheras, hydrangeas, irises and many other ornamentals may also be attacked and show similar symptoms.

Biology. Adults, 1–2 mm long, move in soil and enter plants through small wounds and natural openings (stomata and lenticels). They feed inside plant tissues, moving freely among damaged cells. Females lay many eggs which soon hatch. Development from egg to adult takes only 2–3 weeks at summer temperatures. When infested plants decay, eelworms leave and move through soil to find new host plants. In dry conditions they become dormant and may survive for several years before reactivation by water. Masses of dormant 'eelworm wool' often form on infested bulbs in bulb stores. Various biological races are known, each race attacking a particular group of plants. Interactions of races and plants are complex and not fully understood.

Treatment. Prevent establishment and spread of eelworms by good hygiene, adequate rotation of crops and efficient weed control. Obtain new plants from reputable suppliers. Buying cheap planting material (especially bulbs) is a false economy since a few infested plants will quickly contaminate a whole garden. Destroy soft, discoloured bulbs by burning immediately and keep suspect plants in isolation until they can be checked for eelworm. If infestations develop, despite these precautions, restrict spread by removal of all affected plants. Destroy by burning or by burying in a deep hole well away from main garden. Also collect up leaves and other plant debris and prevent spread by wind, on shoes and on implements and containers. Do not grow susceptible hosts on affected areas for 2–3 years and control weeds, which can maintain eelworm populations.

Onion eelworm attacks may be avoided by growing from healthy seed rather than from sets and by rotating crops so that onions and related vegetables are only grown on the same site once every three years.

Infested stocks of perennial phlox should be propagated by taking true root cuttings since eelworms do not penetrate the roots. Avoid reinfestation by growing on a site well away from infested areas.

Commercial growers use hot water treatment to maintain healthy stocks of bulbs and this treatment is sometimes used by gardeners. Dormant narcissus bulbs are immersed in water maintained at 44.5°C (112°F) for three hours. The temperature is critical as it must be high enough to kill eelworms inside the tissues without killing the bulbs. Fine thermostatic control is therefore essential. Chemicals used by commercial growers for stem eelworm control are too toxic for garden use.

POTATO TUBER EELWORM   *Ditylenchus destructor*
Symptoms. Mainly attacks potatoes but is less serious than potato cyst eelworm (p. 137). Infested tubers crack and wrinkle and internal tissues show dry, mealy, white pockets of eelworm infestation within generally discoloured areas. Plants grown from infested tubers fail to establish and eventually die. Also attacks some bulbous irises: bulbs produce stunted, discoloured leaves and bulb scales show longitudinal dark streaks. Occasionally attacks dahlias, gladioli, and tigridias, producing similar symptoms.
Biology. Generally very like stem eelworm (p. 132) but can survive by feeding on soil fungi as well as on weeds.
Treatment. Do not plant suspect tubers, corms or bulbs and remove suspect plants during the growing season. Destroy immediately by burning or burying. Do not grow susceptible plants on infested areas for at least two years and control weeds. No chemical treatments are safe for garden use.

CHRYSANTHEMUM EELWORM
Aphelenchoides ritzemabosi   **Pl.1**
Also known as leaf and bud eelworm. Particularly damaging on late-flowering outdoor and glasshouse chrysanthemums but can attack more than 150 different plant species. Widespread distribution.
Symptoms. *Ornamentals.* Chrysanthemums are attacked at any stage of growth, from young cuttings to mature plants. Most attacks develop during cool wet summers or in autumn. Leaves are discoloured, hang limply against stems and eventually die, starting with lower leaves and spreading up plants (Pl. 1). Exact symptoms vary but usually some leaves show discrete dark areas between veins where tissues have been killed. Buds are also damaged, checking growth, and if the terminal bud is killed lateral shoots develop. Blooms growing from damaged buds are distorted and stems may be scarred. Cuttings are stunted and usually killed. Poor growth and leaf discolourations are also caused by some diseases and nutritional disorders. Eelworm diagnosis may be confirmed by breaking affected leaves into small pieces and leaving in water in a glass tube for half an hour. Eelworms, if present, will move out of the leaves and collect into a wriggling mass in the bottom of the tube. They should not be confused with leaf hairs which look very like eelworms but do not move. Seek expert advice in doubtful cases. Other ornamentals attacked include asters (both china and perennial), buddleias, calceolarias, dahlias, delphiniums, doronicums, lavenders, paeonies, penstemons, phlox, pyrethrums, rudbeckias, saintpaulias, verbenas, wallflowers and zinnias. Symptoms vary considerably but generally show as stunted growth and discoloured leaves.

Fruit. Strawberries show symptoms in early spring and again in autumn. Leaves are crumpled and distorted and have characteristic white marks along veins on upper surface where eelworms have fed. Severe attacks kill the growing point and multiple crowns develop; flower trusses are killed and fruiting is seriously affected. The same symptoms are also produced by leaf eelworm (p. 136) and both eelworms may be present together. Association of these eelworms with a bacterium causes severe stunting and swelling of leaf and flower stalks which is known as 'cauliflower' disease of strawberries (p. 331). Blackcurrants are also attacked by chrysanthemum eelworm in winter and early spring. Eelworms feed in buds causing distortion of developing leaves and checking growth. Buds are sometimes killed.

Biology. Very similar to stem eelworm (p. 132) but feeds and breeds externally in buds, leaf axils and similar situations as well as internally in plant tissues. Eelworms move over wet surfaces of leaves and stems by swimming in films of water. They tend to move upwards, invading healthy leaves by entering stomata, and move 5 cm or more a night. Development from egg to adult takes 3–4 weeks in favourable conditions and eelworms survive in dead leaves and other plant tissues for about three months. Long-term persistence (up to three years) is possible in dry plant debris but is not typical of natural environments, which seldom stay dry for so long. Survival in soil is not important but survival as breeding populations in weeds is. Many are alternative hosts, especially groundsel, sowthistle, chickweed, goosegrasses, speedwells and buttercups.

Treatment. Avoid introducing infested plants into gardens. Obtain plants from reputable sources and scrutinise all new introductions carefully before planting. If any plants look suspect, grow them in isolation away from the main garden. Despite these precautions, chrysanthemum eelworms may be carried into gardens in wind-blown plant debris and they may also be present on weeds. If established plants show symptoms it is usually best to destroy them immediately by burning or by burying in a deep hole. Valuable stocks of chrysanthemums can be disinfested by hot water treatment of the stools before taking cuttings. Trim plants and wash all soil off stools before treatment then immerse in hot water at 46°C (115°F) for 5 minutes or at 43.5°C (110°F) for 20–30 minutes. Plunge stools into clean, cold water to end treatment and then box up to grow on for cuttings well away from infested material. Alternatively, take tip cuttings from infested stocks and grow them on rapidly in a heated frame or glasshouse. After some weeks take further cuttings from the tips of these cuttings and root them in sterile compost. Resultant plants should be free from infestation since eelworms will have been unable to keep up with rapid new growth. Always keep plants well ventilated and relatively dry to limit spread of eelworms in water on leaf and stem surfaces. Spread of eelworm on larger chrysanthemum plants may also be restricted by applying bands of vaseline or vegetable grease (*not* car grease) around bases of stems. No chrysanthemum cultivars are entirely immune to eelworm but 'Orange Peach Blossom', 'Delightful', 'Amy Shoesmith', 'Covent Garden' and 'New Princess' are partially resistant.

Strawberry runners may be treated by immersing in hot water at 46°C (115°F) for 10 minutes but this treatment is not suitable for other host plants. No chemicals are readily available for eelworm control in gardens and hosts other than chrysanthemum and strawberries must therefore be treated by removing and destroying infested plants and by maintaining good hygiene.

LEAF EELWORM *Aphelenchoides fragariae*
Also known as fern eelworm and strawberry eelworm. Closely related to chrysanthemum eelworm (p. 134). Widespread distribution.
Symptoms. Produces similar symptoms to chrysanthemum eelworm on ornamentals, on strawberries and on mint. Discoloured areas develop between main veins, especially on ferns. Usually most apparent in winter. Begonia leaves turn yellow and transparent, starting at the edges. Similar symptoms develop on coleus, gloxinias, lilies, paeonies, primulas, saintpaulias, violets and many other plants. Buds and terminal shoots are also attacked, with resultant stunting, distorting and scarring of young growth.
Biology. Similar to chrysanthemum eelworm but has not been studied to the same extent.
Treatment. Good hygiene, combined with maintenance of relatively dry conditions, will limit eelworm movements in glasshouses. Infested strawberry, fern, begonia and mint plants may be given hot water treatment. Immerse strawberry runners in water at 46°C (115°F) for 10 minutes. Treat ferns at 43.5°C (110°F) for 10–15 minutes; small pot begonias at 48°C (118°F) for 30 minutes and mint runners at 46°C (115°F) for 10 minutes (which will also kill mint rust disease). These temperatures and times are critical and if they are exceeded plants will be seriously damaged. At the end of the treatment plunge plants into cold water and then grow them in sterile potting compost away from any possible sources of reinfestation.

SCABIOUS BUD EELWORM *Aphelenchoides blastophthorus*
Related to the previous two species, it attacks flower buds of scabious, especially cultivar 'Clive Greaves'. Buds are either killed or produce poor flowers. Also attacks anemones, bulbous irises, cephalarias, narcissus, trollius and violets. Remove and destroy infested plants. No hot water treatment or other specific control measures have been developed.

ROOT-KNOT EELWORMS *Meloidogyne* spp.   **Pl. 1**
At least 17 different species occur in western Europe. Some are native species but others have been accidentally introduced with plants from the tropics and sub-tropics and are now established in heated glasshouses. All attack roots, tubers and corms and more than 800 different plant hosts have been recorded.
Symptoms. Galls develop on roots (Pl. 1). Size of individual galls varies from about 1 mm to 2 cm or more. Galls are sometimes regular but often coalesce into irregular growths. Severe attacks interfere with normal root function, plants show signs of malnutrition, wilt and may die. Pot plants in glasshouses and houses are especially susceptible but plants growing in soil, both under glass and outdoors, are also affected. Plants commonly attacked include begonias, cacti, carnations, chrysanthemums, coleus, cyclamen, gloxinias and plumbagos. Also cucumbers, tomatoes, lettuce, French beans, and occasionally carrots, parsnips and beetroot. Attacks are gen-

erally worst on light soils and at high temperatures. All galls on plant roots are not necessarily caused by root-knot eelworms. Confirm diagnosis in doubtful cases by breaking open galls and examining with a lens for presence of female eelworms (see below), or submit specimens to an expert. **Biology.** Minute eelworm larvae move through soil and invade root tissues. They feed internally for 1–2 months and as they feed the plant cells enlarge and proliferate to form the galls. Some species are parthenogenetic and only produce females. Mature females are pear-shaped, white and 0.5–1 mm long, either embedded in plant tissues or protruding from roots. Each lays 300–1,000 eggs over 2–3 months and these hatch into larvae which extend infestations by moving through the roots or the soil to new feeding sites. Eggs and larvae survive for only a few weeks in dry soil if no suitable host plants are available but can survive for months in moist soil at 10°C (50°F). In parthenogenetic species survival of a single egg is sufficient to start a new infestation.

Treatment. Do not introduce suspect soil or suspect plants into gardens. Examine roots of all newly acquired plants, especially pot plants and young lettuce, cucumber, cabbage and tomato plants. Destroy infested plants immediately by burning or burying and observe strict hygiene to prevent spread of eggs and larvae in soil, potting compost, staging gravel and on pots and implements. Sterilise pots, seed-trays, garden tools and potting composts by steaming, if possible, and grow seedlings and cuttings in sterile compost. There are no chemicals suitable for use in gardens.

YELLOW POTATO CYST EELWORM

Globodera rostochiensis (= Heterodera rostochiensis) **Pl. 1**

Also known as potato root eelworm and sometimes as golden nematode. Widespread in temperate regions and one of the most damaging pests of potatoes. Other very similar species also attack potatoes.

Symptoms. Roots of potatoes and tomatoes are infested by eelworms in the soil. Growth is checked and yields reduced. In severe attacks plants yellow and die, often in conspicuous patches of plants growing on infested areas of soil. White, yellow or shiny brown cysts, about 0.5 mm diameter, are usually present on roots and are easily seen with a hand lens or magnifying glass (Pl. 1).

Biology. Each cyst is the dead, swollen body of a female eelworm and contains 200–600 eggs. These remain dormant within the cysts in the soil for up to 10 years but most hatch within a year or so when potatoes, tomatoes or related weeds grow near them. The roots of these plants release chemicals which diffuse through the soil and stimulate eggs to hatch. Young eelworms then move through the soil, locate and penetrate the roots, and start to feed. Females growing within roots swell and eventually burst through the outer tissues and protrude with their heads still attached to the plant. They are then fertilised by the smaller, thinner males and mature into spherical brown cysts which break away and lie free in the soil. Populations of cysts increase very rapidly in favourable conditions. On main-crop potatoes there may be an overall increase of 25 times the initial number in one growing season, which leaves many millions of cysts to overwinter and infest subsequent crops.

Treatment. Cysts are present in most soils but usually remain at undetectable low levels until susceptible plants are grown. Avoid undue build-up of the pest by not growing potatoes or tomatoes too often on the same site. Tomatoes are best grown in soil-less media without direct contact with the soil and if good hygiene is observed they should not create an eelworm problem. Potato crops must be rotated. Soil and situation have some effect so strict rules can be misleading but main-crop potatoes should not be grown on the same site more than once every five years. Early crops may be taken more frequently, about once every three years, as they take less time to grow and therefore give eelworms less time to multiply. There are no satisfactory chemical treatments for infested soils and the only way to bring seriously affected areas back into use is to refrain from growing potatoes or tomatoes on them for at least six years. Some eelworm resistant potato cultivars are available but should be used under expert professional guidance as they are not immune to all strains of the eelworm and their unsupervised use could actually increase eelworm. 'Maris Piper', 'Maris Anchor', 'Pentland Javelin', 'Pentland Lustre' and 'Pentland Meteor' are resistant to the yellow potato cyst eelworm, which can be recognised by the yellow colour of its cysts in July and early August but these cultivars are not resistant to the other species, the white potato cyst eelworm, which does not produce yellow cysts. Weeds are not of any real importance in maintaining populations of this eelworm although it can occur on wild species of *Solanum*.

PEA CYST EELWORM   *Heterodera goettingiana*

A serious pest of peas throughout Europe. Closely related to potato cyst eelworm.

Symptoms. Pea plants make poor growth and haulms yellow prematurely. Patches of yellowing plants stand out against the normal green colour of healthy plants. Severely infested plants are stunted, with small leaves, and usually die. Cysts are usually present on roots. Similar symptoms develop on broad beans. Runner beans and French beans are not attacked.

Biology. Generally the same as potato cyst eelworm (p. 137) but cysts are lemon-shaped rather than spherical.

Treatment. Rotate crops so that peas and broad beans are only grown on the same site once every four years. If attacks develop, do not grow susceptible crops on affected area for 7–8 years. Weed control has no effect on this pest and no suitable chemical controls are available for garden use.

BRASSICA CYST EELWORM   *Heterodera cruciferae*

Also known as cabbage cyst nematode. Attacks all brassicas and many other plants. Spring cabbages are particularly susceptible. Widespread distribution. Avoid build-up by rotating crops so that brassicas are grown on the same site only once every three or four years.

CARROT CYST EELWORM   *Heterodera carotae*

Attacks carrots and other umbelliferous plants. Limited distribution in U.K.; mainly stunts carrots in E. Anglia. Biology and treatment similar to other species of *Heterodera*.

CACTUS CYST EELWORM   *Heterodera cacti*
Attacks cacti and euphorbias and may be troublesome on pot plants. Cacti can be saved by removing all roots, washing thoroughly and re-potting in sterile compost in clean pots.

FIG CYST EELWORM   *Heterodera fici*
Attacks edible figs and ornamental *Ficus* ('rubber' plant). Healthy *Ficus* can be grown from leaf cuttings taken from infested plants and inserted in sterile compost.

ROOT-LESION EELWORMS   *Pratylenchus* spp.
Various species, but especially *Pratylenchus penetrans*. Widespread distribution outdoors and under glass.
Symptoms. Roots, tubers and rhizomes are attacked by eelworms in soil. Small discoloured longitudinal slits in roots extend and eventually the roots rot after secondary invasion by bacteria and fungi. Affected plants are stunted, yellow, wilt and may die. Anemones, begonias, clematis, delphiniums, hellebores, lilies, narcissus, ranunculus and thalictrums are especially susceptible but many other plants may be attacked.
Biology. Relatively uncomplicated. Eelworms live free in soil and enter roots to feed and breed, leaving them again when tissues start to rot. No resistant stages are known.
Treatment. Easily destroyed by desiccation. Will not survive on bulbs or corms in store if these are dried thoroughly. No chemical treatment of infested soils is possible but some plants, notably African and French marigolds *(Tagetes)*, reduce *Pratylenchus* populations in soil. Hence the often publicised but inaccurate claim that *Tagetes* will control all eelworms.

STUBBY-ROOT EELWORMS   *Trichodorus* spp.
NEEDLE EELWORMS   *Longidorus* spp.
DAGGER EELWORMS   *Xiphinema* spp.
Free-living soil eelworms that feed externally on roots of many plants, causing direct damage to root systems and so stunting growth but they are generally more important as virus vectors. Among important viruses transmitted by these eelworms are tobacco rattle (p. 462) by *Trichodorus*; raspberry ringspot by *Longidorus* and arabis mosaic (p. 465) by *Xiphinema*. Professional diagnosis and advice on treatment is essential.

SLUGS AND SNAILS
(Plate 1)

Soft-bodied, unsegmented gastropod molluscs (Mollusca: Gastropoda), moving on large, slimy, muscular foot. Snails have conspicuous hard shells into which the body retracts but slugs do not have this protection. Most species are widely distributed in Europe.
Symptoms. Effects of various species of slugs and snails are similar since they all feed by rasping plant tissues with a toothed tongue (radula).

Irregular holes are eaten in roots, bulbs, corms, tubers, stems, leaves, buds, flowers, seed-capsules and fruits of many different plants. Most feed at night. Characteristic slime trails persist on plants and soil and indicate the extent of nocturnal activity. Damage to outdoor plants is generally most severe during warm humid periods in spring and autumn but in glasshouses and frames high temperatures and humidity prolong activity and increase damage.

Fruit. Ripening strawberries are eaten by slugs and snails but most other fruits escape attack.

Vegetables. Often seriously damaged. Pea and bean seeds are eaten soon after sowing, especially if they germinate slowly in cold weather. Seedlings, especially brassicas and lettuces, are eaten through at soil level, both outdoors and in seed boxes under glass. Tubers of Jerusalem artichokes and potatoes (Pl. 1) are extensively holed by subterranean slugs and damage is usually worst on main-crop potatoes. Leaf stalks of celery and chicory are often attacked by slugs and both slugs and snails feed on stems and leaves of cabbages, cauliflowers, Brussels sprouts and other brassicas, and on lettuces.

Ornamentals. Campanula zoysii, Omphalodes luciliae and *Phyteuma comosum* are so attractive to slugs that it may be quite impossible to grow these plants successfully. Other ornamental plants regularly attacked include anemones, coreopsis, daffodils and narcissi, dahlias, delphiniums, gerberas, heleniums, hostas, hyacinths, irises, lilies, primulas, rudbeckias, sweet peas, tulips and violas but many others are also attacked occasionally. Most damage is done in early spring by slugs and snails feeding above ground on seedlings, new shoots and crowns of plants and by slugs feeding in the soil on bulbs, corms and tubers. Leaves and stems of tulips are often extensively damaged as they grow up through the soil and similar symptoms may appear on other bulbous plants.

Biology. Individual slugs and snails are truly hermaphrodite, functioning both as male and female, but cross-mating is usual and courtship and mating behaviour is often very elaborate. Clusters of 10–50 spherical, opaque or transparent eggs (Pl. 1) are laid in small cavities in or on soil and each individual lays about 500 eggs during adult life, which may last for months or years. Eggs normally hatch within a month of laying in spring and summer but some species lay in autumn and eggs then lie dormant in soil until the following spring. They are often exposed when gardens are dug in winter. Young slugs and snails are miniature versions of adults and they feed for up to a year before maturing. Young and adults feed on dead, decaying organic matter, such as decaying leaves, as well as on living plants. Slugs are most abundant in soils containing decaying organic matter, such as composts and manures. Snails are commonest on calcareous soils.

The main pest species in gardens are:

FIELD SLUG *Deroceras reticulatum (= Agriolimax reticulatus)*
Commonest and generally most damaging. Widespread in Europe and has been accidentally introduced to other parts of the world. Colour is variable but usually light grey to fawn. Length up to 3–4 cm. Feeds mainly above

ground but will enter soil. Main breeding seasons in April–May and September–October.

GARDEN SLUG   *Arion hortensis*   Pl. 1
Common and widespread in Europe. Dark grey to black. Length up to 4 cm. Feeds above and below ground and breeds in summer.

BLACK SLUG   *Arion ater*
Large black slug, up to 15 cm long. Widespread in Europe and northern hemisphere. Although large, and therefore conspicuous, this species is generally less damaging than smaller species.

KEELED SLUGS   *Milax* spp.
Many different species. *Milax budapestensis* is commonest in northern Europe. Grey, dark brown or black, up to 10 cm long, with distinct ridge (keel) running down back of body. Live in soil but may come above ground to feed. Breed in autumn and early winter. Eggs lie dormant in soil until spring.

GARDEN SNAIL   *Helix aspersa*
Common and widely distributed in Europe. Grey-brown shell up to 3 cm across. Often abundant near dry stone walls and in rockeries, overgrown herbaceous borders and blackcurrant plantations. Adults live for several years and spend winters hibernating in clusters in dry holes in walls and similar situations.

STRAWBERRY SNAIL   *Trichia striolata (=Hygromia striolata)*
Smaller than garden snail and with flatter shell, up to 1 cm across. Shell colour varies from dirty grey to reddish brown.

BANDED SNAILS   *Cepaea nemoralis* and *C. hortensis*
Shells attractive. Broad white, yellow, grey and pink bands alternate with narrower dark bands. Conspicuous where they occur but do not normally cause much damage.

Treatment. Moist soils with high organic content support large populations of slugs and if conditions in a garden are particularly favourable it may be impossible to prevent damage to plants. Some relief is given by cultivating frequently and thoroughly to expose eggs, young and adult slugs to predators and weather; by limiting use of organic fertilisers, such as manures and composts; by restricting use of mulches around plants and by generally maintaining good garden hygiene. In addition, slugs may be trapped by laying old lettuce or cabbage leaves, grapefruit skins, old sacking, bran and various other materials on soil surfaces. Warm humid evenings give best catches. Inspect traps early next morning and kill slugs by dropping into a bucket containing a strong solution of salt or very hot water. Slugs can also be trapped by sinking shallow dishes to soil level and filling them with beer, which attracts them, intoxicates them and drowns them. Beer remains effective for 2–3 nights. Alternatively, use poison baits containing either metaldehyde or methiocarb. These are sold as slug pellets which are placed in strategic positions near susceptible plants. Pellets remain effective for 3–4 days and repeat applications may be necessary. Methiocarb pellets are expensive but generally give best results and 1 kg of pellets should be sufficient to treat about 1 hectare. Follow manufacturer's instructions for use carefully. Keep children and pets away from treated areas and do not use edible crops until at least a fortnight after treatment.

Areas where celery or potatoes are to be grown are best treated before planting to reduce the local slug population so far as possible. Susceptible potato cultivars, especially 'Redskin', 'Maris Piper', 'Desiree', 'Glade', 'King Edward', 'Record' and 'Pentland Crown' should not be grown where slugs are troublesome and main-crop potatoes should be lifted and stored as soon as possible, not left in the soil at the end of the growing season. Protect seedlings of ornamental plants and vegetables growing in seed trays or in the open ground by sprinkling slug pellets, weathered ashes or soot around them and use the same treatment to protect the young growing points of susceptible plants in the spring.

Snails are generally less troublesome than slugs. Where they are pests they can be dealt with in the same way and it is also possible to pick them off plants by hand.

EARTHWORMS

Segmented worms (Annelida : Oligochaeta). Widespread distribution; many species. Generally have beneficial effects on soil fertility and drainage but some species are a nuisance in lawns and sports turf.

Symptoms. Casts of excreted soil appear on soil surface, especially in spring and autumn when earthworms are most active. Casts make mowing difficult, damage cutting edges of cylinder mowers, provide sites for colonisation of lawns by weeds, and make lawns muddy and slippery in wet weather.

Biology. Earthworms live in the soil and feed by ingesting soil and decaying plant material, especially leaves. They do not feed on living plants but are attracted to dead and dying plants and are therefore often found when such plants are uprooted. Most earthworms live in the top 50 cm and populations in grassland often exceed 5 million worms per hectare. In cold or dry weather earthworms move down a metre or more in the soil and remain inactive until conditions improve. Individuals are hermaphrodite, functioning both as male and female, and they mate either on the soil surface or in burrows in the soil. 20–30 eggs are laid in cocoon-like capsules made from mucus secreted by the clitellum, which is the smooth broad band around the middle of the body of mature worms. After 1–6 months, young worms hatch from eggs. They mature about a year later and may then live for some years as adults. The main pest species in lawns are *Allolobophora longa* and *A. nocturna*.

Treatment. Earthworms cannot thrive in acid soils and may therefore be discouraged by managing lawns so that acid conditions (pH 5 or lower) are maintained. This is done by removing all cut grass and by using top dressings of peat, sulphate of ammonia or sulphate of iron. Do not use top dressings containing lime or heavy dressings of organic matter, such as compost. If this treatment is not feasible, it may be possible simply to disperse worm casts by sweeping with a birch or similar broom before mowing. Many different chemicals are recommended for earthworm control and they vary considerably in effectiveness. The safest to use are expellents, such as derris dust or potassium permanganate solution. These bring worms to the surface but do not kill them and the expelled worms must be collected up and destroyed. Derris is usually applied as a 1% dust at the rate of 25 g/sq. metre and must be well watered in after application and potassium permanganate is applied at the rate of 4 g per litre of water to every square metre. Potassium permanganate is especially useful on lawns near ponds and streams as it is not toxic to fish. Both derris and potassium permanganate are non-persistent and repeat treatments may have to be made every six months. Worm killers based on chlordane are much more toxic and persistent and should only be used when there is no adequate alternative. This chemical kills earthworms within the soil, so there is no disposal problem, and although it is relatively expensive, it will control earthworms and other soil pests for a year or more after application. All worm killers are best applied in warm, humid weather in spring and autumn since worms are near the surface then, and label instructions for correct use should always be followed. Keep children and pets off treated areas for at least two weeks after applying chlordane.

ENCHYTRAEID WORMS

Commonly known as pot worms. Closely related to earthworms but very much smaller, up to 2 cm long, thin and white. Often abundant in leaf mould, compost and manure where they feed on decaying organic matter. They cause concern when found in potting composts but they do not feed on living plants and are not pests. No treatment necessary.

MILLEPEDES
(Plate 1)

Segmented arthropods (Diplopoda) with up to 50 articulated segments and two pairs of legs per segment. World-wide distribution with many species. Mostly live in soil and leaf litter.Move slowly and smoothly, like snakes, and curl into tight coils when at rest. Feed mainly on dead and decaying plant material.
(*Note:* Centipedes are often confused with millepedes but have only one pair of legs per segment and move rapidly. They are mainly carnivorous, feeding on small insects, earthworms, slugs and other invertebrates, and are beneficial.)
Symptoms. Although millepedes generally feed on dead plants they sometimes eat seeds and young seedlings and often extend wounds on roots, bulbs, corms and tubers that have been caused by other pests, such as slugs, or by pathogens. Pea and bean seeds and seedlings are especially susceptible in early spring. Other plants commonly affected are straw-berries, carrots, cucumbers, potatoes, lilies, daffodils and tulips. In all cases the plants are attacked at or below soil level and the symptoms are indistinct. Soft tissues are eaten and rots quickly establish. Careful examination of damaged plants and the adjoining soil will usually reveal numbers of millepedes.

The main species in gardens are:

BLACK MILLEPEDES *Cylindroiulus londinensis,*
and *Tachypodiulus niger* **Pl. 1**
Large species with black, shiny, cylindrical bodies up to 6 cm long.

SPOTTED MILLEPEDE *Blaniulus guttulatus* **Pl. 1**
Thin, pale yellow with distinct red spots along each side of body. Up to
2 cm long. The most important millepede pest in gardens, often found in
potatoes and in other tubers, corms and bulbs.

FLAT MILLEPEDES *Brachydesmus superus,* and *Polydesmus
angustatus*
GLASSHOUSE MILLEPEDE *Oxidus gracilis*
Species with distinctly flattened bodies with about 20 segments. They grow
to 2–4 cm long and colour varies from white to brown. *Brachydesmus* and
Polydesmus are common outdoors and *Oxidus gracilis*, which is an introduced tropical species, occurs mainly in heated glasshouses and frames.

Biology. Millepedes breed in spring and summer and females lay 50–100
eggs in small chambers that they excavate in the surface layers of the soil.
Eggs hatch after 2–3 weeks. Young millepedes resemble adults but are
much smaller and have fewer body segments. As they grow they moult
periodically and the number of segments increases until they reach the full
adult complement. Adults remain inactive in the soil in winter and may live
for 2–3 years.

Treatment. Millepedes thrive in moist soils with a high content of organic
matter and are especially favoured by undisturbed accumulations of decaying plant debris and other rubbish. Good hygiene and thorough cultivation
will therefore reduce their numbers and may often be sufficient to prevent
serious damage. Good hygiene is particularly important in glasshouses and
frames where millepedes breeding under seed trays and pots may attack
seedlings growing in them. Susceptible seeds, seedlings, bulbs and corms
may be protected by working a little HCH dust into the soil just before
sowing or planting and established plants can be treated by watering a
spray-strength solution of HCH onto the roots. Use of HCH should
however be kept to a minimum, especially in vegetable gardens where
heavy applications may taint potatoes and other root crops. There is of
course little point in controlling millepedes when they are not the primary
cause of damage. In such cases the pest, disease or disorder responsible for
the initial damage should be identified and treated.

SYMPHYLIDS

Small arthropods (Symphyla), up to 1 cm long, white, with long antennae
and 12 pairs of legs. Related to centipedes and millepedes (p. 143) and are
sometimes confused with them and with springtails (p. 146). Widespread
but localised distribution, mainly in glasshouse soils in northern Europe

but also occur outdoors in mild climates. The glasshouse symphylid, *Scutigerella immaculata* is usually the most troublesome species.

Symptoms. Young and adult symphylids live in soil and eat root hairs and fine rootlets. They may also tunnel into larger roots. Affected plants are stunted, with poor root systems and may wilt and die. Foliage sometimes has a blue or red tinge and, in addition to these direct effects, plants may suffer from secondary invasion of wounds by bacteria and fungi. Tomatoes, lettuces and chrysanthemums are commonly attacked and young tomato 'plants are particularly susceptible. Asparagus, beans, brassicas, celery, cucumbers, parsley, peas, potatoes, strawberries, anemones, lilies, primulas, saintpaulias, and sweet peas may also be affected. When plants are seriously damaged, careful examination of the soil will usually reveal numbers of symphylids. An easy way to detect them is to lift a whole plant and plunge the roots into a bucket of water. Within about 10 minutes symphylids will float to the surface.

Biology. Females lay clusters of about 20 eggs in soil and larvae hatch after 2–3 weeks. At first they have only six pairs of legs but as they feed and grow the number increases until they have the full adult complement of 12 pairs. Development from egg to adult takes at least three months and adults may live a year or more. Larvae and adults feed near the soil surface in late winter and early spring and populations increase most rapidly in spring and summer when soils are moist and warm and food is abundant. When the upper layers of the soil dry out symphylids migrate downwards in natural crevices and earthworm burrows to a depth of 1–2 metres.

Treatment. Symphylids occur relatively infrequently and it is therefore usually impracticable to use preventative chemical treatments. Good hygiene may help prevent their introduction on plants or in composts and manures. If they do appear, susceptible plants may be protected by watering a spray-strength solution of HCH into the roots at a rate of about 8 litres/sq. metre or by working HCH dust or diazinon granules into the soil. HCH dust applied to planting holes will protect young transplants but must be used cautiously since heavy applications may check growth and cause taint in root crops.

WOODLICE
(Plate 1)

Terrestrial crustaceans (Isopoda). Bodies with hard, jointed exoskeleton and up to 7 pairs of legs. Up to 2 cm long. Many species, widely distributed and common in leaf litter, rotting wood and other decaying organic matter. The commonest species in gardens are *Oniscus asellus*, *Porcellio scaber* and *Armadillidium vulgare*. Other species, especially *Armadillidium nasutum*, *Porcellio laevis* and *Androniscus dentiger* are commoner in glasshouses.

Symptoms. Seedlings in seed boxes in glasshouses and frames are eaten off at soil level and irregular holes are eaten in leaves; stems, fruits and roots of older plants sometimes gnawed. Cacti, chrysanthemums, cucumbers, cyclamen, lupins, pansies and petunias are especially susceptible. Roots of ferns and orchids may also be attacked. Most damage is done at night but woodlice can usually be found during the day hiding near damaged plants under seed boxes, pots, stones and pieces of wood.

Biology. Females lay batches of 20 or more eggs in brood pouches under the body. Eggs soon hatch into young woodlice, which are miniature, light-coloured replicas of the adults, and they stay in the brood pouch for the first week or so after hatching. They then disperse and feed for about a year before maturing. Females of most species produce a number of broods in early summer but will reproduce at other times if conditions are favourable. Woodlice cannot survive dry conditions and therefore tend to concentrate in moist, cool situations with an abundance of plant debris and plenty of shelter. They are therefore abundant in overgrown neglected gardens. Their main source of food is dead organic matter, not living plants.

Treatment. First clear up accumulations of plant debris, old seed-boxes, stacks of unwashed pots, piles of rubble and rotting wood and any other possible breeding sites. Any woodlice found in such places may be killed by dusting with HCH or carbaryl or by simply pouring boiling water on them. Baits of bran, dried blood, boiled potato, turnip, cheese or sugar may then be used to attract survivors and HCH or carbaryl dust may be mixed with the bait to kill woodlice when they feed on it. These pests do not move very far from their breeding sites and a concerted campaign against them in a glasshouse or garden in winter or early spring should reduce the local population for the rest of the season.

SPRINGTAILS

Wingless, terrestrial arthropods (Collembola) related to insects. Body cylindrical or globular, 1–5 mm long. World-wide distribution from arctic to antarctic and often extremely abundant in soil and leaf litter. Able to jump by flicking a forked terminal appendage against the ground. Many different species, mostly feeding on fungal spores and mycelium and on dead plant tissues. Some attack living plants and the commonest pests in gardens are the garden springtail, *Bourletiella hortensis*, which is often abundant outdoors in April–June, and species of *Onychiurus, Hypogastrura* and *Orchesella*, which attack glasshouse and house plants.

Symptoms. Very small holes eaten in young leaves near soil, resemble flea-beetle damage but holes are much smaller. Leaves are sometimes reduced to a skeleton of veins. Soft young stems, roots and root-hairs are also occasionally damaged. Springtails are usually present in considerable numbers when damage is severe. Seedlings and plants commonly affected include beans, lettuce, tomatoes, cyclamen, cinerarias, chrysanthemums, conifers, orchids, violets and zantedeschias. Springtails are often noticed on house plants standing in saucers or other shallow containers as they come to the surface of potting composts when plants are watered and also float on water accumulating in saucers. They do not necessarily harm the plants but often cause concern as they are mistaken for fleas.

Biology. Females lay eggs in soil singly or in groups and each female deposits about 100 eggs over two to three weeks. Eggs soon hatch and young springtails, which are miniature versions of the adults, feed for some months before maturing. Development from egg to adult takes 2–18 months, depending on temperature, and breeding continues so long as temperatures are high and other conditions are favourable.

Treatment. Springtails are favoured by wet, acid conditions and may be checked by improving drainage, cultivating thoroughly and applying lime. If they cause appreciable damage outdoors, or in glasshouses and houses, plants and adjacent soil may be sprayed or dusted with HCH, diazinon, nicotine or malathion.

MOLE CRICKETS

Large insects (Saltatoria: Gryllotalpidae), up to 45 mm long, with power-ful fore-legs modified for digging in soil. One species, *Gryllotalpa gryllo-talpa* is widespread in northern Europe but now rare in Britain.
Symptoms. Roots and stems of young plants are eaten at or below ground-level. Potatoes, carrots, tomatoes and sweet corn are especially susceptible but many other plants may be attacked where mole crickets occur, which is usually on well-cultivated loams and sands in low-lying areas near streams, rivers and ponds. Small tunnels are excavated in soil near damaged plants and parts of plants may be dragged into tunnel entrances.
Biology. Young and adults live in tunnels excavated in soil. Each female lays up to 300 eggs from early May onwards in an underground nest chamber and tends young nymphs for first few weeks of development. Nymphs then disperse to feed during summer and autumn, hibernate in tunnels, and mature 1–2 years after hatching. They feed on insects and other small invertebrates as well as on plants. Adult males stridulate by rubbing wings together, making characteristic soft churring sound on warm spring and summer evenings.
Treatment. In Britain mole crickets are so rare that they should be conserved, not destroyed. Consult local naturalists.

CRICKETS

Medium-sized insects (Saltatoria : Gryllidae), up to 25 mm long, with long, thin antennae and wings folded horizontally over broad, flat body. Many species in northern Europe. Seldom pests of plants growing out-doors but the house cricket and some other species occasionally damage plants in heated glasshouses.

HOUSE CRICKET   *Acheta domesticus*
Symptoms. Tops are eaten off young seedlings and stems of young plants may be eaten through at soil level. Most damage is done at night and the easiest way to detect the presence of crickets is to listen for characteristic chirping of males after dark. Cockroaches (p. 149), cutworms (p. 200) and mole crickets (above) cause similar damage.
Biology. Introduced to Britain and northern Europe from warmer cli-mates and now well established in heated buildings and glasshouses. Each

female lays several hundred eggs singly or in small batches in organic debris and soil. Eggs hatch in 1–3 months and nymphs mature after 5–8 months at normal temperatures. Development is much more rapid at high temperatures. House crickets hide during the day and come out to feed at night, eating almost any dead organic matter as well as living plants. Adults fly and males chirp continuously in warm conditions.

Treatment. Reduce available food supplies and hiding places by good hygiene and if house crickets are really troublesome use methods recommended for cockroach control (p. 149).

BUSH CRICKETS

Medium-sized insects (Saltatoria : Tettigonidae), up to 5 cm long, also known as long-horned grasshoppers because of very long antennae. Many different species in Britain and northern Europe, feeding mainly on small insects and other animal matter. Some species occasionally eat plants but damage is sporadic and local and seldom serious enough to justify treatment. If valuable plants are threatened, apply contact insecticides, such as malathion or HCH, as soon as bush crickets are seen on plants.

GRASSHOPPERS

Large insects (Saltatoria : Acrididae), up to 6 cm long, with short antennae, long wings and powerful hind legs adapted for jumping. Grasshoppers and closely related locusts are serious pests of plants in many parts of the world but locusts do not normally reach northern Europe and grasshoppers only cause serious damage to plants in abnormally favourable seasons when populations are unusually high. Contact insecticides, such as HCH or malathion, may be used as sprays or dusts to protect plants, if necessary.

STICK INSECTS

Thin, long insects (Phasmida), generally up to 10 cm long (but some tropical species much longer) and resembling twigs and sticks. Not native to northern Europe but some exotic species have become established locally in glasshouses and in warm situations outdoors. All species feed on leaves of plants but are only likely to be pests in exceptional circumstances.

EARWIGS
(Plate 2)

Insects (Dermaptera), up to 25 mm long. Body ending in horny forceps. Large, delicate hind-wings fold tightly into short wing cases formed from modified fore-wings. The common earwig, *Forficula auricularia* is the only species of general importance in Britain and northern Europe but other species may be locally troublesome in glasshouses and also outdoors in southern areas with warmer climates.

Symptoms. Ragged holes are eaten in petals of chrysanthemums (Pl. 1), cinerarias, clematis, dahlias, delphiniums, pansies, violas, zinnias and some other ornamental flowers. Most damage is done between June and September outdoors. Earwigs hide within damaged flowers during the day and feed mainly at night. Leaves and buds are also damaged and young buds may be killed. Earwigs are often found in cavities in fruits, especially peaches and apples, and in distorted leaves. In such cases they are not usually the main cause of the damage but are simply using these sites as hiding places.

Biology. Females overwinter in soil under stones, wood and other debris. In December and January each female lays 50–100 eggs in a small chamber excavated in the soil and she remains with them until they hatch in February–March. Females then remain with the small, white earwig nymphs, tending them and feeding them with regurgitated food, until they mature and disperse in May–June. Further egg batches may then be laid, giving rise to a second generation of adults by September. Although they have wings, adults do not normally fly and mainly disperse by crawling relatively short distances from their breeding sites. Most of the earwigs present in a garden during the summer are therefore likely to be from local breeding in the previous winter and spring.

Although earwigs undoubtedly damage plants, they also feed on aphids and other small insects and may have some beneficial effect by limiting pest populations.

Treatment. Tidy up accumulations of rubbish, maintain good garden hygiene and cultivate thoroughly to destroy breeding sites. Trap earwigs under old sacking, planks or cardboard placed on soil surface or in upturned flower pots stuffed with straw and positioned on the top of canes or stakes. Treat infested plants by spraying with HCH or malathion and treat known breeding and hiding sites with HCH or carbaryl dust. Insecticides are best applied late in the evening or after dark when earwigs are most active. Pot chrysanthemums that have stood in the open during late summer and early autumn should be examined carefully before they are moved into glasshouses. If earwigs are present, spray before moving or fumigate with HCH as soon as they have been moved into glasshouse.

COCKROACHES

Medium to large insects (Dictyoptera : Blattodea), up to 4 cm long with long antennae and tough fore-wings lying horizontally on flattened body and covering hind-wings. Native European species are small and of no importance as pests but four larger species have been introduced from warmer climates and are now well established in heated glasshouses and in bakeries, kitchens and other warm buildings. Symptoms, biology and treatment of these four species are similar.

COMMON COCKROACH *Blatta orientalis*

AMERICAN COCKROACH *Periplaneta americana*

AUSTRALIAN COCKROACH   *Periplaneta australasiae*

SURINAM COCKROACH   *Pycnoscelus surinamensis*

Symptoms. Tops eaten off seedlings, seeds are destroyed and roots, stems, leaves and flowers of older plants are chewed. Cinerarias, chrysanthemums, pot cyclamen, nicotianas, orchids and schizanthus are particularly prone to attack but many other glasshouse plants may also be damaged, especially when populations of cockroaches are high. Most damage is done at night and the presence of cockroaches can usually be confirmed by inspecting glasshouses 2–3 hours after sunset.

Biology. Cockroaches hide during the day in plant debris, pipe ducts, crevices in walls and woodwork, under pots and seed trays and in soil. They come out to feed at night and eat almost any organic matter. They seldom fly but run quickly and erratically. Females lay eggs in tough, brown, purse-like cases (oothecae) which are either deposited on the ground or carried by females until eggs hatch. Each female produces up to 30 oothecae over 3–12 months and each contains about 15 eggs. Eggs usually hatch in 1–3 months at high temperatures but may not hatch for a year or more at low temperatures. Young cockroaches resemble adults but lack wings and wing cases. They take 9–12 months to mature and develop most rapidly above 21 °C (70 °F).

Treatment. Cockroaches are usually most troublesome in old glasshouses run at high temperatures as these usually contain plenty of food and cover and therefore favour rapid breeding. Limit potential breeding by removing accumulations of rubbish and by repairing old brickwork and woodwork. Then apply dusts of HCH or malathion to pipe ducts, paths, brickwork and other surfaces on which cockroaches will walk at night and renew at intervals of a week or so until the population is reduced. Baits of bran, bread, beer, sugar, treacles and various other materials can be used to trap cockroaches in jars or wide-mouthed bottles sunk flush with soil surface. A little bait in the bottom of each container attracts cockroaches which fall in and are then unable to scramble out. Insecticide may be mixed with bait or trapped cockroaches may be killed by pouring very hot water into containers. Persistent trapping for a few weeks will give good control but eggs may continue to hatch for up to a year and repeated trapping may be necessary.

LACE BUGS
(Plate 2)

Small insects (Hemiptera : Tingidae), up to 6 mm long. Adults have unusual lace-like fore-wings and extensions from the thorax. Nymphs and adults feed on plant sap but only one species is of importance as a pest of garden plants.

RHODODENDRON BUG   *Stephanitis rhododendri*   **Pl. 2**

Symptoms. Upper surfaces of rhododendron leaves mottled yellow and undersides with characteristic rusty brown discolourations (Pl. 2). Nymphs are present on young leaves from May onwards. Foliage of severely infested plants droops and is extensively discoloured. Plants growing in open, dry positions are more susceptible than plants growing in shade and hybrids of *Rhododendron arboreum, R. campanulatum, R. campylocarpum, R. catawbiense* and *R. caucasicum* are generally more susceptible than hybrids of other species.

Biology. Probably a north American species. First recorded in Britain in 1901 and now widely established in southern England and Wales and in parts of northern Europe. Eggs laid in autumn alongside leaf mid-ribs hatch from May onwards. Nymphs feed on undersides of leaves in groups of 10–50. Adults mature from June onwards and females move up onto youngest leaves to lay eggs. Adults have wings but do not fly and colonies therefore remain relatively static on the same plants year after year. Adults die out in early winter but eggs survive.

Treatment. Do not grow susceptible species and hybrids if this pest is known to be troublesome and avoid growing rhododendrons in exposed, dry sunny conditions, if possible. If plants are attacked, prune out and burn affected branches in March, if feasible, to destroy overwintering eggs, and spray with contact insecticides, such as HCH, malathion or nicotine, or with systemic insecticides, such as dimethoate or formothion, on two or three occasions at about three-week intervals from mid-June. Spray thoroughly to contact colonies of bugs on the undersides of leaves.

Light infestations on small plants may be dealt with in spring and summer simply by shaking or picking off nymphs and adults by hand.

CAPSID BUGS
(Plate 2)

Small, active insects (Hemiptera : Miridae), up to 6 mm long. Adults with fore-wings hardened, hind-wings membranous, legs, antennae and proboscis long. Nymphs similar but without wings. Colour varies with species from pale yellow and green to red-brown, often with subsidiary markings in other colours. Many different species are common and widespread in Britain and northern Europe, mostly feeding on plants. Nymphs and adults pierce plant tissues with fine stylets, inject saliva and feed on sap. Saliva of main pest species kills plant tissues. Some species of capsid bugs prey on mites, aphids, caterpillars and other small invertebrates and may have some beneficial effect by reducing pest populations.

Symptoms. Small ragged holes appear in young leaves, followed by characteristic tattering (Pl. 2). Buds and shoots may be killed and flowers developing from damaged buds are deformed (Pl. 2). Careful examination of affected plants may detect capsids but they are elusive and quickly drop to the ground or fly away when disturbed. Most damage is done in late spring and during the summer and many different garden plants are affected.

Injury on apples often shows as bumps and other irregularities on developing fruits (Pl. 2) and other plants that are commonly affected in various ways include pears, plums, black and redcurrants, gooseberries, raspberries, strawberries, potatoes, runner beans, asters, arctotis, buddleias, caryopteris, ceanothus, chrysanthemums, dahlias, forsythias, fuchsias, hydrangeas, magnolias, nasturtiums, pelargoniums, poppies, roses, salvias, sunflowers, venidiums and zinnias.

The main pest species are:

COMMON GREEN CAPSID   *Lygocoris pabulinus*
Nymphs are yellow-green, adults bright green, without brown markings. Common and widespread. Attacks many wild and cultivated plants, including fruits and ornamentals. Eggs laid in young twigs of apple, currants, hawthorn and other woody hosts during autumn overwinter and hatch in the spring. Nymphs feed on new growths and then migrate to herbaceous summer hosts where females mature and lay eggs in stems and petioles during June and July. A second generation of adults matures in autumn and females lay overwintering eggs before dying.

POTATO CAPSID   *Calocoris norvegicus*
Similar to common green capsid both in appearance and biology. Eggs laid during July and August in cracks in woody stems and sometimes in herbaceous plants overwinter and hatch in May and June. Attacks chrysanthemums and other Compositae as well as potatoes.

APPLE CAPSID   *Plesiocoris rugicollis*
Also similar to common green capsid. Eggs overwinter in bark of shoots and branches of apple trees and of some other woody host plants. Nymphs hatch in spring and feed on leaves, blossom trusses and young fruitlets. Adults mature by July and lay overwintering eggs before dying out.

TARNISHED PLANT BUG   *Lygus rugulipennis*
Also known as the bishop bug. Adult colour variable, yellow-green with red-brown markings. Adults overwinter in litter under hedges, in herbaceous borders and in similar situations. Females emerge from hibernating sites in April and May and lay eggs in stems, buds and leaves of many herbaceous ornamental plants and in various weeds. Nymphs soon hatch and feed until June when females mature and lay further eggs. Populations are high in summer and early autumn and adults often migrate off weeds onto garden plants, especially during dry weather when well-watered garden plants are most attractive to them. They often congregate on flowers, especially on arctotis and on *Chenopodium bonus-henricus*. Most adults seek overwintering sites in late autumn but may remain active well into the winter on chrysanthemums and other ornamentals growing in glasshouses and similar protected situations.

Treatment. Capsids tend to be rather elusive and control is not always easy but when damage is expected populations of nymphs and adults should be reduced before attacks develop. This is best done by a combination of chemical and non-chemical methods. In winter tidy under hedges by raking out leaf litter and clear up any other accumulations of plant debris and coarse herbage that may provide overwintering sites for the tarnished plant bug. Eggs of other capsids overwintering on deciduous woody hosts, such as apples, pears, plums, currants, gooseberries and some ornamental shrubs, may be killed by spraying with a winter wash in December or January while the plants are fully dormant but this treatment is not always very effective and it does not protect plants from re-invasion by capsids in the spring. Better control is usually achieved by spraying with systemic insecticides, such as dimethoate or formothion, or with non-systemic insecticides, such as fenitrothion, malathion or nicotine, in spring, summer and autumn, when the pests are active. Spray strawberries, raspberries and loganberries immediately before flowering and treat apples, pears, plums, currants and gooseberries immediately after flowering. Ornamental plants should be treated in spring and early summer, if necessary, and may also need protection later in the year if capsids are abundant.

FROGHOPPERS
(Plate 2)

Medium-sized sap-feeding insects (Hemiptera : Cercopidae), up to 6 mm long. Rather frog-like, with prominent eyes and powerful hind legs, with which adults jump when disturbed. Pale-coloured nymphs feed on stems, leaves and roots of cultivated and wild plants under conspicuous coverings of froth, commonly known as 'cuckoo spit' (Pl. 2). Many species with widespread distribution in Britain and northern Europe but only two are pests in gardens.

COMMON FROGHOPPER *Philaenus spumarius* **Pl. 2**
Symptoms. Froth masses appear on plants from May onwards. Young growths may be distorted and occasionally wilt but damage is seldom severe. Blackberries, raspberries, perennial asters, campanulas, chrysanthemums, coreopsis, geums, lavender, lychnis, phlox, roses, rudbeckias and solidagos growing outdoors are especially susceptible but many other plants may be attacked. Glasshouse plants are sometimes infested, especially chrysanthemums taken under glass after standing outside.
Biology. Females lay eggs in batches of about 30 in stems of shrubs and herbaceous plants during autumn. Eggs overwinter and hatch in May. Nymphs start to feed and protect themselves from desiccation by forming froth from a liquid excreted from the anus. Adults mature by late July and feed on plants until autumn before laying overwintering eggs and dying.
Treatment. Froth and nymphs can be removed from infested plants quite simply by spraying forcibly with water from a garden hose or pressurised sprayer. If insecticidal treatment is considered necessary, spray with malathion or nicotine.

RED AND BLACK FROGHOPPER *Cercopis vulnerata*
Slightly larger than common froghopper and adult is more strikingly marked. Nymphs feed on roots and adults occasionally damage apples, pears, blackberries, raspberries and strawberries by feeding on leaves, producing angular leaf spot symptoms. Damage is seldom severe and routine insecticide treatment for aphid control usually prevents froghopper infestations on fruits.

LEAFHOPPERS
(Plate 2)

Medium-sized sap-feeding insects (Hemiptera: Cicadellidae = Jassidae), up to 5 mm long. Similar to froghoppers but generally less robust. Many different species are widespread in Britain and northern Europe on wild and cultivated plants. Adults are often abundant and conspicuous on plants in summer jumping, flying briefly and resettling on leaves. Some species transmit pathogens, especially mycoplasmas (p. 430).

GLASSHOUSE LEAFHOPPER
Hauptidia maroccana (=Zygina pallidifrons)   **Pl. 2**

Symptoms. Coarse white mottling of upper leaf surfaces on glasshouse and house plants (Pl. 2). Young and adult leafhoppers and cast skins are usually present on undersides of damaged leaves; leaves completely blanched in severe attacks. Primulas, especially *Primula malacoides* and *P. obconica*, are most susceptible but tomatoes, calceolarias, fuchsias, gloxinias, pelargoniums, verbenas and many other plants are commonly affected. Some other leafhopper species cause similar symptoms.

Biology. First introduced into Britain about 1918 and now well established in glasshouses. Females live on plants for 2–3 months and lay about 50 eggs, inserting them singly into leaf veins. Eggs hatch after about a week in summer or a month or more in winter. Nymphs feed on undersides of leaves, moulting five times and maturing to winged adults in 1–2 months. Breeding continues throughout the year in heated glasshouses and in houses but slows down during winter.

Treatment. Remove older affected leaves to reduce numbers of eggs and nymphs, then protect new growth by spraying or dusting with a contact insecticide, such as malathion, nicotine or bioresmethrin on two or three occasions at fortnightly intervals.

RHODODENDRON LEAFHOPPER   *Graphocephala fennahi*
Formerly incorrectly referred to as *Graphocephala coccinea*.

Symptoms. No obvious direct damage to plants but this pest facilitates the spread of bud blast disease (p. 347). Adults with striking red longitudinal bands on green wing cases are present on plants in late summer and early autumn. They sit on leaves, jumping and flying briefly when disturbed but soon resettling.

Biology. Females cut shallow slits in outer scales of rhododendron flower buds in late summer and autumn and insert eggs. Eggs overwinter and hatch from April onwards and nymphs feed on undersides of leaves and mature by the end of July. Nymphs and adults tend to congregate towards the ends of young shoots.

Treatment. Presence of rhododendron leafhoppers does not necessarily mean that plants will develop bud blast disease but where bud blast is known to be prevalent, leafhopper control may help to reduce infections. Spray two or three times in August–September to kill adults before eggs are laid. Use contact insecticides, such as malathion or bioresemthrin, or systemics, such as dimethoate or formothion. Fungicides may also be used to control the disease (p. 347).

ROSE LEAFHOPPER
Edwardsiana rosae (= Typhlocyba rosae) **Pl. 2**

Symptoms. Fine white mottling on upper leaf surfaces (Pl. 2), especially on climbing roses on walls. Adults, nymphs and cast skins are usually present on undersides of leaves (Pl. 2). Severe attacks may develop in dry summers and can cause premature leaf-fall. Damage generally restricted to roses but this species also occurs on apple leaves in summer.

Biology. Eggs laid by females in shoots in autumn overwinter and hatch from May onwards. Nymphs of first generation mature by July and females lay eggs in leaf tissues. These eggs hatch in August and September, producing a second generation of adults which later lay the overwintering eggs.

Treatment. Systemic insecticides, such as dimethoate and formothion, used against aphids, also check leafhoppers. Spray against the first generation in May and June and repeat later in the season, if necessary. Nonpersistent contact insecticides, such as bioresmethrin, malathion or nicotine, may also be used.

Note: Hornbeam and beech hedges are often infested by similar leafhoppers and many other garden plants may also be attacked at some time. Such infestations can generally be treated in the same way as rose leafhopper.

APPLE LEAFHOPPERS
At least three different species of leafhopper occur on apple trees but seldom cause severe damage. Biology and treatment are essentially as for rose leafhopper (above) and they are usually well-controlled by routine sprays applied against aphids.

STRAWBERRY LEAFHOPPERS
Species belonging to the genera *Aphrodes* and *Euscelis* are important pests as they transmit green petal disease (p. 467). Protect runner beds by spraying with dimethoate or malathion in July and repeating twice at fortnightly intervals.

RASPBERRY/BLACKBERRY/LOGANBERRY LEAFHOPPERS
Two species of leafhopper, *Macropsis fuscula* and *M. scotti* transmit the mycoplasma that causes *Rubus* stunt disease (p. 439). Control leafhoppers as soon as they are seen by spraying with a non-persistent insecticide, such as malathion or bioresmethrin, but do not spray during the flowering period.

POTATO LEAFHOPPERS
At least four species of leafhopper feed on potato leaves and may cause some speckling and browning. Infestations are seldom serious and these leafhoppers do not transmit diseases. They can be controlled, if necessary, by applying systemic insecticides, as for aphids.

PSYLLIDS
(Plate 2)

Small sap-feeding insects (Hemiptera: Psyllidae), about 2–3 mm long. Nymphs, commonly known as 'suckers', have wide, flattened bodies, prominent eyes and conspicuous wing buds. Adults have two pairs of wings and can jump and fly. Both nymphs and adults feed by extracting sap from leaves, stems and other parts of wild and cultivated plants. Young buds and shoots are damaged and plants are often fouled with sticky honeydew and sooty moulds. Many species are widespread and often abundant in Britain and northern Europe.

APPLE SUCKER *Psylla mali*
Symptoms. Blossom trusses and leaf buds of apple trees are damaged as they open in April and May. Affected blossom is discoloured and distorted (Pl. 2) and may fail to develop at all if attack is severe. This damage is sometimes wrongly attributed to late frost but close examination will usually reveal psyllid nymphs which are orange–brown or green, with bright red eyes.
Biology. Females lay pale yellow, elongate oval eggs (Pl. 2) near leaf scars and buds on fruit spurs in September. Eggs overwinter on trees and hatch in March and April. Young nymphs immediately crawl into opening buds and start to feed. They mature from May onwards and adults live on trees until autumn, when overwintering eggs are laid and adults die out. There is only one adult generation each year.
Treatment. Overwintering eggs are easily killed by a 5% tar oil wash applied in December or January, when the trees are dormant. This treatment is also used against overwintering aphid eggs (p. 164) but does not control fruit tree red spider mites and can lead to an increase of this pest (p. 249). It may therefore be better to wait until the green cluster stage of bud development and then apply a systemic insecticide, such as dimethoate or formothion, as for aphids (p. 163). If apple sucker attack is severe, spray with HCH as it gives a better kill.

PEAR SUCKER *Psylla pyricola*
Symptoms. Leaf and blossom buds are damaged in spring, as with apple sucker (see above) but damage may continue through the summer as populations increase. This often causes extensive fouling of foliage and fruit with sticky honeydew and black sooty moulds. Fruit buds forming in late summer for the following season may also be damaged, with resultant loss of crop potential, and severe infestations may cause premature leaf-fall.
Biology. Pear sucker overwinters as adults, not as eggs. Adults remain on twigs and branches of pear trees and on other trees and shrubs in gardens and hedgerows. In late March females lay eggs on spurs and shoots of pear trees and these eggs hatch within a few weeks. Young nymphs feed in buds and on young growth and mature by early June. Breeding and feeding then continues until winter sets in. Three generations of adults develop, with

population peaks in June, July and October, but there is much variation in different places and seasons.

Treatment. Spraying with tar oil in winter is not effective against this pest as adults survive in sufficient numbers to re-invade treated trees in spring. Best control is given by spraying with dimethoate, formothion or malathion three weeks after petal fall to kill nymphs of the first generation. If spring infestations are particularly severe, apply this spray immediately after petal fall and repeat three weeks later and if summer populations are high, spray again after harvest to reduce damage to next year's buds.

BAY SUCKER   *Trioza alacris*

Symptoms. Leaves of bay laurel *(Laurus nobilis)* are attacked in late spring and early summer. Leaf edges thicken and curl over and sucker nymphs are present under curled leaves. Nymphs secrete tufts of conspicuous white woolly wax and also excrete sticky honeydew which fouls foliage and encourages growth of sooty moulds. Young plants, up to about five years old, are most susceptible and, when heavily infested, leaves shrivel and fall prematurely and shoots die.

Biology. Bay sucker occurs in Europe and parts of North America. It was introduced to Britain in the early 1920's and is now well established. Adults overwinter in curled leaves, amongst dense foliage, in leaf litter and in similar situations. On warm sunny days in spring they assemble on young bay shoots and feed on the leaves. Leaves then curl and females lay clusters of about six eggs under the edge of each curled leaf. Flat, oval, yellow-brown nymphs hatch from May onwards and feed until mid-October, when they become adult.

Treatment. Prune off and burn severely affected shoots and clear accumulations of leaf litter under plants. Spray with malathion in April–May to kill females before they have laid eggs. Control is more difficult after eggs have hatched as nymphs are protected from insecticides both by the wax that they secrete and by the curled leaves. Systemic insecticides, such as dimethoate or formothion, may therefore give better control of established infestations.

BOX SUCKER   *Psylla buxi*

Symptoms. Leaves at tips of infested box shoots curve inwards and form tight clusters, like miniature cabbages. Severe infestations make box hedges look unsightly and may check growth.

Biology. Females lay eggs in slits in leaf axils and twigs in August. Eggs overwinter and hatch in following April, just as new leaf buds start to develop. Psyllid nymphs then feed in malformed shoots until August, when they become adult.

Treatment. Cut out affected shoots and burn them, preferably before April, to reduce numbers of overwintering eggs. Spray thoroughly with a contact insecticide, such as malathion or HCH, or with a systemic insecticide, such as dimethoate or formothion, in May and June to kill nymphs.

ALDER PSYLLID   *Psylla alni*

Symptoms. Conspicuous tufts of white woolly wax appear on twigs of alder from early June onwards. These resemble tufts of wax produced by woolly aphid (p. 166) but can be distinguished by presence of psyllid nymphs.

Biology. Females lay eggs in crevices in bark during autumn. Eggs overwinter and hatch from April onwards and colonies of nymphs are present on young shoots until the end of June, when they develop into adults. Adults live on plants until September and then die out after laying overwintering eggs.

Treatment. This pest may retard growth of young shoots but is seldom important. If necessary, spray with malathion on one or two occasions in April–June.

BLUE-GUM SUCKER   *Ctenarytaina eucalypti*

Also known as eucalyptus psyllid.

Symptoms. Colonies of nymphs feeding on leaves and young shoots during summer months distort and weaken new growths. Widespread on eucalyptus, especially on nursery stocks.

Biology. Females lay eggs in late summer. Eggs overwinter on plants and hatch in spring. Two or more generations develop during the summer.

Treatment. Frequent applications of malathion dust give best control as eucalyptus leaves are water-repellent and therefore tend to shed sprays.

WHITEFLIES
(Plate 2)

Small sap-feeding insects (Hemiptera: Aleyrodidae), up to 2 mm long. Pure white adults have two pairs of wings folded over back of abdomen when at rest. Young nymphs crawl over plants but soon settle to feed and become immobile scales. Many species occur in Britain and northern Europe on wild and cultivated plants. Most are very similar in appearance and biology.

GLASSHOUSE WHITEFLY   *Trialeurodes vaporariorum*   **Pl. 2**

Symptoms. Adults, eggs and scales are present on undersides of leaves (Pl. 2) and upper leaf surfaces are often fouled with sticky honeydew and sooty moulds (p. 303). Tomatoes and cucumbers are often severely infested but this pest also attacks many other glasshouse plants and house plants, especially abutilons, begonias, calceolarias, chrysanthemums, cinerarias, coleus, dahlias, freesias, fuchsias, gerberas, heliotropes, hibiscus, pelargoniums, poinsettias, primulas, salvias, verbenas and zantedeschias. Sometimes occurs on outdoor plants and weeds in summer but may be confused with other species. Persistent infestations reduce plant vigour and sooty moulds make infested plants look unsightly. Leaves of some plants develop yellow spots and other discolourations where whiteflies have fed.

Biology. Originally a tropical and sub-tropical species accidentally introduced to glasshouses in Britain and northern Europe and now widely established. Adults live for a month or more and each female lays up to 200 eggs on undersides of leaves, often in neat circles. Eggs darken soon after laying and hatch about ten days later. Young nymphs crawl over plants for a few hours then settle to feed. Legs and antennae degenerate and nymphs then become immobile scales which feed for about two weeks before entering a non-feeding pupal stage from which adults emerge. Development from egg to adult takes three weeks at 21°C (70°F); four weeks at 15°C (60°F), and much longer at lower temperatures. Breeding continues throughout the year in heated glasshouses and is mainly parthenogenetic. Outdoor populations on weeds and other plants are killed by severe winters but some hibernating adults may survive mild winters.

Treatment. Prevent serious damage by examining plants regularly. If adult or immature whiteflies are seen, spray thoroughly with an appropriate contact insecticide, such as malathion, HCH, pyrethrum, permethrin or bioresmethrin, or with a systemic insecticide, such as dimethoate or formothion, and repeat on two or three occasions at about weekly intervals. Fumigants, such as HCH or dichlorvos, may also be used to kill adult whiteflies but may have to be used repeatedly at weekly intervals as they have little effect on nymphs, pupae or eggs. In southern England, and probably elsewhere, populations of glasshouse whitefly are now resistant to organochlorine and organophosphorus insecticides and control with these chemicals may fail. If so, pyrethrum, resmethrin, permethrin or bioresmethrin may still be effective.

Biological control by introduction of a small parasitic hymenopteran, *Encarsia formosa*, gives good results at temperatures above 21°C (70°F) and has been used in commercial horticulture. *Encarsia* females lay eggs in whitefly scales and parasitised scales turn black. Adults emerge through holes in the tops of parasitised scales and are almost all females, since reproduction is parthenogenetic.

Encarsia has been widely distributed in Britain and elsewhere during the past sixty years and is likely to be present in established glasshouses. Where it does not occur, it can be established quite easily by introducing a few leaves bearing parasitised scales. Supplies of these can be bought from commercial suppliers and from some other organisations, but may only be available for a restricted period. Enquire about supplies well in advance of intended use as it may be difficult to obtain stocks of the parasite at short notice. Use of insecticides must be kept to a minimum once *Encarsia* has been introduced and persistent insecticides, such as HCH, should not be used.

CABBAGE WHITEFLY *Aleyrodes proletella* Pl. 2

Symptoms. Adults and scales infest undersides of leaves of cabbages, Brussels sprouts and other brassicas (Pl. 2). Leaves are fouled with sticky honeydew and sooty moulds. Infestations are often severe on allotments in southern England where brassica plants at different stages of growth maintain breeding populations throughout most of the year. In such situations adults may be found at any time and may even be seen flying on sunny, frosty mornings in mid-winter when most other insects are inactive.

Biology. This pest resembles the glasshouse whitefly but is biologically quite distinct since it only attacks brassicas and it can survive severe winter weather. Breeding continues through spring, summer and autumn, but ceases in early winter as temperatures fall. Adults, and possibly some pupae, overwinter on infested plants and breeding starts again when temperatures rise in early spring.

Treatment. Pull up and burn severely infested plants after harvest to reduce numbers of whiteflies surviving to the next season and protect young brassica plants by examining them frequently and spraying as soon as any adult whiteflies are seen on them. Thorough high-volume spraying gives best results and the spray must be directed to the undersides of the leaves as that is where most whiteflies congregate. Non-persistent insecticides, such as bioresmethrin, pyrethrum, malathion or nicotine are safest to use but may have to be applied at weekly or fortnightly intervals. Systemic insecticides, such as dimethoate or formothion, give longer protection but should not be applied within a week or so of harvest. Biological control, which works against glasshouse whitefly, is not effective against cabbage whitefly.

RHODODENDRON WHITEFLY   *Dialeurodes chittendeni*
Symptoms. Infestations are most easily detected in June and July when adults congregate on young leaves of rhododendrons. This pest occurs mainly in southern England and attacks rhododendrons but not azaleas. Smooth-leaved rhododendrons seem most susceptible, especially *Rhododendron campylocarpum, R. catawbiense, R. caucasicum, R. ponticum* and hybrids derived from them. 'Pink Pearl', 'Goldsworth Yellow', 'Bagshot Ruby' and 'Mount Everest' seem to be especially prone to attack.
Biology. This pest is probably of Himalayan origin and it was accidentally introduced into Britain early this century. It was first recorded at Chiddingfold in 1926 and is now well established in southern England.

Adult females lay eggs on undersides of young leaves in June and July and then die. Eggs soon hatch and young scales feed for almost a year before developing into adults in the following May. There is therefore only one adult generation a year and populations increase slowly.
Treatment. Spray thoroughly with bioresmethrin, permethrin, resmethrin, pyrethrum, malathion, nicotine or a systemic insecticide, such as dimethoate or formothion, in June and July to kill adults. Two or three applications at fortnightly intervals may be needed to eliminate established infestations.

AZALEA WHITEFLY   *Pealius azaleae*   Pl. 2
Symptoms. Pale green, flat scales infest undersides of leaves and sticky honeydew and sooty moulds accumulate on upper surfaces. Adults are present on young leaves in summer. Locally a pest of evergreen azaleas, mainly *Rhododendron mucronatum* and *R. simsii*.
Biology. First recorded in Britain at Edinburgh in 1931 on azaleas imported from Europe and now widely established. Young scales overwinter on leaves and adults emerge in early summer. Only one generation a year.
Treatment. Spray thoroughly with bioresmethrin, permethrin, resmethrin, pyrethrum, malathion, or nicotine in June–July to kill adults. Systemic insecticides, such as dimethoate or formothion, applied in spring, summer or autumn, will also check this pest by killing adults and young scales.

VIBURNUM WHITEFLY   *Aleurotrachelus jelinekii*   Pl. 2
Symptoms. Only attacks *Viburnum tinus* and, to a lesser extent, *Arbutus unedo*. First sign of attack is usually the appearance of sooty moulds on

upper surfaces of leaves and close examination of undersides will reveal young scales and characteristic pupae, which are black with a white wax frill (Pl. 2).

Biology. Introduced into Britain about 1936 and seems to be restricted to southern England, where it is locally abundant. Scales and pupae over-winter on leaves of host plants and adults emerge in June and July. Eggs are laid on younger leaves and adults then die out. One generation a year.

Treatment. Spray thoroughly with bioresmethrin, permethrin, resme-thrin, pyrethrum, malathion, or nicotine in June and July to kill adults, directing spray to the undersides of leaves. Systemics, such as dimethoate or formothion, may be used at other times of year to kill scales.

Other species of whitefly, such as the phillyrea whitefly, *Siphoninus phillyreae* and the honeysuckle whitefly, *Aleyrodes lonicerae*, also occur on garden plants. Biology and treatment are essentially the same as for the species dealt with above.

APHIDS
(Plate 3)

Small sap-feeding insects (Homoptera: Aphididae) generally 1–5 mm long with soft bodies, relatively long legs and antennae and usually a con-spicuous pair of tube-like structures, the siphunculi, at the end of the ab-domen. Body colour varies between species and sometimes within species and ranges from white through various shades of red, yellow, orange, green, brown and blue to black. Aphid colonies often consist of winged and wingless individuals and the winged aphids have two pairs of broad, trans-parent wings with conspicuous dark veins. They are weak fliers but can be carried in thermals and air currents for hundreds of miles and this is how many species disperse and migrate to new host plants. Aphid mouth-parts contain very fine stylets which are inserted into plant tissues so that plant sap can flow up minute canals in the stylets and so enter the aphid's digestive system. Both young and adult aphids feed almost continuously and colonies are commonly found on leaves, buds, stems, roots and flowers (Pl. 3). Large volumes of sap are ingested by most species and excess sugars and water are excreted as drops of a sticky substance, commonly known as honeydew. This usually falls from the aphids or is flicked away by the hind legs but some aphid species are tended by ants which feed on honeydew and give the aphid colonies some protection from predators and parasites.

More than 500 aphid species occur in Britain and northern Europe. Many of these are common pests of fruits, vegetables and ornamental plants growing outdoors and some are particularly troublesome on glasshouse and house plants. Some species are restricted to one plant species or to a few closely related species or genera but others, such as the peach-potato aphid (p. 167) attack hundreds of different plant hosts, many of which may not be closely related botanically.

Aphids are one of the most important groups of pests attacking garden plants and are well known to gardeners. They are often referred to as blackfly, greenfly or blight and other common names are used locally in different countries and regions. Primary damage to plants results from the effects of colonies feeding on young tissues, which weakens and distorts new growth. Secondary effects, which are often more important, result from fouling of leaves and stems with honeydew, which encourages the growth of sooty moulds (see p. 303), and from the transmission of viruses (p. 429) which are carried from diseased plants to healthy plants on the stylets and in the saliva.

Aphid reproduction is mainly asexual and most aphids seen on plants are parthenogenetic females which are capable of giving birth to live young. At summer temperatures, young aphids mature in about a week and populations can therefore increase very rapidly to make maximum use of their plant hosts while conditions are favourable. The reproductive potential of many common species is so great that a single aphid would give rise to about ten million tons of aphids by the end of 100 days of summer breeding. This theoretical maximum is never achieved since many factors operate to control populations (see p. 18).

The seasonal biology of some aphids is complicated by alternation of sexual and asexual periods of development, combined with migrations from one set of host plants to another. This type of seasonal biology is exemplified by the black bean aphid (p. 171) and by some of the apple aphids (p. 165). Other species, such as the mottled arum aphid (p. 177) may never reproduce sexually and simply continue to breed asexually on the same host plant so long as conditions are suitable.

The symptoms and biology of the main pest species are detailed in the following section and for convenience they are grouped under fruits, vegetables and ornamentals. It should however be realised that the same species may be an important pest on all three groups of plants, as in the case of the peach-potato aphid which attacks peaches, potatoes, lettuces, tomatoes, chrysanthemums, roses, China asters and many other plants.

General methods of treating aphid infestations are summarised below and detailed recommendations are given in the entries for each of the main pest species.

Aphid species	*Main plants attacked*
FRUITS	
Green apple aphid (p. 165)	Apples and some related ornamentals
Rosy apple aphid (p. 165)	Apples
Rosy leaf-curling aphid (p. 165)	Apples
Apple-grass aphid (p. 165)	Apples and some related ornamentals
Woolly aphid (p. 166)	Apples and some related ornamentals
Pear-bedstraw aphid (p. 167)	Pears
Peach-potato aphid (p. 167)	Peaches, potatoes, lettuces, tomatoes and many ornamentals
Leaf-curling plum aphid (p. 168)	Plums, damsons and many ornamentals
Mealy plum aphid (p. 168)	Plums and damsons
Damson-hop aphid (p. 169)	Damsons
Cherry blackfly (p. 169)	Cherries, fruiting and ornamental
Black currant aphid (p. 170)	Black currants and red currants
Currant-sowthistle aphid (p. 170)	Black currants
Red currant blister aphid (p. 170)	Red, white and black currants
Gooseberry aphid (p. 170)	Gooseberries
Small raspberry aphid (p. 171)	Raspberry and loganberries
Large raspberry aphid (p. 171)	Raspberries
Strawberry aphid (p. 171)	Strawberries

VEGETABLES

Black bean aphid (p. 171)	Beans and some ornamentals
Cabbage aphid (p. 172)	Brassicas
Currant-lettuce aphid (p. 173)	Lettuces, currants and gooseberries
Lettuce root aphid (p. 173)	Lettuces and Lombardy poplars
Willow-carrot aphid (p. 174)	Carrots
Pea aphid (p. 174)	Peas and other leguminous plants
Melon and cotton aphid (p. 174)	Cucumbers and melons
Potato aphid (p. 175)	Potatoes, lettuces, tomatoes and many ornamentals
Glasshouse and potato aphid (p. 175)	Potatoes, lettuces, tomatoes and many ornamentals

ORNAMENTALS

Rose aphids (p. 175)	Roses
Chrysanthemum aphids (p. 177)	Chrysanthemums
Mottled arum aphid (p. 177)	Zantedeschias, pot cyclamen and many other glasshouse and house plants
Tulip bulb aphid (p. 178)	Tulips and other bulbs and corms in store
Honeysuckle aphid (p. 178)	Honeysuckle
Viburnum aphids (p. 178)	Viburnums
Elder aphid (p. 178)	Elder
Privet aphid (p. 179)	Privet
Violet aphid (p. 179)	Violets and some other ornamentals
Water-lily aphid (p. 179)	Water-lilies
Lime leaf aphid (p. 179)	Limes
Birch aphids (p. 180)	Birches
Sycamore aphid (p. 180)	Sycamores and some other acers
Beech aphid (p. 180)	Beech
Large willow aphid (p. 180)	Willows
Green spruce aphid (p. 181)	Spruces
Fern aphid (p. 181)	Ferns in houses and glasshouses
Root aphids (p. 181)	Various ornamental house and glasshouse plants and some outdoor plants

General treatment. The aphids that attack garden plants are mostly mobile species and serious infestations often develop after plants have been colonised by a few airborne winged females. These may have come from many miles away, possibly from hundreds of miles away, but they soon become established and breed rapidly when conditions favour them. The main danger period on outdoor plants is from March to October but in glasshouses, houses and other protected situations infestations may develop at any time of year. Non-chemical methods of control are seldom effective in these circumstances. It may be theoretically possible to limit aphid infestations on some plants by reducing the use of nitrogenous fertilisers so that young growths are less attractive to these pests but this is hardly practicable in most gardens. Greater reliance must therefore be placed on chemical control if plants are to be adequately protected.

Many different insecticides may be used to control aphids but to be really effective they must be applied before damaging populations have built up. This calls for frequent and careful examination of plants so that the first signs of attack can be noticed and prompt action can be taken. If this is not done, insecticides may be applied too late, after plants have been seriously damaged and often at the point where aphid populations are in any case about to be dramatically reduced by predators, parasites and other natural control factors. The full impact of the insecticides may then fall on predators and parasites and this may result in a subsequent increase in aphid numbers, rather than the expected reduction.

The chemicals that are used against aphids fall into three main categories: winter washes, non-systemic insecticides and systemic insecticides.

Winter washes based on tar oil are used only on fully dormant woody deciduous plants, such as apples, pears, cherries, currants, birch and willow, during December and January. They are sprayed or brushed onto plants to kill overwintering eggs of aphids and also control some other pests, especially psyllids, scale insects, mealybugs and some moth eggs. Winter washes must be applied very carefully as they may irritate skin and eyes and will kill any green plant tissues that they contact. On old apple trees this latter effect may prove beneficial as they will kill off growths of mosses and lichens on trunks and branches.

Non-systemic insecticides are used on growing plants and they kill aphids mainly by direct contact. They are mostly formulated as water-based sprays or as aerosol packs. Non-persistent insecticides, such as derris, malathion, nicotine, pyrethrum and the newer synthetic pyrethroids, bioresmethrin, permethrin and resmethrin, may be used on edible and ornamental plants at any time when aphids are present. Residues do not persist for more than one or two days and phytotoxic side effects are less likely than with more persistent insecticides. Application to outdoor plants is usually by conventional sprayer but glasshouse and house plants may be treated with aerosols. None of these non-persistent insecticides will protect plants from subsequent reinvasion by aphids and frequent applications may be necessary if plants are to be kept clean.

More persistent non-systemic insecticides, such as diazinon, fenitrothion, HCH, pirimicarb and pirimiphos-methyl give longer protection, usually of one to two weeks after application. Pirimicarb is especially useful since it is a highly selective insecticide which kills aphids rapidly but does not affect beneficial predators and parasites. When these persistent non-systemic insecticides are used on edible plants the label instructions should be observed to ensure that the required minimum period of one or two weeks elapses before harvest and to avoid possible phytotoxic side effects.

Systemic insecticides, such as dimethoate, formothion and menazon, give best control of aphids feeding in protected situations in rolled and curled leaves, in galls, and on higher branches of trees, as they are absorbed into plant tissues and poison the sap on which aphids feed. These three chemicals may be applied either as water-based sprays or as soil drenches of spray-strength solutions watered into the roots. They protect plants for two to three weeks after treatment and minimum periods between application and harvest of edible crops must be observed. Because of their systemic activity these insecticides tend to be phytotoxic to some plants and must be used cautiously (see p. 31).

Aphids on glasshouse plants can be controlled by fumigating with HCH, nicotine or dichlorvos and dichlorvos may also be used to fumigate house plants by sealing them in large plastic bags with dichlorvos resin strips for one to two hours.

APPLE APHIDS   Pl. 3

Five different species commonly infest apples in Britain and northern Europe. Four of these have similar seasonal cycles and are therefore dealt with together while the fifth species, the woolly aphid, is dealt with separately.

ROSY APPLE APHID   *Dysaphis plantaginea*

Infests leaves and blossom in early spring and may cause considerable stunting of fruits. The aphids are pink to grey and are covered with a white wax powder.

ROSY LEAF-CURLING APHID   *Dysaphis devecta*

Infested leaves curl tightly and curled areas turn bright red or yellow. The aphids are bluish grey with a mealy wax covering and mostly occur within the curled leaves. This pest is usually restricted to older trees, especially those with rough bark, and tends to recur on the same trees every year.

APPLE-GRASS APHID   *Rhopalosiphum insertum*

Also known as the oat-apple aphid. Attacks rosette leaves of blossom trusses on apple and may also attack medlar, rowan and hawthorn. Damaged leaves curl and yellow but the effect is only serious when substantial infestations develop. The aphids are yellow-green, rather plump, with short siphunculi.

GREEN APPLE APHID   *Aphis pomi*

Colonies develop in late spring and early summer on young extension growths of fruiting and ornamental apples and pears and on cotoneasters, hawthorn, pyracanthas, rowan and quinces. This is a bright green to yellow-green aphid with contrasting dark brown siphunculi. It is mainly a pest of young plants.

Biology. All four species overwinter as eggs laid on apple and other woody hosts in summer and autumn. Adults do not normally survive the winter. The eggs are black, about 0.5 mm long, and can be seen quite easily on twigs and branches during the winter months (Pl. 3). They hatch from mid-March onwards and young aphids quickly move into opening buds to feed on new leaves and blossom trusses. Peak populations develop by mid-June and most damage is done in the spring months. Rosy apple aphids migrate to plantains and apple-grass aphids migrate to oats, grasses and reeds during the summer but the other two species do not migrate from their main hosts. Overwintering eggs of the rosy leaf-curling aphid are laid under flakes of bark and in crevices as early as mid-June but the sexual egg-laying generations of the other three species do not normally develop until the autumn and most eggs are laid in September, October and early November. Reproduction at all other times is parthenogenetic and viviparous.

Treatment. Most apple trees are attacked by aphids every year and spring infestations of buds, blossom and young shoots may affect growth and yield. Valuable trees should therefore be treated by spraying thoroughly in the pre-blossom period with a systemic insecticide, such as dimethoate, or with contact insecticides, such as fenitrothion or malathion. Trees should be sprayed when they have reached the 'green cluster' stage of development since most of the overwintering eggs will have hatched by then. Dwarf pyramid trees, cordons, espaliers and fans are easily treated with hand sprayers but larger trees require power equipment and should only be sprayed if it can be done efficiently, since inefficient spraying may do more harm than good by killing predators and parasites without killing the aphids.

Aphids on apple and on other deciduous woody trees and shrubs may also be controlled by applying a tar oil winter wash in late December or in January, while the plants are fully dormant. Some gardeners favour this treatment as it cleans up trunks and branches by killing growths of moss, lichen and algae, but it will also scorch evergreen herbaceous plants and must be applied carefully. It may also favour increases in fruit tree red spider mites by killing off some of the predators that normally keep populations of this pest at a low level.

WOOLLY APHID   *Eriosoma lanigerum*   Pl. 3

Symptoms. Colonies of small brown aphids, covered by conspicuous tufts of white woolly wax, infest stems and branches of apple trees, ornamental crab apples, cotoneasters, hawthorn, pyracanthas, sorbus and some other plants. Colonies are most conspicuous in late spring and early summer but are present throughout the year. Infested plants develop irregular swellings on twigs and branches (Pl. 3) and these may later crack open and provide foci of infection by pathogens that cause canker and die-back.

Biology. Accidentally introduced into Europe, possibly from North America, and was first reported from London by Sir Joseph Banks in 1787. Now well established but is still sometimes referred to as 'American blight'.

Young aphids overwinter in cracks in bark and in galls. From March onwards they feed on twigs and branches, especially in leaf axils, on spurs and in wounds. Successive generations of wingless aphids develop during the summer. Some winged aphids develop and fly off to other host plants but this species does not spread so widely or rapidly as other aphids and colonies tend to persist on the same trees or shrubs year after year. Breeding stops in the autumn and immature aphids seek sheltered overwintering sites. No overwintering eggs are laid by this species in Europe.

Treatment. Localised infestations on trunks and branches can be dealt with by applying a spray-strength solution of malathion or HCH with a brush but if infestations are too extensive the affected plants should be sprayed thoroughly after flowering. Systemic insecticides, such as dimethoate or formothion, are most effective but contact insecticides, such as malathion or HCH, will give good results if applied thoroughly. Severely galled branches of apple and of cotoneasters, pyracanthas and other ornamentals should be pruned out and burned before the plants are sprayed as they will never recover and will make complete chemical control more difficult.

Biological control of woolly aphid in southern England was attempted about 1924 by introduction of a small parasitic hymenopteran, *Aphelinus mali*, from North America. This parasite lays eggs in the bodies of live woolly aphids and the parasitic larvae feed within the aphids and eventually kill them. They then pupate within the dead aphids and adult parasites emerge through small holes that they cut in the bodies. *Aphelinus* is now well established throughout southern England and parasitised aphids can usually be found in woolly aphid colonies. If the parasite is known to be active in a garden it should be conserved by limiting the use of persistent insecticides.

PEAR-BEDSTRAW APHID   *Dysaphis pyri*

This is the commonest species affecting pears although some other species may also occur on this host.

Symptoms. Colonies of relatively large pink or pink-brown aphids, covered with white mealy wax, infest young growths in early spring and cause severe leaf-curl and discolouration. Colonies may persist on pears into July and at times these aphids are so numerous that they swarm over trunks and branches of infested trees.

Biology. Overwintering eggs are laid in autumn, mainly on the swollen sections of fruiting spurs. They hatch in early spring and aphids breed on pear trees before migrating to bedstraws from late spring onwards. There is a return migration of aphids to pear trees in the autumn.

Treatment. Prune out and burn old fruiting spurs during the winter to reduce the numbers of overwintering eggs. Apply winter washes and spring and summer sprays as for apple aphids (see p. 165).

PEACH-POTATO APHID   *Myzus persicae*

Symptoms. This aphid attacks more host plants than any other species and is the most important aphid vector of plant virus diseases (p. 429). Colonies feeding on young growths of peaches and nectarines in spring cause severe leaf-curl. On other host plants dense colonies seldom form and the aphids are generally dispersed over the undersides of the leaves. This may cause some distortion of young growths as well as contaminating the upper surfaces of older leaves with sticky honeydew, sooty moulds, and the white moult skins. Glasshouse and house plants may be infested at any time of year but outdoor plants are mainly attacked in spring and summer. Tomatoes, lettuces and potatoes are commonly affected as well as many ornamentals, including antirrhinums, aubrietias, begonias, calceolarias, carnations, China asters, chrysanthemums, cinerarias, hibiscus, hyacinths, lilies, myosotis, nasturtiums, orchids, palms, phlox, primulas, roses, tulips and sweet peas. Among the important viruses transmitted by this pest are bean yellow mosaic, cauliflower mosaic, cucumber mosaic, dahlia mosaic, lettuce mosaic, pea leaf roll, plum pox, potato leaf roll, tomato aspermy and tulip breaking virus.

Biology. Peaches and nectarines are the primary woody hosts of this species and it overwinters on them as eggs. It also survives the winter as actively breeding populations in heated glasshouses and houses and in some protected situations outdoors. Development from nymph to adult is greatly influenced by temperature, taking over a month at an average of 6°C (43°F) but only a week at 24°C (75°F). Winged aphids appear from May onwards and migrate to other hosts and it is at this time that most virus diseases are spread.

The colour of peach-potato aphids varies from green to pale yellow and pink and individuals of different colours may be present together in the same colony.

Treatment. Overwintering eggs on peaches and nectarines may be killed by spraying with a tar oil winter wash, as for apple aphids (p. 165) but this must be done by the end of December as peaches and nectarines break dormancy earlier than apples, especially under glass, or control may be delayed until the spring when systemics, such as dimethoate, applied as soon as flowering has finished, will give best results. Systemics or non-systemics may be used on vegetables and ornamentals providing minimum periods to harvest are observed and phytotoxic effects are avoided (see p. 30).

Some populations of the peach-potato aphid are now resistant to organophosphorus and other insecticides, especially on chrysanthemums and other commercially-grown plants which have been intensively treated with chemicals. Because of the danger of selecting resistant strains, insecticides should only be applied when really necessary and it is best to use a number of unrelated materials rather than rely on continuous use of any one insecticide.

Resistant aphids on commercial chrysanthemums have been controlled biologically by rearing and releasing the small hymenopterous parasite *Aphidius matricariae*. This technique might be adapted for garden use but it requires considerable skill.

PLUM APHIDS

Three species are important pests of plums and damsons and also attack some ornamental plants.

LEAF-CURLING PLUM APHID *Brachycaudus helichrysi*

Dense colonies of small yellow-green aphids with pale siphunculi infest young shoots of plums and damsons causing severe leaf-curl during spring and early summer. From May onwards winged aphids migrate onto China asters, chrysanthemums, cinerarias, myosotis and other herbaceous ornamentals and aphid colonies stunt and distort growth of young plants. Aphids migrate back to plum and damson in the autumn and overwintering eggs are laid on twigs, especially at the bases of buds. They hatch early in the following year, usually by mid-January, and the first generation of aphids develops on the dormant buds. This pest is therefore already well established by the time the buds open.

MEALY PLUM APHID *Hyalopterus pruni* Pl. 3

Tightly packed colonies of green, narrow-bodied aphids with a slightly blue tinge and powdery white wax covering develop on the undersides of plum and damson leaves. Infested leaves are not deformed and the most obvious symptom of attack is the extensive growth of sooty mould on the upper leaf surfaces. Large quantities of honeydew are excreted by this pest and the sooty moulds growing on it often form a thick coating. This affects growth and contaminates fruit.

Eggs overwinter on plums and damsons but do not hatch until April so that colonies seldom build up to damaging levels before mid-summer. Colonies may persist on plum trees until August but winged aphids migrate to reeds from June onwards and there is a return migration to plums and damsons in the autumn.

DAMSON-HOP APHID *Phorodon humuli*

Colonies of shiny yellow-green aphids with dark green longitudinal stripes develop on young growths. Most troublesome on damsons in gardens but is also an important pest of commercial hops. Eggs overwinter on damsons, plums and blackthorn and hatch in February and March. Colonies increase until May when winged aphids migrate to hops. This species has become resistant to many insecticides.

Treatment. Plum aphids may be checked by spraying with tar oil between mid-December and mid-January but the buds of some cultivars, especially 'Belle de Louvain', 'Victoria', 'Yellow Egg' and the gages may be damaged, especially if they are sprayed during frosty or wet weather, and myrobalan is so susceptible to damage that it should not be sprayed with tar oil at all.

Spraying in the spring with systemic insecticides, such as dimethoate or formothion, or with contact insecticides, such as fenitrothion, malathion or nicotine, will generally give equally good results if done thoroughly in March, just before flowering, with a second treatment in May or June if mealy plum aphid is troublesome. By the end of June most aphid colonies on plum will contain coccinellid beetles (ladybirds) and their larvae, hover-fly larvae, anthocorid bugs, hymenopterous parasites and other natural enemies. These reduce aphid populations only after the main damage has been done but they may restrict aphid populations in late summer and autumn and so reduce the numbers of overwintering eggs laid. Use of insecticides on plums and damsons should therefore be limited to spring and early summer.

Plum aphid infestations on ornamental plants may be treated with systemic or non-systemic insecticides (see p. 164) and pirimicarb will give particularly good control of aphids, with little or no effect on natural enemies and hardly any danger of phytotoxicity.

CHERRY BLACKFLY *Myzus cerasi*

Symptoms. Colonies of black aphids infest young leaves and shoots of fruiting and ornamental cherries in spring. Leaves are distorted, growth is checked, and foliage is fouled with sticky honeydew and sooty moulds. *Prunus avium* and *Prunus cerasus* are especially susceptible; *Prunus serrulata* and *Prunus yedoensis* are less susceptible, and *Prunus sargentii* and *Prunus subhirtella* are usually unaffected.

Biology. Eggs laid near buds in October overwinter and hatch in March. Young aphids feed on leaves and shoots and colonies increase until July when winged aphids migrate from cherry trees to bedstraws (*Galium* spp.) and some other wild plants. Colonies on cherries are usually eliminated by parasites and predators towards the end of July but winged aphids migrate back to cherry trees from summer hosts in autumn and lay overwintering eggs before dying out.

Treatment. Tar oil winter wash, applied in late December or early January, will kill overwintering eggs and reduce initial infestations of young growths. Alternatively, wait until early April and spray with a systemic or non-systemic insecticide to kill young aphids (see p. 164).

CURRANT APHIDS

At least seven different species of aphid infest black, red and white currants in Britain and northern Europe. The three main pest species are dealt with below and two other species, the gooseberry aphid and the currant-lettuce aphid, are dealt with below and on p. 173, respectively.

CURRANT-SOWTHISTLE APHID   *Hyperomyzus lactucae*
Symptoms. Colonies of green aphids with swollen siphunculi infest young leaves in early spring causing leaf distortion and yellow discolouration. Mainly on black currants but may also attack red currants.

RED CURRANT BLISTER APHID   *Cryptomyzus ribis*
Symptoms. Colonies of pale yellow aphids with long siphunculi infest the undersides of young leaves in spring and early summer causing characteristic red blistering on red and white currants and similar yellow blistering on black currants.

BLACK CURRANT APHID   *Cryptomyzus galeopsidis*
Symptoms. Colonies of greenish-white aphids infest the undersides of black currant leaves, causing no obvious symptoms initially but as populations increase in early summer infested plants are fouled with honeydew and sooty moulds.

Biology. All three species overwinter as eggs laid near buds in the previous autumn. Eggs hatch as soon as buds open in spring and young aphids quickly infest young leaves. Colonies breed until early summer when winged aphids migrate to various weeds which are the summer hosts. In early autumn winged aphids return to currants and the sexual generation produces the overwintering eggs.
Treatment. Apply a tar oil winter wash in January to kill overwintering eggs and spray with a systemic or non-systemic insecticide immediately before flowering and, if necessary, after flowering to kill any aphids that survive the winter treatment (see p. 164).

GOOSEBERRY APHID   *Aphis grossulariae*
Symptoms. Dense colonies of small grey-green aphids infest tips of gooseberry shoots in spring, causing slight leaf-curl.
Biology. Eggs overwinter on gooseberries and hatch in early spring. Colonies develop on young growths and may persist throughout the season but some winged aphids migrate to willow herbs during the summer.
Treatment. As for currant aphids (see above).

RASPBERRY, LOGANBERRY and BLACKBERRY APHIDS

Two species occur commonly on raspberries in Britain and northern

Europe and both are important vectors of virus diseases. The same or similar species also attack loganberries and blackberries.

LARGE RASPBERRY APHID *Amphorophora idaei*

Symptoms. Colonies of large shiny yellow-green aphids with long legs and antennae and long, slightly swollen siphunculi infest young growths from March onwards and may cause slight leaf-curl.

SMALL RASPBERRY APHID *Aphis idaei*

Symptoms. Small powdery grey-green aphids form dense colonies on young shoots and fruiting laterals in spring and early summer, causing leaf-curl.

Biology and Treatment. Similar to currant aphids. Use winter washes against overwintering eggs and systemic or non-systemic insecticides before flowering, with repeat applications after flowering, if necessary. The various viruses inducing raspberry mosaic can be spread by quite small populations of aphids and it is difficult to obtain complete control of aphids in gardens so some virus transmission may occur despite the use of routine chemical control.

STRAWBERRY APHID *Chaetosiphon fragaefolii*
(= *Pentatrichopus fragaefolii*)

This is the main species on strawberries in Britain and northern Europe although some other species also occur on these plants.

Symptoms. Pale green or yellow aphids with long siphunculi infest the undersides of strawberry leaves. Infestations are most severe on young growth in spring, especially on strawberries grown in glasshouses and houses and under glass or polythene protection outdoors. Aphids may also be present at other times of year and are potentially damaging on runner beds as they transmit several different viruses, especially those associated with yellow edge disease.

Biology. This species lives on strawberry plants throughout the year. Populations are low during the winter but increase as temperatures rise.

Treatment. Spray with systemic or non-systemic insecticides, when necessary, outside the flowering and fruiting period. Do not rely entirely on any one insecticide as that may encourage the development of resistant strains of glasshouse red spider mites and strawberry mites.

BLACK BEAN APHID *Aphis fabae* Pl. 3

One of the best known aphids in gardens. Common and widespread in Britain and northern Europe.

Symptoms. Conspicuous dense colonies of small black aphids (Pl. 3) develop on young shoots of broad, French and runner beans in May, June and July and on dahlias, poppies, nasturtiums *(Tropaeolum)* and some other ornamentals later in the summer. Infested shoots are often stunted, leaves are curled and distorted, and growth is checked. Similar infestations develop on *Euonymus, Viburnum opulus* and *Philadelphus* in early spring.

Many other black aphids occur on cultivated and wild plants so all black aphids seen in gardens are not necessarily *Aphis fabae.*

Biology. Females of the sexual generation developing in the autumn lay many small black shiny eggs near buds of the winter hosts, which are mainly *Euonymus europaeus* and *Viburnum opulus*. Adults usually die out in the winter but in milder southern areas wingless females may overwinter on plants. Eggs start to hatch in April, just as the buds of the winter hosts are opening, and colonies develop on young leaves and shoots. Winged females migrate to beans and other summer hosts from May onwards and as summer temperatures rise the rate of reproduction increases so that in June and July each successive generation may mature in about a week. Most damage is done at this time, especially on beans, but colonies persist on the summer hosts until autumn when winged females migrate to the winter hosts and the sexual generation develops. This aphid is also an important vector of bean yellow mosaic virus.

Treatment. Serious infestations of black bean aphids tend to occur every other year so trouble can be anticipated by observing the relative severity of attacks on the summer hosts each year and by looking for eggs on the winter hosts, especially on *Euonymus europaeus*, in late winter and early spring.

Infestations on broad beans can be checked by pinching out infested shoots in May and early June but in most cases of moderate to severe infestation chemical control is easier and more effective. Low-persistence contact insecticides, such as malathion or nicotine, sprayed in the pre-flowering period will give good control if applied thoroughly but longer lasting protection will be given by systemic insecticides, such as dimethoate or formothion, either sprayed onto plants or watered into the roots. Do not apply insecticides to beans in flower as they may kill bees and other pollinating insects.

Infestations on ornamentals may be treated with systemic or non-systemic insecticides, providing phytotoxic effects are avoided by choosing the correct insecticides for the plants to be treated (see p. 30).

CABBAGE APHID *Brevicoryne brassicae*   Pl. 3

Symptoms. Dense colonies of grey-green aphids covered with a powdering of white mealy wax develop on leaves of many brassicas, especially cabbages (Pl. 3), Brussels sprouts, broccoli, cauliflowers and swedes. Leaves are discoloured and distorted where aphids are feeding and severe infestations check growth and may kill young and weak plants. Infestations build up from July onwards and usually reach a peak in September and October.

Biology. This species remains on cultivated and wild brassicas throughout the year. A sexual generation develops in autumn and overwintering eggs are laid on stems and leaves of brassicas but asexual females may also survive the winter in milder areas. Eggs hatch in April and young aphids feed on leaves and, later in the season, flower buds and flower stalks are infested. During May, June and July winged aphids fly off old brassica plants and establish new colonies on younger plants. Parthenogenetic reproduction of wingless females then continues until autumn when the next sexual generation matures. This aphid is an important vector of cauliflower and turnip mosaic viruses.

Treatment. Limit the carry-over of eggs from one season to the next by pulling up old brassicas as soon as cropping is over and destroying them by burning or burying. If this can be done by mid-April local sources of reinfestation will be reduced but plants will still be open to invasion by winged aphids from more distant sources. This risk is often greatest on allotments where brassicas at different stages of growth are present most of the time. All young plants should therefore be examined regularly from June until September and should be sprayed thoroughly with a suitable insecticide as soon as any cabbage aphids are seen. Malathion or nicotine give good short-term control and systemics give longer protection. Give particular attention to brassica seed-beds and, if necessary, dip transplants in a spray-strength solution of malathion or nicotine just before planting out. Wear rubber gloves when doing this to prevent contact of the insecticide with the skin.

CURRANT-LETTUCE APHID   *Nasonovia ribisnigri*
This is one of three species that commonly attack lettuces.
Symptoms. Colonies of green or yellow-green shiny aphids with darker markings develop on lettuce leaves during the summer outdoors and throughout the year under glass. This species also attacks young shoots of gooseberries and red currants, causing some leaf-curl and checking growth.
Biology. Eggs overwinter on gooseberry and currant bushes and hatch in early spring. Colonies then develop on new growths of these hosts and in May and June winged aphids spread to lettuces and to hawkweeds. In glasshouses and other protected situations this species continues to breed throughout the year.
Treatment. Examine outdoor lettuces regularly during early summer and check aphid infestations by spraying. Non-persistent insecticides, such as malathion or nicotine, can be used to within a few days of harvest but at least a week must elapse between applications of systemic insecticides and harvest. Plants in glasshouses and frames and under cloches may need treatment at other times of year.

On gooseberries and currants this pest is checked by winter washes and pre-blossom sprays applied against other species (see p. 164).

LETTUCE ROOT APHID   *Pemphigus bursarius*   Pl. 3
Symptoms. Colonies of yellow-white wingless aphids develop on roots of lettuces in summer (Pl. 3). Severe infestations kill plants, especially during dry weather.

This aphid also attacks Lombardy poplars (*Populus nigra* var. *italica*), where it lives in pouch-like galls on the leaf stalks. A related species, *Pemphigus spirothecae*, produces spiral pouch-galls on Lombardy poplars.
Biology. Eggs overwinter on Lombardy poplars and hatch in early spring. Wingless females feed on young leaf stalks and galls develop. Dense colonies of aphids then build up inside the galls and when these dry and split in June winged aphids migrate to the roots of lettuces and sowthistles where colonies feed and breed throughout the summer. Winged aphids return to Lombardy poplars in the autumn and lay overwintering eggs but summer colonies on lettuces may persist well into the winter and can survive in the soil until the following season.
Treatment. Some lettuces, such as 'Avoncrisp' and 'Avondefiance', are resistant to this pest but most of the varieties commonly grown in gardens are highly susceptible. Root systems of infested lettuces may be treated by

applying a spray-strength solution of malathion or nicotine as a soil drench but this is only likely to be effective if done as soon as damage occurs. Adequate watering during dry weather, good hygiene, rotation of crops and thorough cultivation all help to prevent serious attacks. If plants are severely damaged, dig up and destroy by burning or burying; cultivate the affected area thoroughly and do not grow lettuces on the same site in the following season.

Lombardy poplars are not normally treated since they are usually too large to spray and they are not appreciably affected by this pest.

WILLOW-CARROT APHID *Cavariella aegopodii*
Symptoms. Medium-sized green aphids with swollen siphunculi infest carrot leaves in late spring and early summer. Colonies are generally inconspicuous but conspicuous secondary symptoms are produced by the viruses that induce carrot motley dwarf disease (p. 441), which are transmitted by this pest.
Biology. Eggs overwinter on willows, mainly *Salix fragilis* and *Salix alba*, and winged aphids migrate to carrots in May and June. Colonies breed on carrots and related plants throughout the summer and there is a return migration to willows in the autumn.
Treatment. Early control of this aphid on carrots by spraying with systemic or non-systemic insecticides may prevent or check spread of motley dwarf disease but this treatment is only necessary if the disease is expected to be troublesome.

PEA APHID *Acyrthosiphon pisum*
Symptoms. Large pale green, pink or yellow aphids with long body, long legs and cauda, and very long siphunculi form smothering colonies on young growths of culinary and ornamental peas and on many other leguminous plants. In addition to direct damage caused by colonies feeding, this pest transmits a number of viruses, including bean yellow mosaic and pea leaf-roll.
Biology. Eggs and some adults overwinter on clovers, lucerne, trefoils and sainfoin and winged aphids spread to peas, beans and other leguminous plants from May onwards.
Treatment. Systemic or non-systemic insecticides applied in late May or in June, before plants come into flower, should give adequate protection to pea crops. Pirimicarb gives good control on sweet peas but is not recommended for use on vegetables.

MELON AND COTTON APHID *Aphis gossypii*
Symptoms. Colonies of relatively small aphids, usually dark green but sometimes yellow or black, infest shoots and leaves of cucumbers and melons in glasshouses and may also attack these and other plants outdoors. Growth is checked, distorted and fouled with honeydew and sooty moulds and this species also transmits cucumber mosaic and other viruses.
Biology. This species can breed continuously on glasshouse plants throughout the year. Reproduction is mainly parthenogenetic and sexual forms of this aphid seldom occur. Development is slow during the winter but colonies increase to damaging levels quite rapidly as soon as temperatures

start to rise in the spring. At summer temperatures aphids mature in about three weeks. Winged aphids spread infestations but there is no regular migration to and from winter hosts, as in many other aphid species.

Treatment. Spray thoroughly with systemic or non-systemic insecticides as soon as infestations are detected (see p. 163), but do not use HCH on cucumbers, melons and related plants as it is toxic to them.

POTATO APHID *Macrosiphum euphorbiae*

Symptoms. Colonies of large, long-bodied aphids with long legs, antennae and siphunculi develop on young shoots. Individual aphids are either green or pink and are rather restless. Potatoes are commonly affected but many other cultivated plants are attacked, both outdoors and in glasshouses and houses. It is often troublesome on tomatoes, lettuces, strawberries, aquilegias, carnations, cinerarias, China asters, chrysanthemums, dahlias, gladioli, hollyhocks, irises, roses, sweet peas, tulips, zantedeschias and zinnias. Because of its restless behaviour this pest is an important virus vector, transmitting narcissus yellow stripe, tulip breaking virus and others.

Biology. This is probably a north American species. It first appeared in Europe about 1917 and is now widely established. Parthenogenetic females overwinter on various plants in glasshouses and in protected situations outdoors and eggs overwinter on roses. Colonies increase from early spring and winged aphids spread infestations.

Treatment. Use appropriate systemic or non-systemic insecticides as soon as infestations are detected (see p. 163).

GLASSHOUSE AND POTATO APHID *Aulacorthum solani*

Symptoms. Medium sized shiny green-yellow aphid with cylindrical tapering siphunculi, each with a dark green pigmented area at the base, infest young leaves, causing some distortion and discolouration. Many different plants are attacked, both outdoors and in glasshouses and houses, and infestations are most troublesome on potatoes, including sprouting tubers, lettuces, tomatoes, anemones, calceolarias, campanulas, carnations, chrysanthemums, dahlias, foxgloves, gloxinias, hyacinths, nasturtiums (*Tropaeolum*), pelargoniums, salvias and zantedeschias.

Biology. Parthenogenetic colonies overwinter in glasshouses and other protected situations. They increase in spring and early summer and winged aphids disperse from June onwards. This species seems to be mainly parthenogenetic in Britain and northern Europe but may occasionally overwinter as eggs produced by a sexual generation in the autumn.

Treatment. As for potato aphid (see above).

ROSE APHIDS

At least seven different species attack roses in Britain and northern Europe. One of these, the potato aphid, is dealt with above and some of the others are noted below.

ROSE APHID *Macrosiphum rosae* Pl. 3

Common and widespread. Large dark green or pink-brown aphids feed on buds, shoots and leaves (Pl. 3). Colonies may persist throughout the year but are most numerous and troublesome in late spring and early summer. Foliage of infested plants is fouled with sticky honeydew and sometimes with sooty moulds and growth may be checked. Some winged aphids migrate from roses to scabious and teasels during the summer and in autumn some overwintering eggs may be laid on shoots and stems of roses.

LESSER ROSE APHID   *Myzaphis rosarum*

Lives mainly in young folded leaves and is usually noticed only when exceptionally abundant. Uniformly green with long swollen siphunculi which are slightly darkened at their ends.   -

ROSE-GRAIN APHID   *Metopolophium dirhodum*

An important pest on cereal crops in the summer and present on roses only in spring and autumn, usually in relatively small numbers and not causing appreciable damage. Uniformly shiny green, long bodied with a dark green longitudinal stripe. This species was unusually abundant in southern England during summer 1979.

SMALL GREEN ROSE APHID   *Chaetosiphon tetrarhodus*

Very small aphids clustering on undersides of leaves and young shoots in May. Can cause severe damage, especially to climbing roses. Colonies persist throughout the year, particularly in sheltered situations, and in some years this species is very common.

ROSE ROOT APHID   *Maculolachnus submacula*

Probably introduced into Britain and now well established. Clusters of eggs are laid on lower stems of roses in autumn and colonies develop on stems and roots in summer. Some cultivars may be killed by persistent infestations. This species is almost always attended by ants.

Treatment. Aphid populations tend to increase most rapidly on soft sappy growths in spring and early summer. Restricted use of nitrogenous fertilisers may limit such growths and so check aphids but cultural methods such as this are of doubtful value and the only certain way of keeping roses free from aphids is to use insecticides. Systemic insecticides, such as dimethoate or formothion, give good results, applied either as sprays or as soil drenches. Rates of application of soil drenches must be varied according to the size of plant and extent of root system. A litre of spray-strength solution may be sufficient to treat a small floribunda but larger shrub roses and climbers will need much more than this.

Many non-systemics give good control if applied thoroughly. Pirimicarb is one of the best and other effective insecticides include HCH, malathion, nicotine and derris. Soapy water may also be effective, if applied with sufficient force.

Chemical control is most necessary in April, May and June and two or three applications of insecticide made at this time before the plants come into full bloom may give adequate control of aphid colonies on new growths. Later in the season natural enemies will limit the increase of any surviving aphids so that further spraying may not be necessary but roses growing in glasshouses may need additional protection at other times of year.

CHRYSANTHEMUM APHIDS

Chrysanthemums growing outdoors or under glass in Britain and northern Europe may be attacked by at least four different species of aphid. Three of these, peach-potato aphid (p. 167), mottled arum aphid (below) and leaf-curling plum aphid (p. 168) have extensive host ranges and are not restricted to chrysanthemums. The fourth species is the chrysanthemum aphid which is dealt with below.

CHRYSANTHEMUM APHID *Macrosiphoniella sanborni*

Glossy brown or green-black aphids with short stout siphunculi form dense colonies on stems and buds and may also affect flowers. Restricted to chrysanthemums.

Treatment. Many different insecticides have been developed for aphid control on chrysanthemums and plants can be kept relatively clean by spraying or fumigating whenever aphids are seen on them. Stools and cuttings growing in glasshouses in winter may be attacked by aphids moving off other glasshouse plants and plants growing outdoors in spring, summer and autumn will be colonised by winged migrants. Fumigation of glasshouses with nicotine or HCH gives good control at temperatures above 16°C (60°F). Systemic insecticides should not be used on chrysanthemums as they may harm some cultivars but non-systemics can be used. Pirimicarb gives good control and can be used on plants in bloom. Spray formulations of malathion, HCH, nicotine or derris are also suitable. Resistance to organophosphorus and carbamate insecticides has developed in strains of some of the aphids that attack chrysanthemums and it is therefore best not to rely on any one insecticide for control.

Biological control of chrysanthemum aphids has been developed on commercial crops by rearing a small parasitic hymenopteran, *Aphidius matricariae*, for release on cuttings before distribution. This ensures that the parasite will be present on young plants and will help to eliminate aphids that may have survived chemical treatments. This technique could possibly be used in gardens but requires more expertise than the use of chemicals.

MOTTLED ARUM APHID *Aulacorthum circumflexum* **Pl. 3**

Common and widespread on glasshouse and house plants.

Symptoms. Colonies of shiny yellow or green aphids with dark U-shaped marks on older individuals infest buds, leaves, flowers and stems (Pl. 3). Plants may be distorted and fouled with honeydew and sooty mould and, in addition to this primary damage, this pest is an important virus vector, transmitting at least 30 different plant virus diseases. Zantedeschias and pot cyclamen are especially susceptible to infestation by this species but many other plants are attacked, including anemones, anthuriums, ornamental asparagus, azaleas, begonias, calceolarias, chrysanthemums, cinerarias, dahlias, freesias, irises, lilies, primulas, roses, saxifrages, schizanthus, tulips and violets.

Biology. Eggs and sexual forms are unknown and breeding is entirely parthenogenetic. Colonies breed throughout the year in glasshouses and houses and generally increase most rapidly between February and May. Populations can reach damaging levels at other times of year if conditions favour them. Winged aphids spread infestations but there are no regular migrations as in some other species.

Treatment. Appropriate systemic or non-systemic insecticides may be used as sprays or fumigants (see p. 163) but HCH is not particularly effective. This seems to be a case of natural resistance or tolerance rather than acquired resistance through over-use of insecticides.

TULIP BULB APHID *Dysaphis tulipae*

Symptoms. Light brown or yellow aphids infest bulbs and corms of tulips, gladioli, irises, crocuses and some other plants. Colonies establish on young shoots and may check growth. Mainly a pest on stored bulbs and corms in winter and early spring but may continue to affect growth after planting.

Biology. Breeds parthenogenetically throughout the year and survives the winter either underground or indoors.

Treatment. Easily controlled by spraying or fumigating as soon as aphids are noticed (see p. 163). Small quantities of bulbs or corms can be treated by enclosing them with dichlorvos strips in polythene bags for 1–2 hours.

HONEYSUCKLE APHID *Hyadaphis foeniculi*

Symptoms. Colonies of dark blue-green aphids with long black inflated siphunculi infest the undersides of honeysuckle leaves in spring and summer. Severe infestations cause extensive fouling of foliage with sticky honeydew and sooty moulds and check new growth, sometimes preventing normal development of the flowers.

Biology. Colonies build up on honeysuckle early in the year and die out during the summer when winged females migrate to umbelliferous wild plants, especially hemlock.

Treatment. Spray thoroughly as soon as aphids are seen. Systemics or pirimicarb give best control and systemics may be applied as a soil drench if overall spraying is not possible (see p. 163).

VIBURNUM APHIDS *Aphis viburni* and *Ceruaphis eriophori*

Symptoms. Young leaves of *Viburnum carlessi*, *Viburnum opulus* and some other species curl tightly early in the season, usually from April onwards, and remain crumpled and unsightly for the rest of the year. Examination of curled leaves in April and May should reveal a few brownish black aphids but these disappear later in the season.

Biology. Eggs overwinter on twigs and branches and hatch as buds open in March and April. Young aphids feed on leaves and colonies persist to some extent until mid-summer.

Treatment. Tar oil applied in late December gives best control by killing overwintering eggs. Systemic or non-systemic insecticides may be used later but must be applied as soon as the buds have opened. Any delay may result in leaves being distorted before the aphids are killed.

ELDER APHID *Aphis sambuci*

Symptoms. Dense colonies of black aphids develop on shoots of elder, *(Sambucus nigra)*, and on some ornamental forms of *Sambucus*, in June and July. Usually attended by ants.

Biology. Eggs overwinter on *Sambucus* and colonies build up until July when winged females migrate to the roots of various wild plants, especially Umbelliferae and docks.

Treatment. Colonies are usually fully exposed and are easily killed by spraying with systemics or non-systemics (see p. 163).

PRIVET APHID *Myzus ligustri*

Symptoms. Colonies develop on young privet leaves in spring and early summer. Affected leaves roll up tightly and, if opened carefully, aphids may be found inside. Extensive infestations make hedges look unsightly and persistent infestations weaken growth and may cause premature leaf-fall.

Biology. This species has not been studied in detail although it is often common in suburban areas. Presumably it overwinters on privet and populations build up on young leaves. Hedge clipping probably encourages this pest by stimulating new growth.

Treatment. Spray thoroughly with a systemic or non-systemic insecticide in spring, if necessary (see p. 163).

VIOLET APHID *Myzus ornatus*

Symptoms. Colonies of small green or yellow aphids with paired dark markings on the thoracic and abdominal segments infest violets, violas and some other plants.

Biology. Similar to the peach-potato aphid (p. 167), which is closely related, but the violet aphid does not go through a sexual generation and breeding is entirely parthenogenetic.

Treatment. Spray thoroughly with systemic or non-systemic insecticides as soon as aphids are seen (see p. 163).

WATER-LILY APHID *Rhopalosiphum nymphaeae*

Symptoms. Colonies of dark green or brown aphids infest flower buds and leaves of water-lilies and of some other aquatic plants in summer, causing distortion and discolouration.

Biology. This species overwinters on *Prunus*, especially on plum and blackthorn, and winged females migrate to water-lilies and other aquatic secondary hosts in the summer.

Treatment. Insecticides should only be applied to aquatic plants if there is no danger of harming fish and other aquatic animals. Use non-persistent insecticides, such as nicotine or malathion, or remove aphids from affected plants by hosing with water.

LIME LEAF APHID *Eucallipterus tiliae*

Symptoms. Distinctive yellow or yellow-green aphids with attractive dark markings infest leaves of lime trees in spring and summer, often in sufficient numbers to cause extensive fouling of foliage with honeydew and sooty moulds. Copious honeydew also falls onto plants, garden furniture and cars under infested trees and may be a great nuisance.

Biology. Restricted to lime trees. Eggs laid in autumn overwinter and hatch in spring. Colonies increase and winged aphids disperse in spring and early summer and breeding slows down as summer temperatures rise.

Treatment. As for sycamore aphid (p. 180).

BIRCH APHIDS

Symptoms. Various aphid species infest buds and undersides of leaves, making foliage sticky and encouraging growths of sooty moulds. Sticky honeydew falling onto plants, garden furniture and vehicles under infested trees causes similar problems to lime aphid (see p. 179).
Biology. Most species overwinter as eggs on twigs and branches. They hatch from March onwards and colonies increase and persist throughout the spring and summer.
Treatment. Apply tar oil in late December or early January to kill overwintering eggs, and use systemic or non-systemic insecticides in spring and summer, if practicable (see p. 163).

SYCAMORE APHID   *Drepanosiphum platanoidis*

Symptoms. Large green aphids with long swollen siphunculi infest buds and undersides of leaves of sycamores and other species of *Acer* in spring and summer. Large quantities of honeydew are produced and foliage is extensively fouled, becoming shiny and sticky in dry weather and with secondary development of sooty moulds. Plants, garden furniture and cars under infested trees are affected.
Biology. Eggs overwinter on trees and hatch from March onwards. Populations increase until July or August and during this period all adults are winged and viviparous. In autumn wingless sexual forms develop and females lay overwintering eggs in November and December.
Treatment. It is usually quite impossible to control this aphid on large trees but small trees can be protected by applying a tar oil winter wash in early January to kill overwintering eggs. Aphids appearing later in the year may be controlled by spraying with systemic or non-systemic insecticides (see p. 163), if feasible.

BEECH APHID   *Phyllaphis fagi*   Pl. 3

Symptoms. Colonies of yellow or green aphids, covered by tufts of white woolly wax, infest the undersides of beech leaves from April onwards. Foliage is fouled with sticky honeydew and sooty moulds and growth may be checked. This pest is particularly troublesome on beech hedges.
Biology. Restricted to beeches. Eggs laid in autumn overwinter on twigs and hatch in April. Young aphids feed on leaves and colonies increase during late spring and early summer. Winged aphids disperse to other beeches and colonies decline in late summer. Sexual forms develop in autumn and lay overwintering eggs.
Treatment. Apply tar-oil in December or early January and use systemics or non-systemics in spring and early summer, if feasible (see p. 163).

LARGE WILLOW APHID   *Tuberolachnus salignus*   Pl. 3

Many different aphid species feed on leaves and shoots of willows but this species is particularly conspicuous because of its size.
Symptoms. Colonies of large aphids, 4–5 mm long, dark brown and covered with fine grey hairs, feed on woody stems and branches of willows. Infestations first appear in late June and populations increase until autumn,

often causing extensive fouling of foliage with honeydew and sooty moulds. Honeydew attracts flies and wasps.

Biology. The seasonal biology of this conspicuous and common aphid is not fully known. Colonies die out in winter and no sexual forms or over-wintering eggs are produced. Small numbers of parthenogenetic females may survive on willows in protected situations and winged migrants may move north from warmer southern areas in early summer.

Treatment. Spray colonies and infested plants with systemic or non-systemic insecticides as soon as they are detected (see p. 163).

GREEN SPRUCE APHID   *Elatobium abietinum*

Symptoms. Small green aphids with distinctive bright red eyes feed on leaves of spruces. At first feeding causes localised yellow discolouration of the leaves but this yellow mottling may be followed by general discolour-ation and premature shedding of leaves. This is most likely on Sitka spruce, *Picea sitchensis*, and on blue spruce, *Picea pungens* 'Glauca', but most species of *Picea* are attacked. Symptoms develop in early spring, especially after mild winters, and infestations tend to be worst on trees that are waterlogged, dry, exposed, or growing in other unfavourable con-ditions.

Biology. This species is active on spruces throughout the year. Wingless females overwinter on the undersides of leaves and can be found quite easily in winter and early spring. They continue to reproduce in mild winters and even after severe winters colonies may start to increase as early as March. Peak populations are usually reached in summer and winged aphids spread infestations.

Treatment. Spray thoroughly with malathion in March–April or at other times when the pest is active.

FERN APHID   *Idiopterus nephrolepidis*

Symptoms. Small dark green or black aphids with whitish antennae and legs and white tips to the siphunculi infest leaves of ferns growing in glasshouses and houses. Infested fronds curl and may be killed. Mainly attacks *Adiantum* and other ferns, especially *Nephrolepis*, *Pteris* and *Poly-podium*, but is also occasionally found on cyclamen and violets.

Biology. Reproduction is entirely parthenogenetic and winged aphids disperse to new host plants when colonies are overcrowded, which is generally in spring and summer.

Treatment. Ferns are easily damaged by insecticides and should only be sprayed when really necessary. Use nicotine, bioresmethrin or pirimicarb.

ROOT APHIDS   Pl. 3

Symptoms. Colonies of white or light-coloured aphids, sometimes cov-ered in white powdery wax, infest roots of some glasshouse, house and outdoor plants (Pl. 3). Severe infestations check growth and plants may wilt. Pot plants are often attacked and root aphid colonies may be dis-covered when plants are being re-potted. Lettuces, auriculas, primulas, car-nations and pinks are especially susceptible but many other plants may be attacked.

Biology. Various aphid species attack plant roots. Sometimes they do this only during one phase of their annual migratory cycle, as in the cases of the lettuce root aphid (p. 173) and the elder aphid (p. 178), but other species, notably the artichoke tuber aphid, *Trama troglodytes*, live on roots continuously. They reproduce parthenogenetically and winged asexual females disperse to new host plants.

Treatment. Treat infested plants with a spray-strength solution of malathion or nicotine applied as a soil-drench in sufficient quantity to permeate the roots. Pot plants may be treated by washing the roots and then dipping them in spray-strength malathion or nicotine before re-potting. Diazinon granules worked into the soil around affected plants may also check infestations.

ADELGIDS
(Plate 4)

Small sap-feeding insects (Hemiptera: Adelgidae), about 1–2 mm long, usually dark brown or black but covered by white wax wool. Closely related to aphids but with shorter antennae, fewer wing veins, and no siphunculi. Restricted to conifers, with many species widely distributed in Northern Hemisphere. Some species cause galls and the biology of many species is complex.

SPRUCE GALL ADELGIDS
Adelges abietis, Adelges viridis **Pl. 4**

Symptoms. Small galls, resembling miniature pineapples, develop at tips of branches of Norway spruce, *(Picea abies)*, and of some other species of *Picea* (Pl. 4). Galls first appear in spring but old galls may persist for years. Galls interfere with normal growth and spoil the shape of young trees. Damage tends to be worst on light soils in cold, dry districts and is particularly severe on Christmas trees.

Biology. Young nymphs of *Adelges abietis* overwinter on spruce trees and mature in early spring. Females then lay batches of about 50 eggs, covered by tufts of white woolly wax and young adelgids hatch within a few weeks. From May onwards they feed at the bases of young leaves at the tips of young shoots. The leaf bases then become abnormally enlarged and form the pineapple-like galls which house the adelgid colonies throughout the summer (Pl. 4). In August and September the galls break open and winged adelgids emerge. Some of these may migrate to other *Picea* trees but there is no migration to other host genera. The biology of *Adelges viridis* is similar but in autumn the winged forms migrate to larches. Wingless nymphs overwinter on larches and in early spring winged forms migrate to spruces.

Treatment. Protect young trees by spraying thoroughly with HCH or malathion during mild weather between November and April to kill off overwintering females before they lay eggs. Remove and burn any galls developing during the summer.

LARCH ADELGIDS *Adelges laricis, Adelges viridis*
and other species
Symptoms. Leaves and shoots are infested by wingless adelgids covered
in tufts of white woolly wax and foliage may be contaminated with sticky
honeydew and sooty moulds. Most conspicuous in summer but adelgids are
also present on leaves and on bark during winter. Severe summer infestations develop on European larch, *(Larix decidua)*, and other species
of *Larix* may also be affected. Adelgids are sometimes associated with dieback but may not be the cause of it.
Biology. Young nymphs of *Adelges laricis* overwinter near buds and
mature in April. Females then lay clusters of small grey or orange eggs from
which nymphs soon hatch to feed on young leaves. Colonies increase
throughout the summer and some winged adelgids may migrate onto
spruces. *Adelges viridis* overwinters on larch but does not cause appreciable
summer infestations.
Treatment. Spray thoroughly with HCH or malathion as soon as infestations are apparent in early spring. Treat summer infestations by
spraying thoroughly with malathion, repeating applications at about
weekly intervals, if necessary.

SILVER FIR ADELGIDS *Adelges nordmannianae,*
Adelges piceae
Symptoms. Colonies of adelgids develop on young shoots, on leaves and
on stems and buds. *Adelges nordmannianae* is a serious pest of common
silver fir, *(Abies alba)*, and may kill young trees. It also attacks Caucasian
fir, *(Abies nordmanniana)*, and some other species of *Abies*. Colonies form
on young shoots and feed on leaves, causing distortion and die-back.
   Adelges piceae, which is known in North America as the balsam woolly
aphid, infests stems and shoots of giant fir *(Abies grandis)*, noble fir *(Abies
procera)* and common silver fir *(Abies alba)*. Infested shoots are stunted
and distorted, sometimes with swellings on woody stems and branches.
Biology. Young nymphs of *Adelges nordmannianae* hibernate at the bases
of leaves on trees. They start to feed in March and mature females lay eggs
on shoots from early April. Eggs hatch as buds open in April/May and
young adelgids then feed on new leaves. During summer some winged
adelgids leave *Abies* and may occasionally establish colonies on *Picea orientalis*, which is an alternative host.
   Adelges piceae also overwinters as nymphs and breeds during spring and
summer. It seldom produces winged forms in Britain and northern Europe
and is only found on species of *Abies*.
Treatment. Spray infested trees with HCH in March/April. Repeat applications two or three times at about fortnightly intervals if infestations are
severe.

DOUGLAS FIR ADELGID *Adelges cooleyi*   **Pl. 4**
Symptoms. Small tufts of white woolly wax appear on undersides of
leaves and branches of Douglas fir, *(Pseudotsuga menziesii)* during summer
(Pl. 4). The tufts conceal small, dark, wingless adelgids. Affected foliage
yellows and is fouled by sooty moulds. Growth of young trees may be
checked by persistent infestations but older trees are not greatly affected.

Biology. Adelgid nymphs overwinter on leaves of Douglas fir and mature in spring. Females lay eggs from which young adelgids hatch to feed on new leaves as they emerge from the buds. Winged forms are produced from June onwards and these migrate to establish new colonies on Sitka spruce, *(Picea sitchensis)*, which is the alternative host, but wingless adelgids continue to feed and breed on Douglas fir throughout the summer.

Treatment. *Adelges cooleyi* is not a serious pest. If necessary, it can be controlled on young trees by spraying thoroughly with HCH or malathion on two or three occasions in April/May.

SCOTS PINE ADELGID *Pineus pini*

Symptoms. Adelgid colonies infest leaves, shoots and bark of Scots pine, *(Pinus sylvestris)*, throughout the year. Often conspicuous tufts of white woolly wax develop. Young trees are most susceptible. Foliage may be discoloured but damage is slight. Other pines are sometimes infested and Weymouth pine, *(Pinus strobus)*, is attacked by a related adelgid, *Pineus strobi*.

Biology. Adelgid nymphs overwinter on pine trees and mature in March. Eggs are laid on stems in March/April and adelgids hatching from them crawl onto new leaves to feed. Some winged forms are produced during the summer and fly off to colonise spruces but wingless adelgids continue to feed and breed on pines.

Treatment. This pest causes little damage but does make trees look unsightly. Infestations on small trees may be treated by spraying thoroughly with HCH or malathion on two or three occasions in March/April or at other times when the pest is seen to be active.

PHYLLOXERIDS

Small or very small insects (Hemiptera: Phylloxeridae), up to 3–4 mm long. Similar to aphids but lacking siphunculi and adults have wings folding flat over body. Life-cycles are complex, with many different forms of each species.

VINE PHYLLOXERA *Daktulosphaira vitifoliae* *(= Viteus vitifoliae)*

A north American species introduced into Europe about 1863. It ravaged European vineyards in the second half of the nineteenth century but was eventually checked by grafting all commercial European grape vines onto resistant rootstocks derived from north American vine species and hybrids. Still present in western Europe but not now a serious pest. Very rare in Britain.

Symptoms. European grape vines growing on their own roots show marked red discolouration of leaves, followed by rapid collapse and death. Colonies of small yellow phylloxerids are generally present on roots of affected plants. Leaf galls may develop on some vines.

Biology. In the full life-cycle on American vines five different types of egg and a dozen different forms of individual phylloxerid develop on roots and

leaves. On European vines the life-cycle is restricted and reproduction is entirely parthenogenetic and subterranean.
Treatment. Notify local agricultural or horticultural officials if suspected phylloxerid infestations develop on grape vines.

OAK PHYLLOXERIDS *Phylloxera glabra* and other species
Symptoms. Yellow and brown spots appear on oak leaves in summer and small yellow phylloxerids and eggs can usually be seen on the undersides of affected leaves. Severe infestations cause extensive browning of foliage and premature leaf-fall.
Biology. Oak phylloxerids probably overwinter mainly as eggs in crevices in bark. Females appear in spring and colonise young leaves. Asexual reproduction then continues until autumn when very small sexual forms develop.
Treatment. Phylloxerids have little effect on older trees but repeated infestations may weaken young trees. They can be protected, if necessary, by spraying, preferably with a systemic insecticide, such as dimethoate or formothion. This is best done in spring to kill off initial colonies before they become established. Contact insecticides, such as HCH, malathion or pirimicarb may also be effective if applied thoroughly.

SCALE INSECTS
(Plate 4)

Small sap-feeding insects (Hemiptera: Coccoidea). Nymphs and mature females produce characteristic white, yellow or brown wax scales up to about 5 mm long covering their relatively featureless bodies. Scales generally remain static on plants and feed by inserting fine stylets into plant tissues but may occasionally move to new feeding sites. Males are 1–2 mm long, with a single pair of wings but are absent or rare in parthenogenetic species. Many species of scale insect are common and widespread in Britain and northern Europe. They are especially troublesome in glasshouses, where introduced tropical and sub-tropical species are widely established, but many indigenous European species attack outdoor plants. Females usually lay hundreds of eggs under wax scales, under coverings of woolly wax or under their bodies. Young nymphs, known as crawlers, hatch some weeks or months later and disperse over plants before settling to feed.
Symptoms. Colonies of nymphs and mature females infest leaves, stems or fruits (Pl. 4). All stages feed on sap and most species excrete honeydew, which makes plants sticky and encourages growth of sooty moulds (p. 303). The extent of this contamination often seems out of all proportion to the number of scales present and a few young nymphs on the undersides of the upper leaves can cause extensive fouling of the older foliage. Persistent infestations weaken growth as well as making plants unsightly. Some fruits are attacked, especially peaches, nectarines and grape vines, but scale insects are mainly pests of ornamental plants. Vegetables are seldom affected.

The most widespread and troublesome pest species are noted below and details of treatment are given on p. 188.

SOFT SCALE   *Coccus hesperidum*   **Pl. 4**
Scales are elongate oval and flat, up to 5 mm long. Colour and shape varies
but the outer edge of the scale is usually light brown and the centre is
darker. Widespread and troublesome on glasshouse and house plants and
also infests outdoor plants, especially in warmer southern areas and in
warm situations against walls and buildings. Scales are usually on undersides of leaves, lying alongside veins. Main ornamental plants attacked are
abutilons, bay laurel *(Laurus nobilis)*, camellias, citrus, clematis, escallonias, ferns, ficus, geraniums, hibiscus, hippeastrums, hollies, ivies, oleanders, poinsettias and stephanotis but many others may also be infested.
Females live about three months and produce about 1,000 live nymphs
(crawlers) which crawl over plants for 2–3 days before settling to feed. At
18–25°C development to maturity takes about two months. Breeding is
almost entirely parthenogenetic and continues throughout the year if conditions are favourable.

HEMISPHERICAL SCALE   *Saissetia coffeae*   **Pl. 4**
Scales are almost perfectly hemispherical up to 3 mm long and dark brown
when mature. Size and shape is influenced by host plants and may be
markedly restricted on hosts with narrow stems and leaves. This species is
common in heated glasshouses, especially at higher temperatures. It often
attacks asparagus ferns, begonias, carnations, clerodendrums, codiaeums,
cycads, ferns, ficus, oleanders, orchids and stephanotis but may also infest
many other ornamental plants. Breeding is almost entirely parthenogenetic
and continues throughout the year if temperatures are high enough. Each
female lays up to 2,000 eggs under the dome-shaped scale and then dies.
Crawlers hatch some weeks or months later and disperse over plants before
settling to feed.

BROWN SCALE   *Parthenolecanium corni*
Also known as peach scale as it is an important pest of peaches and
nectarines under glass and outdoors. Similar to hemispherical scale but
normally occurs in unheated glasshouses and on outdoor plants. Colonies
develop on woody stems as well as on shoots and leaves and mature female
scales may be more than 5 mm long. Peaches, nectarines, currants and
grape vines are the main fruits affected but this pest has been recorded from
more than 300 different ornamental plants, including acacias, brooms,
ceanothus, chaenomeles, clematis, cotoneasters, flowering currants
(Ribes), escallonias, honeysuckle, laurel, magnolias, oleander, ornamental
cherries, pyracanthas, roses, spiraeas and wistarias. It also attacks hazel and
may affect nut production. Males are very rare and females reproduce
parthenogenetically, laying up to 2,000 eggs each under the scale in
May–July. Eggs hatch within a month and crawlers disperse and settle to
feed on undersides of leaves. In autumn the small nymphs move from
leaves to stems and branches where they hibernate. They feed again in
spring and mature from April to July. There is only one generation a year
and this species is therefore easier to control than species that breed
continuously.
    A closely related species, *Parthenolecanium pomeranicum*, attacks yew.

OLEANDER SCALE *Aspidiotus nerii*
Scales look like miniature fried eggs up to 2 mm diameter. Colonies develop on undersides of leaves of many glasshouse and house plants, especially on acacias, ornamental asparagus, aucubas, azaleas, pot cyclamen, dracaenas, ericas, fatsias, ficus, laurels, oleanders and palms. Occurs outdoors in warmer southerly areas.

WOOLLY CURRANT SCALE *Pulvinaria ribesiae*
WOOLLY VINE SCALE *Pulvinaria vitis* **Pl. 4**
Scales feed mainly on stems and females produce conspicuous egg sacs of white wax wool in May and June. Currants and grape vines are the main fruits attacked and some ornamental plants, especially alder, birch, hawthorn, cotoneasters and willows, are also susceptible. Generally only attacks outdoor plants. Breeding is parthenogenetic with a single generation a year.

HORSE-CHESTNUT SCALE *Pulvinaria regalis*
Closely related to the woolly currant scale but only established in southern England and north-west France since about 1960. Dense colonies form on branches and trunks of horse chestnut trees and on limes, elms, sycamores, magnolias, maples, cornus and some other ornamental plants. Little apparent damage is done, despite the presence of large colonies. Eggs hatch in June–July and nymphs feed on leaves until autumn when they move onto branches and trunks to overwinter. Females mature in spring and reproduce parthenogenetically.

CUSHION SCALE *Chloropulvinaria floccifera*
Young scales infesting undersides of leaves of camellias, orchids and some other plants look very like soft scale but in April and May females produce conspicuous white wax egg sacs up to 10–15 mm long. Usually one generation a year but may be more in warm glasshouses and in warmer southern areas.

BEECH SCALE *Cryptococcus fagisuga*
(= *Cryptococcus fagi*) **Pl. 4**
Mature females are only about 1 mm long but colonies containing females, eggs and nymphs form conspicuous areas of white wax powder on trunks and branches of beech trees. Reproduction is parthenogenetic and most eggs are laid between June and October. Eggs overwinter and colonies may persist on the same tree for many years. This pest predisposes trees to infection by fungi of the genus *Nectria* which cause a serious and widespread condition that is known as beech bark disease (p. 311).

MUSSEL SCALE *Lepidosaphes ulmi* **Pl. 4**
Small mussel-shaped brown scales, up to 3 mm long, form dense encrusting colonies on woody stems and branches of ceanothus, cotoneasters, ornamental and fruiting apples and pears, roses and various other trees and shrubs. Infestations are most severe on old plants lacking vigour and the scales are often so densely packed that they look like a normal feature of the

bark. Eggs laid under the scales in late summer overwinter and hatch in the following May–June.

A closely related species, *Lepidosaphes machili*, infests stems and leaves of cymbidiums in glasshouses.

SCURFY SCALE *Aulacaspis rosae*
Round, flat, opaque white scales, up to 2–3 mm across, form dense colonies on woody stems of certain species of rose, both outdoors and under glass. Males are present in this species and the male scales are elongate white fluted scales which are smaller than the females. Females lay eggs in July–August. Nymphs soon hatch and settle to feed before hibernating.

WILLOW SCALE *Chionaspis salicis*
Similar to scurfy scale but female scales are whiter and more irregular, often covering woody stems so densely that they look as if they have been whitewashed. Attacks alder, ash, brooms, ceanothus, elms, lilacs, privet, willows and some other ornamental shrubs and trees.

EUONYMUS SCALE *Unaspis euonymi*
Female scales are similar to mussel scales, 1–2 mm long, and form dense colonies on woody stems of *Euonymus japonica*. Male scales are white, elongate, with three longitudinal ridges, and mainly develop on the undersides of leaves. Only of local importance in southern England but more widespread in France. *E. japonica* is the only host plant seriously affected although other species of *Euonymus* are sometimes infested.

FLUTED SCALE *Icerya purchasi*
Originally an Australian species that is now widely established throughout sub-tropical and mediterranean areas in the southern and northern hemispheres where it is a serious pest of citrus crops and of other plants. Not an important pest in Britain and northern Europe but it occasionally appears on glasshouse and house plants and its striking appearance causes concern. Female scales are brown, oval, and up to 3 mm across and produce large white egg sacs with longitudinal grooves. Each egg sac may be up to 1 cm long and contains 600 or more eggs. Citrus, acacias, cytisus, mimosas and roses are commonly infested but many other plants may be attacked.

Treatment. Scale insects are relatively immobile pests. Crawlers move about to a limited extent and may occasionally be blown around by wind or transported on flying insects or on birds but most scales found in gardens will have been introduced on plants. All newly acquired plants should therefore be examined very carefully and should be treated, if necessary, before establishing them in the garden, glasshouse or house.

Some glasshouse and house plants can be treated quite effectively by wiping scales off leaves and stems with a soft rag, sponge, or brush dipped into soapy water. Aspidistras, ornamental citrus, codiaeums, dracaenas, ficus, ivies, orchids, sansevierias and other plants with relatively robust leaves may be treated in this way.

Chemical treatments are most effective when applied to kill young crawlers before they have settled and started to form their protective scale. They

are easily killed by thorough application of non-persistent contact insecticides, such as bioresmethrin, diazinon, malathion or nicotine, but
timing is critical. The main period of crawler activity on outdoor plants is in
late spring and early summer but in heated glasshouses and in houses
crawlers may hatch at almost any time. By using a magnifying glass or hand
lens crawlers can be seen quite easily. Infested plants should be treated with
one of the above insecticides when crawlers are active and, since some of
them will escape treatment, a second application of insecticide should be
made about two weeks after the first.

If insecticide treatment cannot be timed to coincide with egg hatch and
crawler activity, systemic insecticides, such as dimethoate or formothion,
may be used to kill maturing females but these chemicals may have harmful
phytotoxic effects on some ornamental plants (see p. 30).

Fully dormant deciduous woody plants may be sprayed or brushed with
tar oil in December or January to kill overwintering scales and eggs and this
is a particularly useful treatment against scale insects on grape vines,
currants and apples and against beech scale on beech tree trunks and
branches. Tar oil must be applied carefully to avoid damage to other garden
plants (see p. 40).

MEALYBUGS
(Plate 4)

Small sap-feeding insects (Hemiptera: Pseudococcidae), up to about 4 mm
long. Males winged but relatively uncommon. Females wingless, with soft,
rounded bodies covered by white wax powder and filaments (Pl. 4). Many
different species on wild and cultivated plants but usually most troublesome on glasshouse and house plants.

GLASSHOUSE MEALYBUGS   Pl. 4
Pseudococcus obscurus is the commonest species in glasshouses in Britain.
Planococcus citri, *Pseudococcus longispinus* and *Pseudococcus calceolariae* are
also common and many other species may occur locally. The four main pest
species are tropical and sub-tropical and have been accidentally introduced
to glasshouses in Britain and northern Europe. They are restricted to
glasshouse plants in colder northern climates but will infest outdoor plants
in the warmer areas of southern Britain and Europe. All species are very
similar in appearance and biology and in the effects that they have on
garden plants.
Symptoms. Mealybug colonies (Pl. 4) develop on leaves, stems, buds,
flowers, fruits and other aerial parts of plants. Roots are sometimes infested
by root mealybugs (see below). Colonies contain young and mature mealybugs clustered together and protected by white wax powder and filaments,
which often completely cover the insects, especially when they congregate
in leaf axils, in the necks of bulbs and on the spines of cacti. Persistent
infestations weaken plants, especially when growing points are attacked,
and foul plant surfaces with honeydew and sooty moulds. Many different
host plants are attacked and the following are particularly susceptible:
vines, currants, sprouting potatoes, abutilons, anthuriums, asparagus fern,

begonias, cacti, ceanothus, chrysanthemums, codiaeums, coleus, crassulas, dracaenas, eucharis, ferns, ficus, fuchsias, gardenias, hippeastrums, hoyas, jasmines, laburnums, musas, oleanders, orchids, pelargoniums, robinias, saintpaulias and palms.

Biology. Each female lays about five batches of 100–150 eggs and each batch is protected by a covering of woolly wax. At relatively high temperatures, eggs hatch in a few days and young mealybug nymphs then crawl over plants for a few hours before settling to feed. They are immobile for long periods when feeding but can withdraw their feeding stylets from plant tissues to move during their development. At 28°C (82°F) development from egg to adult takes about a month. Breeding can continue throughout the year in heated glasshouses and in houses but most mealybugs mature in summer and populations are generally highest in autumn and early winter. Winged males develop from delicate white cocoons and may appear in large numbers at times but females can reproduce parthenogenetically when males are not present and some species are almost entirely parthenogenetic.

Treatment. Mealybugs are often difficult to control as they tend to live in inaccessible situations on plants and they are well protected from insecticides by their wax coverings. To eradicate established infestations, first remove as many colonies as possible by cutting out and burning infested shoots and branches, by washing plants with powerful jets of water or by removing mealybugs with a household paint brush. This latter treatment can be made more effective by dipping the brush in a spray-strength solution of malathion or nicotine. Methylated or surgical spirit may also be used. Brushing is often the only way of removing mealybugs from cacti and is also a useful way of treating dormant grape vines.

Once the main colonies have been cleared up, spray plants with malathion, nicotine, or one of the systemic insecticides, such as dimethoate or formothion. The systemics may also be applied as soil drenches to be taken up by the roots and translocated within the plant to the sites where mealybugs are feeding. This is often the best way to control infestations on cacti, succulents and bulbs but may cause phytotoxic side-effects on some plants (see p. 30).

Avoid introducing mealybugs into clean collections of plants by examining all new introductions carefully and by keeping them in quarantine for a month or so, if possible. Treat if any sign of mealybug infestation is observed and repeat the treatment on two or three occasions at about fortnightly intervals to make sure that the pests are eradicated.

Since most glasshouse mealybugs have been introduced from tropical and sub-tropical countries, and since they are relatively immobile pests, it should be possible to control them by introducing parasites and predators into the glasshouse environment. This type of biological control has been used successfully against mealybug pests of citrus and other crops in the sub-tropics but there has not been any sustained attempt to develop biological control of mealybugs in glasshouses in Britain and Europe.

ROOT MEALYBUGS Pl. 4

Very similar to glasshouse mealybugs described above but they mainly feed on roots. A number of species of the genus *Rhizoecus* occur in Britain and northern Europe. *Rhizoecus falcifer* is the most widespread and trouble-some pest species on glasshouse and house plants and other species affect outdoor garden plants.

Symptoms. Colonies of root mealybugs develop on roots (Pl. 4) and on bases of stems of many different plants. Root mealybugs may be confused with root aphids (p. 181) and with springtails (p. 146). Severe root mealy-bug infestations interfere with normal root function and plants do not grow well and may wilt, especially if they are growing in pots. Damage is generally worst when plants are growing in dry potting composts or in dry borders. Abutilons, acacias, cacti and succulents, cassias, correas, dianthus, dracaenas, gardenias, grevilleas, olearias, pelargoniums, saintpaulias and stephanotis are particularly susceptible.

Biology. The biology of root mealybugs, so far as it is known, resembles that of the glasshouse mealybugs (see above).

Treatment. Water plants regularly to keep the root zone moist and so limit root mealybug populations by creating unfavourable conditions. Drench infested pot plants with a spray-strength solution of a contact insecticide, such as malathion or nicotine, or a systemic insecticide, such as dimethoate or formothion, but take care to avoid phytotoxic effects (p. 30). If root mealybugs are discovered when re-potting plants, wash the infested roots in soapy water or in a spray-strength solution of weak insecticide. Then re-pot in sterile compost.

THRIPS
(Plate 4)

Small, elongate, cylindrical insects (Thysanoptera), up to 3–4 mm long. Adults, commonly known as 'thunder flies', have two pairs of very nar-row wings fringed with long, fine hairs. Colour varies with species from white through shades of yellow to brown and black. Larvae resemble adults but lack wings and are generally lighter coloured. More than 150 species occur on plants in Britain and northern Europe. Many species are abundant and widespread and some are cosmopolitan. Larvae and adults of most species feed on plants, especially on flowers and buds, but some prey on mites, aphids and other small invertebrates. Mouth-parts of larvae and adults are adapted to pierce plant or animal tissues superficially and to extract cell contents. Direct damage to garden plants is caused by large numbers of thrips feeding on leaves, flowers and buds, producing charac-teristic light mottling and silvering (Pl. 4) and some distortion. Some species may cause indirect damage by transmitting tomato spotted wilt virus disease although this is now less serious than it once was.

Females live for about a month and lay 60–80 eggs which are either inserted into plant tissues with a saw-like ovipositor or are simply deposited on plants. Eggs soon hatch at summer temperatures and larvae feed for a few weeks before passing through quiescent, non-feeding prepupal and pupal stages. Winged adults emerge about a week after pupation and are often carried long distances in thermals and air currents before settling on plants. Development from egg to adult takes about a month at 20°C (68°F) but lasts much longer at lower temperatures. Development is usually sexual but some species are parthenogenetic. Adults and immature stages over-winter in soil, leaf litter and similar situations.

The main species attacking garden plants are noted below and general methods of treatment are summarised on p. 193.

ONION THRIPS   *Thrips tabaci*
A cosmopolitan species that attacks many different plants. Mainly a pest of glasshouse and house plants in Britain and northern Europe but also damages some outdoor plants.

Symptoms. Characteristic fine, light flecking and silvering of leaves, buds and flowers, with numerous yellow or brown larvae and adults usually present. Onions, brassicas and tomatoes are the most susceptible vegetables and begonias, carnations, chrysanthemums, cinerarias, cyclamen, dahlias, gerberas, gloxinias, orchids, pelargoniums, sweet peas and zantedeschias are often attacked. Transmits tomato spotted wilt virus.

Biology. Males are rare and breeding is almost entirely parthenogenetic. Females lay eggs in plant tissues and larvae feed for 2–3 weeks before pupating in the soil. There are usually only two adult generations a year outdoors but breeding may continue during most of the year in glasshouses and other protected situations.

Treatment. See p. 193.

PEA THRIPS   *Kakothrips pisivorus* (= *Kakothrips robustus*)   **Pl. 4**
Symptoms. Flowers, leaves and pods of peas and broad beans misshapen, discoloured, and usually with characteristic silver sheen. Yellow larvae and yellow-brown adults are present on affected plants in May–July, especially during hot, dry weather.

Biology. Adults emerge from pupae in soil during late spring and early summer and fly up onto plants. Females lay eggs in flower stamens and larvae hatch after a week or so, feed for 2–3 weeks and then descend to the soil to overwinter. There is only one adult generation a year and popu-lations reach a peak between mid-June and the end of July.

Treatment. See p. 193.

GLADIOLUS THRIPS
Thrips simplex (= *Taeniothrips simplex*)   **Pl. 4**
A serious pest of gladioli, first introduced into Britain about 1950 and now well established.

Symptoms. Leaves and flowers of gladioli and related plants show charac-teristic light flecking where thrips have fed. Yellow larvae and darker adults are usually present, especially under leaf bases, in flowers and in flower buds. Severe attacks ruin flowers which wilt, brown and die (Pl. 4). Most severe attacks develop on growing plants during July and August but corms may be attacked in store during winter and early spring by thrips feeding on the scales.

Biology. Thrips overwinter on corms in store and some may also survive mild winters outdoors in soil and leaf litter. When temperatures rise above 10°C (50°F), females lay eggs in plant tissues and larvae soon hatch. Breeding continues so long as conditions are favourable and damaging populations build up rapidly in summer.

Treatment. Control is relatively easy and effective since the annual cycle can be broken by dusting corms with gamma HCH before they are stored and before they are planted out in spring. HCH may also be applied to growing plants if thrips are present or a less persistent contact insecticide, such as malathion, or nicotine may be used to check early infestations on susceptible plants.

ROSE THRIPS *Thrips fuscipennis*
Symptoms. Flowers of roses and of some other plants are flecked with numerous light spots and streaks, which later darken and rot. Leaves may also show silver flecking and brown thrips are usually present on plants. Roses grown under glass may be severely damaged by infestations developing early in the season.
Biology. Adults overwinter in cracks in brickwork, woodwork, canes, stakes and similar situations and some thrips may continue to breed through the winter in protected situations. Females lay eggs from February onwards in young petals and other soft tissues and breeding continues until autumn.
Treatment. See below.

CARNATION THRIPS *Taeniothrips atratus*
Produces typical thrips symptoms on carnations in glasshouses and also attacks many other plants outdoors.

GLASSHOUSE THRIPS *Heliothrips haemorrhoidalis*
Generally not as common in glasshouses as the onion thrips (see p. 192). Adults have dark brown bodies with light yellow tips to the abdomen and attack azaleas, citrus, ferns, fuchsias, orchids, zantedeschias and many other plants growing in glasshouses and houses. Infested plants are marked with small red and brown globules of liquid that are deposited by feeding thrips.

PRIVET THRIPS *Dendrothrips ornatus*
Yellow larvae and dark-bodied adults, with alternating dark and light bands on wings attack privet and lilac, causing characteristic silvering of leaves. Established plants are not appreciably damaged but young plants may be checked by severe infestations causing premature leaf-fall.

General Treatment. Severe thrips attacks on glasshouse and house plants are often associated with poor growing conditions resulting from underwatering and overheating. Regular watering and maintenance of a cooler, more humid atmosphere can therefore help to prevent infestations. Similarly, infestations on outdoor plants are usually worst during hot dry periods and thrips populations are reduced in cool, wet weather.

Most species are easily controlled with contact insecticides, if necessary, and HCH, malathion or nicotine are generally effective. These are usually applied as sprays, but dusts or fumigants may be used when appropriate (see p. 29). If insecticides are used, they should be applied as soon as thrips or symptoms are seen, with a repeat application after 2–3 weeks if damage continues.

CATERPILLARS
(Plate 5)

Caterpillars are larvae of moths and butterflies (Lepidoptera) and all have a generally similar structure which resembles that of some sawfly larvae (see p. 224). The head is well developed and has a pair of strong mandibles which are used to bite and chew plant tissues; the thorax is three-segmented, usually with a pair of jointed legs on each segment, and the abdomen consists of ten segments bearing up to five pairs of fleshy, non-jointed legs (prolegs).

Size (5–100 mm) and colour pattern vary with species and more than 2,500 species are recorded from Britain and northern Europe. Caterpillars of most species feed on wild plants but at least fifty species are widespread pests of garden plants.

Biology is varied but relatively uncomplicated. Adults feed only on nectar and other sugary fluids or do not feed at all. Reproduction is normally sexual and, after mating, females lay a few hundred eggs on or near plants, either singly or in batches. Caterpillars generally hatch within a few weeks and feed for 1–2 months before pupating on plants, in soil and plant debris or on fences, buildings and other structures. Adults usually emerge from pupae after a few weeks. They fly during daytime (most butterflies) or at night (most moths) and some species migrate over long distances, often reaching Britain by non-stop flights from northern Africa or central Europe.

Many important families of Lepidoptera have similar habits and are known by their family names (pyralids, tortricids, tineids, noctuids, geometrids, hepialids, pierids, etc.). Larval habitats include all parts of plants (leaves, stems, buds, flowers, fruits and roots) and symptoms are correspondingly varied. Most caterpillars feed openly on leaves, buds and shoots but some produce silk webbing which they use to draw leaves together into protective tent-like coverings. Some caterpillars live in soil and a few are able to tunnel in woody branches and trunks of trees.

The symptoms and biology of the main pest species are detailed in the following section and for convenience they are grouped under fruits, vegetables and ornamentals. To some extent this is an artificial division since many species commonly attack all of these groups of plants.

General methods of treating caterpillar pests are summarised below and specific recommendations are given under the main pest entries.

Caterpillar species	*Main plants attacked*
FRUITS	
Winter moths (p. 196)	Apples and other fruit trees and many ornamental trees and shrubs
Codling moth (p. 197)	Apples, pears and quinces
Fruit tortrix moths (p. 198)	Apples
Plum fruit moth (p. 198)	Plums and damsons
Magpie moth (p. 198)	Gooseberries, currants and some ornamentals

Raspberry moth (p. 199)	Raspberries
Leopard moth (p. 199)	Apples, plums, pears, cherries and some ornamental trees and shrubs

VEGETABLES

Cutworms (p. 200)	Brassicas, carrots, lettuces, potatoes and many ornamentals
Swift moths (p. 201)	Carrots and other root crops, lettuces and many ornamentals
Cabbage caterpillars (p. 201)	Cabbages and other brassicas
Pea moth (p. 203)	Peas
Silver Y moth (p. 203)	Beans, cabbages, lettuces, potatoes and many ornamentals, especially chrysanthemums
Tomato moth (p. 204)	Tomatoes and various ornamentals, especially carnations and chrysanthemums
Leek moth (p. 204)	Leeks

ORNAMENTALS

Angleshades moth (p. 205)	Chrysanthemums and many other ornamentals
Vapourer moth (p. 205)	Roses and other ornamental trees and shrubs
Buff-tip moth (p. 205)	Roses and other ornamental trees and shrubs
Grey dagger moth (p. 206)	Roses and other ornamental trees and shrubs
Brown-tail moth (p. 206)	Roses and some other ornamentals
Garden tiger moth (p. 206)	Herbaceous ornamentals
Hawk moths (p. 207)	Apples, privet, lilac and fuchsias
Pine looper moth (p. 207)	Pines
Carnation tortrix moth (p. 207)	Carnations and many other glasshouse and house plants
Green oak tortrix (p. 208)	Oaks
Lackey moth (p. 208)	Roses and other ornamental trees and shrubs
Hawthorn webber moth (p. 209)	Hawthorns and cotoneasters
Juniper webber moth (p. 209)	Junipers
Small ermine moths (p. 209)	Hawthorns, euonymus and some other ornamental trees and shrubs
Lilac leaf miner moth (p. 210)	Lilacs and privets
Laburnum leaf miner moth (p. 210)	Laburnums
Azalea leaf miner moth (p. 211)	Azaleas grown as house and glasshouse plants
Goat moth (p. 211)	Willows, oaks and some other trees
Pine shoot moth (p. 211)	Pines

General Treatment. Non-chemical methods of control can be used against some caterpillars, especially when they occur in small numbers on a few plants. The simplest and most effective method is to examine plants regularly and crush eggs and caterpillars whenever they are seen. This technique may seem crude and cruel but it kills eggs and caterpillars more certainly and swiftly than insecticides. Species that overwinter as caterpillars or pupae in the soil or in plant debris may be partly controlled by good hygiene and thorough cultivation during autumn and early spring and species that overwinter as conspicuous egg batches can be restricted by pruning out and destroying the eggs in winter. Grease-banding is a special non-chemical technique that is only effective against the winter moth group and sack-banding is occasionally used against codling moth (see p. 197).

Contact insecticides are widely used against caterpillars, generally as sprays but sometimes as dusts. Their effectiveness is enhanced if they have some residual activity so that they act as internal poisons when ingested on treated plant tissues, but this is not essential. Systemic insecticides, which are so effective against aphids and other sap-feeding insects, are not equally effective against caterpillars although some of them do have sufficient contact activity to be of use.

Non-persistent contact insecticides, such as derris, bioresmethrin, resmethrin or malathion, are quite adequate for treatment of caterpillars feeding in exposed situations on plants and can be used safely on vegetables and fruits within a few days of harvest. More persistent insecticides, such as HCH, permethrin, carbaryl or trichlorphon, may give better control of caterpillars feeding in protected situations on plants and in soil. Further details of these and other insecticides are given on p. 33.

WINTER MOTHS   Pl. 5

Three main species are grouped under this heading: the winter moth, *Operophtera brumata*; the March moth, *Alsophila aescularia*, and the mottled umber moth, *Erannis defoliaria*. All three are widespread and important pests of fruit trees and ornamental plants in Britain and northern Europe.

Symptoms. Leaves, buds and young shoots are eaten during spring and early summer by green, yellow-green or brown 'looper' caterpillars (Pl. 5). Characteristic irregular holes are eaten in leaves, often before they have expanded from the buds, and these symptoms persist throughout the growing season (Pl. 5). Extensive damage to fruit blossom and foliage may affect cropping and leaf damage on ornamental trees, shrubs and hedges makes them look unsightly. Persistent attacks weaken plants and facilitate the development of bacterial and other diseases. Apples, pears, plums and cherries are the main fruit trees attacked and many ornamentals, especially beech, flowering cherries, cotoneasters, crab-apples, dogwoods, elms, hawthorns, hazels, hornbeams, limes, rhododendrons, roses, sycamores and willows, are commonly affected.

Biology. Wingless female moths emerge from pupae in soil during winter or early spring, mate with winged males, and then crawl up trunks and stems of trees and shrubs to lay about 200 eggs each. Winter moths emerge in October–December; mottled umbers emerge in October–March and March moths emerge from January to April. Winter moth and mottled umber females lay eggs singly or in small clusters near buds and in crevices in bark and March moths lay distinctive bands of eggs around twigs. Eggs hatch from March onwards and caterpillars feed on buds and young leaves as they develop. Caterpillars of all three species are similar in structure but differ in colour. Winter moth caterpillars are usually green with three yellow longitudinal stripes; mottled umber caterpillars are dark brown with yellow lateral longitudinal bands (Pl. 5) and March moth caterpillars are green with light stripes. All grow to a length of about 30 mm and most finish feeding from mid-May to the end of June. They then go to the soil to pupate and are often seen at this time of year hanging from trees and shrubs on long silken threads. There is only one adult generation a year and pupae remain dormant in the soil until the next adults emerge in the following winter and spring.

Treatment. Protect valuable fruit and ornamental trees by applying grease bands to the trunks in October to stop females crawling up to lay eggs. Use specially formulated grease (*not* car grease) applied either direct to the bark or onto strips of paper to form a band about 100 mm wide about 1–2 metres above soil level. (Ready-made grease bands can be bought from gardening shops.) Tie bands tightly onto trunks after removing any loose or rough bark and make sure that female moths cannot by-pass the bands by crawling under them or by crawling over them on dead leaves or other plant debris. Inspect regularly and renew if necessary.

Where grease bands cannot be used effectively, as on shrubs and hedges, use insecticides. Tar oil winter washes applied to control aphids (p. 161) will give some protection by killing overwintering eggs but contact insecticides such as trichlorphon, fenitrothion, malathion, permethrin, or derris, applied as soon as buds have opened in spring, will give better control, especially on fruit trees.

CODLING MOTH   *Cydia pomonella*
(= *Laspeyresia pomonella*)   **Pl. 5**
A common and world-wide pest of apples.

Symptoms. Caterpillars eat into flesh of maturing apples, making them inedible (Pl. 5). Pears and some other fruits are occasionally attacked but apples are the main host plants. Tortrix caterpillars (p. 198) and apple sawfly larvae (p. 224) produce similar symptoms but do not tunnel so extensively within the mature fruit.

Biology. Adults are active on warm nights in June/July but are small and inconspicuous. Females lay flat, translucent eggs singly on fruits and leaves and caterpillars hatch from them after about two weeks. They immediately tunnel into developing fruits, often entering through the eye (calyx) so that there is no apparent entry hole. Caterpillars feed within fruits for a month or more, eating down through the flesh to the core, and when they have finished feeding they leave the damaged fruits and spin cocoons under loose bark, under tree ties, and in similar situations. Most caterpillars remain in cocoons until the following season but some pupate in August to produce a second adult generation in September.

Treatment. Chemical control is difficult since timing must be sufficiently accurate to ensure that young caterpillars are killed before they can enter fruits and thorough spraying may only be possible on smaller trees. If possible, spray apple trees with fenitrothion or permethrin after blossoming, first about mid-June with a second application three weeks later.

Sacking or corrugated cardboard tied around branches and tree trunks by mid-July can be used to trap caterpillars seeking sites in which they can form cocoons but is of doubtful value since the fruits will have been damaged before the caterpillars are trapped. Trapping is also unlikely to reduce numbers of egg-laying females in the following season since they can easily fly into gardens from adjacent untreated trees.

FRUIT TORTRIX MOTHS

Three species of tortrix caterpillar attack fruits: the fruitlet mining tortrix, *Pammene rhediella*; the summer fruit tortrix, *Adoxophyes orana*, and the fruit tree tortrix, *Archips podana*. All three produce similar symptoms.

Symptoms. Attacked fruits show various types of superficial scarring, distortion and tunnelling. Ripening apples are commonly affected but fruits of pears, plums and cherries may also be attacked. Green caterpillars, up to 25 mm long, may be found feeding on damaged fruits and on leaves and buds. They spin silk to draw leaves into protective covers and if these are opened carefully caterpillars will characteristically wriggle backwards as they are disturbed.

Biology. Small, inconspicuous adult moths fly at night during the summer. Females lay eggs singly or in small batches on leaves and caterpillars feed for about a month before spinning cocoons and pupating. A second generation of adults emerges in autumn and the young caterpillars hatching from eggs laid by females of this generation hibernate in cocoons on infested plants. They become active again in the following spring and feed on young buds and leaves before pupating.

Treatment. Damage is often slight and tortrix activity is generally too unpredictable for chemical controls to be used effectively. Sprays applied to control winter moths (p. 196) will give some incidental control of tortrix moths and, if tortrix moths are known to be particularly troublesome, a post-blossom fenitrothion spray applied in mid-June, with a second application a fortnight later, will limit damage to fruits.

PLUM FRUIT MOTH   *Cydia funebrana (= Laspeyresia funebrana)*

Symptoms. Small white, pink or red caterpillars, up to 15 mm long, enter plum and damson fruits in July/August, causing rotting and premature ripening. Similar symptoms are produced by the plum sawfly (p. 224).

Biology. This is similar to that of the codling moth (p. 197), which is a closely related species.

Treatment. Collect and destroy affected fruits in July/August before caterpillars leave them.

MAGPIE MOTH   *Abraxas grossulariata*

Symptoms. Black, white and yellow 'looper' caterpillars, up to 40 mm long, eat leaves of many plants in April/June. Gooseberries and currants may be stripped of leaves and other plants, especially apricots, cherries, crab apples, euonymus, hawthorns, hazels, laurels and plums are similarly affected. Gooseberry sawfly (p. 225) produces similar symptoms on gooseberries and currants.

Biology. Adult moths are active at night in July/August and females lay eggs singly or in small batches on undersides of leaves. Caterpillars hatch about a fortnight later and feed on leaves until autumn. The caterpillars, which are still small, overwinter in cracked bark, leaf litter and similar situations and start to feed again in the following spring, from about April onwards. After feeding voraciously for a month or so, conspicuous black and yellow banded pupae develop in thin cocoons attached to leaves and twigs of host plants or to fences, walls and other structures. Adults emerge from these in July/August to complete the cycle.

Treatment. Look for caterpillars on susceptible host plants in early spring, paying special attention to the centres of gooseberry and currant bushes, where attacks often develop. Remove caterpillars by hand or spray with non-persistent contact insecticides (p. 195). Spraying with tar oil (p. 197) may help to limit infestations by killing some of the overwintering caterpillars.

RASPBERRY MOTH *Lampronia rubiella*

Symptoms. Young shoots of raspberries, loganberries and blackberries shrivel and die in April–June and small pink to red caterpillars, up to 15 mm long, are usually present in tunnels in affected shoots and buds. This pest tends to be most troublesome in northern areas, especially Scotland.

Biology. Small, inconspicuous adult moths are active in May/June and females lay eggs in flowers. Caterpillars hatch about a week later and feed in cores of developing fruits. They do not cause any real damage at this stage and as fruits ripen the partly-grown caterpillars leave them and spin cocoons in soil, leaf litter, cracks in canes and similar situations. Caterpillars remain in cocoons until the following growing season and eventually leave them in April/May to tunnel into buds and shoots. They then feed for about a month before pupating either within the galleries that they have eaten out or in suitable crevices on plants or adjoining structures.

Treatment. Reduce numbers of overwintering caterpillars by pruning out old canes, by removing accumulations of leaf litter and other debris and by applying a tar oil winter wash in January (see p. 197).

LEOPARD MOTH *Zeuzera pyrina*   Pl. 5

Symptoms. Terminal leaves on branches of young trees suddenly wilt and die when caterpillars tunnel in wood (Pl. 5). Holes and frass on the surface of branches indicate the presence of caterpillars within. Attacks are local and sporadic but can cause serious damage to young apple, pear and cherry trees and to many ornamental trees and shrubs, especially ash, birch, cotoneasters, elms, hawthorns, horse chestnuts, lilacs, maples, sycamores and rhododendrons.

Biology. Adult moths are large (wing-span up to 60 mm) but fly at night and are seldom seen. Females lay eggs in bark during June/July and caterpillars spend their first autumn and winter feeding just beneath the bark. In the following spring they tunnel into the wood and each caterpillar eats out a gallery up to 20 cm long. Caterpillars feed for up to three years before pupating in galleries and when adults emerge the empty pupal cases are often left protruding from the exit holes.

Treatment. This pest is so sporadic and local that it is difficult to prevent attacks but if symptoms are recognised soon enough it may be possible to save branches by killing caterpillars. First locate holes, clear them with a stick or wire, introduce a fumigant, such as paradichlorbenzene, or an insecticide, such as HCH, and seal the holes with plasticine or putty. Branches that have been seriously damaged will not recover and should be pruned out and burned.

CUTWORMS   Pl. 5

Caterpillars of various noctuid moths live in the upper few centimetres of the soil and feed at the surface, mainly at night. They attack young plants at ground level, often eating right through the stems, and are therefore commonly known as cutworms. Most species are widespread and abundant.

Symptoms. Stems of young plants are eaten at ground level and are often completely severed. Damaged plants wilt and die. Young vegetable plants, especially lettuces and brassicas, are especially susceptible and small numbers of cutworms can cause considerable damage by working along rows. Roots, tubers, corms and leaves are also eaten. Carrots, celery, beet and potatoes are often attacked but strawberries are usually the only fruit plants likely to be damaged. Many ornamental plants are attacked, especially China asters, chrysanthemums, dahlias, marigolds, primulas, zinnias, and young trees and shrubs in nurseries. Plants may be damaged at almost any time of year and attacks are often severe on light soils during dry summers. Drab cutworm caterpillars (Pl. 5) may be found by searching the soil near damaged plants and adult moths are sometimes disturbed from leaf litter and other plant debris.

Biology. Most species have similar life-cycles. Adults emerge from pupae in the soil in June/July and females lay eggs in batches of 30–50 on leaves and stems of cultivated plants and weeds and on dead leaves and other plant debris. Each female lays up to 1,000 eggs over a period of 1–2 weeks and eggs hatch about two weeks after laying. During summer the caterpillars feed for 1–2 months before pupating and a second adult generation emerges in August/September. Caterpillars that develop later in the season over-winter in the soil, feeding whenever the weather is favourable, and finally pupate in the following spring. There is considerable variation in the time taken for development from egg to adult.

The main pest species found in gardens are:

LARGE YELLOW UNDERWING MOTH   *Noctua pronuba*

Fat, soft, brown, yellow or green caterpillars, up to 50 mm long, are often seen when land is being dug in winter and early spring. Colour and markings are variable but there is usually a line of dark markings on each side of the body. Adult moths have brown fore-wings and bright orange-yellow hind-wings, with dark borders. Wing-span is about 40–60 mm.

TURNIP MOTH   *Agrotis segetum*   Pl. 5

This is a smaller species. Caterpillars are grey-brown with fine dark spots on the body (Pl. 5).

HEART AND DART MOTH   *Agrotis exclamationis*

This species is closely related to the turnip moth and both larvae and adults are similar. The fore-wings of the adult have characteristic dark 'heart and dart' markings.

Treatment. Cultivate soil thoroughly in winter to expose cutworm caterpillars to weather and to birds and other predators. Keep plots free from weeds as they encourage egg-laying. Protect susceptible young plants during the early stages of growth by applying bromophos, chlorpyrifos or other insecticides to the soil around the stems. On a small scale it may be possible to locate and destroy caterpillars by hand, especially in small nursery beds and in frames.

SWIFT MOTHS   Pl. 5

Various species of hepialid moth, especially the garden swift moth, *Hepialus lupulinus*, and the ghost swift moth, *Hepialus humuli*, are widespread pests of garden plants. Symptoms, biology and treatment are essentially the same for all species.

Symptoms. White caterpillars, up to 65 mm long, with shiny brown heads (Pl. 5) live in soil and feed on roots, tubers, bulbs, corms and rhizomes of many plants. Unlike cutworms, swift moth caterpillars feed almost entirely below ground. Strawberries and lettuces may be seriously damaged by caterpillars eating up into stems from below and many ornamentals are susceptible to attack, especially anemones, perennial asters, auriculas, chrysanthemums, colchicums, convallarias, daffodils, dahlias, delphiniums, gladioli, irises, lilies, paeonies and perennial phlox. Damage may occur at almost any time of year and caterpillars are often seen when old herbaceous borders, old strawberry beds, grassland and weedy or uncultivated areas are being dug in winter.

Biology. Adult moths fly at dusk during June–August and hover over borders, beds, lawns and patches of weeds. Females release eggs while in flight and each female may lay up to 700 eggs. These fall to the soil and hatch about a fortnight later. Young caterpillars then burrow into the soil and start to feed on fine roots of cultivated plants and weeds. They continue to feed throughout autumn and winter, attacking major root systems, bulbs, corms, tubers and other underground parts of plants, and either pupate in the following spring or continue to feed for a further year before pupating.

Treatment. Female swift moths are attracted to overgrown weedy areas to lay eggs and do not normally lay over bare soil. Good weed control and regular cultivation can therefore prevent attacks by limiting egg-laying. Cultivation also exposes caterpillars to predators and weather and thorough winter digging is particularly effective since young caterpillars die as a direct result of exposure or indirectly as a result of infection by disease-causing fungi. Perennial plants, such as strawberries and herbaceous ornamentals, should be lifted and replanted regularly and, if necessary, susceptible plants may be protected by applying HCH, bromophos, or other insecticides to the soil around them, preferably in autumn to kill young caterpillars. Newly cultivated land is always likely to contain caterpillars and should therefore be thoroughly cleaned and cultivated before planting

CABBAGE CATERPILLARS   Pl. 5

Caterpillars of four different species of Lepidoptera are widespread and common pests of cabbages and other brassicas and some other species also occasionally attack these plants.

Symptoms. Caterpillars eat holes in leaves and tunnel into hearts. Damage encourages secondary rotting and plants are often fouled with excreta produced by caterpillars. Caterpillars of some species feed in exposed positions on leaves (Pl. 5) but others feed under silken webbing or in cavities eaten in heart leaves. Plants may be attacked at almost any time of year but the most serious damage is generally that done to mature plants in summer, autumn and early winter.

The main pest species are:

LARGE WHITE BUTTERFLY   *Pieris brassicae*   **Pl. 5**
Also known as the cabbage white. Adults emerge from overwintered pupae in April/May and each female lays a few hundred yellow, conical eggs in batches of 10–20 on leaves of brassicas and of garden nasturtiums (*Tropaeolum*). Eggs hatch about a fortnight after laying and yellow and black, hairy caterpillars (Pl. 5), up to 50 mm long, feed on plants for a month or more. Fully-fed caterpillars then leave plants to pupate in sheltered situations on fences, walls and buildings. A second generation of adults emerges in July/August and caterpillars developing from eggs laid by females of this generation are often abundant in August/September. A further adult generation may develop in September/October but most pupae developing in autumn overwinter to produce the first adult generation of the following year. Adult large white butterflies migrate and local populations are often augmented by invasions of adults from distant sources, which may be hundreds of miles away.

SMALL WHITE BUTTERFLY   *Pieris rapae*   **Pl. 5**
Biology and host range are similar to those of the large white but eggs are laid singly and caterpillars are smaller and velvety-green (Pl. 5). Adults emerge from overwintered pupae in February/April; a second generation of adults appears in June/July, and peak populations of caterpillars develop in August/September, often outnumbering large white caterpillars.

Another closely related species, the green-veined white, *Pieris napi*, is similar but does not attack garden plants to the same extent.

CABBAGE MOTH   *Mamestra brassicae*
This is a noctuid moth, related to the cutworms (p. 200), but caterpillars behave quite differently. Adults emerge in May/June from pupae that have overwintered in the soil and females lay batches of 20–100 hemispherical eggs on the undersides of leaves of cabbages and other brassicas and also on lettuces, onions, sweet corn, tomatoes and various ornamental plants. Small, green caterpillars hatch about a fortnight later and feed on leaves for a month or more. Full-grown caterpillars are up to 50 mm long and their colour varies from light green, through brown to black. They pupate in the soil and some adults emerge in August/September to produce a second generation of caterpillars. The remaining pupae, together with pupae of this second generation, overwinter and these pupae may be seen in the soil during winter digging.

DIAMOND-BACK MOTH   *Plutella xylostella*
This species has a world-wide distribution but tends to be a local and
sporadic pest with severe attacks suddenly developing after long-range
migrations by adult moths. It is much smaller than the other species
described above and the inconspicuous adults are generally active at night
from May onwards. Small, green caterpillars, up to 12 mm long, feed for
about a month under silk webbing on the undersides of leaves and then
pupate in cocoons under this webbing. Two or three generations of adults
develop during the growing season and the species probably overwinters
mainly as cocoons on plants.

Treatment. Inspect susceptible plants weekly throughout their growth
and crush any eggs or caterpillars seen. If necessary, dust plants with derris
or carbaryl or spray with derris, malathion or trichlorphon. Apply insecticides thoroughly so that they make good contact with caterpillars on
both sides of leaves and in growing points and try to eliminate all caterpillars from cabbages before the plants start to heart up.

PEA MOTH   *Cydia nigricana* (= *Laspeyresia nigricana*)   **Pl. 5**
Symptoms. Small caterpillars, up to 6 mm long, with pale yellow bodies
and black heads, feed on peas inside ripening pods (Pl. 5). Damage is
usually most severe in July/August.
Biology. Small, inconspicuous moths emerge from cocoons in soil from
early June onwards and females fly to pea plants to lay small flat eggs singly
or in clusters on leaves and stipules. Adults are most active during warm,
sunny weather in July. Eggs hatch about a week after laying and caterpillars
eat into young pods to feed on developing peas for about a month. Caterpillars then eat their way out of pods and descend to the soil where they
overwinter in cocoons. There is only one adult generation a year.
Treatment. Since egg-laying is mainly restricted to June/July, serious
infestations can be avoided by sowing early or late, so that peas are not in
flower and pod during this critical period. If possible, do not sow peas in
March/April for June/July cropping but sow earlier and later than this. If
crops are in flower during June/July they may be protected by chemicals
but correct timing is essential since the young caterpillars must be killed
before they enter the pods. Apply fenitrothion or permethrin spray about a
week after first flowering, and make a repeat application about a fortnight
later. Thorough winter cultivation may also help to check this pest by
exposing overwintering caterpillars to weather and predators.

SILVER Y MOTH   *Autographa gamma*   **Pl. 5**
Symptoms. Leaves of beans, cabbages, lettuces, potatoes and many ornamental plants are eaten by bright green to dark olive caterpillars, up to
40 mm long (Pl. 5). Infested plants may be extensively defoliated and large
pellets of frass, deposited on leaves, often indicate the presence of caterpillars, which hide during the day and feed mainly at night. This pest is often
troublesome in glasshouses where it feeds on many ornamentals, especially
pot azaleas, chrysanthemums, coleus, daturas and pelargoniums.
Biology. The silver y moth is a migrant species and is unable to survive
northern winters, although some caterpillars may overwinter in heated

glasshouses. In May/June adult moths migrate north and west from permanent breeding areas around the Mediterranean and some mes reach as far north as Iceland and Finland. They are relatively small, with a wingspan of 35–45 mm, and the dark fore-wings are marked with a distinctive small white Y. Females lay up to 500 eggs each, either singly or in small batches, on leaves of host plants. Caterpillars hatch about ten days later and feed for about a month before pupating in loose cocoons on infested plants. Populations of adults often reach a peak in autumn when dozens of moths may be seen feeding at dusk on buddleias, sedums and other flowers. By late autumn many of these adults will have migrated south to the Mediterranean and north Africa where breeding continues.

Treatment. Watch glasshouse and house plants carefully at all times and locate and destroy caterpillars if symptoms appear. Apply contact insecticides (see p. 195), if necessary, to indoor and outdoor plants.

TOMATO MOTH *Lacanobia oleracea*

Symptoms. Holes are eaten in leaves and fruits of tomatoes during summer and autumn by yellow-green to brown caterpillars, up to 40 mm long, with yellow bands and white spiracles. Caterpillars mainly attack tomatoes and some other plants, especially carnations and cucumbers, growing in glasshouses but will occasionally attack brassicas growing outdoors.

Biology. Adults emerge from overwintered pupae in May/June outdoors and earlier in glasshouses. Females lay up to 1,000 eggs each in batches of 60–300 on the undersides of leaves. Caterpillars hatch about ten days later and at first feed in groups, eating away the surfaces of leaves. Later they disperse and feed voraciously on the leaves, often eating them down to a skeleton of veins. Full-grown caterpillars pupate in cocoons on walls, woodwork and plant debris and a second generation of adults emerges about 2–3 weeks later. During July–September long-term pupae develop and these remain dormant until the following season.

Treatment. Remove and destroy egg masses and caterpillars whenever they are seen and use contact insecticides (see p. 195), if necessary.

LEEK MOTH *Acrolepiopsis assectella* (= *Acrolepia assectella*)

This pest is of only local importance in Britain but is widely distributed in Europe.

Symptoms. Small yellow-green caterpillars, up to 10 mm long, tunnel into leaves of leeks and related plants and eat through leaves into stems. Damage is particularly severe in warm, dry summers and is often extended by secondary rotting. Similar symptoms are caused by onion fly (see p. 219).

Biology. Small, inconspicuous adult moths overwinter in plant debris and survive quite severe winter weather. Adults become active as temperatures rise in spring and females lay about 100 eggs each on host plants, generally depositing them near ground level. Caterpillars hatch about a week later and mine leaves before eating down through leaves into growing points and stems. After feeding for about a month they crawl back to the upper leaves and pupate in cocoons. Two or three generations develop during the growing season in northern Europe but up to six generations develop in southern areas when conditions are favourable.

Treatment. Destroy severely infested plants and collect and burn accumulations of leaf litter and other debris in winter to reduce numbers of overwintering adults. Use contact insecticides (see p. 195) in spring to check populations before they reach damaging levels.

ANGLE SHADES MOTH *Phlogophora meticulosa* Pl. 5

Symptoms. Leaves, buds and flowers of outdoor and indoor plants are eaten by caterpillars feeding at night. Caterpillars are soft and plump, green to brown, minutely spotted with white, with a light band along the back and grey V-shaped markings on the sides (Pl. 5), and they grow to a length of 40–50 mm. Adults are sometimes seen in plant debris and on sides of glasshouses and are marked with a characteristic camouflage pattern of fawn, green and pink.

Biology. Adults are active from May to October outdoors and longer in heated glasshouses. Females lay clusters of 50–100 eggs on leaves and caterpillars hatch about ten days later. They feed on leaves, buds and flowers for 1–2 months and then pupate in cocoons in the soil. Two generations of adults develop outdoors, usually in May/June and September/October, but breeding may continue throughout the year in heated glasshouses.

Treatment. Inspect plants regularly and destroy egg batches and caterpillars. Apply contact insecticides (see p. 195), if necessary.

VAPOURER MOTH *Orgyia antiqua* Pl. 5

Symptoms. Leaves of many outdoor plants are eaten in summer by conspicuous hairy caterpillars, up to 25 mm long, marked with red and yellow dots and lines (Pl. 5). Roses, rhododendrons, flowering cherries, crab apples, hawthorns, pyracanthas, ceanothus, buddleias and heathers are particularly susceptible and this pest tends to be most troublesome in urban and suburban gardens. Outbreaks are usually local and damage is seldom severe, but caterpillars have urticating hairs which may cause skin irritations.

Biology. Eggs overwinter on cocoons attached to twigs and branches and hatch sporadically from May onwards. Caterpillars feed until July/August and then construct cocoons and pupate. Wingless female moths emerge in August/September and, after mating with winged males, lay 200–300 eggs each on or near their cocoons. Adults die out in autumn and eggs overwinter.

Treatment. Remove and destroy cocoons and eggs in winter and apply contact insecticides (see p. 195) in spring and summer, if necessary. Wear gloves when handling caterpillars and cocoons to avoid contact with urticating hairs.

BUFF-TIP MOTH *Phalera bucephala*

Symptoms. Leaves of oaks, limes, elms, hazels, willows, roses and other trees and shrubs are eaten in August/September by groups of a dozen or so yellow and black caterpillars, up to 50 mm long. These caterpillars feed voraciously and strip leaves but damage is usually restricted to one or two branches and the overall effect on plants is slight.

Biology. Adult moths emerge from overwintered pupae in July/August.

They are nocturnal but may sometimes be seen resting on trees or in plant debris, where they look like small pieces of broken stick. Females lay eggs in small batches on the undersides of leaves and caterpillars hatch about a week later. They then feed for about a month before going to the soil to pupate. Pupae overwinter in the soil and adults emerge in the following summer.

Treatment. Remove caterpillars by hand and destroy them or transfer them to wild host plants. Use contact insecticides (see p. 195), if necessary.

GREY DAGGER MOTH   *Acronicta psi (= Apatele psi)*

Symptoms. Distinctive caterpillars with an erect process on the first abdominal segment and a clear yellow back flanked by red spots and edged with black, feed on many plants in summer and are often found on birch, cotoneasters, hawthorn, limes, ornamental cherries, hazels, crab-apples and roses.

Biology. Pupae overwinter in cocoons in soil and leaf litter and adult moths emerge in April–June. Females lay eggs on leaves and caterpillars feed for about a month before pupating in the soil. A second generation of adults emerges in July–September and gives rise to overwintering pupae.

Treatment. Pick off caterpillars by hand or use contact insecticides (see p. 195).

BROWN-TAIL MOTH   *Euproctis chrysorrhoea*

This is not a common pest but may be very abundant in some localities, especially in south-east England, and is a public health hazard as the caterpillars shed large numbers of hairs which can cause skin irritation and allergic reactions.

Symptoms. Dark brown caterpillars, up to 40 mm long, with long tufts of brown hairs and white scales and with red spots at the end of the abdomen, feed on roses, hawthorn, sloes and some other plants. Colonies of caterpillars construct protective canopies ('tents') by spinning silk and drawing leaves together. Leaves are mainly eaten in spring and summer but 'tents' may be present throughout the year.

Biology. Caterpillars hatch from eggs in August/September and construct small 'tents' in which they overwinter. In the following spring they feed on new growths and construct larger 'tents' in which they spin cocoons and pupate in June/July. Adults emerge in August/September and females lay eggs from which overwintering caterpillars hatch about a week later.

Treatment. Prune out and burn 'tents' whenever they are seen but use gloves and avoid contact with irritant hairs. Use contact insecticides (see p. 195), preferably against young caterpillars in autumn as they may not be fully effective against larger caterpillars feeding in 'tents' in spring.

GARDEN TIGER MOTH   *Arctia caja*

Symptoms. Very hairy large, dark brown caterpillars, commonly known as 'woolly bears', feed on leaves of many cultivated plants and on various weeds, especially during May/June. Caterpillars are 70–80 mm long when fully fed and are often seen wandering over paths, walls and lawns in June.

Biology. Caterpillars hatch in August/September and feed for a few weeks before seeking suitable hibernation sites on plants, in plant debris, and in walls, fences and other structures. They emerge in early spring and feed for about a month before pupating in cocoons in June. Adults emerge in July/August and lay eggs on plants.

Treatment. Remove caterpillars from plants by hand and use contact insecticides (see p. 195), if necessary. Caterpillars have urticating hairs which may irritate skin and gloves should be worn when handling them.

HAWK MOTHS

Caterpillars of some species of hawk moth occasionally attack garden plants and cause consternation because of their large size (up to 10 cm long) and strange appearance. They can consume large quantities of leaves quite quickly but seldom occur in any numbers and are not important pests. The main species likely to be found in gardens are the privet hawk moth, *Sphinx ligustri*, on privet and lilacs; the elephant hawk moth, *Deilephila elpenor* on fuchsias and impatiens, both outdoors and in glasshouses and houses, and the eyed hawk moth, *Smerinthus ocellata*, on apples, willows and poplars.

Treatment. Remove caterpillars by hand and transfer them to suitable wild hosts to complete their development. The use of insecticides against hawk moth caterpillars is seldom, if ever, justified.

PINE LOOPER MOTH   *Bupalus piniaria*

This is a serious pest of pines and other conifers growing in forests and may sometimes attack trees in gardens, especially if they are in forest areas.

Symptoms. Green 'looper' caterpillars, up to 30 mm long, eat pine needles in summer, causing severe discolouration and defoliation. Damaged leaves wilt and turn brown and caterpillars hang from infested trees on long silk threads. Attacks are commonest on sandy soils in low rainfall areas and the effects are often most apparent in September. Douglas firs and larch may also be affected.

Biology. Adult moths emerge from overwintered pupae in May/June and females lay eggs on the leaves of the higher branches of trees. Eggs hatch about a fortnight later and caterpillars feed from June to September. They then go to the soil to pupate and the pupae overwinter.

Treatment. This pest feeds high up in the canopy of larger trees and is beyond the range of most garden spraying equipment. If caterpillars do appear on smaller trees, use contact insecticides (see p. 195).

TORTRIX MOTHS

Many species of tortrix moth attack ornamental plants and some species damage fruit (see p. 198). The carnation tortrix and the green oak tortrix, which are dealt with below, are typical of most of the species that occur in gardens.

CARNATION TORTRIX MOTH

Cacoecimorpha pronubana **Pl. 5**

This is the commonest and most troublesome species attacking glasshouse and house plants but some other species produce similar symptoms.

Symptoms. Small yellow or green caterpillars, up to 20 mm long, feed in leaf axils, in the tips of young shoots, in buds and in flowers and characteristically draw the affected parts together with silk webbing to form a

protective covering enclosing the caterpillars (Pl. 5). Caterpillars can usually be found by pulling this webbing apart. They wriggle backwards when
disturbed and drop from plants on silk threads. Plants are damaged directly
by caterpillars eating leaves and other tissues and are also affected indirectly
by the webbing, which restricts young growths. Many glasshouse and
house plants are attacked, especially acacias, carnations, cytisus, daphnes,
grevilleas, and hypericums. Outdoor plants may also be damaged, particularly euonymus, ivies, laurel and privet.

Biology. This species is native to southern Europe but is now well established in glasshouses and outdoors in Britain and northern Europe. Small,
inconspicuous adult moths have grey-brown fore-wings and copper-
orange hind-wings. They are active at night and each female lays about
eight batches of 10–200 flat, oval, overlapping eggs, protected by a mucila-
ginous covering. Eggs hatch after 2–3 weeks and young caterpillars wander
over plants to find suitable feeding sites. They then spin silk webbing to
protect themselves and start to feed on plant tissues. After feeding for 1–2
months, caterpillars pupate under webbing on plants and adults emerge
about a month later. There are usually two adult generations a year but the
length of larval development is so variable that there is no distinct gap
between them. Caterpillars overwinter and may be present in heated
glasshouses at all times of year.

Treatment. Examine glasshouse and house plants regularly and crush
caterpillars by hand. This is much more effective than using contact insecticides but sprays and fumigants may be used against adults and young
caterpillars (see p. 195).

GREEN OAK TORTRIX MOTH   *Tortrix viridana*

Symptoms. Small grey-green caterpillars, up to 20 mm long, draw young
leaves of oak trees together with silk webbing and feed on them from May
onwards. In some seasons mature oaks are extensively defoliated and many
caterpillars may be seen hanging from branches on long silk threads. The
pedunculate oak, *(Quercus robur)*, is generally more severely damaged than
the sessile oak, *(Quercus petraea)*, possibly because the latter has hairy
leaves and bud scales and comes into leaf earlier, so that leaves are tougher
and less attractive to young caterpillars. Many other plants are also
attacked.

Biology. Eggs overwinter on twigs and branches and hatch in April–June,
as buds open. Caterpillars feed on young leaves for about a month and then
pupate in webbing on damaged trees. Adults emerge about 2–3 weeks later
and females lay eggs which remain on the trees until the following spring.

Treatment. Various parasites and predators feed on the caterpillars when
they are abundant and, as a result, years when serious damage occurs are
usually followed by years when this pest is relatively scarce. Young trees
may be protected by applying contact insecticides (see p. 195), if necessary,
but larger trees cannot be sprayed adequately with normal garden spraying
equipment.

LACKEY MOTH   *Malacasoma neustria*

Symptoms. Groups of blue-grey caterpillars, up to 50 mm long, with
longitudinal white and orange stripes feed on leaves and twigs of outdoor

plants in summer and spin protective silk webbing over affected parts to form 'tents'. Similar webbing is also produced by caterpillars of the brown-tail moth (p. 206) and by the hawthorn webber and ermine moths. Lackey caterpillars attack many different ornamental plants, especially alders, birch, cotoneasters, elms, eucryphias, hawthorn, lilacs, roses, pyracanthas and willows, and occasionally attack apple and other fruit trees.

Biology. Bands of 100–200 eggs, laid on twigs and branches in the previous season, hatch in April/May. Caterpillars feed for 1–2 months and gradually construct communal 'tents', which may be up to 30 cm long. Caterpillars pupate in cocoons on bark and between leaves and adults emerge from July to September. Females lay overwintering eggs and adults die out in autumn.

Treatment. Cut out and destroy egg bands in winter and spray thoroughly with contact insecticides (see p. 195) in spring and early summer, if necessary. Cut out and burn large communal tents found in late summer as insecticides cannot penetrate them effectively.

HAWTHORN WEBBER MOTH   *Scythropia crataegella*
Symptoms. Colonies of small red or yellow-brown caterpillars, up to 15 mm long, feed on leaves of hawthorn and cotoneasters from May onwards and spin silk webbing over infested branches. Outbreaks are local and are most likely in southern areas.

Biology. Female moths lay eggs on plants in July/August and young caterpillars eat out mines in leaves in autumn. Mined leaves remain attached to plants and caterpillars overwinter in them. From March onwards caterpillars spin protective webbing and feed on leaves, eventually pupating in cocoons on webbing in June/July.

Treatment. Remove mined leaves from plants in winter and destroy them and spray with contact insecticides (see p. 195) in spring to kill young caterpillars.

JUNIPER WEBBER MOTH   *Dichomeris marginella*
Symptoms. Foliage of junipers turns brown and dies. Close examination may reveal small brown caterpillars, up to 20 mm long, and small pupae in silk webbing amongst dead leaves. Webbing is often extensive on older junipers.

Biology. Inconspicuous female moths lay eggs on juniper leaves in July/August. Eggs soon hatch and caterpillars feed until the following June/July before pupating in the webbing that they have produced. Adults emerge about 2–3 weeks later to complete the cycle.

Treatment. Spray forcibly with contact insecticides (see p. 195) in late summer or early autumn to kill young caterpillars.

SMALL ERMINE MOTHS   *Yponomeuta* spp.
At least four different species of *Yponomeuta* attack garden plants. They are all very similar in appearance and biology but have preferences for different host plants.

Symptoms. Small grey-green caterpillars, up to 20 mm long, feed in colonies on leaves of many garden plants and spin silk webbing which is similar to that produced by caterpillars of the lackey moth (p. 208) and hawthorn webber (above). Apples, euonymus, hawthorns, willows and bird cherries are particularly susceptible but some other garden plants may also be attacked.

Biology. Caterpillars hatch in August/September but remain under egg scales until the following April/May. They then become active and feed on leaves and flowers until June, when they pupate in cocoons on infested plants. Small, pure white moths with black spots on their wings emerge in July/August and females lay batches of flat eggs covered by a hard coat which protects the overwintering young caterpillars.

Treatment. Cut out and burn webbing and caterpillars, if possible, or spray forcibly with contact insecticides (see p. 195), if necessary.

LEAF MINERS

Some species of caterpillar tunnel in leaf tissues, producing leaf mines which resemble leaf mines produced by some fly maggots (see p. 222). Three of the most damaging species are dealt with below but there are many others that may occur in gardens.

LILAC LEAF MINER *Caloptilia syringella*

Symptoms. Large brown blistered areas appear on leaves of lilac and privet from June onwards. Each blister contains one or more small white or green legless caterpillars. Later in the season some leaves are rolled and tied with silk by caterpillars that have left their mines.

Biology. Caterpillars overwinter in cocoons at the bases of lilac trees, privet hedges, fences and other structures. Small silvery-white moths emerge in May/June and females lay batches of eggs on undersides of young leaves. Caterpillars hatch about a week later and feed inside leaves to make blister mines. When they have almost finished feeding, they come out of the mines, roll and tie leaves with silk threads, and then continue to feed within the protection of the rolled leaves. They then pupate in cocoons in leaf axils, and on branches and stems, and a second generation of adult moths emerges in July/August. Caterpillars developing from eggs of this generation are fully fed by September/October and then form overwintering cocoons.

Treatment. Remove and destroy affected leaves, if possible, or spray with HCH, nicotine or trichlorphon in May/June and again later in the season if new mines appear.

LABURNUM LEAF MINER *Leucoptera laburnella*

Symptoms. Unsightly spiral and blotch mines appear on laburnum leaves from June onwards. Each mine contains a single small green-white caterpillar but more than one mine may develop on each leaf. Severely damaged leaves turn brown and shrivel.

Biology. Pupae overwinter in cocoons on stems and branches and in leaf litter under plants. Inconspicuous adult moths emerge in May/June and females lay eggs on the undersides of young leaves. Caterpillars soon hatch and eat through the bottom of the egg into the leaf tissues where they eat out spiral mines. As the caterpillars grow they make more extensive blotch mines which obliterate the original spiral mines. Fully-fed caterpillars spin cocoons on damaged leaves and a second generation of adults emerges from

them in August. More mines develop and caterpillars leave them in September/October to spin overwintering cocoons.
Treatment. Remove and destroy affected leaves, if possible, and spray with insecticides, if necessary, as for lilac leaf miner (p. 210).

AZALEA LEAF MINER *Caloptilia azaleella*
Symptoms. Small yellow-green caterpillars, up to 15 mm long, make blister mines in leaves of pot azaleas growing in heated glasshouses and houses and also curl young leaves by drawing them together with silk threads. Symptoms usually appear in winter after plants have been brought into glasshouses and houses to flower. Infested plants have a scorched, unsightly appearance and may shed leaves prematurely. This pest may also attack azaleas growing outdoors in warmer, southern areas.
Biology. This pest is native to Japan and was introduced into Britain about 1925. It is now widely established in Britain and northern Europe. Small, inconspicuous female moths lay oval, flattened eggs on the undersides of leaves, mostly alongside veins. Eggs hatch after about a week and young caterpillars tunnel into leaves and feed for about a month in blister mines. They then leave the mines and pupate in silken cocoons at the tips of the shoots. In heated glasshouses breeding may continue throughout most of the year and 4–5 adult generations develop but breeding outdoors is restricted to the summer months.
Treatment. Examine pot azaleas regularly and remove affected leaves. Spray with a contact insecticide (see p. 195) in October/November, if necessary, to kill caterpillars and moths before plants come into flower.

GOAT MOTH *Cossus cossus*
This is an uncommon pest but it causes great concern when it does occur in gardens.
Symptoms. Large pink and red-brown caterpillars, up to 10 cm long, tunnel in trunks of oaks, elms, ash, beech, poplars, willows and many other trees. Holes up to 12 mm diameter appear at bases of trunks and small heaps of sawdust accumulate beneath them. A hundred or more caterpillars may develop in a single tree, severely weakening or even killing it. Similar damage is caused by leopard moth caterpillars (p. 199) but they generally attack smaller branches of young trees rather than the trunks of older trees.
Biology. Adult goat moths are active at night in June/July and females lay eggs in cracks and wounds in bark on tree trunks. Caterpillars hatch about two weeks later and tunnel into the wood where they feed for up to four years. They smell like goats – hence the common name – but are said to be edible! Fully-fed caterpillars leave their tunnels in April/May and pupate in cocoons from which adults emerge in June/July.
Treatment. Inject fumigants, such as paradichlorbenzene, into galleries and seal them to kill caterpillars inside. If substantial damage has already been done before the caterpillars are killed, tree surgery may be needed to remove damaged tissues and to ensure that the tree is safe.

PINE SHOOT MOTH *Rhyacionia buoliana*
Symptoms. Young shoots of pines and some other conifers wilt and die in spring and early summer and the brown, dead shoots stand out against the

dark green background of healthy foliage. Small caterpillars or pupae may be found inside dead shoots if they are opened up carefully. Damage to shoots and to buds may severely distort the subsequent growth of stems and branches (but also see pine twisting rust (p. 282)).

Biology. Inconspicuous adult moths are active in July/August and females lay eggs on shoots near terminal buds. Caterpillars hatch within a few weeks and feed at the bases of the leaves before tunnelling into one of the side buds. They overwinter inside the buds and in the following spring they tunnel into the terminal shoots. In June/July the fully-fed caterpillars pupate in the galleries that they have eaten out and adults emerge about a month later.

Treatment. Prevention of damage is difficult but valuable young trees can be protected by applying a contact insecticide (see p. 195) early in August and again three weeks later to kill young caterpillars before they enter the buds. Shoots that have been attacked should be cut out and burned before the caterpillars complete their development.

Caterpillars of many other moths are likely to be found attacking plants in gardens and the species dealt with above represent only a small selection. Other caterpillars that are occasionally troublesome include those of the delphinium moth, *Polychrisia moneta*, damaging buds and flowers of delphiniums, larkspur and monkshood; the peppered moth, *Biston betularia* , feeding on leaves of beech, birch and other trees and occasionally on chrysanthemums and roses; and the yellow-tail moth, *Euproctis similis*, eating leaves of beech, birch, hawthorn, oak, ornamental cherries, crab-apples, roses and viburnums.

FLIES
(Plate 6)

Adult flies (Diptera) have only one pair of functional wings, the hind wings being reduced to a pair of club-shaped balancing organs (halteres). This clearly distinguishes them from whiteflies, greenflies, butterflies and other groups of insects which, despite their common names, are not true flies. Size of adult flies varies from very small (1–2 mm long) to large (3–4 cm long) and detailed structure also varies considerably. The head is prominent, with large eyes, a pair of antennae, and mouth-parts modified for piercing and sucking rather than biting and chewing. More than 5,000 species occur in Britain and northern Europe. Some of these feed on blood (mosquitoes, clegs, biting midges); others prey on insects and other invertebrates (hover-flies, robber-flies) or are parasites (tachinid flies) and many feed on decaying animal and vegetable matter (house-flies, flesh-flies, moth-flies). In most cases the larva is the main feeding stage and adults only feed casually on fluids, such as nectar, exudations of sap, animal secretions and similar materials. Larvae always lack legs and are mostly colourless, relatively featureless maggots. Sexual reproduction is usual and females generally lay eggs on or near suitable food for the larvae. Larvae feed for a few weeks or months and then pupate, often inside a characteristic hard, dark puparium (Pl. 6). The larvae of some species feed directly on living plant tissues and may be important pests of cultivated and wild plants. The main species attacking garden plants in Britain and northern Europe are dealt with in detail in the following section.

Fly species	Main plants attacked
FRUITS	
Pear midge (p. 213)	Pears
Raspberry cane midge (p. 214)	Raspberries
Black currant leaf midge (p. 214)	Black currants
VEGETABLES	
Letherjackets (p. 215)	Young brassica and lettuce plants and many ornamentals
Sciarids (p. 215)	Cucumbers, mushrooms and many glasshouse seedlings
Frit fly (p. 216)	Sweet corn
Beet leaf miner (p. 217)	Beetroot and spinach beet
Celery fly (p. 217)	Celery and parsnips
Carrot fly (p. 217)	Carrots, parsnips, celery and parsley
Cabbage root fly (p. 218)	Cabbages and other brassicas and some ornamental plants
Onion fly (p. 219)	Onions, leeks and shallots
Bean seed fly (p. 219)	Beans and peas
ORNAMENTALS	
Chrysanthemum gall midge (p. 220)	Chrysanthemums
Violet leaf midge (p. 220)	Violets
Yew gall midge (p. 220)	Yews
Hawthorn button top midge (p. 221)	Hawthorns
Large narcissus fly (p. 221)	Daffodils, narcissi and some other bulbs
Small narcissus flies (p. 222)	Daffodils, narcissi and some other bulbs
Chrysanthemum leaf miner (p. 222)	Chrysanthemums and some related plants
Holly leaf miner (p. 223)	Hollies
Chrysanthemum stool miner (p. 223)	Chrysanthemums
Carnation fly (p. 223)	Carnations, pinks and sweet williams

PEAR MIDGE *Contarinia pyrivora* **Pl. 6**

This gall midge (Cecidomyiidae) is one of two species that occasionally attack pears in Britain and northern Europe. The other species is the pear leaf midge, *Dasineura pyri*, which causes characteristic leaf-rolling on young shoots but is not an important pest.

Symptoms. Young pear fruitlets that have been attacked by pear midge larvae swell rapidly and then fail to develop normally. Many small yellow-white larvae, up to 4 mm long, may be present inside cavities in blackened, distorted fruitlets (Pl. 6) and affected fruitlets fall prematurely in May/June. Attacks are usually local, often affecting one particular tree every year, and severe attacks reduce yields by limiting the numbers of fruits developing. Pears flowering in mid-season are most susceptible to attack and varieties flowering early or late often escape the main period of midge activity.

Biology. Larvae overwinter in small silk cocoons in soil under pear trees and pupate in spring. Small, inconspicuous adult midges, up to 4 mm long, emerge in March/April and females fly up onto pear blossom to lay eggs in unopened flowers. Eggs hatch about a week later and larvae enter young fruitlets to feed for about a month. Fully-fed larvae either leave fruitlets while they are still on the tree or fall with them to the ground. They then enter the soil, spin cocoons, and remain dormant until the following spring. **Treatment.** Reduce the numbers of larvae entering the soil by picking off affected fruitlets as soon as they are seen and by collecting fallen fruitlets. Destroy them immediately, preferably by burning. In addition, cultivate the soil under infested trees in summer and apply HCH dust to the soil in spring. Spraying with fenitrothion in spring to kill emerging adults may check attacks but timing must be correct if the females are to be killed before they lay eggs. Best results are usually given by spraying just before the blossom opens.

BLACK CURRANT LEAF MIDGE   *Dasineura tetensi*

Symptoms. The youngest terminal leaves on black currant shoots fail to open and remain folded and twisted. Small white or orange larvae, up to 3 mm long, may be found by carefully unrolling affected leaves. Plants are attacked from April to August and persistent infestations check growth and may encourage the development of lateral shoots. 'Goliath', 'Seabrook's Black', 'Baldwin' and 'Wellington XXX' are more susceptible to attack than other cultivars.
Biology. Larvae overwinter in cocoons in soil and pupate in spring. Inconspicuous adult midges, up to 3 mm long, emerge from April onwards and females fly to the tips of young shoots to lay eggs on the youngest unopened leaves. Eggs hatch after about a week and larvae feed for about a month before going to the soil to pupate. Three or four generations develop during the growing season and increasing numbers of larvae form cocoons in the soil and remain dormant until the following spring.
Treatment. Spray the tips of shoots with dimethoate as soon as symptoms are seen but do not spray during the flowering period when pollinating insects will be visiting the plants.

RASPBERRY CANE MIDGE   *Resseliella theobaldi*
   (= *Thomasiniana theobaldi*)

Symptoms. Small pink or red larvae, up to 4 mm long, feed in cracks and under raised bark of raspberry canes in summer and autumn. Affected areas of bark peel away and cane blight, due to a fungal infection (see p. 322) may develop, causing die-back of shoots. 'Bath's Perfection', 'Malling Enterprise', 'Malling Jewel' and 'Malling Promise' are particularly susceptible to attack.
Biology. Larvae overwinter in cocoons in soil near raspberry canes and pupate in the spring. Adults emerge in May/June and females lay eggs in cracks in bark and in wounds. Larvae hatch about a week later and feed under the bark for about a month before going to the soil to pupate. A further three or four generations develop between July and September and increasing numbers of larvae go to the soil to overwinter.
Treatment. Spray canes with fenitrothion during the first week in May

and repeat about a fortnight later. In addition, cultivate the surface layers of the soil in winter to expose overwintering larvae to weather and predators.

LEATHERJACKETS   Pl. 6
Leatherjackets are the larvae of crane-flies (Tipulidae), also known as daddy-long-legs. Many species occur in grassland and the commonest are *Tipula paludosa*, *Tipula oleracea* and *Nephrotoma maculata*. They are mainly pests in newly cultivated land and in lawns.

Symptoms. Grey-brown larvae, up to 5 cm long (Pl. 6), live in soil and feed on roots and stems of many different plants. Most damage is done to young plants in spring but attacks may develop at other times during the growing season. Affected plants turn yellow, wilt and may die and symptoms often resemble those produced by cutworms (p. 200) and certain root-infecting fungi. Young cabbages and other brassicas, lettuces, strawberries and various ornamental plants are susceptible to attack and a few leatherjackets can cause considerable losses of seedlings and young plants. They are occasionally introduced into seed boxes and pots in unsterilised potting composts. High populations of leatherjackets in lawns produce yellow patches during dry weather and affected lawns may be further damaged by starlings and other birds that probe and loosen the turf to extract leatherjackets from the soil.

Biology. Adult crane-flies emerge from pupae in soil during August/September and, after mating, females lay up to 300 eggs each in soil near plants. Eggs hatch about a fortnight later and larvae feed during the autumn and again in the following spring and summer before pupating. There is usually only one generation a year. High populations of leatherjackets develop after prolonged damp, warm weather in summer and early autumn and are mainly found in old meadow land and in poorly maintained lawns.

Treatment. Leatherjackets are usually only troublesome during the first few years after new areas of garden have been brought into cultivation. Thorough and regular digging and hoeing and a summer fallow period should be used to reduce populations of larvae before plants are grown. Susceptible plants can be protected by working a little HCH dust into the soil around them but over-use of this insecticide should be avoided, especially on vegetable plots where root crops may be tainted. Applications of bromophos, diazinon or chlorpyrifos granules to the soil around plants may give some protection and the use of methiocarb pellets against slugs may also help to check leatherjackets. Affected areas of lawn can sometimes be treated by thoroughly soaking them with water and then placing a tarpaulin or black polythene sheet on the affected area overnight. Leatherjackets will then work to the surface by the following morning and may then be swept up and killed.

SCIARID FLIES
Also known as mushroom flies. Larvae of many different species feed in larger fungi and also feed on fungal mycelium in soil, in decaying plant tissues and in similar situations. Most species are of no importance as pests but a few do occasionally attack garden plants and some may damage cultivated and wild mushrooms.

Symptoms. White larvae, up to 25 mm long, with distinct black or brown heads live in soil and in potting and rooting media where they may feed on root-hairs and roots of seedlings, in the bases of cuttings and in the stalks and caps of mushrooms. Carnation, chrysanthemum and pelargonium cuttings and seedlings and young plants of cucumbers, sweet peas, orchids, primulas and freesias are especially susceptible to attack. Adult sciarids are small, dark, midge-like flies, up to 5 mm long, with relatively long, narrow antennae and they are often seen in glasshouses and on pot plants in houses as they swarm over the surfaces of moist soil and potting media, characteristically running rapidly with wings vibrating.

Biology. Adults are attracted to soil and potting media, especially if the organic content is high. Females deposit about 100 eggs each and these hatch after about a week. Larvae feed for about a month before pupating in the soil and adults emerge after a further week or so. Breeding continues throughout the season. Populations are often highest in heated glasshouses, where high temperatures favour rapid and continuous breeding, but adults and larvae are also abundant outdoors when conditions are favourable.

Treatment. The presence of sciarid adults or larvae near plants does not necessarily mean that the plants will be damaged, since many of the harmless species will be attracted to potting media, manures and plant debris. If established plants do show signs of attack, water a spray-strength solution of malathion or nicotine around them but first note possible phytotoxic effects on some plants (see p. 30). Seeds, seedlings and cuttings of susceptible plants may be protected by a light dusting with HCH or malathion dust and adult flies in glasshouses can be killed by fumigating with HCH, nicotine or dichlorvos. Correct watering of seedlings, cuttings and pot plants may also help to limit attacks since sciarid larvae tend to thrive on plants that are under or over-watered.

FRIT FLY   *Oscinella frit*

This is mainly a pest of oats and other cereal crops but it also attacks sweet corn in gardens.

Symptoms. Light yellow longitudinal stripes appear on young leaves of sweet corn plants and affected leaves later disintegrate into frayed strips. Infested plants are often stunted and this early check in growth, which usually occurs in June/July, may reduce the number of cobs produced later in the season. In severe attacks the growing point is killed and plants die. Close examination of damaged plants may reveal small white larvae, up to 5 mm long, feeding in the stems and at the bases of leaves.

Biology. Larvae overwinter on grasses and winter cereals and pupate in the spring. The small, inconspicuous adults emerge in May/June and females lay eggs on or near the bases of young sweet corn and other host plants. Eggs hatch about a week later and larvae enter the young shoots to feed for 2–3 weeks. They pupate in late June/early July, either within damaged plants or in the soil, and a second generation of adults emerges in July/August. Females of this generation mainly lay eggs in developing seed-heads of oats and wheat. In autumn a third generation of adults emerges and females lay eggs on grasses and winter cereals. Larvae hatch and then develop slowly on these hosts during the winter and pupate in the following spring.

Treatment. Protect sweet corn seedlings during June/July by dusting with HCH. In addition, provide good growing conditions so that plants can become established rapidly.

BEET LEAF MINER *Pegomya hyoscyami (=P. betae)*
This pest is also known as the beet fly or the mangold fly.
Symptoms. White larvae, up to 10 mm long, tunnel in leaves of beetroot and spinach beet and occasionally attack some other vegetables, making extensive blotch mines. Severe attacks on young plants may check growth but older plants are only slightly affected. Most damage is done in early summer but attacks may continue into the autumn.
Biology. Pupae overwinter in the soil and adults emerge in April/May. Females lay batches of eggs on the undersides of leaves and larvae hatch about a week later. They soon eat into the leaves and feed for 2–3 weeks before dropping to the soil to pupate. Two or three generations develop during the growing season and puparia overwinter.
Treatment. Cultivate thoroughly in winter to reduce the numbers of overwintering puparia and grow susceptible plants under favourable conditions in well-prepared soil so that they can establish themselves quickly. If necessary, spray plants thoroughly with dimethoate, formothion or trichlorphon as soon as mines appear, repeating later in the season if attacks persist.

CELERY FLY *Euleia heraclei (= Philophylla heraclei)*   **Pl. 6**
This pest is also known as the celery leaf miner. Symptoms resemble those of the beet leaf miner (see above) but these two pests are not closely related and they affect different plants.
Symptoms. White larvae, up to 7 mm long, tunnel in leaves of celery and parsnips, producing conspicuous yellow-brown blotch mines (Pl. 6). Symptoms first appear in May but attacks may continue throughout the growing season, often with peak activity in summer and early autumn. Severe attacks check plant growth.
Biology. Pupae overwinter in the soil and adults emerge during April–June. They are small, up to 5 mm long, and have conspicuous dark markings on the wings. Females lay about 100 eggs each in leaves and eggs hatch about a week after laying. Larvae then eat into leaves and feed for about a month before pupating either within the leaf mines or in the soil. A second generation of adults emerges in late summer and a third generation may develop in autumn.
Treatment. Protect young plants by spraying with malathion, trichlorphon or dimethoate as soon as symptoms appear or as soon as females are seen on plants. Light attacks on established plants can be checked by crushing larvae and pupae within the mines.

CARROT FLY *Psila rosae*   **Pl. 6**
Symptoms. Cream-white larvae, up to 10 mm long, feed on roots of carrots and also attack parsnips, celery and parsley. In early summer the finer roots are attacked and this causes reddening of the leaves and some stunting of growth, which resembles symptoms produced by the willow-carrot aphid (p. 174) and by motley dwarf disease (p. 441). Later in the

growing season carrot fly larvae tunnel into the main roots and eat out irregular, discoloured galleries (Pl. 6). Light infestations may have little effect on well-grown crops but serious attacks may make roots unusable.

Biology. Pupae and some larvae overwinter in soil and in roots left in the ground and the small, shining-black adults emerge in May/June. Females lay clusters of small white eggs in soil near carrots and other host plants and eggs hatch after about a week. Larvae feed for about a month before pupating in the soil. A second generation of adults emerges in August/September and larvae from this generation continue to feed well into the winter before pupating.

Treatment. This pest is common in most gardens where carrots and related plants are grown regularly and it may prove difficult to control. Low levels of injury may be tolerated, since it is relatively easy to cut away affected parts when roots are being prepared for cooking, but persistent infestations should be checked by a combination of chemical and non-chemical methods. Clean and dig vegetable plots thoroughly during the winter to reduce populations of overwintering pupae and larvae and clear overgrown adjoining areas, especially nettle beds, which may provide cover for adult flies in spring and summer. If possible, avoid the first generation attack by delaying sowing until late May or early June and sow thinly to limit the need for hand thinning later. If thinning is necessary, remove thinnings immediately, consolidate the soil around the plants and water thoroughly to disperse the odour of crushed plants which may attract egg-laying females. Lift roots as soon as they are ready and store them rather than leaving them in the soil. Applications of bromophos, chlorpyrifos or diazinon granules to seed drills may give some protection against early attacks and a spray-strength solution of trichlorphon watered in as a soil drench on two or three occasions in August/September may limit damage on established plants.

CABBAGE ROOT FLY   *Delia brassicae*
(= *Erioischia brassicae*)   **Pl. 6**
This is a serious pest of cultivated brassicas throughout Europe and N. America.

Symptoms. White larvae, up to 8 mm long, feed on roots of cabbages, cauliflowers, Brussels sprouts, calabrese, radishes, turnips, swedes and some ornamental plants, especially wallflowers. Severe attacks during April–July kill young plants and transplanted brassicas are especially susceptible (Pl. 6). The first signs of attack often show on outer leaves which wilt and develop a blue-red tinge but similar symptoms are caused by club root (p. 328) and other types of root damage. Established plants can usually survive slight to moderate infestations by cabbage root fly but growth may be checked and yields reduced. Larvae feeding in the main roots of radishes, swedes and turnips and inside Brussels sprouts make them inedible. The most severe damage is usually done in late spring and early summer but this pest remains active throughout the growing season and can continue to damage plants well into the autumn.

Biology. Pupae overwinter in the soil and adult flies emerge in late April/early May. Adults resemble small house-flies and are relatively inconspicuous. Females lay eggs in soil near host plants and occasionally lay some eggs actually on host plants. Eggs hatch about a week after laying and larvae feed for about a month before pupating in the soil. A second generation of adults emerges in June/July and a third generation develops in autumn.

Treatment. Thorough winter cultivation and good hygiene will reduce numbers of overwintering pupae but adult flies are mobile and may move into gardens from adjoining areas. Calomel dust, which is sometimes used to combat club-root disease (p. 328) may give partial control of cabbage root fly and on a small scale it may be possible to protect young brassica plants by cutting felt, cardboard or plastic discs to fit around the bases of stems to stop females laying eggs. If serious attacks are expected, protect susceptible plants by incorporating granules of bromophos, chlorpyrifos or diazinon in the soil at each planting position just before the plants are set out. This is especially advisable in spring when the main period of egg-laying usually coincides with the flowering of hedge parsley *(Anthriscus sylvestris)*. If attacks develop on established plants, water a spray-strength solution of trichlorphon into the soil to drench the roots. Cabbage root flies are resistant to certain insecticides in some areas and expert advice should be obtained if the above treatments fail to give satisfactory results.

ONION FLY   *Delia antiqua*   Pl. 6

Symptoms. White maggots, up to 8 mm long, feed in stems and bulbs of onions and may also attack shallots and leeks. Young plants wilt and die (Pl. 6) and tissues of older plants soften and rot. Most damage is done in June/July but plants may also be attacked at other times during the growing season. Identical symptoms are produced by the bean seed fly (p. 219) and similar symptoms are caused by stem eelworm (p. 132) and by whiterot disease (p. 367).

Biology. Pupae overwinter in soil and adults emerge in May. Females lay eggs on young leaves and stems or in the soil near them and larvae hatch 3–4 days later. They then feed in plants for 2–3 weeks before leaving them to pupate in the soil. Two or three further generations may develop during summer and early autumn but increasing numbers of pupae remain dormant in the soil until the following spring.

Treatment. Cultivate thoroughly during the winter to disturb over-wintering pupae and dust established seedlings with HCH in June/July if damage is expected. Avoid over-use of HCH since it can affect growth and may cause taint (see p. 37). Applications of bromophos, diazinon or chlorpyrifos granules to the soil before sowing should also give some protection against this pest.

BEAN SEED FLY   *Delia platura (= Delia cilicrura)*   Pl. 6

This pest is very similar to the onion fly. It occasionally attacks onions but is more important as a pest of peas and beans. Larvae attack germinating seeds and seedlings early in the season and losses may be considerable, especially if germination is slowed by cold weather. Protect seeds with a combined HCH/captan seed-dressing or apply bromophos, diazinon or chlorpyrifos granules to the soil just before sowing.

CHRYSANTHEMUM GALL MIDGE
Rhopalomyia chrysanthemi (= Diarthronomyia chrysanthemi)
This is a potentially serious pest of chrysanthemums but it has been well-controlled by commercial growers in recent years.
Symptoms. Small thorn-like galls, up to 2 mm long, protrude obliquely from leaves, buds and stems of autumn-flowering chrysanthemums. Galls may be sufficiently numerous to distort and stunt growth. Stems swell and twist, flowers are malformed and cuttings are checked and weakened. Most attacks develop on glasshouse chrysanthemums between July and December.
Biology. This pest was introduced into Europe from N. America on cuttings from about 1927 and is now well established. Small, inconspicuous females, up to 2 mm long, lay up to 150 orange-red eggs each on young leaves, and on leaf and flower buds. Eggs hatch about a week later and small larvae feed on tissues for weeks or months. Galls develop around them and larvae eventually pupate within the galls. Development from egg to adult takes less than a month in summer but extends to 4–5 months during the winter.
Treatment. Spray affected plants with HCH or nicotine to kill adults and young larvae. Two or three applications may be needed to check established infestations since older larvae and pupae will survive treatment.

VIOLET LEAF MIDGE *Dasineura affinis*
Symptoms. Leaves of cultivated and wild violets thicken abnormally and roll upward and inward. Young leaves are attacked throughout the growing season and persistent infestations check growth and limit flowering. If thickened leaves are unrolled carefully, many small white to orange larvae, up to 4 mm long, and clusters of white silk cocoons may be found inside.
Biology. Larvae overwinter in cocoons inside galled leaves or in the soil. Adults emerge from April onwards and females lay eggs on the youngest leaves. Eggs hatch within a week and larvae feed on the developing leaves, which react by curling and thickening to produce the galls. Larvae feed inside galls for 2–3 weeks before pupating and adults emerge about two weeks later to repeat the cycle. Four or five generations develop during the growing season and larvae of the last generation overwinter.
Treatment. Pick off affected leaves and destroy them, preferably by burning. If necessary, spray with HCH, fenitrothion or dimethoate.

YEW GALL MIDGE *Taxomyia taxi* Pl. 6
Symptoms. Characteristic leafy galls, resembling miniature young globe artichokes, develop at the tips of young shoots of yews (Pl. 6). Galls persist on affected yews for two or more years and are often so common that they may be mistaken for a normal feature of the plants. The overall effect is not particularly harmful but infested hedges and specimen trees look unsightly and persistent infestations interfere with normal growth.
Biology. Almost the entire life-cycle of this pest is spent within the galls. Inconspicuous adults, up to 4 mm long, emerge from galls in May/June and females lay eggs in the tips of young shoots. Larvae soon hatch and galls develop around them as they feed. Each gall contains a single larva but there may be many galls on one plant. Larvae mostly remain within galls

until the following season when they pupate and give rise to the next generation of adults.

Treatment. Chemicals are unlikely to be effective against this pest except during the brief period in May/June when the adults are active. Spraying with HCH, fenitrothion, or other contact insecticides during this period may check infestations and slight attacks may be dealt with by picking off and destroying galls as soon as they develop.

HAWTHORN BUTTON TOP MIDGE   *Dasineura crataegi*

Symptoms. The growth of young hawthorn shoots is checked and rosettes of leaves develop at their tips. These rosette galls are most conspicuous on hedges that are regularly trimmed and which therefore produce many new shoots for the midge to attack.

Biology. Larvae overwinter in cocoons in the soil and adults emerge from May onwards. Females lay eggs in the tips of young shoots and each rosette gall contains many larvae which feed for 2–3 weeks before going to the soil to pupate. Breeding continues until September.

Treatment. Cut out and burn galls as soon as they appear and, if necessary, spray with HCH or some other contact insecticide in June to kill egg-laying females of the first generation. Dimethoate or formothion sprayed later in the season may kill larvae within galls.

LARGE NARCISSUS FLY   *Merodon equestris*   Pl. 6

Symptoms. Narcissus, hyacinth and some other bulbs produce narrow, yellow, distorted leaves in spring and fail to flower. If affected bulbs are lifted and opened, single, large, dirty grey-brown, fleshy larvae, up to 20 mm long, may be found inside rotting tissues (Pl. 6). Larvae of small narcissus flies (see below) may also be found in rotting bulbs but are smaller and more numerous. The large narcissus fly mainly attacks bulbs growing in open, sunny situations and severe attacks may kill bulbs completely so that no foliage or flowers are produced in spring. Similar symptoms are produced by stem eelworm (p. 132).

Biology. Adult flies emerge from pupae in the soil during April–June and are active on warm, sunny days. They are about 15 mm long, resemble small bumble bees, and make a distinct humming sound in flight. Females lay about 40 eggs each near the necks of bulbs, often entering the holes left in soil above bulbs as the leaves die down. Larvae hatch about a week later, crawl to the bases of the bulbs, and tunnel through the base into the interior. Usually only one larva develops in each bulb and it feeds for six months or more before leaving to pupate in the soil in the following spring.

Treatment. This is a difficult pest to control since symptoms do not appear until after the damage has been done. Bulbs bought from reputable suppliers are unlikely to contain narcissus fly larvae since they are killed by commercial hot water and chemical treatments. Bulbs obtained from other sources should be carefully examined before planting and any suspect bulbs should be destroyed, preferably by burning. In areas where this pest is known to be troublesome, healthy stocks of bulbs can be preserved by planting them in shade or by providing artificial shade with muslin rolls or similar materials as the foliage dies down. Egg-laying females can also be discouraged by raking or sifting soil over plants to fill the holes that are left

as the leaves die down. HCH dust applied over plants in June/July may also
be partially effective.

SMALL NARCISSUS FLIES *Eumerus tuberculatus* and
Eumerus strigatus

Larvae of these two species feed in rotting narcissus and other bulbs. They
are only about half the size of large narcissus fly larvae, are lighter in colour,
and usually feed in groups of six or more. They may occasionally cause
primary damage to otherwise healthy plants but are much more likely to
have invaded plants that have already been damaged by other pests, such as
stem eelworm (p. 134), or by diseases. Affected bulbs should be destroyed,
preferably by burning, and the initial cause of the trouble should be
diagnosed and treated.

CHRYSANTHEMUM LEAF MINER
Phytomyza syngenesiae   **Pl. 6**

Symptoms. Sinuous white mines appear in the leaves of chrysanthemums
(Pl. 6), cinerarias, calendulas, lettuces and some other cultivated plants and
weeds. Attacks generally develop during May/June on plants growing out
of doors but occur at most times of year on glasshouse and house plants. In
severe attacks leaves are extensively mined and plants may collapse and die.
Chrysanthemum cultivars 'Galaxy', 'Vibrant', 'Tuneful', 'Long Island
Beauty', 'Ice cap', 'Salmonchip', 'Silverstone', 'Criterion', 'Favourite',
'Corsair', 'Yellow Iceberg', 'Yellow Shasta', 'Dark Delmar', 'Cerise
Shoesmith' and 'Delightful' are particularly susceptible.

Biology. Adult chrysanthemum leaf miners are small, inconspicuous, dark
flies, about 3 mm long, with a light grey bloom on the thorax. They make
short, jerky flights amongst leaves of their host plants and females pierce
leaf tissues with their ovipositors to feed on sap that exudes from the
wounds. This produces characteristic white spots, which are the first
indication of the presence of this pest on plants. Each female lays up to 100
eggs in similar incisions and larvae hatch about a week later. They then
tunnel into the leaves and as the larvae grow, the mines get wider and
meander towards the mid-rib. Larvae feed for 2–3 weeks and then pupate
within the mines, where they can be seen as small, dark bumps. At 16–21 °C
(60–70 °F) adults emerge from pupae after 9–12 days but at lower temperatures the pupal stage lasts much longer. In heated glasshouses and in houses
breeding continues through most of the year but out of doors it is restricted
to the summer months.

Treatment. Examine susceptible plants regularly and treat them as soon
as feeding punctures or mines are seen. If only a few leaves are attacked,
pick them off and destroy them by burning. If necessary, spray plants with
HCH, nicotine or trichlorphon to kill eggs and young larvae, repeating
after two weeks if attacks persist. Control weeds, especially groundsel and
sowthistle, as these may support populations of this pest.

Note: Another species, the American serpentine leaf miner, *Liriomyza trifolii*, has been introduced into Britain and northern Europe in recent years,
especially on chrysanthemums, and may become established as a pest.

HOLLY LEAF MINER *Phytomyza ilicis*   **Pl. 6**
Symptoms. Blotch mines develop in leaves of wild and cultivated hollies
(Pl. 6). Few hollies are ever completely free from this pest.
Biology. The small, inconspicuous adult flies emerge from pupae from
May onwards and lay eggs on the undersides of the leaves near the mid-rib.
Larvae soon hatch and tunnel into the leaves, first making a straight and
narrow mine but later eating out a larger blotch mine as they grow. Larvae
feed throughout the growing season and remain in the mines until the
following spring, when they pupate. Only one generation of adults develops each year.
Treatment. If only a few leaves are affected, pick them off and destroy
them, preferably by burning. If necessary, spray with HCH once or twice
in May/June to kill egg-laying females. Once larvae are inside leaves they
are not easily killed but systemic insecticides, such as dimethoate or formothion, might be effective.

CHRYSANTHEMUM STOOL MINER *Psila nigricornis*
Symptoms. Tough-skinned, yellow-white larvae, up to 6 mm long, tunnel in roots, underground stems and shoots of chrysanthemums in autumn
and early winter. Young shoots often wilt and die and older shoots are
weakened. Effects are most marked on stools of early-flowering chrysanthemums brought into heated glasshouses for the production of cuttings
and the cultivar 'Favourite' seems to be especially susceptible to attack.
Biology. Adult chrysanthemum stool miners are small, brown-black,
shiny flies, similar to the closely related carrot fly (p. 217). They first appear
out of doors in May/June and females lay eggs in soil near chrysanthemum
plants. Eggs hatch about two weeks later and larvae feed for 1–2 months
before pupating in the soil or in the tunnels that they make in plants. A
second generation of adults emerges during August–October and it is the
larvae of this generation that cause most damage. They feed for 3–4 months
before pupating and many of the pupae overwinter.
Treatment. Protect stools reserved for the production of cuttings by
watering a spray-strength solution of HCH or nicotine into the roots before
lifting or immediately after lifting and boxing up.

CARNATION FLY *Delia cardui*
Also known as the carnation leaf miner but is not closely related to other
dipterous leaf miners.
Symptoms. Cream-white larvae, up to 10 mm long, tunnel in leaves and
stems of carnations, pinks and sweet williams, making extensive mines.
Affected plants may wilt and die. Attacks develop from September onwards
and persist into the winter months.
Biology. Larvae feed throughout the winter and pupate in the following
spring. Adults emerge in June but are not reproductively mature until
September, when females lay eggs on leaves and in leaf axils of host plants.
Larvae hatch 1–2 weeks later and feed on plants throughout the winter,
eventually leaving them to pupate in the soil.
Treatment. Spray plants with HCH or dimethoate as soon as mines
appear on leaves in September.

SAWFLIES
(Plate 7)

Adult sawflies are relatively small, usually up to 10 mm long, and generally inconspicuous insects (Hymenoptera: Tenthredinoidea), with two pairs of wings and usually with dark bodies and legs. They resemble flying ants but can be distinguished by the absence of a distinct waist-like construction between thorax and abdomen. Females have a saw-like ovipositor which is used to insert eggs in cuts made in soft plant tissues. Larvae resemble caterpillars of moths and butterflies (p. 194) but differ in having at least six pairs of fleshy prolegs on the abdominal segments. Adults feed mainly on pollen but larvae eat leaves and other plant organs. At least 400 different species attack wild and cultivated plants in Britain and northern Europe and some of these are important pests of garden plants. Reproduction is usually sexual, with both males and females being produced, but males are rare in some species and reproduction is then parthenogenetic.

APPLE SAWFLY *Hoplocampa testudinea*   **Pl. 7**
Symptoms. Cream-white larvae, up to 15 mm long, tunnel into developing apple fruitlets in May/June and eat out large cavities, which are filled with wet red-brown frass (Pl. 7). Similar but distinct symptoms are produced by codling moth caterpillars (p. 197) but they attack later in the season. Many of the fruitlets damaged by sawfly larvae fall prematurely in June/July but some fruits that are only superficially damaged survive until harvest and show characteristic scars and distortions (Pl. 7). 'Worcester Pearmain', 'Charles Ross', 'James Grieve' and 'Ellison's Orange' are generally more susceptible to sawfly attack than other cultivars.
Biology. Larvae overwinter in cocoons in soil and adults emerge in spring. Peak activity generally coincides with the main flowering period of apples in April/May. Females lay about 30 eggs each in the open blossom and eggs hatch about a fortnight later, just as the petals are falling. Young larvae then tunnel into the fruitlets, eventually penetrating to the core and eating the developing seeds. They then leave damaged fruitlets to enter healthy ones and during their feeding period of about a month each larva feeds on 2–3 fruitlets. Fully-fed larvae drop to the soil in June/July and spin cocoons in which they overwinter. Most larvae pupate in the following spring but some may remain dormant in the soil for a further year before completing their development to adults.
Treatment. Spray trees thoroughly with HCH, malathion, dimethoate or nicotine about a week after the petals have fallen. Timing is critical as the young larvae must be killed before they tunnel into the fruitlets. If hive bees are visiting blossom, defer spraying until late evening and use non-persistent insecticides, such as malathion or nicotine.

PLUM SAWFLY *Hoplocampa flava*
This species is closely related to the apple sawfly but only attacks plums and damsons.
Symptoms. Cream-white larvae, up to 15 mm long, eat into developing fruitlets of plums and damsons in spring and early summer. Dark, sticky

frass exudes from holes in damaged fruits as larvae tunnel into the flesh. Fruits fall prematurely and severe attacks reduce crops. 'Czar', 'Belle de Louvain' and 'Victoria' plums are generally more susceptible to attack than other cultivars.

Biology. Larvae overwinter in cocoons in soil and adults emerge in spring. Females lay eggs in blossom during April/May and larvae hatch after about a week. Young larvae tunnel into fruitlets but the initial entry holes are inconspicuous. Larvae move from damaged to healthy fruits during their development, attacking about four fruits each in a month, and it is the older, larger larvae that cause the most obvious damage. Fully-fed larvae leave damaged fruits and go to the soil to overwinter in cocoons.

Treatment. Attacks are sporadic and years of severe attack may be followed by years in which little damage is done. In areas where this pest is known to be troublesome, spray with fenitrothion about a week after petal-fall.

PEAR AND CHERRY SLUGWORM   *Caliroa cerasi*   **Pl. 7**

Symptoms. Small, shiny black, slug-like larvae (Pl. 7), up to 15 mm long, feed on upper surfaces of leaves of pears, cherries, apples, almonds, hawthorns and rowans during June–September. Leaf tissues are characteristically eaten away from the upper surfaces, leaving the veins and lower epidermis intact (Pl. 7). These symptoms are sometimes referred to as 'window-paning' or 'skeletonising'.

Biology. Larvae overwinter in cocoons in soil and small, black adult sawflies emerge in May/June. Females lay eggs in slits in leaves. Eggs hatch about two weeks later and larvae feed on leaves for about a month before going to the soil to pupate. A second generation of adults emerges in July/August and larvae feed until autumn before going to soil to overwinter in cocoons.

Treatment. This pest is easily controlled by spraying with non-persistent insecticides, such as derris, malathion or bioresmethrin, as soon as young larvae are seen on plants.

COMMON GOOSEBERRY SAWFLY   *Nematus ribesii*   **Pl. 7**

This is the main species attacking gooseberries but related species also occur on gooseberries and currants.

Symptoms. Leaves of gooseberries are eaten in spring and summer by green larvae, up to 40 mm long, with black heads and black spots (Pl. 7). Magpie moth caterpillars cause similar symptoms but differ in colour and form (p. 198). Leaves attacked by sawfly larvae are often reduced to a skeleton of veins and this type of severe defoliation affects yields.

Biology. Larvae overwinter in cocoons in soil and small black and orange-yellow adults emerge from April onwards. Females lay eggs in slits made alongside the main leaf veins on the undersides of leaves and eggs hatch about a week later. Larvae feed for about a month and then pupate in the soil. A second generation of adults emerges about three weeks later and the cycle is repeated. Three adult generations develop during the growing season and larvae of the third generation overwinter in cocoons in the soil.

Treatment. Inspect gooseberry bushes carefully every week from May onwards, paying particular attention to the centres of bushes, since females generally lay most of their eggs there. If caterpillars are seen, spray immediately with a non-persistent insecticide, such as derris or malathion, and repeat at intervals of a week or so if damage continues.

ROSE SAWFLIES

Cultivated and wild roses are particularly susceptible to attack by many different species of sawfly. Most of these feed only on roses but related species occur on other cultivated plants.

ROSE SLUG SAWFLY *Endelomyia aethiops* Pl. 7

This species is also known as the rose slugworm or rose skeletonizer.

Symptoms. Soft, yellow-green, slug-like larvae, up to 15 mm long, eat away leaf tissues, exposing veins and usually leaving the epidermis intact on one side (Pl. 7). Symptoms may appear at any time between May and September. Plants are seldom severely affected but damaged leaves look unsightly.

Biology. Larvae overwinter in cocoons in soil and pupate in early spring. Females emerge in May/June and lay eggs in the edges of leaves. Eggs hatch about a week later and larvae feed for about a month before going to the soil to pupate. A second generation of females emerges in August/September and larvae from this generation go to the soil to overwinter in cocoons. Males are extremely rare and reproduction is therefore almost entirely parthenogenetic.

Treatment. Winter and early spring cultivation of soil under roses may reduce populations of overwintering larvae by exposing them to weather and predators. If attacks develop in spring and summer, spray thoroughly with a contact insecticide, such as malathion, derris or bioresmethrin, as soon as larvae are seen on plants, and repeat treatment after a week or so, if attacks persist.

LEAF-ROLLING ROSE SAWFLY *Blennocampa pusilla* Pl. 7

Symptoms. Rose leaves roll up tightly, with the edges rolled downwards and inwards along their length (Pl. 7). Symptoms first appear quite suddenly in May. Leaves remain rolled for the rest of the season and, although there is little direct effect on the vitality of infested plants, the rolled leaves look unsightly. This pest is most troublesome in southern areas of Britain and northern Europe and some rose cultivars, especially 'Peace', 'Albertine', 'Frensham', 'Golden Wings', 'Grand'mère Jenny', 'Masquerade', 'Mischief', 'Mme Butterfly', 'New Dawn', 'Queen Elizabeth' and 'Sutter's Gold' are more susceptible to attack than others.

Biology. Larvae overwinter in cocoons in soil and pupate in early spring. Adults emerge in May/June and females probe young leaves with their sawlike ovipositors before laying eggs. They only lay eggs in some of the probed leaves but any leaf that has been probed rolls up within a few hours. Larvae hatch from eggs after about a week and feed on leaf tissues within rolled leaves until July/August, when they go to the soil to overwinter in cocoons.

Treatment. Cultivation of soil under roses in winter and early spring will help to reduce the numbers of overwintering larvae and slight infestations developing on plants in May/June can be easily dealt with by picking off and burning all affected leaves. Insecticides are not particularly effective against this pest but spraying with contact insecticides, such as HCH or trichlorphon, may check egg-laying females in May/June and systemic insecticides, such as dimethoate or formothion, may give some control of larvae later in the season.

LARGE ROSE SAWFLY   *Arge ochropus*
Symptoms. Young shoots and flower stalks of roses turn black and die in June/July after female sawflies have laid eggs in them and colonies of blue-green larvae, with black spots, feed on leaves and shoots later in the season.
Biology. Larvae overwinter in the soil and adults emerge in June/July. Females lay eggs in lines of incisions on young shoots and flower stalks and eggs hatch a few weeks later. Larvae then feed for about a month before going to the soil to pupate. A second generation develops in September/October and larvae of this generation overwinter.
Treatment. Cut out and burn affected shoots in June before eggs have hatched. Otherwise, spray larvae with a contact insecticide, such as HCH, trichlorphon or malathion, whenever they are seen on plants.

BANDED ROSE SAWFLY   *Allantus cinctus*
Symptoms. Yellow-brown larvae with slightly tapered bodies feed on rose leaves in summer and characteristically hold their bodies in the air while clinging to the edges of leaves.
Biology. Larvae overwinter in tunnels in rose stems and adults emerge in June. Females lay eggs in slits in the undersides of leaves and larvae hatch about a week later. They then feed on leaves for about a month before tunnelling into pithy stems through pruning snags and dead shoots. A second generation of adults may develop later in the season.
Treatment. Remove larvae by hand or spray with a contact insecticide.

SOLOMON'S SEAL SAWFLY   *Phymatocera aterrima*
Symptoms. Blue-grey larvae, up to 20 mm long, with black heads, feed in groups on leaves of Solomon's seal and other polygonatums during the summer. Leaves are rapidly stripped when larvae are abundant.
Biology. Larvae overwinter in cocoons .in soil and adults emerge in May/June. Females lay eggs in leaf stalks and larvae hatch about a week later. They then feed for about a month, mainly on the undersides of leaves, before going to the soil to overwinter. There is only one generation a year.
Treatment. Spray with contact insecticides, such as HCH, trichlorphon or malathion as soon as larvae are seen on plants.

IRIS SAWFLY   *Rhadinoceraea micans*
Symptoms. Blue-grey larvae, up to 20 mm long, feed along leaf edges of cultivated and wild irises, especially *Iris pseudacorus* and *Iris laevigata*, during the summer.
Biology. Larvae overwinter in cocoons in soil and adults emerge in June/July. Females lay eggs on leaves and larvae feed for about a month before pupating. A second generation may develop later in the season.
Treatment. Spray with contact insecticides, as for Solomon's seal sawfly (see above).

WILLOW BEAN-GALL SAWFLY   *Pontania proxima*   **Pl. 7**
This is one of many species of *Pontania* that cause various types of leaf galls on willows.
Symptoms. Conspicuous, hard, raised, red or green galls, up to 2 cm long, develop on leaves of willows (Pl. 7) in summer. Sawfly larvae eat out cavities within galls but cause little direct damage to plants, although the galls may sometimes be considered unsightly.
Biology. Galls develop after female sawflies have laid eggs in young leaves during May and they are usually first noticed in June. Larvae feed within galls until July and then leave them to pupate in cocoons on bark or in soil. A second generation of adults emerges in September/October and larvae of this generation overwinter. Males are relatively rare and reproduction is therefore mainly parthenogenetic.
Treatment. Pick off and destroy affected leaves, if necessary and feasible. Chemical control is seldom, if ever, justified.

PINE SAWFLIES

At least twelve different species of sawfly attack pines in Britain and northern Europe. The two commonest are usually *Neodiprion sertifer* and *Diprion pini*.
Symptoms. Yellow-green larvae, up to 25 mm long, feed together in parties of a dozen or more on shoots of pines. Leaves are eaten and shoots may be rapidly defoliated. *Diprion pini* larvae feed on leaves of both the current and the past year's growth but larvae of *Neodiprion sertifer* only attack the older leaves and do not feed on new growth of the current season. *Pinus contorta* is especially susceptible to attack but other pines and some other conifers, such as spruces, are also attacked.
Biology. The seasonal biology of these two species differs quite markedly. *Diprion pini* larvae overwinter in cocoons in the soil and adults emerge in May/June. Females then lay eggs in slits in older leaves. Larvae hatch about two weeks later and feed until July when they spin cocoons and pupate either on the trees or in the soil. A second generation of adults emerges in July/August and the cycle is repeated, with larvae going to the soil in autumn to overwinter. *Neodiprion sertifer* overwinters as eggs which are laid in the leaves in September/October. These hatch in the following May and larvae feed on leaves until June/July before going to the soil to pupate. Adults emerge in autumn and females lay overwintering eggs before dying out.
Treatment. Spray larvae with contact insecticides, such as HCH or malathion, as soon as they are seen on plants.

SPIRAEA SAWFLY   *Nematus spiraeae*

Symptoms. Groups of green-yellow larvae, up to 20 mm long, with pale brown heads, feed on undersides of leaves of *Aruncus sylvester* from May onwards. Leaves are often reduced to a skeleton of veins when larvae are abundant.
Biology. Larvae overwinter in the soil and adults emerge in April/May. Males are very rare and reproduction is normally parthenogenetic. Females lay eggs on the undersides of leaves during May/June and larvae hatch after about a week. They then feed for a month before pupating in the soil. A

second generation of adults emerges in July/August and a third generation may develop in September.

Treatment. Apply contact insecticides, such as HCH or malathion, as soon as larvae are seen on plants.

HAZEL SAWFLY *Croesus septentrionalis*

This species is also known as the birch sawfly.

Symptoms. Green-blue larvae, up to 25 mm long, feed in groups on edges of leaves of birch, acers, hornbeam, hazel, ash, poplars, willows and some other trees from May onwards. Young plants in nurseries may be defoliated and this can check growth.

Biology. Larvae overwinter in cocoons in soil and adults emerge in May/June. Females lay eggs in leaf veins and larvae hatch a week or so later. They feed for about a month before pupating in the soil and a second generation of adults emerges in August. Larvae developing from eggs of this generation overwinter.

Treatment. Apply contact insecticides, such as HCH or malathion when larvae are seen on plants.

GALL WASPS
(Plate 7)

Small, inconspicuous insects (Hymenoptera: Cynipidae), up to 8 mm long, with two pairs of wings, long antennae and elongate dark body. Despite their common name, they do not resemble wasps. Females generally lay eggs in young buds, leaves and other plant tissues and larvae develop within characteristic galls that are produced by the host plant. Each gall wasp species induces development of a particular gall or series of galls and many of these are conspicuous and well known. Oaks are especially liable to attack by gall wasps and at least 100 different species have been recorded in Britain and northern Europe. Although the symptoms are often spectacular, galls seldom cause severe damage to plants, but persistent infestations on young plants may check growth. The biology of many gall wasp species is complicated by alternation of sexual and asexual generations which develop in quite distinct galls, often on different parts of the same host plant.

MARBLE GALL WASP *Andricus kollari*   Pl. 7

Widespread and especially common on young scrubby oaks. Large, spherical, smooth, green or brown woody galls, up to 25 mm across, develop from buds (Pl. 7) after female gall wasps have laid eggs in them in spring. Each gall contains a single small larva which feeds in an internal chamber. Adults emerge through holes in galls in September/October and brown, empty galls remain attached to trees for years.

OAK-APPLE GALL WASP *Biorhiza pallida*

Also common and widespread on oaks and is sometimes confused with the preceding species. Large, pithy, slightly irregular, smooth galls develop from buds in May. At first they are green or yellow but they soon turn red and look like small apples. Each gall contains many small chambers in which larvae feed. Adults emerge in June/July and females crawl down stems and trunks to lay eggs on roots. Root galls then develop and house larvae of the overwintering generation. Adults emerge in spring and move up to the young buds to lay eggs and complete the cycle.

ARTICHOKE GALL WASP *Andricus fecundator*

Oak buds develop into galls that resemble small artichokes. These galls appear in June after females have laid eggs in young buds and they persist for a year or more. Female gall wasps emerge in the following spring and lay eggs in male flower buds. Hairy galls then develop on male catkins and adults of the sexual generation emerge from them in May/June. After mating, the females lay eggs and give rise to the next crop of artichoke galls.

SPANGLE GALL WASPS *Neuroterus* species   **Pl. 7**

Four different species of *Neuroterus* induce the development of small, circular, 'spangle' galls, up to 4 mm diameter, on the undersides of oak leaves in late summer and autumn. These are the common spangle gall wasp, *Neuroterus quercusbaccarum*, the silk-button spangle gall wasp, *Neuroterus numismalis*, the cupped spangle gall wasp, *Neuroterus tricolor*, and the smooth spangle gall wasp, *Neuroterus albipes*. Each gall contains a single gall wasp larva in a small chamber. Galls fall off leaves in autumn, often in great numbers which carpet the soil under infested trees. Larvae overwinter in them and adults emerge in the following spring. Parthenogenetic females then lay eggs on catkins or leaves and various specifically distinct pea and blister galls develop before the sexual generation of adults emerges in June/July to give rise to the next crop of spangle galls.

Treatment. Oaks are seldom seriously affected by spangle galls and it is generally unnecessary to try to control them. Young trees may be protected to some extent, if necessary, by applying a contact insecticide, such as HCH, during June/July when adults of the sexual generation are active.

ACORN CUP GALL WASP *Andricus quercuscalicis*

Causes 'knopper' galls on acorns which develop conspicuous irregular green sticky growths. Established in southern England since 1960 and extremely abundant in recent years. Mostly noticed in September – October when galled acorns fall from trees.

BEDEGUAR GALL WASP *Diplolepis rosae*

This species occasionally attacks wild and cultivated roses. Galls develop from affected buds in June/July and take the form of bright red and green, moss-like outgrowths, up to 10 cm across, which are sometimes known as 'Robin's pincushions'. The central woody core of the gall contains up to 50 separate larval chambers and adults emerge from them in May of the following year. Little damage is done to affected plants but if galls are considered unsightly they should be cut out and destroyed.

ANTS

All ants are social insects (Hymenoptera: Formicidae) and they live in nests of varied types and sizes which generally contain large numbers of wingless workers. Winged males and females are produced at certain times, usually during warm weather in summer, and these flying ants swarm in mating flights. Fertilised females (queens) then shed their wings and either enter existing nests or establish new nests. Individual queen ants live for several years and maintain colonies by continuing to lay eggs throughout this long reproductive period. The detailed biology of ants is very complex and interesting. Some species are carnivorous and some harvest pieces of plants on which fungi are cultured to create fungus gardens inside their nests. Some colonies keep slave ants, as well as maintaining various inquilines which inhabit their nests, and many species tend aphids, scale insects and mealybugs, feeding on the honeydew that they produce and affording them some protection from parasites and predators. The most troublesome species in gardens are the common black ant, *Lasius niger* and the mound ant *Lasius flavus*. In glasshouses and houses various introduced species occur, especially Pharoah's ant, *Monomorium pharaonis* and the Argentine ant, *Iridomyrmex humilis*.

Symptoms. Symptoms vary according to the situation and the species of ant concerned. Occasionally ants eat small pieces of tissue out of leaves but this type of injury is seldom important, except on choice pot plants in glasshouses. More damage is caused, both in glasshouses and outdoors, by ants collecting seeds from seed boxes, pots and open ground and carrying them away to their nests where they are used as food for larvae. Seeds with a high oil content are often selected and those of buddleias, meconopsis, rhododendrons and violas are often taken. Plants may also be damaged by ants tunnelling in soil under their roots and in dry borders and rock gardens on light soils root function may be seriously affected so that plants wilt and die. The mound ant causes similar trouble in lawns and rough grass by throwing up heaps of soil over nests. These mounds are often as large as mole hills and can interfere with mowing. Much indirect damage is also caused by ants tending aphids, mealybugs and scales since this can encourage the spread of these pests and the maintenance of healthy colonies.

Biology. Nests are usually constructed in soil, under paving stones, in walls, woodwork and other structures, and even in flower pots. Each nest contains at least one egg-laying female (the queen) and the nest is serviced by many wingless worker ants which construct galleries, tend eggs, rear larvae, forage for food and protect nests from invaders. Nests persist for many years and each year they produce mating flights of winged males and females which extend infestations.

Treatment. Direct treatment of nests is the best method of control as it eradicates these pests at source. If nests are accessible, treat them by watering a spray-strength solution of HCH or trichlorphon into the entrances, using sufficient solution to permeate the whole nest. If nests are inaccessible they may be poisoned indirectly by putting down baits or dusts in areas where workers are active. Proprietary ant baits usually consist of sugar or other attractants mixed with a persistent insecticide. These are applied to the soil or are placed in shallow lids or other containers to which ants have easy access. Workers then carry particles of poisoned bait back to the nest. Baits should be renewed regularly so long as ants continue to visit them and it may take weeks of baiting before there is any marked reduction of ant populations. Many different baits are available but those that contain trichlorphon or propoxur generally give best results.

WASPS
(Plate 7)

Wasps are relatively large insects (Hymenoptera: Vespidae), up to 20 mm long, and with conspicuous yellow and black markings (Pl. 7). Six species of wasp and the similar, but larger hornet, *Vespa crabro*, occur in Britain and northern Europe.

Symptoms. Adult wasps are usually most troublesome in summer and early autumn when they feed on ripening fruits, especially apples, pears, plums and grapes, often extending damage caused by other pests, such as birds, and by diseases and disorders (Pl. 7). Large cavities are eaten in fruits and people are sometimes stung by wasps when picking fruit. These pests may also damage plants such as dahlias by scraping the stems.

Biology. The two commonest species, the common wasp *(Paravespula vulgaris)* and the German wasp *(Paravespula germanica)*, usually construct nests in cavities in soil, walls and buildings but some of the other species make nests amongst twigs and branches of shrubs and trees. Nests are constructed from a paper-like material which the wasps make by chewing small pieces of wood and mixing them with saliva. Fertilised female wasps (queens) overwinter under loose bark on trees, in garden sheds, in outbuildings and in similar dry, protected situations and establish new nests in the spring. Initially the nests are small, with only a few open hexagonal cells in which the queen rears the first brood of larvae, feeding them on a high-protein diet of caterpillars and other insects. When the first worker wasps of the new season emerge in early June they take over nest-building and feeding functions from the queen and by mid-summer the nests are greatly enlarged, often containing thousands of workers. At this stage males and females are reared and the young fertilised queens seek suitable sites in which they can overwinter. Nests continue to thrive until the first hard frosts kill them out in late October and early November.

Treatment. Wasp populations fluctuate considerably from year to year and in years when conditions favour them they may be so numerous that it may be almost impossible to protect all susceptible plants from attack. Nest destruction may have some effect but workers may often fly into gardens from nests situated some distance away. Nests found within gardens may be destroyed by applying carbaryl or derris dust to the entrances at night when most of the workers will be inside. It is best to wait until at least an hour after sunset. Wasps leaving and entering treated nests will be contaminated by the dust and the nests will eventually die out. Wasps visiting fruit trees can be trapped in great numbers by baiting jam jars or other containers with a little jam and then partly filling them with water and detergent. Prevention of bird damage to fruit (p. 255) will also reduce wasp damage by limiting the initial injuries that the wasps find so attractive. Wasp stings should be treated with a mild antiseptic and with cold compresses and anti-histamine creams. Individuals who are hypersensitive to stings should seek medical advice.

LEAF-CUTTER BEES
(Plate 7)

These solitary bees are closely related to honey bees but do not live in communal nests. *Megachile centuncularis* is the commonest species in gardens but other species also occur.

Symptoms. Neat semi-circular to oblong pieces are cut out of leaves of roses (Pl. 7), laburnums, lilacs, privet, rhododendrons and other ornamental plants during the summer by small, hairy bees which are about 10 mm long. Damage to plants is seldom severe but the sudden appearance of the unusual symptoms may cause concern.

Biology. Female bees cut pieces out of leaves with their mandibles and use them to construct series of thimble-like cells in cavities in decaying wood, old brickwork, light soils, or even in flower pots. The female bee provisions each cell with a mixture of pollen and honey before laying a single egg and then sealing the top with a neat, circular piece of leaf. About six cells are built end to end in each series. Larvae feed within the cells during the summer and then overwinter in them. They pupate in the following spring and adults emerge in June, eating their way out of the cells in an orderly fashion so that adults developing from the first eggs laid are the last to leave.

Treatment. Prevention of damage is difficult and is seldom necessary since plants are not normally seriously affected. If necessary, plants can be protected by spraying or dusting with HCH and nests can be located and destroyed by following the bees when they are working.

BUMBLE BEES

Bumble bees (*Bombus* species) occasionally eat into the corolla tubes of some flowers, especially those of runner beans, sweet peas, antirrhinums and aquilegias, to feed on nectar. Damage is generally slight and is completely outweighed by the beneficial effects of pollination.

BURROWING BEES

The tawny burrowing bee, *Andrena fulva*, is sometimes common in gardens and other species may also occur. Small underground nests are excavated by bees working in soil during April/May and miniature spoil heaps are thrown up at the entrances. These may be a nuisance on lawns and paths but they soon disappear once the breeding season is over.

BEETLES
(Plate 8)

Beetles (Coleoptera), comprise the largest order of insects with more than 250,000 species known and at least 5,000 of these recorded from Britain and northern Europe. The fore-wings of adult beetles are characteristically modified to form a pair of hard wing-covers (elytra) which hinge over the abdomen and over the membranous hind-wings. The head is generally relatively small but well developed, with prominent eyes, a pair of antennae, and strong mouthparts capable of biting and chewing hard or soft food. Size of pest species varies from very small (1 mm long) to large (3 cm long). Most species are black, grey or brown but some are brightly coloured. Reproduction is normally sexual but a few species are parthenogenetic. Females lay eggs on or near food, depositing them singly or in clusters. Larvae vary considerably in structure from active, predaceous types with three pairs of well developed thoracic legs and hard, pigmented bodies to relatively inactive, colourless, legless, maggot-like larvae with soft bodies. Most larvae have a well developed head with biting mouthparts. Both adults and larvae feed and their feeding habits are varied. Predators (ground beetles, ladybirds) feed on other insects and small invertebrates and may therefore be beneficial in gardens. Some groups are aquatic (dytiscid and whirligig beetles); others feed mainly on dead wood (furniture, longhorn and death-watch beetles) and many species feed on living plant tissues and are pests of cultivated and wild plants. The most important pests of garden plants are listed below and are dealt with in detail in the following section.

Beetle species	Main plants attacked
FRUITS	
Chafers (p. 235)	Raspberries, strawberries, fruit trees, lettuces and other vegetables, ornamental plants and lawns
Wireworms (p. 236)	Strawberries, potatoes and some ornamentals
Strawberry beetles (p. 237)	Strawberries
Raspberry beetle (p. 237)	Raspberries, loganberries and blackberries
Apple blossom weevil (p. 241)	Apples
Leef weevils (p. 241)	Apples, pears and ornamental shrubs and trees
Nut weevil (p. 244)	Hazel nuts
Shot-hole borers (p. 244)	Apples, pears, plums and cherries
VEGETABLES	
Flea beetles (p. 238)	Young brassicas and some ornamentals
Colorado beetle (p. 239)	Potatoes
Asparagus beetle (p. 239)	Asparagus
Turnip gall weevil (p. 242)	Brassicas and turnips
Pea and bean weevils (p. 242)	Peas and beans

ORNAMENTALS

Water-lily beetle (p. 239)	Water-lilies
Lily beetle (p. 240)	Lilies
Willow and poplar leaf beetles (p. 240)	Willows and poplars
Vine weevil (p. 242)	Pot cyclamen and begonias and other ornamentals
Elm bark beetles (p. 244)	Elms

CHAFERS   Pl. 8

The larvae of at least five different species of chafer beetle (Scarabaeidae), also known as white grubs, live in soil, usually in old grassland and overgrown gardens, and both adults and larvae attack cultivated plants.

Symptoms. Soft-bodied, white, C-shaped larvae, up to 40 mm long, with well developed brown head and three pairs of thoracic legs (Pl. 8) feed on roots, corms, tubers and stems of fruits, vegetables and ornamental plants and also attack lawns. Raspberries, strawberries, potatoes, lettuces, young trees and various herbaceous ornamentals are especially susceptible. Affected plants wilt and die when the main root is severed. Similar symptoms are produced by vine weevil larvae (p. 242) but they can be distinguished by the absence of any thoracic legs. Adult chafers also cause some damage by feeding on leaves, buds, flowers and fruits of apples, roses and some other plants. When lawns are attacked by larvae, brown, withered areas appear during dry weather and chafer larvae can usually be found by lifting affected turf. Similar symptoms are produced by leatherjackets (p. 215) and other factors (p. 365).

Biology. The biology of the five main chafers found in gardens is basically the same. Females lay eggs in soil near plants during summer and larvae hatch a few weeks later. They then feed underground on roots, corms and tubers or in rotting wood until they are ready to pupate, which is generally within a year of hatching for the smaller species but may take up to five years for the larger cockchafer. Before pupating, larvae construct pupal cells in the soil at depths down to about 60 cm. These cells may be exposed during winter digging and, if opened carefully, will be found to contain a chafer larva, a pupa, or even a fully grown adult ready to emerge in the following spring.

The main chafer species found in gardens are:

COCKCHAFER *Melolontha melolontha*   Pl. 8

Also known as the May bug. Large adults, up to 30 mm long, have redbrown wing cases, lamellate, clubbed antennae, and white triangular markings on the sides of the abdomen. They sometimes swarm on apples, oaks and other trees, especially on warm nights in May/June. Larvae are common and widespread and are the largest of the chafer grubs, growing to a length of 40 mm (Pl. 8).

GARDEN CHAFER *Phyllopertha horticola*

This is a smaller species, up to 15 mm long, and it is often abundant on light soils overlying chalk or limestone. The head and thorax are metallic blue-green and the wing cases are brown.

SUMMER CHAFER   *Amphimallon solstitialis*
This is about the same size as the garden chafer but is uniformly red-brown, without metallic colouring. It is generally less common but sometimes swarms on elms and poplars.

BROWN CHAFER   *Serica brunnea*
This species is locally common, especially in well-wooded areas. Adults are about 15 mm long and are entirely red-brown.

ROSE CHAFER   *Cetonia aurata*
This is a striking metallic-green beetle, up to 20 mm long. It is locally common and may be seen feeding on rose buds, flowers and leaves in June.

Treatment. Chafer larvae are mainly pests in newly-cultivated land and are soon eliminated by thorough cultivation and good weed control. Valuable plants may be protected by working a relatively persistent insecticide, such as HCH, bromophos or diazinon, into the soil around the roots, preferably before planting. Infested lawns benefit from heavy rolling in late spring to kill pupae and emerging adults and, if necessary, HCH dusts or sprays may be used.

WIREWORMS   Pl. 8
Wireworms are the larvae of various species of click beetle (Elateridae), which are well known for their ability to flick themselves into the air when lying on their backs. This acrobatic click jump puts them back onto their legs after a fall. The larvae mainly develop in grassland but are occasionally troublesome in gardens, the main species being the garden click beetle, *Athous haemorrhoidalis*, and the common click beetle, *Agriotes lineatus*.
Symptoms. Elongate, cylindrical, white to golden yellow, tough-skinned larvae, up to 25 mm long, with three pairs of thoracic legs (Pl. 8) live in soil and feed on roots, corms, tubers and stems of many plants. Larvae tunnel into mature potato tubers, leaving small holes, 2–3 mm across, which may be enlarged later by slugs and millepedes. Wireworms also attack strawberries, cabbages and other brassicas, beans, beetroot, carrots, lettuces, onions, tomatoes and many ornamental plants, especially anemones, carnations, chrysanthemums, dahlias, gladioli and primulas. Most damage is done in spring but wireworms are present throughout the year and may attack plants at other times.
Biology. Female click beetles lay eggs in soil during June/July and larvae hatch about a month later. They feed for up to five years, with the main periods of active feeding in March–May and September/October. Fully fed larvae construct oval pupal cells about 20 cm below the soil surface before pupating. Adults emerge from pupae in late summer and early autumn but remain within the cells until the following summer when they come to the surface, mate, and continue the cycle.
Treatment. Thorough cultivation and good weed control should soon eliminate wireworms from newly cultivated ground and, if susceptible plants need temporary protection, the insecticides recommended for chafer control should be adequate (see above). Compost heaps and heaps of turf stacked to produce loam are sometimes infested by wireworms and should

be checked before use. If many wireworms are seen to be present, the compost or loam should be exposed to weather and birds to reduce the numbers of wireworms before use. Main-crop potatoes should always be lifted early when wireworms are likely to be present since damage will be greatest when they are left too long in the soil after maturing.

STRAWBERRY BEETLES

Most ground beetles (Carabidae) are predators on insects and other small invertebrates but a few species, notably the strawberry seed beetle, *Harpalus rufipes*, and the strawberry ground beetles, *Pterostichus madidus* and *Pterostichus melanarius*, are omnivorous or specialist seed-feeders and may damage ripening strawberry fruits.

Symptoms. Black beetles, up to 20 mm long feed on ripening strawberries, removing seeds and eating out pieces of flesh. Similar damage is caused by birds and slugs but when strawberry beetles are the cause they can usually be found by looking under the straw beneath damaged fruit trusses.

Biology. Adults and larvae overwinter in burrows in soil, in accumulations of plant debris, and in overgrown areas of gardens. They become active in spring and females lay eggs in the soil from May onwards.

Treatment. Clean up strawberry beds in autumn by removing old straw and other plant debris and also clear adjacent areas of rough grass or other overgrowth that could harbour beetles. In early spring, reduce beetle populations by sinking jam jars level with the soil surface to act as pit-fall traps. Examine these regularly and kill any strawberry beetles found in them. Methiocarb pellets, used to control slugs, may have some effect on strawberry beetles but should not be used during the harvesting period since ripe fruit could easily be contaminated.

RASPBERRY BEETLE *Byturus tomentosus* Pl. 8

Symptoms. Yellow-brown larvae, up to 8 mm long, with brown heads and brown dorsal markings, feed in and on ripening fruits of raspberries (Pl. 8), loganberries and blackberries, causing direct damage by feeding on the fruits and reducing the palatability of picked fruit by their presence, which is often first noticed when the fruit is being prepared for eating. Adult raspberry beetles may also cause some damage by feeding on buds and flowers earlier in the season.

Biology. The small, inconspicuous adult beetles, which are up to 4 mm long, overwinter in small earthern cells in the soil near their host plants and emerge from hibernation in April/May. They then fly to hawthorn, apple blossom and other flowers to feed and move onto raspberry, loganberry and blackberry blossom as it opens later in the season. In June/July females lay eggs on the stamens and other floral parts of these host plants. Larvae hatch about two weeks later and feed for about a month by tunnelling into the developing fruits. Fully fed larvae pupate in earthern cells in the soil and adults emerge in the autumn but remain within the pupal cells until the following spring.

Treatment. This is a common species and most cultivated raspberries, loganberries and blackberries are likely to be attacked by adults flying into gardens from adjacent hedgerows and uncultivated areas where wild blackberries and raspberries thrive. Under these circumstances, only chemical control can effectively check this pest and correct timing is critical since the larvae must be killed before they tunnel into the fruit. Fenitrothion is probably the best insecticide currently available for use in gardens, with malathion or derris as possible alternatives. One or other of these insecticides should be applied as a spray or dust as soon as flowering has finished. Reduce risks to pollinating insects by applying insecticides in the late evening and do not harvest fruit until at least a week after application of insecticides.

FLEA BEETLES   Pl. 8

Many species of flea beetle (Chrysomelidae) attack cultivated plants. The main species in gardens are the small striped flea beetle, *Phyllotreta undulata*, the large striped flea beetle, *Phyllotreta nemorum*, the turnip flea beetles, *Phyllotreta atra* and *Phyllotreta cruciferae*, and *Phyllotreta nigripes*. These *Phyllotreta* species are mainly pests of brassicas and other cruciferous plants but other garden plants may be attacked by similar species belonging to other genera. Some species transmit radish mosaic and turnip crinkle viruses (p. 440).

Symptoms. Adult flea beetles attack germinating brassica and other seedlings as they first emerge from the soil in spring. Small holes and pits, 2–3 mm across, are eaten out of the young leaves (Pl. 8) and sustained attacks check growth and may kill plants. Most damage is done in April/May, especially during dry weather, and at this time the small adult beetles, up to 3 mm long, may be seen jumping like fleas on and near affected plants. Larvae also damage some plants by feeding on roots or mining leaves. Radishes, swedes, turnips, cabbages, kale, beet and also various ornamentals, especially alyssums, anemones, drabas, godetias, irises, nasturtiums *(Tropaeolum)*, stocks and wallflowers are especially susceptible.

Biology. Adult flea beetles hibernate in accumulations of plant debris under hedges, in grass tussocks, under loose bark on trees, and in similar situations. They emerge from hibernation in early spring and move onto young plants to feed. Adults may fly for a kilometre or more from their hibernation sites and are especially attracted to brassicas and related plants. Females lay eggs in soil near plants during May/June and larvae feed in leaf mines (large striped flea beetle) or on plant roots (other species). Larvae pupate in the soil and the next generation of adults emerges in August–October. These adults feed for a few weeks and then seek hibernation sites to overwinter.

Treatment. Clear accumulations of plant debris to limit numbers of adults overwintering and protect susceptible young plants in spring by applying a combined insecticide/fungicide seed-dressing to seed before sowing. If necessary, protect plants after germination by applying an insecticidal dust, such as derris or HCH, to the young leaves and to the soil around plants. The effects of flea beetle attack may also be reduced by thorough preparation of the soil before sowing, to encourage rapid establishment and healthy growth, and by watering young plants in dry weather.

COLORADO BEETLE   *Leptinotarsa decemlineata*   **Pl. 8**
This species is native to N. America where it feeds on wild plants that are
related to cultivated potatoes. It was accidentally introduced into France
about 1921 and has gradually become established throughout most of
western Europe. Adults are often found in Britain, mainly on imported
horticultural produce, but constant vigilance and strict control measures
applied by the Ministry of Agriculture have prevented the establishment of
breeding populations here. Any suspected outbreak found in Britain must,
by law, be reported immediately to the Ministry of Agriculture.
Symptoms. Conspicuous black and yellow striped adult beetles, up to
12 mm long (Pl. 8) and equally conspicuous bright red larvae feed on leaves
of potatoes, tomatoes and related plants, often reducing them to a skeleton
of bare stems. Infested plants are characteristically fouled by black excre-
ment produced by the beetles and their larvae. Severe infestations cause
substantial or complete loss of crops.
Biology. Adult beetles hibernate in the soil at depths down to about 25 cm.
They emerge from hibernation on warm days in spring and early summer
and make short flights of a few kilometres in search of potato plants. When
they find these or other suitable host plants, they settle to feed on the leaves
and each female lays about 500 orange-yellow eggs in small batches on the
undersides of the leaves. Eggs hatch about a week later and larvae feed for
about a month before going to the soil to pupate. A second generation of
adults may lay eggs and produce further larvae in late summer but by the
autumn most beetles will have gone to the soil to hibernate.
Treatment. Any infestation discovered in Britain must be reported
immediately to the Ministry of Agriculture, which will then take responsibility for investigation and treatment of the affected area. In continental Europe contact insecticides (see p. 33) applied to potato plants in
June /July may give adequate control.

ASPARAGUS BEETLE   *Crioceris asparagi*
Symptoms. Conspicuous adult beetles, up to 6 mm long, with distinc-
tively marked yellow and black wing cases and grey-black, hump-backed
larvae feed on leaves and stems of asparagus from May onwards. Persistent
attacks defoliate plants, damage stems and may check growth.
Biology. Adults hibernate under stones, in plant debris, in soil, and in
similar protected situations. They emerge in May to feed on young asparagus shoots and females lay small batches of brown eggs on the leaves in
June. Eggs hatch about a week later and larvae feed for about two weeks
before going to the soil to pupate in cocoons. One or two further generations develop during July–October and towards the end of this period
adults seek hibernation sites.
Treatment. Limit overwintering populations by clearing up accumulations of plant debris and spray or dust asparagus with a non-persistent
insecticide, such as derris or malathion, if beetles or larvae appear in spring.
Repeat this treatment later if infestations persist.

WATER-LILY BEETLE   *Galerucella nymphaeae*   **Pl. 8**
Symptoms. Small, brown, adult beetles, up to 6 mm long, and soft-
bodied, brown-black larvae, eat irregular, narrow furrows and small holes

in the upper surface of water-lily leaves in spring and summer (Pl. 8). Affected leaves rot and disintegrate.

Biology. Adult beetles hibernate in coarse vegetation and plant debris at the sides of pools. They emerge in May/June and move onto water-lily leaves to feed. Females then lay batches of eggs and the larvae, which hatch about a week later, feed for 2–3 weeks before pupating on the leaves. Two or three further generations develop during the growing season and adults leave the water-lilies in autumn to seek hibernation sites.

Treatment. Insecticides should not be used against this pest if fish are present in pools or if there is any possibility of insecticides contaminating water-courses. In these circumstances the best method is to knock adults and larvae off leaves with a powerful jet of water from a garden hose. Fish will then be able to feed on them in the water and the treatment can be repeated as often as is necessary. If there is no danger to fish, non-persistent insecticides, such as derris, malathion or nicotine, may be used, either as dusts or sprays.

LILY BEETLE   *Lilioceris lilii*   Pl. 8

Symptoms. Conspicuous, scarlet-red adult beetles, up to 6 mm long (Pl. 8) and reddish-yellow, hump-backed larvae, covered with black slime, feed on leaves, stems and seed pods of lilies and also attack fritillarias, nomocharis and polygonatum (Solomon's seal) from May onwards. Both adults and larvae feed voraciously and cause considerable damage when conditions favour them.

Biology. Adult beetles overwinter in plant debris and coarse vegetation and move onto lilies and other host plants in May/June. Females lay 200–300 eggs each on leaves and stems and larvae, which hatch about a week later, feed for about a month before going to the soil to pupate in earthern cells.

Treatment. Apply a contact insecticide, such as malathion, resmethrin or HCH, as soon as beetles or larvae are seen in the spring and repeat the treatment later if infestations persist.

WILLOW and POPLAR LEAF BEETLES

Various beetles attack the leaves of willows and poplars, causing similar damage. The three commonest are the brassy willow beetle, *Phyllodecta vitellinae*, the blue willow beetle, *Phyllodecta vulgatissima*, and the willow flea beetle, *Chalcoides aurata*.

Symptoms. Small, metallic blue, red or green beetles, up to 5 mm long, and groups of black-spotted larvae feed on leaves of willows and poplars in spring and summer, stripping away the tissues and exposing the leaf veins. Persistent infestations make affected plants look unsightly.

Biology. Adults hibernate under dead bark, in plant debris and in similar situations and move onto willow and poplar leaves in May/June. Females lay small batches of eggs on leaves and larvae, which hatch after about two weeks, feed on the undersides of the leaves in small groups, gradually extending the area of attack. After feeding for 2–3 weeks the larvae pupate in the soil and a second generation of adults appears in August/September.

Treatment. Spray adults and larvae with a contact insecticides such as malathion, nicotine or resmethrin, as soon as they are seen on plants.

WEEVILS
Pl. 8

Weevils belong to a remarkably successful family of beetles (Curculionidae) with more than 50,000 world species. At least 500 of these are recorded from Britain and about 1,000 from northern Europe. They vary in size from small seed weevils, less than 2 mm long, to the large pine weevils, 20–25 mm long. All weevils have a characteristic snout (rostrum) projecting forward from the head, with terminal mouthparts and many have elbowed antennae. The long-nosed weevils have an exceptionally long rostrum, which may exceed the length of the rest of the body, but in most weevils it is much shorter. Adults and larvae of all species feed either on living or on dead plant tissues and many species are important pests of cultivated plants. Larvae are relatively featureless, with a well-developed head, no thoracic legs, and a soft, white body (Pl. 8). The main pest species are dealt with below but there are many other species that may occur in gardens as minor pests.

APPLE BLOSSOM WEEVIL   *Anthonomus pomorum*   Pl. 8
Symptoms. Apple blossom fails to develop normally in spring. Flowers remain closed, with dead petals attached (Pl. 8), and careful examination of affected flowers may reveal weevil larvae, pupae or adults inside. Similar symptoms are caused by frost (p. 482).
Biology. The small, black to brown adult weevils, up to 4 mm long, hibernate under loose bark and in dead leaves and other accumulations of debris near apple trees. In early spring they crawl up trees and feed on young leaves. Females then lay up to 20 eggs each in flower buds and the larvae, which hatch after 1–2 weeks, feed on the stamens and other parts of the young flowers and prevent the normal opening of the petals. Usually a single larva develops in each affected flower and, after feeding for a few weeks, the larvae pupate in the flowers, under the dead petals. Adults emerge in June/July and feed on the leaves for about a month before seeking hibernation sites.
Treatment. This pest does not normally cause appreciable damage since it tends to act as a natural thinning agent, reducing the amount of fruit setting. If it is known to be locally important, protect blossom by spraying trees with fenitrothion or HCH just before the flower buds open, to kill female weevils before they lay eggs.

LEAF WEEVILS   *Phyllobius* spp.
Small, (5–10 mm long), dark brown to black weevils, densely covered by metallic gold or greenish-bronze scales, feed on leaves of apples and other fruits and on leaves of birch, limes, flowering cherries, crab-apples, poplars and rhododendrons, eating small holes in the leaves and occasionally damaging blossom. The main species are the brown leaf weevil, *Phyllobius oblongus*, the silver-green leaf weevil, *Phyllobius argentatus*, and the common leaf weevil, *Phyllobius pyri*, but some other species also occur in gardens. These weevils are usually seen on plants in May–June. They seldom cause severe damage and can easily be controlled, if necessary, by spraying with a contact insecticide, such as malathion or resmethrin.

TURNIP GALL WEEVIL   *Ceutorhynchus pleurostigma*   **Pl. 8**
Symptoms. Characteristic bumpy round galls develop on roots of turnips, cabbages (Pl. 8) and other brassicas. They resemble symptoms of club-root disease (p. 328) but the galls are hollow and contain weevil larvae. Extensive galling stunts growth and may make turnips unusable.
Biology. The small adult weevils, up to 3 mm long, are active from March onwards. Females lay eggs in holes that they make in the roots of their host plants and as the larvae develop the plant tissues form galls around them. Larvae feed in the galls until the following spring and then leave them to pupate in the soil.
Treatment. Remove and destroy affected roots of old brassica crops in early winter and cultivate the soil thoroughly. Protect spring and early summer sowings of cabbages and turnips by dressing the seed with a combined insecticide/fungicide seed-dressing before sowing. Discard any brassica transplants that show signs of attack.

PEA and BEAN WEEVIL   *Sitona lineatus*   **Pl. 8**
This is one of several different species of *Sitona* that feed on cultivated plants. Similar damage is caused by these weevils on roses, lupins, cytisus, sorbus, laburnum and carnations, among others. Some species transmit certain viruses that infect broad beans (p. 438).
Symptoms. Small, semi-circular pieces are eaten out of the edges of pea and broad bean leaves in spring and summer, producing a characteristic scalloped effect (Pl. 8). Small, short-snouted weevils, up to 5 mm long, may be seen on affected plants but often drop off when disturbed.
Biology. Adult weevils overwinter in plant debris and coarse vegetation and move onto peas, beans and related plants in early spring. Females lay eggs in the soil during warm weather and the larvae, which hatch about two weeks later, feed on the root nodules for about a month before pupating in the soil. Adults emerge in June/July and feed on various plants until the autumn, when they seek hibernation sites. There is only one generation a year.
Treatment. Encourage good germination and rapid growth by preparing the ground thoroughly before sowing and by treating the seed with a combined insecticide/fungicide seed-dressing. If necessary, protect young plants by dusting the leaves with derris or HCH. Older plants are not greatly affected by this pest and need not be treated unless damage is excessive.

VINE WEEVIL   *Otiorhynchus sulcatus*   **Pl. 8**
This is the most important species of *Otiorhynchus* attacking garden plants in glasshouses, houses and out of doors but two other species, the clay-coloured weevil, *Otiorhynchus singularis*, and the strawberry root weevil, *Otiorhynchus rugifrons*, attack some plants out of doors. The clay-coloured weevil is particularly associated with damage to apples, currants, gooseberries, raspberries, yew, roses, rhododendrons, polyanthus and clematis and the strawberry root weevil is mainly a pest of strawberries. The biology and treatment of both of these weevils is essentially the same as for the vine weevil.
Symptoms. Legless, white larvae, up to 8 mm long, feed on roots, corms

and tubers of many plants. They look like miniature chafer larvae but lack thoracic legs (Pl. 8). They are especially troublesome on pot plants and container-grown plants in houses and glasshouses but also affect plants growing outdoors in borders, rock gardens and similar situations. Pot cyclamen, primulas and begonias are most susceptible to attack but this pest also affects crassulas, ferns, gloxinias, saxifrages, sanseverias, camellias, orchids and many others. Destruction of roots by larvae checks growth and may cause sudden wilting and collapse of shoots and leaves. Adult vine weevils are seldom seen since they are mainly nocturnal and hide during the day. Their presence is generally indicated by irregular notches and holes eaten out of leaves and by the death of young shoots following ring-barking. Camellias and rhododendrons are most susceptible to this damage, especially in woodland plantings.

Biology. The biology of the vine weevil is unusual. Most other beetles reproduce sexually but virtually all vine weevils are female and they must therefore reproduce parthenogenetically. Male vine weevils have been found occasionally but are very rare. Each female lays up to 1,000 eggs in soil or potting media over a period of 3–4 months during spring and early summer and, although many of these eggs are infertile, this means that a single female has the potential to start a serious infestation. The eggs are usually laid close to suitable host plants and the larvae, which hatch after about two weeks, feed for three months or more before pupating in small earthern cells in the soil or other growing medium. Adults mostly emerge in the autumn under glass but not until the following spring out of doors. There is basically one generation a year but, because of the staggered emergence of adults, there is some overlap in late winter and early spring when eggs, larvae, pupae and adults may all be present at the same time, especially in glasshouses and houses. Adult weevils are unable to fly but crawl into glasshouses through doors and ventilators or are introduced on plants. They hide at soil level during the day, in leaf litter, loose brickwork and woodwork and similar situations, and crawl up onto plants after dark.

Treatment. In glasshouses damage can be prevented by regular inspection, good hygiene, and the use of chemicals. Remove all accumulations of plant debris that could provide cover for adults and, if necessary, fumigate or spray with HCH to kill adults before they lay eggs. Protect susceptible plants, such as cyclamen, primulas, and begonias, by incorporating HCH dust in the potting medium when potting up. This will protect young plants while they are establishing and additional protection may be given later by working HCH dust into the surface layer of the potting medium or by watering in a spray strength solution of HCH. During re-potting destroy any larvae, pupae or adults seen and use the above treatments to eliminate survivors.

Protection of plants growing out of doors is not so easy since the damage is usually done before the presence of the pest is realised. Saxifrages and sedums growing in light soils are especially susceptible and may be protected by working HCH dust into the soil around them in spring and early summer. Leaf feeding by adult weevils seldom causes severe damage although it may make rhododendron and other foliage look unsightly. If necessary, adult weevils could be killed by spraying affected plants with HCH in May/June.

NUT WEEVIL   *Curculio nucum*
Larvae of this long-snout weevil feed on kernels of hazel nuts and emerge through neat exit holes that they cut in the shells. Eggs are laid in young nuts from June onwards. Damage is seldom severe but if weevils are seen on plants during the summer it might be possible to reduce their numbers by spraying with a contact insecticide, such as HCH.

ELM BARK BEETLES   *Scolytus scolytus* and *Scolytus multistriatus*   **Pl. 8**
Symptoms. These beetles are serious pests of elms since they are mainly responsible for spreading Dutch elm disease, which causes sudden yellowing of leaves on affected branches and subsequent death of most infected trees (p. 307). If the dead branches and trunks of affected trees are examined, bark beetles and their larvae will be found in characteristic tunnels under the loose bark. These tunnels consist of a main gallery, about 25 mm long, in which the female beetle lives, and a characteristic series of radiating side galleries, in which the larvae develop (Pl. 8). The beetles are 3–6 mm long, with brown wing cases and shining black thorax.
Biology. Adults, pupae and larvae overwinter in dead trees and branches and adults emerge from May onwards. Development is greatly influenced by the weather and may vary considerably from year to year, from place to place, or even from one side of a log to the other. Adults fly to the tops of healthy trees to feed on the smaller twigs. Spores of the Dutch elm disease fungus are carried on their bodies and mouthparts and are introduced into the tissues of healthy trees as the beetles feed on them. Transmission of the disease is mainly due to this feeding activity of adult beetles and infections therefore tend to spread most rapidly in warm springs and summers when the beetles are most active.
Treatment. There is no really effective way of controlling elm bark beetles. Destruction of infected trees removes potential breeding sites and limits local populations but the beetles can probably fly for several kilometres so that these measures will only have an effect if used over large areas. Even when this is done, beetles may survive in sufficient numbers to initiate further infestations. In North America large-scale spraying from the ground and from the air has failed to eradicate *Scolytus scolytus* and it seems unlikely that chemical control of the beetles will ever be really effective. Biological control has been suggested as a possible alternative solution to the problem, both in North America and in Europe, but will require a substantial research input if it is to have any hope of succeeding.
    Timber from infected trees is not seriously affected either by the disease or the beetles. It may therefore be used but all branch wood, trimmings and bark should be removed during the winter and should be burned by the end of March to kill the beetles before they emerge.

SHOT-HOLE BORER   *Xyleborus dispar* (= *Anisandrus dispar*)
Symptoms. Small holes, about 1–2 mm across, appear in the bark on trunks of trees. Fine wood dust may be produced from some of these holes and on some host plants, notably plums and cherries, a sticky gum may exude from them. Shot-holing often develops on plums, cherries and

peaches that are already dying because of disease or unfavourable growing conditions but primary attacks may develop on otherwise healthy trees, especially on apples, pears and apricots and, if left unchecked, may eventually kill them.

Biology. Adult beetles, which are only about 3 mm long, overwinter in galleries that they excavate in their host plants. They emerge in the spring when temperatures rise and, after mating, the females disperse and establish new galleries. Eggs are laid in May/June and when the larvae hatch they feed on the mycelium of a fungus ('ambrosia') that grows in the galleries from spores brought from old galleries on the bodies of the female beetles. Larvae pupate and adults emerge in the tunnels during late summer but remain in them until the following spring.

Treatment. This pest is only exposed outside its host plant for a brief period in spring and is therefore difficult to control. Severely infested trees should be cut down and burned in early winter to limit the numbers of adults dispersing in the following spring or should be treated by a professional tree surgeon, who may be able to prune away damaged areas and conserve the trunk. Susceptible trees can be protected to some extent by painting the trunks with a spray-strength solution of HCH at fortnightly intervals from mid-April to mid-May to prevent the establishment of dispersing females.

MITES
(Plate 9)

Mites differ from insects in having four, not three, pairs of jointed legs in the nymphal and adult stages and in lacking any clear division of the body into head, thorax and abdomen. They are generally small (1–2 mm long) or very small (less than 0.5 mm long) and are extremely abundant in most terrestrial habitats. Many species are general scavengers, feeding in soil, in decaying organic matter and in similar situations; others prey on small invertebrates, including other mites and many species of insect, and some feed on healthy plant tissues and may therefore be pests of cultivated and wild plants. Reproduction is generally sexual and the main stages of development are: egg, larva, various nymphal stages and adult. Males are absent or rare in some species and reproduction is then parthenogenetic.

Despite their relatively small size, mites are often serious pests of cultivated plants since they are able to breed rapidly when conditions are favourable and they are often less susceptible to chemical control than many insects. The main pest species in gardens are:

Mite species	*Main plants attacked*
FRUITS	
Glasshouse red spider mites (p. 247)	Strawberries, peaches, vines and also many vegetables and ornamentals, especially cucumbers, beans, tomatoes, chrysanthemums, fuchsias, dahlias and roses growing in glasshouses and outdoors
Fruit tree red spider mite (p. 249)	Apples, pears, plums and related ornamental trees and shrubs
Bryobia mites (p. 250)	Gooseberries, apples and various ornamentals
Strawberry mite (p. 251)	Strawberries, cyclamen, Michaelmas daisies and some other ornamentals
Black currant gall mite (p. 252)	Black currants
Nut gall mite (p. 253)	Hazels
Pear leaf blister mite (p. 253)	Pears
ORNAMENTALS	
Conifer spinning mite (p. 249)	Spruces and other ornamental conifers
Lime mite (p. 250)	Lime trees
Broad mite (p. 251)	Begonias, chrysanthemums, pot cyclamen, dahlias, fuchsias and other ornamentals in glasshouses
Bulb scale mite (p. 252)	Narcissus and other bulbs
Nail gall mite (p. 254)	Lime trees
Sycamore gall mite (p. 254)	Sycamores and hedge maples
Witches' broom gall mites (p. 254)	Birches and willows
Broom gall mite (p. 254)	Genistas and cytisus
Yew gall mite (p. 254)	Yews
Bulb mite (p. 254)	Bulbs, corms and tubers of various ornamental plants

RED SPIDER MITES   Pl. 9

These are small mites (Acarina: Tetranychidae), up to about 1 mm long. They move relatively quickly and resemble small spiders, to which they are distantly related. Colonies develop on leaves and other aerial parts of plants and usually contain all stages, from eggs to adults (Pl. 9). Young and adult mites feed by extracting sap and cell contents from plant tissues, producing characteristic fine light mottling and other discolourations (Pl. 9). Many cultivated plants are attacked by various species of red spider mite and serious damage is done when populations increase to astronomical numbers in heated glasshouses and during warm summer weather outdoors. Glasshouse and house plants may be almost covered by seething masses of mites and the webbing that they spin and this may even happen on large trees, such as limes, which may have vast numbers of mites swarming over leaves, branches and trunks during hot dry weather in summer.

The red spider mites dealt with here are those that feed on plants but there are other groups of red, spider-like mites that feed on decaying organic matter or prey on small invertebrates. These are sometimes seen in or on soil, on walls and paths, or on tree trunks and branches and may be mistaken for the plant-feeding red spider mites (especially since the latter are often not particularly red-coloured). If in doubt, seek expert advice.

GLASSHOUSE RED SPIDER MITE *Tetranychus urticae* and CARMINE SPIDER MITE *Tetranychus cinnabarinus* **Pl. 9**

These are two of the most troublesome pest species found in glasshouses and houses. *Tetranychus urticae* is generally commoner and more widespread than *Tetranychus cinnabarinus*.

Symptoms. Infestations first show as a very fine light speckling or as localised pale yellow spots on the upper surfaces of leaves (Pl. 9). Careful examination of the undersides of affected leaves, preferably with a hand lens or magnifying glass, should reveal colonies of mites, with numerous minute spherical eggs and, if the mites have been present for some time, scurfy accumulations of egg shells and of mite moult skins will also be present, especially at the sides of the leaf veins. As mite numbers increase, the damage to leaves, buds and flowers is extended. Leaves usually become progressively discoloured, often showing characteristic bronzing, and in severe attacks the leaves may wither and die. Mites then swarm over plants on fine silk webbing and when populations are exceptionally high ropes of webbing, swarming with mites, hang off the tips of leaves. Adult female mites are about 0.5 mm long and the males are slightly smaller. They have four pairs of relatively short legs and the body colour during most of the growing season is usually yellow or green with darker green markings. Hibernating females of *T. urticae* turn bright red as they stop feeding in autumn and active females of *T. cinnabarinus* are reddish brown during most of the year.

The worst red spider mite infestations develop between June and September, both outdoors and under glass, but damage may occur at other times in heated glasshouses. Many different plants are attacked, including strawberries, peaches and nectarines, vines, cucumbers, tomatoes, beans, aubergines, capsicums, cacti, carnations, chrysanthemums, fuchsias, impatiens ('busy Lizzie'), orchids, pelargoniums, poinsettias, primulas, roses and violets.

Bryobia mites are sometimes confused with red spider mites but are quite distinct (see p. 250).

Biology. Female red spider mites live for a month or more and lay up to 100 eggs each, mostly on the undersides of leaves. Eggs hatch after 3–30 days, depending on temperature, and the six-legged larvae feed for a few days before moulting and transforming into the first nymphal stage (protonymph). The protonymph also feeds for a time before transforming into the second nymphal stage (deutonymph) which later develops into the adult. Development from egg to adult takes only about 8 days at an average temperature of 26.5°C (80°F) but more than two months at 10°C (50°F). Breeding continues throughout spring, summer and early autumn but as the days shorten in September females of *T. urticae* stop laying eggs, change colour and leave plants to seek hibernation sites in cracks in woodwork, in straw and leaf litter, under loose bark on stakes, in canes, in soil and potting media and in similar situations outdoors and in glasshouses and houses. They then remain dormant until the weather warms up in March and April when they gradually leave their hibernation sites and seek plants on which to lay eggs and establish the first infestations of the new season.

Some infestations may also remain active throughout the winter and this is especially the case with *T. cinnabarinus*.

Treatment. The simplest way of checking glasshouse red spider mite infestations is to spray susceptible plants with a fine mist of plain water twice a day. The high humidity depresses mite activity and, provided the plants are not affected, this treatment can be used effectively on glasshouse and house plants, and possibly to some extent outdoors on tomatoes and beans. Red spider mite attacks are often most severe on plants that are growing under unfavourable conditions or on plants that are ageing and lacking vigour and pot plants growing at high temperatures in dry, over-crowded glasshouses are especially susceptible. Provision of good growing conditions and frequent propagation of new plants, followed by rigorous discarding of old infested plants, will help to prevent serious attacks. Any severely infested plants present in glasshouses and houses in early autumn are best removed and destroyed, preferably by burning immediately. This will kill off many female mites that would otherwise carry over to the next season in hibernation and numbers of mites can be further reduced by tidying up accumulations of plant debris, straw, old canes, plant ties and other materials that may provide hibernating sites. If glasshouses can be cleared of plants, they should be thoroughly washed down and sterilised.

Chemical control of glasshouse red spider mites is generally unsatisfactory and is best avoided, if possible. Few of the chemicals available to gardeners are specifically active against these pests and, even when they are, mite populations can soon become resistant to most of the main groups of chemicals (see p. 30). If chemical control is attempted, the main choice is between derris, dicofol, malathion, dimethoate, formothion or diazinon, applied as sprays. Derris is usually the safest chemical to use on delicate plants but is of such short persistence that it has to be applied more frequently than other materials.

A biological method of controlling glasshouse red spider mites has been developed in recent years and overcomes some of the difficulties of chemical control. It involves introducing a predaceous species of mite that feeds on all stages of the glasshouse red spider mites. This red spider mite predator is *Phytoseiulus persimilis*, a species that is native to Chile but is now maintained in cultures in Europe and N. America. Adult mites are slightly larger than red spider mites, with longer legs and an orange-red body. They move rapidly over plants, searching for their prey, and each individual can kill about 5 adults or about 20 immature stages a day. Females lay eggs amongst red spider mite colonies and when temperatures are high eggs soon hatch and predator populations increase, often reaching sufficiently high numbers to eliminate all red spider mites. When this happens, the predator eventually starves and dies out, so must be reintroduced if new red spider mite infestations develop. *Phytoseiulus* is mainly used on commercial crops of cucumbers and tomatoes and is sold by commercial suppliers in batches of 100 or more adult mites. Best control is given when the predator is introduced into glasshouses early in the season. It should be established on infested plants throughout the glasshouse since it does not spread very quickly if left to its own devices. The subsequent use of chemicals to control other pests must be restricted as they would kill out the predators and when mite populations have already reached damaging levels it is best to spray with a non-persistent chemical, such as malathion or derris, a week before introducing the predator. Further details of this method of biological control can be obtained from the Ministry of Agriculture and from the Royal Horticultural Society.

FRUIT TREE RED SPIDER MITE *Panonychus ulmi* **Pl. 9**
This is the most important red spider mite attacking apple and plum trees but the similar glasshouse red spider mites (see above) also attack some fruits, especially peaches, nectarines and strawberries.
Symptoms. Mite colonies build up on leaves from April onwards, causing progressive light speckling, bronzing and desiccation. Severe attacks cause extensive damage to foliage in June/July and may reduce fruit yields both in the year of attack and in the following year. The general appearance of mite colonies on the undersides of leaves is very similar to that described for glasshouse red spider mites (see above). Fruit tree red spider mite mainly attacks apples, pears, plums, damsons, cotoneasters, hawthorns and rowans but may also affect some other related plants.
Biology. Small, spherical red-brown eggs, about 0.15 mm diameter, overwinter on twigs and branches of host plants (Pl. 9), often in such great numbers that the reddish colour of the eggs on the branches can be seen from 20–30 metres away. Eggs hatch during April–June and mites feed and breed on the undersides of the leaves throughout the growing season, producing 4–5 successive generations, which results in rapid increases of populations. When temperatures start to fall in September the females lay overwintering eggs and then die out.
Treatment. Examine susceptible host plants carefully in December/January to determine the extent of autumn egg laying. If eggs are abundant, occurring in clusters of a dozen or more rather than as isolated single eggs, apply a soaking spray of DNOC/petroleum to fully dormant plants in January/February, paying strict attention to the instructions for safe use. Otherwise wait until after blossoming and examine leaves for the presence of mite colonies. If they are detected, spray in May or early June with malathion, derris, or other suitable acaricide to check these colonies before they reach peak numbers. Dinocap, benomyl and some other fungicides are partially effective against fruit tree red spider mite and their regular use against diseases may be sufficient to keep this pest under control.

Populations of fruit tree red spider mites have become resistant to organophosphorus acaricides, such as malathion, and to other groups of chemicals, and expert advice should be sought if resistance is suspected. Fruit tree red spider mites can also become troublesome if tar oil winter washes are used regularly to control aphids and other insects since tar oil reduces the numbers of predaceous insects feeding on overwintering red spider mite eggs.

CONIFER SPINNING MITE *Oligonychus ununguis* **Pl. 9**
Symptoms. Colonies of this mite build up on conifers, especially dwarf conifers, in May, June and July. Affected foliage turns yellow (Pl. 9) and in severe attacks the damaged leaves may be shed prematurely. Spruces,

cedars, cypresses, junipers, thujas and pines are susceptible to attack.
Biology. Eggs laid on trees in autumn overwinter and hatch in early spring. Mite colonies then become established on young shoots and breeding continues until the autumn.
Treatment. Small ornamental conifers can be treated by spraying thoroughly on two or three occasions in early spring with malathion or dicofol to check the initial colonies.

LIME MITE   *Eotetranychus tiliarius*

Symptoms. Mite colonies develop on the undersides of leaves in summer and infested leaves may shrivel and fall prematurely. In September vast numbers of mites may swarm over tree trunks and the fine silk webbing that they spin glistens in bright sunlight and makes the tree trunks look as if they have been tightly covered with thin polythene sheeting.
Biology. The detailed biology of this pest is not well known but it probably overwinters as eggs laid on trees during late summer and early autumn. Colonies develop from spring onwards and reach peak populations in late summer.
Treatment. Attacks are sporadic and, although the appearance of large numbers of mites in autumn may cause concern, it is then too late to prevent damage.

BRYOBIA MITES   *Bryobia* spp.

Various species of mites belonging to the genus *Bryobia* attack some fruit and ornamental plants, producing symptoms that resemble those produced by red spider mites.
Symptoms. Mites feed on leaves in spring and summer causing progressive discolouration which may result in general bronzing and desiccation, followed by premature leaf fall. Adult bryobia mites can be distinguished from red spider mites by their relatively long fore legs and by the absence of any long bristles on their bodies. They also prefer to feed on the upper surfaces of leaves and do not produce silk webbing. Apple, gooseberry, pear, polyanthus, ivy and various alpine plants, especially campanulas, dianthus, gentians and saxifrages, are susceptible to attack. In autumn large numbers of bryobia mites move off cultivated plants and weeds and may be seen swarming over paths and invading garden sheds, glasshouses and buildings.
Biology. Most species of *Bryobia* reproduce parthenogenetically and males are unknown. Some species overwinter as eggs on host plants but others survive as adults on plants or in buildings, walls and similar protected situations. Eggs hatch and adults become active from March onwards and up to five successive generations develop during the growing season. Some species breed slowly but others reproduce rapidly, giving rise to high populations in summer and early autumn.
Treatment. Chemicals applied to control fruit tree red spider mites and aphids on apples and other fruits will check bryobia mites. If local infestations on other plants justify control, use malathion sprays, repeating applications after a week, if necessary.

TARSONEMID MITES
Pl. 9

Tarsonemid mites (Tarsonemidae) are very small (up to 0.25 mm long), relatively slow moving mites with elongate, rounded, shiny cream to light brown bodies. They generally feed in buds, under leaf sheaths, in the tips of young shoots and in similar situations on plants, and are therefore more difficult to detect than are the red spider mites, which feed openly on leaves. The main stages of development are: egg, larva, quiescent pre-adult stage (sometimes inaccurately known as a pupa), and adult. Males occur regularly in most species but reproduction may be parthenogenetic for much of the time. Females lay about 50 eggs each, either singly or in small clusters, on or near young buds, leaves and flowers. Eggs hatch after a few days and larvae feed for a week or more before entering the quiescent pre-adult stage. The rate of development from egg to adult is greatly influenced by temperature and at 75°F (24°C) the complete cycle may take less than a week. Mites continue to breed throughout the year but the rate of development slows down considerably during the winter. Females extend infestations by crawling over plants and are sometimes assisted by males which carry female 'pupae' on their backs to new feeding and breeding sites. Such movements are generally restricted so that infestations tend to be relatively local. Effects on plants vary but symptoms mostly result from mites feeding on young tissues. New growth is checked and distorted; buds and flowers may be killed, and damaged stems and leaf bases often show characteristic light brown, russet-like scarring (Pl. 9). Leaves of infested plants are often puckered, curled, slightly thickened and rather brittle and damaged flowers are generally discoloured and distorted (Pl. 9). Damage is often out of all proportion to the numbers of mites present since localised death of plant tissues in attacked buds results in extensive symptoms on the leaves and flowers that later develop from them. Both indoor and outdoor plants are attacked by tarsonemids but damage is often greatest in heated glasshouses and in houses, where higher temperatures encourage rapid and continuous breeding.

The main species attacking garden plants are:

STRAWBERRY MITE
Tarsonemus pallidus
Pl. 9

This species is also known as the cyclamen mite and is especially troublesome on strawberries, pot cyclamen (Pl. 9) and perennial asters (Michaelmas daisies). It also attacks many other plants, including antirrhinums, aphelandras, azaleas, begonias, chrysanthemums, cissus, columneas, crassulas, delphiniums, fatshederas, ficus, fuchsias, gerberas, gloxinias, ivies, kalanchoes, pelargoniums, saintpaulias, verbenas and many others.

BROAD MITE
Polyphagotarsonemus latus

This is very similar to the cyclamen mite and can only be positively distinguished by microscope examination of adults. It is mainly a pest in glasshouses where it attacks cissus, begonias, fatshederas, fuchsias, impatiens and pelargoniums. Some of these host plants may also be infested by cyclamen mite and it is possible to have both species present on the same plant at the same time. The condition of begonias that is sometimes known as 'begonia rust' is caused by broad mite infestations and not by a fungus.

BULB SCALE MITE   *Steneotarsonemus laticeps*   **Pl. 9**
This is mainly a pest of narcissus and hippeastrum bulbs, especially when
these are forced at high temperatures in heated glasshouses and in houses.
If infested narcissus bulbs are cut transversely with a knife, characteristic
brown marks may be seen on the bulb scales where mite colonies have
developed (but also see p. 133). On hippeastrums conspicuous red streaks
show where mites have fed on leaves and stems.

Treatment. None of the chemicals currently available for garden use is
likely to give effective control of tarsonemid mites. Pot plants may be
protected to some extent by dusting with powdered sulphur if temperatures are above 65°F (18°C) but this leaves an unsightly deposit on foliage
and flowers. Spraying with malathion or one of the other organophosphorus acaracides is sometimes recommended but there is evidence that use
of these chemicals can actually increase mite populations, probably by
eliminating natural predators and also by altering the nutritional value of
the plants so that they have a beneficial effect on the mites. Apart from
chemical control, good hygiene can prevent serious outbreaks, especially in
glasshouses. Do not introduce suspect plants into glasshouses, houses or
gardens and isolate or destroy affected plants as soon as they are noticed.
Avoid undue disturbance of infested plants by handling or re-potting since
infestations may be spread by dropping onto other plants and onto staging.
Heat treatments may sometimes be used to eliminate mites from certain
plants but must be carefully applied. Strawberry runners, saintpaulias and
cyclamen are treated by immersing whole plants in water maintained at
46°C (115°F) for 7 minutes and forced narcissus bulbs can be treated by
sudden cooling, which can be achieved quite simply by putting them
outside on one or two frosty nights about a week after forcing has started.

GALL MITES   **Pl. 9**
Many different species of eriophyid mites (Eriophyidae), commonly known
as gall mites, induce galls and other malformations on wild and cultivated
plants. Adult mites are very small, usually less than 0.25 mm long, with
narrow white or yellow bodies and only two pairs of legs. Adults and
immature stages feed on plant tissues, usually inside buds or in galls, and
colonies of gall mites often contain many thousands of individuals. Although the galls caused by these mites are often conspicuous, most species
cause little or no permanent damage to their host plants. The only real
exception to this is the black currant gall mite, which transmits the agent
causing reversion disease.

The main species in gardens are:

BLACKCURRANT GALL MITE
  Cecidophyopsis ribis   **Pl. 9**
This is an important and widespread pest of black currants.
Symptoms. Gall mites infest buds which then become swollen and rounded and fail to develop normally (Pl. 9). This condition is commonly known
as 'big bud' and is most apparent on infested black currant bushes in winter
and early spring. Direct damage to buds is generally less important than

secondary effects resulting from the transmission by the mites of the agent that causes reversion disease (see p. 445).

Biology. Mites breed inside buds in summer and autumn and many mites overwinter in them. Populations increase from January onwards and when the buds start to open in March/April mites disperse from them and invade healthy young dormant buds that will develop in the following season. Mites can only crawl over short distances on plants but may be carried passively over longer distances by wind and rain or on flying insects, such as aphids. Most dispersal occurs during warm, humid weather in May/June and by July the mites are established and breeding in the dormant buds, where colonies persist until the following spring.

Treatment. If possible, remove enlarged buds from bushes during the winter and early spring to limit the numbers of mites carrying over from one season to the next. Destroy these affected buds immediately, preferably by burning, so that the enclosed mites are killed before they can disperse. Chemical control is difficult since the only material available for garden use is $\frac{1}{2}$–1 per cent lime sulphur. This should be sprayed thoroughly 2–3 times in May/June and must be used strictly according to the label instructions to avoid scorching sensitive cultivars. Benomyl fungicide also gives some protection against this pest. Once reversion disease has become established the affected plants will have to be replaced by healthy plants on a new site (see p. 21). To limit the spread of mites and the disease from diseased to healthy plants it is best to lift and burn all infested plants during the winter, well before the new plants are brought into the garden.

NUT GALL MITE *Phytoptus avellanae*

This species resembles the black currant gall mite and produces similar enlarged buds on hazels. It is a distinct species and cannot transfer from hazels to black currants. Damage to hazels is slight and does not justify the use of any control measures, other than picking off and destroying affected buds.

PEAR LEAF BLISTER MITE *Eriophyes pyri* **Pl. 9**

Symptoms. Irregular pink or yellow blisters appear on young pear leaves in spring and, as the leaves age during summer and autumn, these blisters turn dark brown and black (Pl. 9). Severely infested leaves may be killed and growth of young shoots may be checked. Some other host plants, especially apples, cotoneasters, mountain ash, whitebeams and wild service trees are occasionally attacked by this or by closely related species. These symptoms may be confused with those of pear leaf curl disease (p. 332).

Biology. Adult mites overwinter under bud scales and feed on young leaves, causing blistering. Mite colonies then develop within the tissues and breeding continues throughout the summer. Mites move out of infested leaves just before leaf-fall in autumn and seek overwintering sites in dormant buds.

Treatment. If possible, remove and destroy affected leaves as soon as they are seen. Otherwise, treat persistent and severe infestations by spraying with 5% lime sulphur towards the end of March, just as the buds begin to open.

NAIL GALL MITE　*Eriophyes tiliae*　　　　　　　　Pl. 9

Small, conical red galls appear on the upper surfaces of the leaves of lime trees in May/June (Pl. 9). A single leaf may bear hundreds of galls and their bright red colour contrasts sharply with the fresh green colour of the young leaves. This unusual and dramatic appearance often causes concern to gardeners but healthy trees are not seriously affected and control measures are not necessary.

SYCAMORE GALL MITE　*Eriophyes macrorhynchus*

The effects of this species are similar to those of the nail gall mite but it only affects sycamores and hedge maple. Numerous small red galls, each about the size of a pin head, develop on the upper surfaces of young leaves from April onwards, producing characteristic rashes of red pimples. These galls tend to be common on young trees, especially those growing in hedges, but do little harm to them.

BROOM GALL MITE　*Aceria genistae*

Irregular light green growths develop on stems of genista and cytisus during late spring and early summer and may persist for some years. These growths develop from buds that have been invaded by the mites and they should be cut out and destroyed as soon as they are seen. This is a local pest which seems to be commoner in polluted urban areas than in rural areas.

WITCHES' BROOM GALL MITES

Various species of gall mite are associated with twiggy and leafy witches' broom galls on birches and willows. Although these galls are common, conspicuous and widespread, it seems that the role played by gall mites in their development is not clearly understood. Some fungi, viruses and other organisms are known to cause similar symptoms (see p. 334).

YEW GALL MITE　*Cecidophyopsis psilaspis*

Terminal and lateral buds of yew become enlarged after mites have invaded them and subsequent growth is spindly and twisted. Young hedges and nursery stock are generally most susceptible. Clippings from infested hedges should be removed and burned immediately to stop mites moving back onto plants. 2 per cent lime sulphur, sprayed on warm spring days, will also help to limit infestations.

BULB MITES　*Rhizoglyphus callae* and *Rhizoglyphus robini*

These are not primary pests but they can cause appreciable secondary damage to bulbs, corms and tubers by extending wounds caused by other pests and by mechanical injury and this encourages the establishment of fungal and bacterial diseases. Narcissus, tulip, hyacinth and lily bulbs, gladiolus and freesia corms and dahlia tubers are particularly susceptible. Adult mites are about 1 mm long, shiny white, with long setae. They are abundant in most soils and rapidly colonise damaged plant tissues. Mites on affected bulbs, corms and tubers can be killed by fumigating in boxes with paradichlorbenzene but it is generally best to discard infested plants and to diagnose and treat the cause of the primary damage that instigated the bulb mite attack.

BIRDS
(Plate 9)

Most birds are welcome visitors to gardens but a few species are destructive pests of fruits, vegetables and ornamental plants and are especially troublesome because of their mobility and their ability to adapt quickly to changing circumstances. Their rate of reproduction, compared with that of insects and mites, is relatively low. Each female lays only a dozen or so eggs a year but this low egg count is compensated for by parental care of the young, by the relatively long life-span of individuals, and by alert awareness of danger and consequent ability to avoid hazards. The main pest species are noted below and general advice on the prevention of damage is given on p. 258.

BULLFINCH *Pyrrhula pyrrhula*   **Pl. 9**
Three distinct races of this species occur in Europe: the British bullfinch, which is confined to the British Isles; the Northern bullfinch of Scandinavia and the Baltic, and the Continental bullfinch, which occurs in France, the Netherlands and other countries of Western Europe.
Symptoms. During winter and early spring the flower and leaf buds of various fruit and ornamental trees and shrubs (Pl. 9) are attacked by marauding parties of bullfinches working surreptitiously in groups of about 6–12 individuals. They nip off buds systematically, working from the tops to the bottoms of shoots and branches, and taking up to 30 buds a minute. The succulent centres are eaten and the harder outer bud scales are dropped and accumulations of dropped scales and other debris may be noticed under attacked plants, especially on frosty mornings or after falls of snow. Persistent attacks seriously affect subsequent flowering, fruiting and growth. Plums, fruiting and flowering cherries, gooseberries, apples, pears, forsythias, wistarias, viburnums and lilacs are particularly susceptible and in spring the damaged branches have bare lengths devoid of buds.
Biology. Bullfinches are shy birds and they generally frequent the fringes of woodland, coppices, thickets, overgrown scrubland and similar habitats that provide them with adequate cover. Females lay up to three successive clutches of 4–5 eggs during May/September in insubstantial nests made of fine twigs, roots and hairs, usually in hedgerows and thickets. Parties of birds forage together during late summer, autumn and winter, feeding mainly on seeds of various weeds, grasses and trees. By mid-winter most of these food sources are exhausted and birds then feed on buds of wild and cultivated plants until the spring when the parties split up as birds pair and nest. Most bullfinches remain within about a 16 km (10 mile) radius of the nest in which they were reared and, despite occasional reports to the contrary, there are no substantial long-range movements of birds.
Treatment. Repellents are sometimes recommended but are seldom really effective and the only certain way to prevent damage is to erect permanent fruit cages or temporary netting, which should be in place by the end of November (see p. 258). Trapping and shooting are unlikely to have any marked effect on populations and are illegal in most areas.

WOOD PIGEON *Columba palumbus* Pl. 9

Wood pigeons are mainly pests of agricultural crops but can also cause substantial damage in gardens. Collared doves (p. 257) occur in gardens in some areas and may cause similar damage.

Symptoms. Leaves of brassica plants are torn and eaten and are often reduced to a skeleton (Pl. 9). Young transplants may be uprooted and killed and seeds of peas, beans and other crops are taken. Buds, leaves and fruits may be stripped from black currants and other fruit bushes and branches are sometimes broken. Wood pigeons cause most damage in spring and early summer when a few individuals making regular raids into gardens, usually in the early morning, can have quite distastrous effects. Adult wood pigeons are mainly blue-grey with a distinct white patch at the base of the neck and white bars visible on the wings in flight. The presence of wood pigeons near gardens is often indicated by their characteristic call, which is a soft 'coo-cooo-coo, coo-coo'.

Biology. Wood pigeon nests consist of loose platforms of twigs, usually constructed in the lower branches of conifers, hawthorns and other wood-land trees, often in thickets and hedges. Nests may be occupied for most of the year and females rear a succession of clutches, which average about two eggs to a clutch. Eggs are incubated for about three weeks and young birds leave the nests after about another month. The peak nesting period is during August/September and it is at this time that most young birds appear. Many of these die through starvation during the winter and, although predators take some toll, wood pigeon populations are mainly regulated by severe winter weather and food shortages. When weather is particularly severe in western Europe, wood pigeons may migrate south and west in search of food, but they do not undertake regular large-scale migrations.

Treatment. Netting or fruit cages provide the only certain protection for susceptible plants although scaring devices may have some deterrent effect (see p. 258). Trapping and shooting are unlikely to have any permanent effect on populations.

HOUSE-SPARROW *Passer domesticus*

House-sparrows are closely associated with houses and other buildings and therefore tend to be particularly destructive in gardens, especially in urban and suburban areas.

Symptoms. Flowers, leaves and buds of many garden plants are attacked and torn to pieces in winter, spring and early summer. House-sparrows mainly feed on seeds and much of the damage that they do in gardens seems quite wanton as it is not directly connected with normal feeding. It is probably a displacement activity that the birds indulge in when they are not fully occupied with essential activities. Plums, currants and gooseberries are disbudded in late winter and early spring, although similar damage is also caused by bullfinches (p. 255), and many spring and early summer flowers, especially crocuses, primulas and polyanthus, may be torn to pieces and even uprooted. Carnations, chrysanthemums, lettuces, beetroot, peas, onion sets and other plants are similarly attacked and, in addition, house-sparrows eat seed and take dust baths in newly sown lawns.

Biology. Nests are built from straw and dried grasses and are sited in holes in walls, under eaves, in drain pipes, in ivy and other climbers, and in similar situations on or near buildings. The main breeding season lasts from May to August and 2–3 successive batches of about 4–6 eggs are hatched and reared. Adults and immature birds stay near buildings for most of the year but during late summer and early autumn large flocks fly off to feed on maturing cereal crops on farm land.

Treatment. Protect susceptible plants with netting or cages whenever possible. Black cotton or other scaring devices and chemical repellents may prove partially effective (see p. 258). Trapping or shooting will have little effect since cleared areas are soon reinvaded by roving flocks of young birds looking for new territories. Putting out food for birds in gardens attracts house-sparrows and probably gives them more free time to attack plants.

Other bird species that occasionally cause damage are listed below. See p. 258 for treatment.

BLUE TIT *Parus caeruleus*   **Pl. 9**
Blue tits damage ripening fruits, especially apples and pears, by pecking small pieces out of them. These wounds are quickly infected by fungal and bacterial rots (p. 338) and may also be extended by wasps (p. 232). Blue tits also peck camellia, magnolia and other flower buds and flowers in early spring. Damage done by blue tits is counterbalanced to some extent by the fact that they also feed on insects, especially aphids and caterpillars, and may therefore help to reduce pest populations.

BLACKBIRD *Turdus merula*

SONG-THRUSH *Turdus philomelos*

FIELDFARE *Turdus pilaris*

REDWING *Turdus iliacus*
These four related species all eat fruits and berries. Blackbirds and song thrushes are resident in the British Isles but fieldfares and redwings are winter migrants which move into Britain from northern Europe in large flocks, especially during severe winter weather. During spring, summer and autumn blackbirds and thrushes attack strawberries, apples, pears and other ripening fruits and during late autumn and winter all four species attack ornamental berries, especially hollies, berberis, pyracanthas and cotoneasters. Netting is the only certain protection although scaring devices may have some effect.

STARLING *Sturnus vulgaris*
Parties of starlings often work over lawns, pecking at the turf and some-times leaving holes and loose tufts of grass. They are usually feeding on leatherjackets (p. 215) or chafer grubs (p. 235) and their activities are therefore mainly beneficial. In some areas they may cause damage to fruits, especially cherries.

COLLARED DOVE *Streptopelia decaocto*
This attractive dove has spread through western Europe during the past twenty years and is now well established in Britain. It nests in cedars and

other large trees and may occasionally damage garden plants, but is possibly blamed at times for damage that has been done by wood pigeons (see p. 256).

JAY *Garrulus glandarius*

Jays raid gardens from adjacent woodland and scrub. They peck conspicuous V-shaped pieces out of developing pods of peas and beans and eat the seeds. If jays are known to occur locally, protect peas and beans with temporary or permanent netting.

General Treatment. In Britain all birds are protected by law and it is illegal to kill them by shooting, trapping or other means or to take their eggs and destroy their nests. Only authorised persons are excluded from this general provision so far as pest species are concerned. Gardeners must therefore rely on preventive measures. Many different techniques have been developed to prevent or reduce bird damage on agricultural and horticultural crops and on garden plants but man's ingenuity in devising them has often been matched by the ability of birds to avoid or circumvent them.

The main methods worth considering for garden use are:

A. *Nets and cages*. In areas where birds regularly attack plants, the best solution is to use temporary or permanent netting to exclude them. The erection of fruit cages to contain soft and top fruits is especially worthwhile since this can prevent both winter attacks on buds by bullfinches and sparrows and summer attacks on ripening fruits by tits, blackbirds and other species of bird. It also gives protection against squirrels (p. 260). A wide range of wire, cord and plastic nettings is available, as well as ready-made fruit cages, and it should be fairly easy to select materials to suit individual requirements. The smallest meshes, 20 mm ($\frac{3}{4}''$) rigid or 12 mm ($\frac{1}{2}''$) pliable, should be used to exclude blue tits but larger meshes can be used against other birds. Nets should be closed by early November, to prevent winter damage to buds, and should be inspected regularly to ensure that they are secure. They should be removed temporarily in spring to allow free access of pollinating insects, but should be replaced as soon as fruits have set.

B. *Scaring devices*. Bird-scaring devices range from traditional scare-crows to modern recorded transmissions of distress calls. Distress calls, explosive devices and other techniques that rely on noises to scare the birds are used in horticulture and agriculture but are generally unsuitable for use in gardens, especially in small suburban gardens where neighbours may object. Techniques that are better suited to garden use include:

    i. Black cotton, string or similar materials strung above plants from canes, stakes or sticks.

    ii. Glittering strips of foil or similar material hung from lines or canes.

    iii. Small windmills, sometimes with rattles, mounted on stakes.

    iv. Scare-crows simulating human forms.

Birds soon become accustomed to a particular device and these techniques therefore give best results if they are changed or moved at least once a week.

C. *Chemical repellents*. Various chemicals with unpleasant tastes are used either as dusts or sprays applied to plants to repel birds from them. Alum and anthraquinone are probably the most widely used active ingredients in proprietary repellents but other materials are also available. These chemical repellents may be partially effective but are limited in use and erratic in performance. They cannot be used to protect ripening fruits, since the fruits will be tainted, and they are not very effective in preventing winter damage to buds by bullfinches and other birds since severe weather quickly removes deposits of repellents from buds and, in any case, starving birds may not be sufficiently deterred from feeding.

Other methods of bird control, such as the use of stupefying baits, trapping, shooting or nest destruction are not recommended for garden use although they may sometimes be used in agriculture, horticulture and forestry. To be effective, they must be used consistently and efficiently over wide areas and care must be taken to avoid illegal killing of non-pest species.

MAMMALS

Some of the smaller mammals, such as mice, voles, moles and rabbits, are often common pests of garden plants, even in urban and suburban areas, and larger mammals, such as deer, are occasional or regular pests in rural areas.

All mammals give birth to live young and these are protected by the parents during the early weeks and months of growth. Populations never reach the astronomical numbers commonly achieved by many insects and mites but the adaptable behaviour of most mammals, combined with their powerful senses of sight, smell and hearing, may make it difficult to prevent damage to plants.

RABBIT   *Oryctolagus cuniculus*
Symptoms. Young and adult rabbits nibble and graze many different wild and cultivated plants. They eat young shoots and leaves, strip bark from young trees, especially in severe winters, and feed on flowers, fruits and seeds. In gardens lettuces, hostas, heathers and lilies and many herbaceous perennials are particularly susceptible to attack and may be seriously damaged in spring and early summer.
Biology. Rabbits are burrowing animals. They live in tunnels excavated in hedge banks, under thickets of hawthorn and bramble, and in similar situations and they emerge at night to feed on plants. Each rabbit eats about 0.5 kg (1 lb) of greenstuff a day and large colonies of rabbits therefore have a marked effect on vegetation growing near their warrens. Each female produces 2–4 litters a year, mostly between January and June, and each litter contains 3–6 young rabbits. These mature and start to breed within a year and populations can increase rapidly when conditions are favourable. This high reproductive potential may be checked by predators, such as foxes and weasels, and by diseases, such as myxomatosis. This virus disease was introduced into Europe in the early 1950's and was initially very virulent. Fleas and mosquitoes transmitted it from unhealthy to healthy

rabbits and rabbit populations were rapidly reduced by as much as 90 per cent. In Britain rabbits almost disappeared and in 1955 it was estimated that the value of cereal crops increased by £15 million following reduction of rabbit populations by myxomatosis. Since then the disease has become less virulent and rabbits have become immune to it so that numbers have increased again, but in some areas there may be an annual cycle with the disease reaching a peak in late summer and early autumn.

Treatment. When rabbits are abundant on land adjoining gardens, the only way to prevent persistent damage to plants is to erect and maintain a rabbit-proof fence. This should be made from $1\frac{1}{4}$–$1\frac{1}{2}$ m (4–5 ft) widths of 18 or 19 gauge wire netting with a 2.5–3.0 cm (1–$1\frac{1}{4}''$) mesh. About 30 cm ($12''$) of this width is buried in the soil to stop rabbits burrowing under it and the remaining 1–$1\frac{1}{4}$ m (3–4 ft) must be well supported by posts and straining wires. Fences should be inspected regularly to ensure that there are no holes through which rabbits can pass.

If rabbits are not particularly abundant, valuable trees, shrubs and other plants may be adequately protected by netting, spiral tree-protectors or other devices. Chemical repellents may give some protection, especially when applied to the bark of trees to prevent bark-stripping, but are not always effective. Some plants are unattractive to rabbits and a list of these may be obtained from the Royal Horticultural Society. Trapping, shooting and gassing are not suitable for general garden use.

HARES
Hares are related to rabbits but are less numerous and therefore seldom cause appreciable damage. Brown hares, *Lepus capensis,* occasionally occur in gardens, especially when these adjoin farmland, and may damage young fruit trees and ornamental trees and shrubs by stripping bark from the stems. Spiral tree-protectors or wire netting should be used to prevent such damage and susceptible young trees may need protection for many years.

GREY SQUIRREL *Sciurus carolinensis*
Symptoms. Grey squirrels usually enter gardens from adjacent deciduous woodland and may cause considerable damage by digging up and eating bulbs, especially crocuses, lilies and tulips, by taking berries and fruits, especially strawberries, hazel nuts and walnuts, and by eating bark off stems and branches of trees, especially ash, beech, sycamore and spruce. They also eat young shoots, buds and flowers of various trees and shrubs in spring and the cones of conifers, such as pine and larch, at other times of year. They have also been known to gnaw hose pipes and plastic or metal plant labels.

Biology. The grey squirrel is a N. American species that was introduced into zoos and menageries in Europe about a century ago. It escaped from captivity and became widely established, eventually replacing the indigenous red squirrel in much of Britain and parts of Europe. Grey squirrels mainly frequent woods containing broad-leaved trees, such as oaks, and during the winter they live together in large nest-like structures (dreys) made from leafy branches and constructed high in the branches of trees.

The does live alone in large breeding dreys or in holes in trees and produce litters of 3–4 young, either in late February or in July.

Treatment. Squirrels can be well controlled on woodland estates by shooting, trapping and the destruction of dreys but these measures cannot usually be used in gardens. Bulbs can be protected with wire netting until they are through the ground and squirrels may also be discouraged from digging them up if lengths of prickly gorse stem are buried amongst the bulbs at planting time. Fruit cages and netting used to protect fruits from birds (see p. 258) will also deter squirrels and temporary netting can be used to protect particularly valuable ornamental shrubs. Stem barking is best prevented by using spiral tree-protectors or small mesh wire netting.

RED SQUIRREL   *Sciurus vulgaris*

This species is now absent from most of central and southern England but is still common in the wilder areas of Scotland, Wales and East Anglia. It is mainly associated with coniferous forests and is not normally a pest in gardens. It may however occasionally cause similar damage to that done by the grey squirrel (see above).

WOOD MOUSE   *Apodemus sylvaticus*

Also known as the long-tailed field mouse.

Symptoms. This species will eat almost anything but in gardens the main damage is to bulbs and corms, especially those of lilies, narcissus and tulips, and to seeds, particularly seeds of peas and beans which are dug up and eaten after they have been sown. In autumn and winter this mouse enters garden sheds and glasshouses and can cause considerable damage to apples and other stored fruits and to packeted seeds. It is much more common in outbuildings than the house-mouse, with which it is sometimes confused. Populations of long-tailed field mice fluctuate and may be exceptionally high in some years.

Biology. Adults are brown with greyish-white underparts and a relatively long tail. They live in burrows in hedge banks and similar situations and breed from March until October. They do not hibernate but remain active throughout the winter and it is at this time that many individuals move into outbuildings. They are preyed on by owls, weasels, cats and other pre-dacious birds and mammals.

Treatment. Traditional mouse-traps should be set in garden sheds, seed stores, glasshouses and other buildings in autumn and winter to catch mice before they cause any damage. Carrot, apple, melon seeds or pieces of potato should be used as bait. Traps can also be set outdoors but should then be covered with cloches or some other form of protection that will prevent birds and pet animals from getting to them. Seeds of peas and beans should be dressed with a combined insecticide/fungicide seed dressing before sowing, as that makes them less palatable to mice, and seed beds and bulbs may also be protected with 1 cm ($\frac{3}{8}''$) mesh wire netting. Anti-coagulant (warfarin) and other proprietary mouse poisons may be used with baits but precautions must be taken to lay them so that domestic and other animals are not affected.

SHORT-TAILED VOLE *Microtus agrestis*
Symptoms. This pest causes serious damage when it eats the bark off young trees and shrubs at about soil level but it also eats seeds, bulbs and roots of outdoor plants. It does not normally enter buildings.
Biology. Short-tailed voles, also known as field voles, are the commonest small rodents in Britain. They have short stout bodies with small ears and short tails and look like miniature guinea-pigs. They prefer to live in unkempt grassy areas where they make extensive networks of runs near the surface of the soil. These runs may include chambers provisioned with short lengths of grass stem, shoots eaten off bulbs, or other succulent pieces of plants. Short-tailed voles do not hibernate but are inactive for short periods during very cold weather. Plague populations tend to develop in some years and up to 500 voles may then be found per 0.2 ha ($\frac{1}{2}$ acre).
Treatment. Clear overgrown areas of grass, especially around trees, and protect the bases of valuable trees and shrubs with spiral tree-protectors or with wire netting. Trapping and baiting are unlikely to be effective when populations are high since cleared areas would soon be recolonised.

WATER VOLE *Arvicola amphibius*
This is a much larger species than the short-tailed vole with adults growing to a length of about 20 cm (8"). It is sometimes mistakenly referred to as the water rat. Water voles tunnel in banks of streams and pools and may be a nuisance in gardens that include or adjoin water. Apart from physically undermining banks, they also feed on plants near their burrows. Populations are usually small and it may be possible to eradicate them by trapping which should be done by an expert. Failing this, wire netting may be used to protect banks of pools.

MOLE *Talpa europaea*
Symptoms. Small mounds of loose earth (molehills) appear on lawns, rough ground and cultivated areas where moles are active. Light, well-drained soils are most affected but moles can thrive in a variety of situations ranging from lowland suburban gardens to upland moors. Mole activity in gardens is usually greatest in late winter and early spring but new tunnels and molehills can be made at any time of year. Moles do not feed directly on plants but damage them indirectly by undermining the roots, which can seriously affect the growth of young plants, especially vegetables and strawberries. Production of molehills also makes lawns unsightly and difficult to mow.
Biology. Moles spend virtually all of their lives underground in a ramifying system of tunnels and chambers that they excavate with their unusual shovel-like fore limbs. Adults are about 15 cm (6") long, with dense dark brown fur. The main breeding season runs from February to June, and each female rears a litter of about four young in a subterranean nest, sometimes known as the fortress. A single family may inhabit a tunnel system extending over 0.2 ha ($\frac{1}{2}$ acre) and individuals patrol the tunnels regularly to feed on earthworms and other small invertebrates that find their way into them. Young moles leave the parental tunnel system after a few months and either establish new systems or enter old systems that are unoccupied.

Treatment. The main methods used to check mole damage are trapping, poisoning and the use of repellent devices or substances.

Two types of mole trap are approved for general use and can be bought from some garden shops and hardware stores. These are the Duffus half-barrel trap and the caliper trap. Permanent mole runs must first be located by careful observation, probing and excavation and traps are then set so that moles are killed instantaneously as they pass along the runs. The Duffus trap is generally easiest to use and gives best results when inserted in relatively straight lengths of run within 15–20 cm (6–8″) of the soil surface. Such runs often follow the lines of fences, walls, buildings and paths and can be located by probing with a thin cane or a metal rod. Once located, the run is carefully exposed, using a trowel or similar tool, the trap is set and inserted and earth and turf are placed over it so that light is excluded but operation of the trap is not obstructed. Caliper traps are inserted in a similar way and are generally easier to use in deeper runs. Traps should be examined at least once a day and if they fail to catch any moles within four days they should be moved to new sites. The best period for trapping is usually in late winter and early spring when mole activity can be easily observed by looking for new molehills and surface runs but traps can be set at any time of year, and continuous trapping may often be necessary if cleared runs are invaded by new moles moving into gardens from adjoining land.

Poisoning is not suitable for use by gardeners since the poisons used are highly toxic and can only be obtained on licence. It may be the best method for clearing large gardens but should then be done by an expert.

Many different repellents are claimed to prevent invasion of gardens by moles or to drive them away once they have become established. Most rely on smell (mothballs, creosote, smoke cartridges, garlic, etc., inserted into tunnels) and some rely on vibrations (empty bottles or tins sunk in the soil so that wind produces vibrations as it blows across them). Moles are certainly sensitive to unusual odours and to vibrations in the soil and it is therefore possible that some of these repellents may have a brief deterrent effect. It is also true that moles can quickly block off their tunnels temporarily so that the effects of repellents could be restricted to small sections of the whole system. It therefore seems that they are unlikely to have a permanent effect on mole activity in a garden.

DEER

Roe deer, *Capreolus capreolus*, and fallow deer, *Cervus dama* are the two species that are most likely to be troublesome in gardens although other species, such as the sika and muntjac may occur in some areas.

Symptoms. Plants in gardens are mainly damaged in March, April and May when deer browse on new shoots and leaves of many shrubs and trees. They also strip bark from conifers and other trees during the winter and bucks may fray the stems and branches of trees and shrubs in spring when they are marking out their territories. Deer feed at dusk and dawn and are seldom seen in gardens but their tracks may be found in soft ground and their presence should be suspected where shoots and stems have been cut through cleanly on one side but left with a frayed edge on the other side.

This characteristic symptom results from the fact that deer lack teeth in the upper jaw.

Biology. Roe deer are usually solitary but may sometimes form small herds. The rutting period is generally from mid-July to mid-August and each doe gives birth to a pair of fawns in the following April–July. These remain with the mother until the following season and become sexually mature at about 15 months. Adult roe deer are only about 0.6 m (2 ft) high at the shoulder and are able to work their way through small gaps in hedges and fences. They can also jump over a height of about 1.5 m (5 ft).

Fallow deer are slightly larger. They form separate herds of bucks and of does with fawns and a herd breaking into a garden overnight can cause a great deal of damage to shrubs and trees. In summer their chestnut coloured coats are generally marked with cream white spots, which are lost in winter, and they have a distinctive black and white tail.

Treatment. In districts where deer are common and damage to plants in gardens is persistent, the only adequate solution is to have deer-proof fences and hedges. Fences must be at least 1.8 m (6 ft) high and should be made from large mesh wire netting, such as two widths of sheep fencing well overlapped. Plastic netting is not suitable for this type of fence. In addition to fencing and hedging the perimeter, deer-proof gates on drives and paths may also be necessary.

If complete exclusion of deer is not possible, susceptible plants should be protected by wire netting around the stems and trunks to limit bark stripping and fraying. Chemical repellents and scaring devices are seldom really effective but temporary protection may be given by tinsel, tin cans, twine, coloured rags or feathers strung across pathways that deer are likely to use. Leading shoots of valuable plants may also be protected by winding lengths of wool around them in spring to enable them to grow out of reach of deer.

Local Deer Control Societies operate in some areas, under the auspices of the British Deer Society, and may be willing to give advice to gardeners.

DOGS *Canis familiaris*

Domestic dogs damage plants by urinating on them, which kills leaves and produces brown discoloured patches on the foliage of dwarf conifers and low-lying shrubs and on lawns. Usually the damage is confined to the owner's garden but on open-plan housing estates passing dogs can become a nuisance. Vulnerable plants should be sited away from pavements and paths where dogs are a problem or should be protected by netting or low fencing. Chemical repellents may prove effective in some situations.

CATS *Felis catus*

Cats are attracted to newly cultivated areas of soil which they use as toilet areas and they often scratch up seed beds and newly sown lawns. They prefer dry soil and can therefore be discouraged by keeping affected areas well watered. Pepper dust and other repellents may also have some deterrent effects.

Diseases

Diseases are malfunctions of plants caused by fungi, bacteria, viruses or virus-like organisms. Any such organism causing a disease is known as a pathogen and the science of their study is plant pathology. In this book, diseases have been mainly grouped together on the basis of similarity of symptoms as different types of causal organism can have essentially similar effects and require similar treatments. It is appropriate to consider briefly therefore some general aspects of fungal and bacterial biology (viruses are dealt with separately on pp. 429–434) as they relate to pathogenicity. There are probably at least 100,000 species of fungi and they are usually considered to be closely related to, if not actually members of, the plant kingdom. They differ from almost all other plants however in lacking the chlorophyll that enables plants to photosynthesise organic nutrients from the raw materials of carbon dioxide gas and water. Fungi must therefore obtain their nourishment by other means and they do this by growing on dead or living organic matter which they digest extracellularly; that is, by secreting enzymes into the material and subsequently absorbing the digested products.

Varying degrees of specialisation have evolved among fungi and it is generally thought that, within any particular group, the ability to live only on dead matter (saprophytism) is less specialised than the ability to live on either dead or living matter (facultative parasitism). Exclusive dependence on living matter (obligate parasitism) is believed to be the most highly specialised. Most of the fungi described in this book are facultative parasites which can thus live on dead or dying organic matter and use this as a springboard to attack healthy plants or healthy tissues. Relatively few of the fungi important in causing garden diseases are obligate parasites and the rusts and powdery mildews are probably the best known of them.

The basic structural units of almost all fungi are microscopic tubular threads termed hyphae. En masse, such hyphae constitute a mycelium and this, in varying degrees of aggregation, is responsible for the form of all the familiar types of fungal structure, ranging in complexity from toadstools downwards. Fungal mycelium grows and ramifies through the soil, through decaying compost and vegetation and through the diseased tissues of plants. It is usually only visible to the naked eye when it forms reproductive bodies and other relatively large structures. Unlike flowering plants, fungi do not produce seeds but spread and multiply by much simpler structures termed spores. Whereas all seeds are the result of sexual reproduction however, spores may be produced by sexual or asexual means. Most fungi do both and the killing of spores or the prevention of their germination is the means of action of many fungicidal chemicals.

Spores produced by different types of fungi or in different ways have
different names but the term conidium (plural, conidia) is applied to
virtually all asexually produced spores and is used repeatedly throughout
this section of the book. The types of spore produced by different fungi are
used as the basis for much of the classification of the group. Fungi that
produce only asexual conidia and for which no sexual reproductive process
is known constitute the group Deuteromycetes, sometimes called the fungi
imperfecti. Fungi that produce spores termed ascospores in microscopic,
enclosed sac-like structures, by sexual reproduction, constitute the Ascomycetes. This is the largest group of fungi and familiar macroscopic examples are the cup-fungi such as *Peziza*. Fungi that produce sexual spores
on unenclosed surfaces are the Basidiomycetes and these spores are termed
basidiospores. All the familiar mushrooms and toadstools as well as rusts
and smuts belong to this group which is generally thought the most advanced fungal type. Sometimes, most notably in the rusts, other spore types
are produced also but it is the presence of basidiospores that remains the
important classificatory feature. All remaining fungi are placed in an as-
sorted group termed the Phycomycetes in which are classified such widely
diverse organisms as the causal agents of clubroot, potato blight, downy
mildews and potato wart. They have no particular spore type in common
although many of the important pathogenic forms produce free-swimming
swarm spores which move in films of water in the soil or on plant surfaces.

In temperate climates, all fungi face the problem of survival during the
winter when the low temperatures do not favour their growth and when
their host plants may have died down or entered a dormant state. Some
survive in sheltered sites as mycelium but most produce specially toughened structures to resist the adverse conditions. Often these structures are
spores but many fungi also develop toughened masses of mycelium (usually
known as sclerotia) which serve the same purpose. The success of much
disease control depends on preventing the formation, survival or germination of these over-wintering bodies, so breaking the life cycle.

As with all other living organisms, fungi are named for scientific purposes using the Latin binomial system which also serves to indicate affinities
between related forms. Again, as with other groups, these names are
changed from time to time, usually as detailed study reveals close relationships between types previously thought different. With some fungi
however a special problem arises. The Deuteromycetes were referred to
above as having no known sexual reproductive process, the emphasis being
on the word known. Detailed research often reveals however that a fungus
classified and named as a Deuteromycete is in fact only the asexual state of
another fungus, named or classified quite separately in another group,
usually the Ascomycetes. Thus the same organism has two names. According to internationally agreed rules, the name of the sexually reproducing (or
'perfect') state takes priority and the old Deuteromycete name thus becomes virtually redundant. In this book however, the Deuteromycete
names have been retained in those instances where the name is already very
familiar to gardeners and/or where it is the asexual state of the fungus that is
largely responsible for the disease symptoms. Thus the familiar name
Botrytis cinerea will be found for the grey mould fungus and not the more
strictly correct *Botryotinia fuckeliana*, which is its ascospore producing

state. In a few instances where both names may be familiar, both are given.

The effects of parasitic fungi on the plants they infect range from the mild, through chronic to acute and are produced in a wide variety of ways. They may be localised, resulting in restricted damage such as leaf spots. Alternatively, they may be systemic, spreading throughout the entire tissues of an affected plant which may then display a general distortion, rotting or other debilitating effect. Obligate parasites develop specialised hyphae that are able to drain nutrients from host cells without killing the plant; for if this were to happen, they would die also. Facultative parasites on the other hand are much more destructive and grow into and between host cells, producing extensive damage by enzymic or sometimes toxic or growth regulatory (hormonal) action.

There are fewer species of bacteria, (probably about 1600), than there are of fungi and in general they are less significant as pathogens although there are a number of important exceptions. Of these, the effects of the soft-rotting bacteria are those with which gardeners are most likely to be familiar.

Structurally bacteria are much simpler than fungi, each bacterium normally comprising a single microscopic cell which may be rod-shaped, spherical or more or less thread-like. Some bacterial cells are capable of movement and some can produce spores although many do not. They reproduce by simple binary fission – by dividing themselves in two, and can multiply with incredible rapidity, especially under warm conditions. Bacteria occur almost everywhere in the environment but relatively few species ever cause any harm to man, animals or plants.

The classification of bacteria is of little importance to the gardener but it is worth stressing that, once established, bacterial diseases often prove difficult to control for relatively few fungicides or other chemicals are effective against them. Both the symptoms and the gross effects of bacterial infections are similar to those produced by fungi.

RUSTS
(Plate 10)

Basidiomycete fungi (Order Uredinales) which produce rust-like areas of fungal growth on leaves, stems or other plant organs. Rusts affect a wide range of vegetables, fruit, trees and ornamentals and several types are very common in gardens. They have a world-wide distribution with about 4500 species; more than 200 in Britain and W. Europe, the majority in the genus *Puccinia*. Many species have complex, highly sophisticated life cycles and appreciation of some of the basic variations will make it easier to understand and to deal with the harmful effects of the diseases on garden plants. Up to five different spore types may be produced by a rust fungus during its development.

Pycniospores. Produced in small, often flask-shaped structures usually on the upper surfaces of leaves. Pycniospores are usually surrounded by a sweet secretion attractive to insects which transfer them to other plants, where they fuse with special structures known as receptive hyphae and so bring about fertilisation. Pycniospores are not easily seen with the naked eye.

Aeciospores. Produced in aecia which in many cases develop after the spermatia have fused with the receptive hyphae. Aecia are normally cup-like and embedded in host plant tissues with an opening to the surface, through which the spores escape. They are often known descriptively as cluster cups and are the most easily seen and distinctly recognisable of rust symptoms.

Urediniospores. Produced in pustules and are probably the most familiar spore type. Urediniospores are often called summer spores and are typically light coloured although many appear dark brown en masse. They spread rust diseases rapidly in the summer, often after initial infection by aeciospores.

Teliospores. Produced later in the season than urediniospores and in late summer the two types may be intermingled. They also develop in pustules and are commonly black or very dark coloured. Unlike most urediniospores, teliospores (also known as winter spores) normally need a period of dormancy before germination and are the form in which many rusts survive the winter. A few rust species overwinter as urediniospores but rather more do so by developing mycelium within the tissues of some perennial part of the host plant.

Basidiospores. Produced from germinating teliospores and bring about initial infection of host plants before spermatial production.

Many rusts regularly alternate between different host plants. Pycniospores and aeciospores are then produced on one host species and the remaining spore types on another. The two host plants are often quite unrelated botanically and some rust diseases may therefore be controlled by eradicating one of the alternate hosts to benefit the other.

The effect of rust infection on plants is a general debilitating one. As highly specialised parasites, most develop hyphae within the host tissues and by means of special feeding structures are able to drain nutrient from the plant without killing it.

Although it is often claimed that excessive nitrogenous fertilisers aggravate the effects of many plant diseases some of the rusts are among the few for which positive evidence of this has been obtained. A number of fungicides are fairly effective in controlling rust diseases. Copper compounds have been widely used in the past but dithiocarbamates, such as maneb and zineb, as well as sulphur in various forms, are valuable for the treatment of certain rust diseases. The systemic fungicide oxycarboxin is highly effective against many rusts but has not been available long enough for detailed treatments to be worked out for all plants. It is certainly worth trying, applied at fortnightly intervals, on ornamentals if other treatments fail. Varieties and cultivars of many plants are resistant to some strains or races of rusts but few plants are resistant to all strains of any particular rust.

It is a usual convention to describe and name rust fungi with reference to the host on which the teliospores are developed. This has not always been followed here however and the various rusts are described under the host plant which is considered most significant from a gardener's point of view.

In common parlance, certain rusts are sometimes named according to the most obvious symptoms on the most important host plants. Thus cluster cup rusts are those that produce only aeciospores on the major host plant and urediniospores and teliospores (if at all) on other plants that are horticulturally insignificant whereas blister rusts produce masses of powdery aecia on swellings on woody hosts.

ANTIRRHINUM RUST *Puccinia antirrhini*

First discovered in California, and first definitely reported in England in 1933 and on the Continent soon afterwards. It is commonest in hot, dry seasons and is now the most serious antirrhinum disease.

Symptoms. Chocolate-brown pustules, usually with a surrounding yellow halo, develop on the undersides of leaves with corresponding pale yellowish spots on the upper surfaces. In severe attacks the stems, capsules and sepals are also diseased. The leaves may die and the flowers are either distorted or fail to develop.

Biology. The pustules normally produce urediniospores; occasionally teliospores, but no other spore types are known. Teliospores may also be produced in separate pustules from the urediniospores but their role is uncertain and it seems that the rust can overwinter as mycelium on plants in glasshouses or other protected situations. It may also be carried in infected seed.

Treatment. Many antirrhinum cultivars, long thought to be resistant to rust in Britain, have, in recent years, developed the disease. More than one race of the fungus exists in North America and it seems that this must now be the case here. Spraying every two or three weeks from early July to late September with maneb and zineb or oxycarboxin provides some protection. Diseased plants should be destroyed and antirrhinums not left in the ground over the winter unless absolutely necessary nor grown on land that has carried a diseased crop during the previous two years. Storing seeds for more than one year may eliminate rust from that source.

AQUILEGIA RUST *Puccinia recondita* f.sp. *agrostidis*

Occasional on garden columbines but its distribution is unknown. Small groups of orange cluster cups develop on the undersides of leaves in summer and the fungus overwinters on bent grasses (*Agrostis* spp.). No treatment is normally necessary although badly diseased plants are best uprooted and burnt.

ARMERIA RUST *Uromyces armeriae*

Fairly common in gardens both on thrift (*Armeria maritima*) and other *Armeria* spp. All five spore types are produced on the same host plant but the urediniospores are those most often seen. They develop in pale brown pustules on purple spots on either or both leaf surfaces. There are no established control measures but odd diseased plants should be destroyed. If whole beds are affected, spraying with dilute Bordeaux mixture or oxycarboxin fortnightly from the time that symptoms first appear until early autumn might be effective.

ASPARAGUS RUST *Puccinia asparagi*

Although considered scarce at present, this disease remains a threat to asparagus growing and has at times reached almost epidemic proportions in parts of Britain. It is occasional elsewhere in northern Europe.

Symptoms. Rust coloured powdery pustules, often in enormous numbers, develop on the feathery shoots or 'bowers' during summer, followed by darker brown streaking on stems, or more rarely, phylloclades ('needles'). Premature death of the shoots may follow. Yellow-orange cluster cups form on young shoots but are often overlooked.

Biology. The development of aeciospores, although common, occurs inconspicuously in yellow-orange aecia on the 'buds' in spring but both urediniospores and teliospores are also produced, the former causing the cinnamon-brown powdery covering on plants and the latter, which later persist and overwinter on the ground, being formed in the dark streak-like slits on the stems. Dew is important for summer spread of the disease which can result in few 'buds' being produced in the following spring.

Treatment. Avoid damp and poorly ventilated sites for beds. Cut back and burn the feathery shoots at the first signs of symptoms in summer in order to prevent the development of overwintering teliospores. As young asparagus is normally not cut until the third year, plant debris lying on the ground in young beds may enable teliospores to persist there undisturbed. These may infect the young buds on which aeciospores form, thus enabling an enormous disease potential to develop. Protect plants by spraying early in the season with zineb. If the rusty spore stage is seen early enough, dust with very fine sulphur. The first application of dust should be about three weeks after the last shoots have been cut, with repeat applications at monthly intervals. Sulphur dusting is claimed to be more effective in hot dry weather.

BEAN RUST *Uromyces appendiculatus*

Infrequent in Britain but more common elsewhere in northern Europe infecting runner and French beans. Dark brown pustules of urediniospores and teliospores on leaves, stems or pods are the most obvious symptoms although white cluster cups form occasionally. Sulphur-based fungicides are sometimes effective if sprayed at ten day intervals from the time of appearance of the first symptoms but the addition of maneb improves them. Zineb is also claimed to be useful if similarly applied.

BEET RUST *Uromyces betae*

Yellowish cluster cups and, more commonly, yellow-brown urediniospore and teliospore pustules occur widely on all above ground parts of red beet and spinach beet, but very rarely cause much harm although plants may wilt in severe cases. Infection begins in spring from germinating teliospores that have survived the winter on dead foliage, so it is important to remove all such debris when the crop is harvested. In severe cases spray fortnightly with thiram or zineb but on spinach beet do not cut the leaves within a fortnight of spraying and wash them thoroughly before cooking.

BERBERIS RUSTS *Puccinia graminis* and *Puccinia poae-nemoralis*

The common barberry *(Berberis vulgaris)* and two species of *Mahonia, M.*

aquifolium and *M. bealii* are alternate hosts for cereal black rust, *Puccinia graminis*, an important disease of wheat. In North America the wholesale destruction of barberries has been carried out in important wheat growing areas but in Britain this is unnecessary as the strain of *P. graminis* found on barberries here probably always has grasses, not wheat, as its alternate host. The symptoms on barberry are bright orange spots on the upper leaf surfaces and masses of orange aecial cups on purple-red spots on the undersides. Aecia may also form on fruits, especially on *Mahonia*. The disease is not normally serious on its ornamental hosts, but an early summer spray with thiram is sometimes necessary.

Puccinia poae-nemoralis produces urediniospores and teliospores on oat grass and also infects barberry, on which witches' brooms develop. It is rare in Britain but sometimes serious elsewhere in northern Europe. Tubular cluster cups develop on the undersides of the leaves on the witches' brooms which should be cut out and destroyed.

BIRCH RUST   *Melampsoridium betulinum*
Common and widespread on common and silver birches and also, especially in Scotland, on the alternate hosts, larches.
Symptoms. The undersides of diseased birch leaves are covered with masses of yellow pustules. Sometimes the leaves die prematurely and the trees are consequently weakened although this effect is usually less serious on young plants. Infected larch needles are distorted, drop prematurely and bear rows of tiny red-orange cups with rough white margins. These cups are virtually indistinguishable from those of another larch rust, which has alder as its alternate host or from those of a number of other larch rusts which alternate on willows and poplars. All these rusts are unimportant as pathogens to larch trees and need not cause concern.
Biology. The yellow masses on birch leaves are urediniospores; teliospores are produced on fallen leaves and are carried by the wind to infect larches on which the aeciospore-producing cups develop. Larches do not seem to be essential for the continuance of this disease however as fungal mycelium seems to survive the winter on leaves that remain attached to the birch seedlings, and possibly in the buds also.
Treatment. Protect birch seedlings by fortnightly sprayings with Bordeaux mixture from the time of bud burst to the end of the summer. Treatment of larches is not necessary.

BLACKBERRY COMMON RUST   *Phragmidium violaceum*

BLACKBERRY STEM RUST   *Kuehneola uredinis*   **Pl. 10**
Both diseases are common and widespread on wild and cultivated blackberries and on cultivated loganberries.
Symptoms. Common rust causes deep red to purple spots on the upper leaf surfaces and orange-yellow and/or black pustules on the undersides. Severe infections distort the leaves and cause premature leaf-drop. Orange slit-like lesions may also appear on the stems. Stem rust produces yellowish spots on the canes, and on the upper and lower sides of the leaves.
Biology. The orange-yellow common rust leaf pustules produce aeciospores and urediniospores; the dark ones produce teliospores. The slit-like

lesions on the stems also produce aeciospores. The pustules on the upper leaf surfaces are aecia of stem rust while those beneath are either also aecia or produce urediniospores or the unusual light-coloured teliospores. Stem rust teliospores germinate on the leaves and the rust overwinters there as mycelium. Common rust teliospores remain dormant through the winter however and do not germinate until the following spring, when they infect young leaves and shoots.

Treatment. Collect and burn fallen leaves in late autumn and early winter, if practicable. Neither disease affects fruit production seriously but, if special protection is necessary, spray with Bordeaux mixture as the leaves unfold in spring.

BLUEBELL RUST *Uromyces muscari*

Quite common on bluebells. Other *Endymion* spp. as well as scillas and muscaris, are affected elsewhere in northern Europe, but only rarely in Britain. Symptoms appear on leaves and stems as small dirty yellow or brown spots bearing brown teliospore pustules, often grouped together in concentric clusters. The outer leaf tissues may rupture to form tiny slits. The disease is not serious and yellowing or death of the leaves only occurs in severe cases. No treatment is necessary.

BOX RUST *Puccinia buxi*

Dark brown or purple-brown teliospore-producing pustules develop quite frequently on both upper and lower leaf surfaces although no other spore types are known. The symptoms appear first in early autumn and persist from September through to spring but are rarely severe enough to damage the plants seriously. Clipping off and removal of affected shoots as a precaution is usually all that is needed.

BROAD BEAN RUST *Uromyces viciae-fabae*

Occasional and widespread but rarely severe in Britain. The same disease is sometimes troublesome on peas in parts of northern Europe.

Symptoms. Very small white or pale yellow cluster cups develop on similarly coloured spots on the undersides of leaves in spring and again in autumn but are easily overlooked. Pale brown pustules are more common on most parts of the plants during summer and defoliation sometimes follows severe attacks. The pustules gradually darken, especially in hot and dry seasons, this effect being most noticeable with those on the stems.

Biology. The aecial cups are relatively unusual on broad beans although quite frequent on related wild host plants. The pale brown urediniospores are much the commonest spore type and are often the only ones formed on leaves although on stems their transition to dark teliospore pustules may be normal in some years. The effect on the plants is not usually serious, partly because the disease does not really get hold until late in the season. In damp and cold years when teliospores appear to be absent, it is possible that urediniospores survive the winter and initiate new infections in spring.

Treatment. No control measures should be necessary but destruction of badly infected plants at the end of the season is probably worth doing to lower the amount of potential infection in the following spring. In severe cases, fortnightly zineb or thiram sprays are effective if applied from the time when symptoms first appear.

CARNATION RUST *Uromyces dianthi* **Pl. 10**
Introduced into England around 1890 on plants imported from Europe. Normally much more severe on glasshouse than outdoor carnations and occasionally epidemic but usually little more than locally troublesome. Rare on other *Dianthus* spp.; the rust on sweet william *(D. barbatus)* is caused by a different fungus (p. 286).
Symptoms. Small yellow blisters develop on leaves and burst during summer to become reddish-brown pustules. Similar symptoms may occur on stems, particularly of cuttings or young rooted plants. The disease usually starts near the base of plants and spreads upwards and when the attack is very severe the plants may appear distorted or stunted. It is often troublesome in glasshouses in winter (unless extra heating is supplied) as the disease is favoured by cool, damp conditions.
Biology. The brown pustules contain urediniospores and occasionally teliospores also. The alternate host of carnation rust, the spurge *Euphorbia gerardiana* does not occur in Britain, where the only natural means of spread and overwintering is by urediniospores.
Treatment. Some older cultivars of carnations are very susceptible but most modern types have some resistance or tolerance. It is essential, especially with glasshouse flowers, to water carefully and avoid splashing spores onto healthy leaves. Similarly, the humidity should not be high enough for dew to form on plants and adequate ventilation must therefore be provided. Cuttings should be taken only from healthy plants but when the disease is very slight, carefully snipping off odd infected leaves may give adequate control. Many different fungicides have been used successfully. Spray established plants at ten-day intervals with zineb, maneb, oxycarboxin or thiram from September to February but take care not to use directly onto blossoms as some discolouration may result. Soil drenches with oxycarboxin around established plants are also very successful. Spray cuttings similarly every ten days during the winter and start spraying plants at other times of the year as soon as symptoms appear. It may be wise not to grow carnations for two years on soil where diseased plants have been found.

CENTAUREA RUST *Puccinia cyani*
Common and widespread as brownish urediniospore pustules on the undersides of the leaves and somewhat darker teliospore pustules both on leaves and stems. A perennial mycelium forms within the plants but flowering is not affected. Moderate attacks may often be checked by spraying every three weeks with thiram or oxycarboxin but in severe instances plants may have to be uprooted and burned. Another rust, *Puccinia hieracii* also occasionally produces reddish or dark brown spots and pustules on *Centaurea* spp. but is rarely serious. It has also been found affecting endives but in England it is confined to the South West on this host.

CHRYSANTHEMUM RUST   *Puccinia chrysanthemi*   **Pl. 10**
Long known in Japan; first found in England in 1895, spreading from there throughout Europe and now fairly common.

Symptoms. Usually commonest late in the summer on indoor chrysanthemums as dirty brown pustules on the undersides of leaves. Yellowishgreen spots appear correspondingly on the upper surfaces and severe attacks may cover and damage large areas of leaf. Defoliation and reduction in flower numbers can also result. The disease is relatively unusual on outdoor plants but annual chrysanthemums are sometimes attacked by other less important rust fungi.

Biology. The pustules produce urediniospores; other spore types are virtually unknown in Europe and it is as urediniospores or possibly as mycelium on old leaves that the disease overwinters. Its significance in many parts of the world is now eclipsed by that of the much more serious chrysanthemum white rust (see below).

Treatment. Many modern chrysanthemum cultivars have at least some resistance although a strict routine of hygiene is still essential. All infected plant material should be burnt immediately and, as a precaution, lower leaves stripped from all cuttings, both when initially taken and when transplanted. Care is also needed to avoid wetting the leaves when watering. Spraying with dilute Bordeaux mixture is claimed to give satisfactory protection but some cultivars are damaged by it and zineb, oxycarboxin or sulphur applied at fortnightly intervals from the first appearance of symptoms until the end of the season is probably better. The hot water treatment used for eelworm control (p. 135) will check rust disease also.

CHRYSANTHEMUM WHITE RUST
Puccinia horiana   **Pl. 10**
This most serious disease was confined to China and Japan until 1963 when an outbreak occurred in England on imported plants. Since then, it has been found here many times and in many other European countries also although a vigorous eradication campaign has probably prevented its establishment in Britain.

Symptoms. Spots, varying from yellow to pale green, on the upper leaf surfaces on glasshouse chrysanthemums are often the most obvious first symptoms. There are usually corresponding dirty-buff and gradually whitening pustules on the undersides. The entire spots may eventually become brown and necrotic in the centre and the plants decline gradually.

Biology. Teliospores germinate to form basidiospores without being released, this change causing the overall white appearance of the pustules. These are the only spore types known and they spread the disease with astonishing speed. There are considerable differences in susceptibility between cultivars but the disease remains an ominous threat to commercial chrysanthemum growing.

Treatment. No control should be attempted and any known or suspected outbreak of chrysanthemum white rust must be reported to officials of the Ministry of Agriculture. The diseased plants will then be destroyed and efforts made to establish the source of the outbreak.

CINERARIA RUST   *Puccinia lagenophorae*
Only known in Europe since 1961 but now very common in Britain on groundsel and related weeds of the genus *Senecio*, as well as on cinerarias (*Senecio cruenta*). The symptoms on cineraria are readily distinguished from those of the other rust infecting this and related hosts (see below) as cineraria rust forms distinct blisters on the leaves (including the upper surfaces) and stems on which vivid orange and yellowish-white aecial cups develop. No other rust attacking cineraria develops cluster cups nor affects the upper leaf surfaces. The symptoms are nonetheless not as dramatic as on groundsel on which dark brown or black teliospore pustules are also common and which may be virtually killed by the infection. The disease may be avoided by keeping cineraria beds clear of groundsel and by applying a monthly spray of zineb, thiram, oxycarboxin or maneb as an additional precaution.

COLTSFOOT RUST   *Coleosporium tussilaginis*
Very common and widespread, existing as many different strains. The bright orange urediniospore pustules or red clusters of teliospore pustules are most likely to be seen in late summer on groundsel (but should not be confused with cineraria rust; see above) but cultivated plants affected include *Senecio pulcher*, various cinerarias, marigolds and (affected by a different strain) *Campanula* spp., especially *C. glomerata*, *C. persicifolia*, *C. rapunculoides* and *C. planifera alba*. Eradication of wild *Senecio* hosts, such as groundsel, will help control as will the avoidance of high humidity through too close planting. A protective spray of dilute Bordeaux mixture or oxycarboxin should be applied where it is known that the disease is common. Alternate hosts of the rust are two-needled pines, especially Scots pine, on the needles of which pale yellow cluster cups develop. The trees are not seriously affected and no control is necessary on them.

COMFREY RUST   *Melampsorella symphyti*
Occasional and probably widespread on several *Symphytum* species. Urediniospores and teliospores form in yellow-brown or yellowish pustules on the undersides of leaves. No other spore types form in Britain or northern Europe. Mycelium is perennial in the rhizome and the disease will therefore appear year after year. Affected plants should be uprooted and destroyed.

CONVALLARIA RUST   *Puccinia sessilis*
Rare on *Convallaria* in Britain; not uncommon on this and other liliaceous hosts elsewhere in Europe. Orange cluster cups develop on lower leaf surfaces and infection originates with teliospores from the alternate host, reed canary grass. Spray fortnightly with zineb or oxycarboxin throughout the season, from the time of first appearance of symptoms.

CROCUS RUST   *Uromyces croci*
Fairly frequent in Holland on *Crocus*; virtually unknown elsewhere in northern Europe. Infections on corms almost invariably result from teliospores in the soil. Early in the season dark teliospore pustules appear on the membranous sheaths although later these are obscured by the darkened

sheaths themselves. The fungus penetrates systemically into the corm and mycelium may grow from there into the daughter corms which show darkening of the basal tissues and teliospore pustules within. No other spore types are known. No chemical treatment is available for garden use and in any event would be useless against systemic infections. Hot water treatment (p. 134) might be effective but has been little tried.

DILL RUST *Puccinia nitida*
Unknown on this host in Britain (although found rarely on parsley) but occasional elsewhere in northern Europe where dill is an important crop. Appears as brown urediniospore and blackish teliospore pustules with consequent defoliation and/or death. No effective treatment is known and care is needed with any fungicides used as they may persist on the dried plant when used as a kitchen herb.

ERYTHRONIUM RUST *Uromyces erythronii*
Rare in Britain and first introduced in 1936 on plants from France, where it is much commoner. Distinguished by yellowish or deep-brown leaf spots producing aeciospores and teliospores respectively in powdery pustules. No treatment should be necessary.

FIR RUSTS
Firs, particularly silver fir, are hosts for a number of rust fungi, most notably aecial forms of several species with ferns as the alternate hosts. The symptoms are typically small white cluster cups on the undersides of the needles. None are particularly common, however, and none serious as far as the trees are concerned.

More dramatic symptoms are cankers or witches' brooms developing from infection by *Melampsorella caryophyllacearum*. Aeciospores are produced on these deformities and can infect the alternate hosts, chickweeds and stitchworts. However, this rust is also infrequent and cutting out the witches' brooms is a satisfactory treatment.

FUCHSIA RUST *Pucciniastrum epilobii*
Very infrequent in Britain on indoor fuchsias and, although it could become a serious problem if established in commercial glasshouses it seems improbable that it will be of importance to gardeners. The same rust does however, affect certain willowherbs which are common garden weeds and the symptoms on fuchsia are worth keeping an eye open for. They appear as ill-defined pale areas on the upper leaf surfaces with corresponding pale orange urediniospore pustules on the undersides. The lower leaves are usually worst affected and may wither and die.

GERANIUM RUST *Uromyces geranii*
Common and widespread on several wild geraniums, some of which are commonly grown in gardens but does not occur on pelargoniums (see p. 282). Orange cluster cups, light brownish urediniospore pustules and/or dark brown teliospore pustules develop on the undersides of leaves, sometimes causing leaf distortion . Rarely severe but spray twice at fortnightly intervals with oxycarboxin to check serious outbreaks, taking particular care to cover the undersides of leaves with spray.

GOOSEBERRY CLUSTER-CUP RUST *Puccinia caricina* var. *pringsheimiana*

Common and widespread but much more important in some years than others. Although most significant on gooseberry, the disease can also occur on black or more rarely, red currants.

Symptoms. Deep orange or red pustules develop on the lower leaves, fruits and, (rarely), on shoots during spring and early summer. As the summer progresses, yellow-rimmed cluster cups develop on these pustules and as they do so the leaves may curl and swellings develop on the stems.

Biology. Aeciospores produced by the cluster cups are unable to re-infect gooseberries but, when carried by wind, do infect sedges, the alternate hosts. There urediniospores and teliospores develop and the latter re-infect fruit bushes the following summer, having survived the winter. Dry weather in March is often the prelude to severe attacks as it delays the germination of teliospores and, when conditions become sufficiently moist later, the spores all germinate at a time when there is plenty of new foliage available to infect.

Treatment. Elimination of sedges from the vicinity of fruit bushes is essential. The disease has been serious at times in Norfolk where a mulch of sedges has sometimes been used on soft fruit. Copper-based fungicides are effective and a single spray applied about a fortnight before flowering is recommended. Do not use copper fungicides on the gooseberry cultivars 'Careless', 'Early Sulphur', 'Freedom', 'Golden Drop', 'Leveller', 'Lord Derby', 'Roaring Lion' and 'Yellow Rough' however as they may be phytotoxic. Maneb, thiram or zineb may be used instead.

HART'S TONGUE FERN RUST *Milesina scolopendrii*

Quite common throughout northern Europe and found in western parts of England, Scotland and Wales. Urediniospores and teliospores are produced on brownish areas on the undersides of the fronds, particularly those that are sterile. Young plants are most frequently infected but no treatment should be needed beyond cutting off affected fronds. Fir trees are probably the alternate hosts (see p. 276) but natural infection is not known for certain to occur on them. Similar symptoms on other ferns are caused by related species of *Milesina*, which also probably infect firs.

HAWTHORN RUST *Gymnosporangium clavariiforme*

Occasional in Britain on both native hawthorn species and also on amelanchiers. In other parts of northern Europe, pears and other rosaceous hosts are affected. Juniper, the alternate host, quite commonly shows symptoms.

Symptoms. Very vivid red and orange discolouration of swollen twigs, leaves and fruits. The symptoms are most apparent from July to September and on the cylindrical or elongate swellings a yellowish-brown pustular mass develops. On junipers elongate yellow-brown swellings develop on the branches. In moist conditions these swellings expand and become soft and bright yellow. From them develop yellowish-brown protuberances up to 1 cm long.

Brownish spots with a yellow margin, mainly on the underside of hawthorn leaves but also on the calyx and fruit are caused by another, much rarer rust fungus, *Gymnosporangium confusum*.

Biology. The symptoms on hawthorns are caused by the production of aeciospores which infect the alternate host juniper where the unmistakable teliospore-producing protuberances develop. Urediniospores are unknown but in junipers a perennial mycelium develops and this can produce new teliospores year after year.

Treatment. Despite its very dramatic appearance, this disease does little damage to either of the two main hosts but pruning off and burning affected shoots may be necessary in bad cases.

HELIANTHUS RUST   *Puccinia helianthi*

Unknown in Britain but common elsewhere in Europe on sunflower and Jerusalem artichoke and sometimes causes considerable damage through inducing premature defoliation. Cluster cups, dusty-brown urediniospore pustules and dark brown teliospore pustules all appear on the under surfaces of the leaves. Lime sulphur or zineb and sulphur sprays applied fortnightly from the time of the first appearance of symptoms are often effective.

HOLLYHOCK RUST   *Puccinia malvacearum*   Pl. 10

First recorded in South America in the middle of the 19th century and soon spread to Europe where it is now extremely common. Mainly attacks hollyhocks but can also occur in gardens on a number of malvas, on tree mallow and on sidalceas.

Symptoms. Raised orange-brown pustules develop on stems, bracts and particularly on the undersides of leaves, with corresponding small yellowish areas on the upper leaf surfaces. The pustules gradually darken and in severe attacks the leaves shrivel and characteristically hang down on the stem. Rarely almost complete defoliation may occur and the disease can sometimes reach almost epidemic proportions.

Biology. All of the pustules contain teliospores; the only spore type produced. Those formed early in the season can germinate immediately and spread the disease to nearby plants. Spores produced later in the year do not germinate until the following spring, surviving the winter in the pustules produced on the lower leaves and stem.

Treatment. Collect and burn diseased plant material in the autumn and repeat this operation in spring to help prevent the all-important first new infections as later, control is very difficult. Also cut off all plants to ground level in autumn and remove the first new leaves in spring as a precaution against their becoming infected. Fortnightly spraying with thiram or oxycarboxin from early spring onwards may be successful and sulphur dust is also sometimes effective. Hollyhock leaves, being hairy, are not always easy to wet and a few drops of wetting agent may be added to fungicide sprays.

HONEYSUCKLE CLUSTER CUP RUST   *Puccinia festucae*

Common and widespread on several *Lonicera* spp. Small pale brown or yellow spots form on the undersides of leaves during the summer and yellow-white aecial cups develop on them. The alternate hosts are fescue grasses, on which urediniospores and teliospores are produced. The disease is rarely serious on honeysuckle and no control measures should be necessary.

HOUSELEEK RUST *Endophyllum sempervivi*
Occasional and probably widespread on sepervivums and echeverias, persisting in the rootstocks as a perennial mycelium. Small orange cluster cuplike dimples on the leaves produce teliospores, while scattered among them are tiny brownish dot-like bodies which produce pycniospores. Leaves of affected plants are generally less fleshy and the rosette habit may also be partly lost. Such plants are best uprooted and burnt and neighbouring healthy stock sprayed with Bordeaux mixture or oxycarboxin.

HYPERICUM RUST *Melampsora hypericorum*
Many *Hypericum* spp. can be attacked; tutsan *(H. androsaemum)* most commonly. Orange cluster cups are produced on the undersides of leaves with corresponding yellow or orange spots on upper surfaces. Teliospores are also produced in reddish or dark brown pustules on the undersides of the leaves. The disease is fairly common in the south-west of Britain but rarely serious.

IRIS RUST *Puccinia iridis*
Widespread but fairly infrequent. The fungus causes pale-coloured leaf spots bearing more or less concentrically arranged brown slit-like urediniospore pustules or darker teliospore pustules. In more severe attacks, the leaves may wither back from the tips. Although most English cultivars are susceptible, many bearded and continental irises seem resistant to at least two of the known races. The alternate hosts are not known for certain but are probably nettles. Chemical or other control treatment is rarely necessary in Britain but elsewhere sulphur dusts are usually effective.

LEEK RUST *Puccinia allii (= Puccinia porri)*   Pl. 10
Common and widespread and probably increasing in frequency in Britain. Most important on leeks but also affects onions, chives (especially in the South), garlic and other alliums.
Symptoms. Dusty reddish-orange slit-like pustules develop on the leaves, sheaths and stems during summer. In severe attacks the leaves may turn prematurely yellow and die, the bulb being reduced in size as a result.
Biology. On leeks only urediniospores are produced; on other hosts teliospores also occur while aeciospores are formed elsewhere in Europe but have never been found in Britain. It is now believed that the leek-infecting rust may be a distinct species from that attacking other host plants. In some areas, wild *Allium* species and crop debris can probably be important however, as refuges for the survival of rust after leeks have been pulled.
Treatment. The disease normally declines with the cooler autumn weather and foliage developing late in the season usually remains healthy but may be at risk in mild autumns. It is often said to be worse on nitrate-rich soils and it is claimed that good supplies of potash will help to suppress it. Other and more reliable control measures are the destruction of infected plants, removal of potential *Allium* weed hosts (although these are infrequently important in Britain), the use of a long rotation (four to five years

if possible) and good drainage. Spraying may be worthwhile if severe attacks develop and fortnightly applications of zineb, up to two weeks before pulling, are sometimes successful.

LETTUCE RUST *Puccinia opizii*

Sporadic outbreaks of this disease in England probably originate from imported plants. It is more common elsewhere in northern Europe. Yellow cluster cups develop on the under surfaces of leaves with corresponding orange-yellow spots on the upper sides. Fungicidal control has apparently never been attempted but two year rotations are successful in southern Europe, where the disease is more important.

LILY RUST *Uromyces aecidiiformis*

Rare in Britain, commoner elsewhere in Europe on lilies and fritillarias. Rusty brown pustules often spreading rapidly and darkening develop on yellow spots on leaves and leaf stalks. Remove and destroy diseased leaves and apply 2–3 sprays of Bordeaux mixture or oxycarboxin at fortnightly intervals.

MAHONIA RUST *Cumminsiella mirabilissima*　　　　Pl. 10

A North American disease introduced into Britain in 1922 and now common and widespread throughout Europe. It causes angular, deep red-purple spots throughout the year on the upper leaf surfaces of *Mahonia aquifolium* and *M. bealii*. On the undersides are corresponding powdery brown pustules which produce urediniospores and teliospores. During May and June cluster cups may occasionally occur on greenish-black thickened areas under the leaves. Prune out and destroy diseased foliage, if necessary to ground level, and spray new shoots with Bordeaux mixture or oxycarboxin at 2–3 week intervals.

MINT RUST *Puccinia menthae*　　　　Pl. 10

Common, widespread and sometimes serious on garden mint, peppermint, marjoram, savory and several other less familiar mints and related genera. **Symptoms.** Affected shoots and leaves are pale-coloured, swollen or otherwise distorted early in the season. Masses of rather indefinite dirty orange cups are produced on the stems and undersides of the affected leaves followed by first yellow and later black pustules on previously healthy parts. Large areas of leaf tissue may die and sometimes plants are defoliated.

Biology. On garden mint, a perennial mycelium is established and the fungus is thus able to survive the winter. On all hosts the aecial cups are followed by pale masses of urediniospores and darker teliospores, the latter being the means of overwintering on all plants except garden mint. Teliospores persist in the soil, and are believed to become lodged among bud scales as the new shoots emerge in spring. Several different races of *Puccinia menthae* are now known to occur.

Treatment. It is rare for a mint bed to be wholly infected but the disease is nonetheless troublesome. Rapid burning in the autumn, perhaps with a flame gun, to remove old debris and kill the spores by scorching the soil is often completely successful. Heat treatment of rhizomes of forcing mint is used in commercial glasshouses to kill any perennial mycelium and may be adapted for garden use where it proves difficult to obtain healthy stock. Before autumn planting, wash plants thoroughly and immerse in water at 44°C (111°F) (no higher) for ten minutes; remove, swirl in cold water and then plant. Take care to keep the water well stirred and, if the treatment is done in a large saucepan on a kitchen cooker, remember that the bottom of the pan can easily reach a higher and damaging temperature. A thermostat and heaters in a tank would give more reliable temperature control.

MOUNTAIN ASH RUST   *Gymnosporangium cornutum*
Fairly common where mountain ash and the alternate host, juniper grow together. Occasionally occurs on apple in Europe, but not in Britain. In autumn mountain ashes can be so severely affected that all the leaves appear discoloured and shrivelled, although the normal autumn appearance of mountain ash leaves is similar so appearances can be deceptive. Masses of orange-yellow aecia on the undersides of infected leaves produce spores that carry the disease to junipers. No urediniospores are formed but teliospores, which reinfect mountain ash, are produced on junipers as rich orange-brown masses on elongated swellings on young shoots or, rarely, on leaves. Few attempts have been made at treatment of affected mountain ashes but Bordeaux mixture at monthly intervals during summer would probably be effective.

OAK RUST   *Uredo quercus*
Occasional in southern England and elsewhere in northern Europe. Small orange urediniospore pustules on the leaves, particularly of coppice shoots, are almost invariably the only symptom seen. Causes no damage and no treatment is necessary.

PEA RUST   *Uromyces pisi-sativi*
Rare in Britain, commoner in other parts of northern Europe where the alternate spurge hosts occur. Appears as rusty brown urediniospore pustules on the under surfaces of leaves and dark teliospore pustules on leaves, stems and leaf stalks. The disease also occurs on laburnum, brooms and other leguminous plants and in Britain at least it probably overwinters as urediniospores without the need for spurges to be present. Destruction of wild spurges, especially *Euphorbia cyparissias* may however reduce the disease while early plantings of peas often escape the worst attacks. Do not grow late-maturing peas in areas where this disease is most troublesome.

PEAR RUST   *Gymnosporangium fuscum*
Rare in Britain, fairly common and sometimes serious elsewhere in Europe. Red spots develop on the upper leaf surfaces in spring, gradually darkening and often having brighter margins. Rough brown aecia develop beneath the leaves and wind-blown aeciospores carry the disease to savin *(Juniperus sabina)* on which the branches swell and ooze orange teliospore masses in spring. Basidiospores developing from them re-infect pears. Removal of all savin bushes within a radius of about 100 metres is the only really effective remedy.

PELARGONIUM RUST *Puccinia pelargonii-zonalis* **Pl. 10**
Known in South Africa since the early years of the century. First recorded
in Europe in 1962, reaching Britain in 1965 and now spreading rapidly
throughout the rest of the world. Ivy-leaved and variegated pelargoniums
are not affected but the disease can be serious on both pot and bedding zonal
types.
Symptoms. More or less circular brown, typically rusty pustules surrounded by a necrotic patch, develop on the undersides of the leaves of
affected plants and may occasionally occur on stipules also. When attack is
severe, the leaves may fall and this can gradually result in a debilitated plant
producing fewer flowers, fewer shoots and consequently few cuttings.
Biology. The pustules produce urediniospores (other types of spore
being unknown in Britain) which can lie dormant on apparently healthy
leaves or stipules until the onset of moist conditions prompts germination.
They can also remain viable on dead leaves for up to three months.
Treatment. On indoor plants, remove and destroy diseased leaves as soon
as they are noticed and take care that leaves do not remain wet for long
periods. On outdoor plants, such hygiene is scarcely practicable but fortnightly sprays with zineb, or alternatively with zineb and thiram will
afford good protection. A number of systemic fungicides, including benomyl, thiophanate-methyl, oxycarboxin and triforine are also effective
although probably no more so than zineb.

PERIWINKLE RUST *Puccinia vincae*
Occasional. Known in Britain since the end of the last century but remaining of uncertain importance. Diseased plants often fail to produce flowers,
develop an erect habit and may become distorted. All spore types are
produced but the aeciospores which are formed, unusually for this type of
spore, in brownish pustules rather than typical cups on the undersides of
leaves are those most commonly seen. A perennial mycelium may develop
and permeate the rootstock so any diseased plants should be uprooted and
destroyed.

PINE RESIN-TOP RUST *Endocronartium pini*
Common on Scots pine in north east Scotland, eastern England and elsewhere in northern Europe. Infection probably takes place through the
needles, and masses of white blister-like aecia subsequently develop in May
and June on large black stem or branch cankers and later burst to release
orange spores. Resin is produced copiously by the cankers and dieback
ensues when the stem is girdled. In most parts of Britain no other host
plants seem to be affected but occasionally paeonies, nasturtiums and,
elsewhere in Europe, a number of other herbaceous plants are attacked by a
rust fungus which is sometimes called *Cronartium flaccidum* but is a race of
the same organism. Unlikely to be a problem other than in forest plantations but diseased trees, once girdled, are better felled.

PINE TWISTING RUST *Melampsora populnea*
Fairly common and widespread and although mainly affects Scots pine,
other two-needled species are susceptible. Causes pale yellow pustules in
early summer on the developing shoots and golden cluster cups develop

on them. The shoot may be killed but typically only one side is damaged and, as the remainder of the shoot grows normally, this results in a permanent S-shaped kink in the stem. This symptom may be confused with damage by the pine shoot moth (p. 211) but the rust causes no hollowing of the stem. Aeciospores infect the undersides of the leaves of the alternate host, aspen (or occasionally white or grey poplars) on which they are indistinguishable from other poplar rusts (below) and the best means of avoidance of this disease is by not planting Scots pine in the vicinity of aspen.

Two-needled pines are also alternate hosts for coltsfoot rust (p. 275) but no damage results.

PLUM RUST *Tranzschelia discolor*   **Pl. 10**
Common and widespread on plums and, less commonly, on apricots, peaches, nectarines and almonds. The alternate hosts are anemones on which considerable damage can occur in some areas.
Symptoms. Very vivid small yellow spots appear on the upper surface of the leaves from July onwards and corresponding brown powdery patches develop on the undersides. As the summer progresses, the patches progressively darken and by autumn are usually completely black. Severe attacks may result in total yellowing of the leaves and, in some cases defoliation which if repeated over a number of years, gradually weakens the trees. Diseased anemones develop an abnormally erect habit, with distortion of the leaves and reduction in the number of leaf lobes. The under surfaces of the leaves bear tiny pale yellow-brown cluster cups with whitish margins and the plants usually fail to flower.
Biology. The brown powdery patches on plum leaves are urediniospore pustules and the gradual darkening is due to their replacement by black teliospores. The latter germinate in spring and infect the alternate hosts; cultivated *Anemone coronaria* and other anemone species. The resulting infection extends into the rootstock and can persist there, aeciospores from the cluster cups on the leaves subsequently reinfecting plum trees. Because the fungus perennates in anemones, diseased plants serve as a permanent reservoir of spores for infecting plums. There is some doubt concerning the significance of the anemone host as far as plum trees are concerned, since severe attacks may develop in plum orchards far removed from any anemones and the fungus can probably persist on plum trees as mycelium.

In Scandinavia (particularly Norway) shot-hole of plum leaves caused by another rust, *Pucciniastrum areolatum*, is sometimes serious. This disease is of minor importance on bird cherries *(Prunus padus)* in Britain.
Treatment. Rust is not usually severe on plums in gardens although spraying the trees with zineb when the fruit is half grown and again after picking may occasionally be necessary. Although the spraying of diseased anemones with zineb may be partially successful, their continued presence, as sources of spores near to plum trees means that they are better uprooted and destroyed.

POPLAR RUST *Melampsora* spp.
A number of rusts of the genus *Melampsora* produce yellow urediniospore and dark teliospore pustules on the undersides of the leaves of several

poplar species. None are normally serious on this host although some defoliation may result and in some instances increased susceptibility to dieback (p. 322) may follow rust infection.

PRIMULA RUST   *Puccinia primulae*
Common on primrose and, outside Britain, on related *Primula* species but probably most familiar in the north of England and in Scotland. Groups of cluster cups develop on yellow spots under the leaves in spring followed by brown urediniospore and teliospore pustules later in the season. Control is difficult but rarely necessary.

RASPBERRY RUST   *Phragmidium rubi-idaei*
Common and widespread but rarely serious. The first sign is the appearance of groups of yellow pustules on the upper leaf surfaces which produce aeciospores in early summer. Later, very small orange urediniospore pustules and, later still, black teliospore pustules develop on the lower surfaces. Defoliation may result, but is infrequent. Protective spraying with Bordeaux mixture might be worthwhile in the year following a severe attack. If practicable, collect and burn the leaves, on which the overwintering teliospores survive.

RHODODENDRON RUST   *Chrysomyxa ledi* var. *rhododendri*
First appeared in Britain in 1913 and is erratic in occurrence there. Common and serious elsewhere in northern Europe affecting a considerable number of *Rhododendron* species and hybrids, including the naturalised *R. ponticum*. The alternate host is Norway spruce, or, rarely, Sitka spruce.
Symptoms. Orange spots and subsequently dark red-brown patches develop on the shoots and undersides of the leaves in spring. Young soft shoots on pruned bushes are particularly susceptible. Somewhat similar rust-like symptoms on rhododendron leaves are also produced by rhododendron bug (p. 150). On spruces, pale yellow bands appear on the young needles in early summer and small white pustules then develop on the undersides of these needles which drop in late autumn.
Biology. On the orange spots urediniospores are produced (although these are rare on *R. ponticum*), the teliospores appearing later on the brownish patches. These infect the alternate spruce host on the needles of which the white aecia develop, the aeciospores subsequently reinfecting rhododendron. On rhododendrons other than *R. ponticum*, the rust can continue to live indefinitely, producing only urediniospores and with no need to alternate on the spruce host. Infection of *R. ponticum* only occurs in the vicinity of spruces however, and the fungus seems unable to survive on it in their absence.
Treatment. Remove and burn affected leaves; spray plants at three week intervals with Bordeaux mixture or oxycarboxin and use a wetting agent as rhododendron foliage is highly waxy and water-repellent.

ROSE RUST   *Phragmidium tuberculatum*   **Pl. 10**
Common and widespread throughout Britain and northern Europe, although the precise identity of the rust fungus responsible is still disputed.
Symptoms. The first symptoms which arise in spring are bright orange
pustules on leafstalks, branches, the undersides of leaves (especially on the
veins) and any persisting fruit. More familiar however, is the development
of yellow-orange pustules on the undersides of the leaves during summer,
and these become increasingly interspersed with black pustules as autumn
approaches.
Biology. The first formed orange pustules produce aeciospores which,
after germination and infection, give rise to the second yellow-orange
pustule type in which urediniospores develop. The teliospores produced in
the black pustules survive the winter on fallen leaves and germinate in
spring to produce basidiospores which, spread by wind or rain splash,
reinfect the bushes.
Treatment. Although several species and most cultivars of roses are
susceptible, 'Allgold', 'Arthur Bell', 'Fragrant Cloud', 'Lilac Rose', 'Pink
Favourite', 'Queen Elizabeth', 'Rose Gaujard', 'Sarabande' and 'Super
Star' show some resistance in certain areas. The cultivars 'Fashion', 'Hec-
tor Deane' and 'The Doctor' are almost always very susceptible. Ideally,
gardeners should collect up and burn rose leaves in autumn but anyone
with more than two or three plants, will rightly dismiss this as academic
advice. Zineb, maneb, oxycarboxin, thiram and some other fungicides are
all fairly reliable when sprayed fortnightly from April to September. Some
gardeners prefer to use Bordeaux mixture (although it may disfigure some
cultivars) in the early part of the season and thiram from June onwards.
The effectiveness of fungicides against black spot (p. 419) and powdery
mildew (p. 300) should be considered if these diseases are also troublesome.

SAXIFRAGE RUST   *Puccinia pazschkei*
Occasional and probably widespread in gardens, appearing as dark red-
brown pustules bursting through the upper leaf surfaces. Damage is very
rarely severe, but it would be wise to destroy badly affected plants.

SPRUCE NEEDLE RUST   *Chrysomyxa abietis*
First appeared in the early years of this century and now quite frequent in
Britain, particularly Scotland, on Norway, Sitka and red spruce.
Symptoms. Light yellow transverse bands appear on those needles that
are not more than one year old and deepen to a bright golden-yellow as the
season progresses. In the following May or June, golden-yellow pustules
burst through the outer tissues of the lower surfaces of the needles which
then drop.
Biology. The second year needles bear teliospores, the only spore type
known for this rust, and these are discharged when the pustules burst out in
spring and are blown to young developing needles which they then infect.
Norway spruce and, less commonly, Sitka spruce are also the alternate
hosts for the common rhododendron rust *C. ledi* var. *rhododendri* (see p.
284) with which this disease, in the early stages, might be confused.
Throughout the remainder of Northern Europe, a closely related rust
fungus with *Ledum palustre* (a rare plant in Britain) as its alternate host, is

the most important spruce needle rust. Its biology is generally very similar to that of the common rhododendron rust.

Treatment. Although trees may lose vigour after severe attacks, the disease rarely strikes in two years consecutively and so any such effect is shortlived. No treatment is necessary.

SWEET WILLIAM RUST *Puccinia arenariae*

Common on sweet williams in most seasons, especially in the south of England, and can also affect gypsophilas and some other garden plants. Brown teliospore-producing pustules, often arranged concentrically, develop on the undersides of the leaves and corresponding pale green areas appear on the upper surfaces. Flowering is sometimes reduced. Other spore types are unknown. Cultivars with dark red flowers and red-green foliage are often resistant and moving sweet williams to poorer soil, thus giving less vigorous leaf growth, is sometimes satisfactory. A spray with Bordeaux mixture, zineb or thiram is only likely to be successful if the attack is light or if the treatment is applied sufficiently early.

TANSY RUST *Puccinia tanaceti*

Fairly uncommon, but has been found on tansies and pyrethrums as well as on most *Artemisia* spp. with the exception of tarragon, *(Artemisia dracunculus)*. The disease probably occurs throughout northern Europe. Yellowish urediniospores are produced in pale brown pustules on the under surfaces of leaves while darker teliospores occur on similar spots on the upper surfaces and sometimes on stems. No treatment should be necessary but spray with Bordeaux mixture or oxycarboxin in bad cases.

VALERIAN RUST *Uromyces valerianae*

Fairly common on *Valeriana officinalis* and *V. dioica* as yellow-white cluster cups which appear on most green parts of the plants in early summer and as yellow-brown urediniospore and darker brown teliospore pustules which arise later in the season. Damage is rarely severe enough to warrant treatment but spray at three week intervals with oxycarboxin if necessary.

VIOLA RUST *Puccinia violae*

Common in Britain and probaby widespread on wild *Viola* species but less frequent in gardens. The symptoms are pale greenish, often swollen, spots on all parts of the plants and all five rust spore types are formed. Orange and white cluster cups are the symptom most commonly seen but brown urediniospore and teliospore-producing pustules can also be found. The rust on pansies was believed to be a different fungus but is now thought to be merely a variant although on that host a perennial mycelium becomes established in the rootstock. Diseased pansies are best uprooted and destroyed and new plants placed in different parts of the garden. On other *Viola* species removal of all above ground parts in autumn should be adequate.

WHITE PINE BLISTER RUST *Cronartium ribicola*

All five-needled pine species are affected by this devastating disease which has virtually prevented the commercial planting of Weymouth pine *(Pinus strobus*) in Europe. The alternate hosts are black currants and other *Ribes* species.

Symptoms. Small yellowish bark swellings develop in spring, followed in the summer of the next year by large white and orange blisters. Large cankers gradually form which can kill small branches and, as they spread to larger branches or the main stem, the entire tree may die. On currants and gooseberries small orange pustules and later, masses of brown bristle-like structures develop on the undersides of the leaves.

Biology. Fresh crops of orange aeciospores are produced annually from the perennial mycelium in the bark blisters and are blown to the alternate hosts on which the yellow urediniospores are first produced, to be followed by brownish teliospores on the tall bristle-like bodies. Black currants are by far the most important alternate host but the effect on them is slight, although some defoliation can occur in severe cases. Teliospores reinfect pine needles and can be carried considerable distances by wind.

Treatment. Nothing need normally be done to protect currant bushes and nothing can be done for badly diseased pines, which should be felled and burned, although in the case of slight branch attack cut off and destroy such branches first as this may protect the rest of the tree.

WILLOW RUSTS *Melampsora spp.*
A number of closely related rust fungi are common and widespread on willows but none are likely to be of any consequence in gardens. The symptoms vary but most typical are orange or yellowish spore pustules on the undersides of leaves. Certain species produce cankers on the stem and one, *Melampsora amygdalinae* is a serious problem in osier beds as it renders the stems useless for basket work. Treatment in gardens should not be necessary, although a fortnightly protective spray with Bordeaux mixture might be worthwhile in severe cases on small trees.

SMUTS
(Plate 10)

Basidiomycete fungi (Order Ustilaginales), closely related to the rusts and most producing black, soot-like spore masses. Over 800 species exist, nearly half in the genus *Ustilago*, and have a worldwide distribution. They cause important diseases of cereals and grasses but relatively few are serious on garden plants. Sweet corn, leeks, onions, salsify and scorzonera are the only non-ornamental garden species attacked and very few smuts affect woody plants. The symptoms are unlikely to be mistaken for any other disease although the superficial growth of sooty moulds (p. 303) may at first sight seem similar. Infected organs may be distorted, as in sweet corn, but some smuts, particularly those of the genus *Entyloma*, which is most important on dahlias in gardens, cause little more than a leaf spotting.

Although related to the rusts, the life cycles of smuts are much less complicated. Smut infections occur on the young parts of plants and those described here either develop on seedlings from spores on the seed or in the soil, or on actively growing points such as shoots or flowers from spores blown or splashed from soil or plant debris. The spores are known as

teliospores (also called chlamydospores or brandspores) and burst en masse
from stems, leaves, flowers, seeds or rarely, from roots. They form in
bodies termed smut balls, always contained within special pustular structures known as sori. These sori form from the mycelium in certain discrete
parts of the host plant. On germination the teliospores do not infect directly
and, as in the rusts, basidiospores (usually called sporidia in smuts) are
often produced first and it is usually these that actually start the infections.

Some smut diseases, sweet corn smut for example, do not spread far
through the plant from the original point of infection. Others, like anther
smut of scillas, develop systemically and permeate almost the entire host
tissues. Such systemic smuts can continue to infect new shoots annually as
they are produced from some perennial infected organ, such as a rootstock.

The smut diseases important in gardens are generally difficult to control.
Teliospores, once in the soil, can persist for several years and most fungicides offer little protection although Bordeaux mixture is useful in some
cases. Destruction of the plants and choice of a fresh site for new stock is
usually the only satisfactory measure.

ANEMONE SMUT *Urocystis anemones*

Common in Britain on anemones and globeflowers. Elsewhere in Europe,
aconites, wild ranunculus species and possibly other genera are also
affected. Several fungal races occur but it is not known if cross infection
takes place between all host plants. Dark streaks and blisters appear on
stems or leaves and dark brown spores burst from these blisters to produce
sporidia which can persist in debris or reinfect other plants at any time of
the year. The fungus is probably localised within the host plant so removal
of affected organs may limit its spread, although total destruction is better
unless the plants are particularly valuable. Take care not to shed spores
onto the ground when removing diseased plants and do not grow anemones, globeflowers or other host plants on contaminated soil for several
years.

CALENDULA SMUT *Entyloma calendulae*

Common and widespread. Symptoms appear first on the older leaves as
circular pale yellow spots which gradually darken, merge and spread to
disfigure the entire plant. This is unlike most smuts as the spores produced
in the lesions are very pale yellow and can germinate to produce sporidia *in
situ* on the leaves and thus bring about rapid spread of the disease within a
season. Spores can also persist in soil and plant debris and the disease is
therefore common where marigolds are allowed to self-sow. Choose a fresh
site for marigolds each year and destroy plant debris at the end of the
season.

CARNATION ANTHER SMUT *Ustilago violacea*

Common and probably widespread on both outdoor and glasshouse carnations and also on wild species of several caryophyllaceous genera. The
flower stalks become stunted and the anthers distorted and filled with a
purplish mass of spores. The fungus is probably systemic throughout the
plant and possibly spreads internally from the lateral buds. Destroy diseased plants before the flower buds open, never take cuttings from such

plants and thoroughly disinfect glasshouses where the disease has occurred. Avoid growing carnations in contaminated soil for at least five
years. Weekly spraying with zineb, maneb or thiram may check established
infections.

COLCHICUM SMUT *Urocystis colchici*

Occasional throughout Europe on colchicums and bulbocodiums. Blister-
like swellings develop on the leaves, corms and less frequently on the
flowers and flower stalks and later burst to release dark brown spores.
Flowering is apparently not affected. No control measures have been
developed but destroy diseased plants to prevent further contamination of
the soil and avoid growing susceptible host plants on the same site for as
many years as possible.

DAHLIA SMUT *Entyloma calendulae* f. *dahliae*

Originated in South Africa and now very common throughout Europe.
Symptoms. Circular or elliptical pale yellow-brown spots arise on the
leaves, normally from July onwards; these gradually enlarge, darken and
merge but usually retain a yellow margin. Darker areas within the lesions
can sometimes be seen in a strong light. It appears first on the lower leaves,
then spreads upwards and large areas of leaf tissue may ultimately die.
Biology. The causal fungus is a form of the species that produces calendula
smut (p. 288) and spores produced in the lesions may germinate *in situ* to
form sporidia which spread the disease to other leaves or plants. It persists
as spores in the soil but transmission does not occur in seeds or tubers.
There is some variation in cultivar susceptibility.
Treatment. Burn diseased foliage in autumn and remove any leaves
adhering to the tubers before storing. Do not grow dahlias in contaminated
soil for at least five years. Three or four fortnightly sprays of Bordeaux
mixture or captan may check the disease on growing plants, if applied from
early in the season.

GAILLARDIA SMUT *Entyloma compositarum*

Common in many parts of Europe since 1960 and found several times in
Britain. More or less circular pale brown, darkening leaf spots develop,
especially on the basal leaves, and in severe attacks the leaves wither and
shrivel. The biology of the disease is similar to that of dahlia and calendula
smuts (see above). Destroy affected plants and choose a different site for
new stock.

GLADIOLUS SMUT *Urocystis gladiolicola*

Occasional in Britain; commoner elsewhere in Europe. Dark brown leaf
blisters burst to liberate the spores and similar lesions develop on the corms
which may be almost totally destroyed in severe attacks. Corms are infected
from spores in the soil and the fungus spreads to the above ground parts
when corms are planted in the second year. It is wise to destroy affected
plants although valuable corms can be saved by hot water treatment.
Immerse them at $47°$ C ($117°$ F) for 45 minutes then plant in uncontaminated
soil.

MYOSOTIS SMUT *Entyloma fergussoni*

Occasional throughout Europe. Small off-white spots form on the leaves of both wild and cultivated *Myosotis* species. The biology of the disease is similar to that of dahlia and calendula smuts (q.v.) and similar treatments may be effective.

ONION SMUT *Urocystis cepulae*

Originally a North American disease, reaching Europe around 1880 and now occurs occasionally. Worse in cool seasons and Northern areas.

Symptoms. Dull streaking or spotting develops beneath the leaves and scales of young onions, chives, garlic, leeks and shallots. Affected leaves are swollen, twisted or otherwise distorted and the dark streaks eventually burst out as a black powdery mass. Some plants are killed; others survive to maturity but remain stunted or malformed.

Biology. Spores may persist in the soil for up to 15 years in the absence of host plants. When they germinate, they usually infect the growing points so that all new leaves in turn develop the disease. The initial infection is always on the first leaf but soon becomes more or less systemic.

Treatment. Because only the first leaf can be the site of initial infection, seedlings from a disease-free seed bed can safely be transplanted into contaminated soil. Care is needed however to avoid the transfer of infested soil (on tools for example), to clean sites. Slightly affected sets are a common source of the disease although the sale of these is illegal in Britain under the Sale of Diseased Plants Orders (see p. 41). Formaldehyde applied to the seed drill is sometimes effective if applied at the rate of 7 ml. formalin (= 40% formaldehyde) to 1 litre water per 10 metres of drill. Apply it with a watering-can immediately after the seed is sown and close the drill afterwards; the seeds are not damaged unless the soil is very dry. Formaldehyde may however be unreliable and seed treatment with thiram using methyl cellulose wallpaper paste as an adhesive gives better results.

SALSIFY SMUT *Ustilago tragopogonis-pratensis*

Rare in gardens in Britain although fairly common on related wild species; occasional elsewhere in northern Europe. Diseased plants at first develop more rapidly but the young flower heads are filled with masses of black spores. Very similar symptoms on scorzonera are caused by *U. scorzonerae*. Destroy affected plants and do not grow salsify on the same land for at least five years.

SCILLA ANTHER SMUT *Ustilago vaillantii*

Common on scillas including *S. bifolia* and *S. taurica* as well as on chionodoxas and *Muscari botryoides* but apparently not on grape hyacinth *(M. atlanticum)* or other *Muscari* species. Brown-black masses of spores appear in the anthers and sometimes in the ovaries, thus disfiguring the flowers. The fungus occurs systemically, passing from parent to young bulbs and affected plants should therefore be destroyed and a fresh site chosen for new stock.

SWEET CORN SMUT *Ustilago maydis* **Pl. 10**

Long known and common throughout the warmer parts of Europe, where high temperatures favour spore germination. Assumed significance in Britain for the first time in the hot summers of 1975 and 1976 on sweet corn and on ornamental and forage maize.

Symptoms. Swelling and other, sometimes very extensive, malformation arises on cobs, inflorescences, stems and occasionally on leaves and the entire plant may be distorted. The swellings are often ashen or whitish and contain a dark powdery mass which later bursts out.

Biology. Spores released from diseased organs can either germinate directly or persist in soil and crop debris. The fungus does not develop systemically in host plants but produces localised lesions.

Treatment. Some cultivars are resistant but not to all fungal races, which complicates the problem and makes choice difficult. Destroy diseased plants and crop debris and avoid growing sweet corn on the same land for at least five years.

VIOLA SMUT *Urocystis violae*

Common on several viola species. Elongated dark swellings develop on the leaf stalks, leaves and, rarely, on the floral parts from which dark spore masses burst. These produce sporidia which infect other plants either immediately or after lying dormant in the soil. Destroy affected plants and do not replant the same site with violas for at least five years.

DOWNY MILDEWS
(Plate 11)

Phycomycete fungi (Family Peronosporaceae); all are obligate plant parasites and are closely related to the common pathogenic genera *Phytophthora* and *Pythium*. Unrelated to the powdery mildews, with which they are sometimes confused and, with a few exceptions, they are a very understudied group biologically. Few very important garden diseases are caused by downy mildews although that on lettuce is commonly troublesome in wet seasons. Most garden species are in the genus *Peronospora* with a few in *Bremia*, *Plasmopara* and *Pseudoperonospora*. They produce off-white or faintly purplish mould in damp conditions, which, despite the name, may differ little from powdery mildew growth in general appearance. Seen under a lens, downy mildews appear as many individual spore-bearing heads whereas powdery mildews are denser and velvety with the individual heads hard to distinguish. When wiped off with the thumb, yellowing of the plant tissues is commonly seen under downy mildew growth. In the genera *Bremia* and *Peronospora*, the 'spores' seen on the mould growth are conidia and these germinate to form a specialised hyphal thread, the germ tube, which infects the host plant. In the genera *Plasmopara* and *Pseudoperonospora* the spores germinate first to produce swimming swarm spores but these also ultimately give rise to germ tubes and infect. Spread of the diseases during the growing season takes place by further spore production. Resistant oospores are formed by many downy mildew fungi and although in many they are probably the main means of overwinter survival, in some they appear to be functionless. It seems probable that each species of downy mildew is only able to infect one particular plant genus and that within these mildew species, further sub-divisions can be made of types capable of infecting only single plant species. Some downy mildews also exist as races, each with a further host restriction to particular ranges of cultivars.

Traditionally, copper-containing fungicides have been used for downy mildew control but at present the dithiocarbamate, zineb is the most effective chemical available for garden use. Most systemic fungicides available to gardeners are of little value.

BEET DOWNY MILDEW   *Peronospora farinosa* f.sp. *betae*

SPINACH DOWNY MILDEW   *Peronospora farinosa* f.sp. *spinaceae*

These two similar diseases are caused by closely related fungi; one occasional and widespread on beet but common and serious on sugar beet and the other common on spinach. Affected beet leaves are pale and toughened with curled edges and ultimately become blackened and necrotic. Grey-purple mould growth develops on the undersides in moist conditions. Spinach leaves are less severely affected but develop pale yellowish patches above and similar grey-purple mould beneath. Resistant oospores form in diseased leaves and persist in the soil or in plants left unpulled at the end of the season and it is important therefore to remove all plant debris each year and maintain as long a rotation as possible between host crops. The disease is also possibly seed-borne on occasions. Spray established attacks with zineb or Bordeaux mixture at fortnightly intervals although the essential cover required on the undersides of spinach leaves is often difficult to achieve. Do not spray within two weeks of cutting.

CRUCIFER DOWNY MILDEW
Peronospora parasitica   **Pl. 11**
Common and widespread on vegetable, ornamental and wild crucifers and found frequently on cabbages, cauliflowers, Brussels sprouts, radishes, stocks, swedes, turnips, wallflowers and watercress, among others. Sometimes a problem on bedding alyssums. This disease is often seen, together with white blister (p. 304), on the weed, shepherd's purse.
Symptoms. Most common on seedlings; yellowish specks develop on the upper surfaces of the leaves and cotyledons, with off-white patches of mould growth beneath in damp conditions. Severely diseased seedlings are stunted or killed but on mature plants, brown-yellow angular patches develop on the upper leaf surfaces with, in humid conditions, white mould beneath. The upper leaf surface is sometimes blistered around the patches and when attack is severe, the leaves turn yellowish and die. There may be pale brown discolouration of cauliflower curds, with black streaks and spots in the stem, although these symptoms can also arise from various other causes. Black internal spotting and streaking are also commonly found in radish.
Biology. Soil-borne; infects seedlings through the roots and the fungus sometimes develops systemically within. Conidia, produced on the mould, are rapidly spread by wind and carry the disease to neighbouring plants. Resistant oospores are sometimes produced and are probably the method of overwintering.

Treatment. Provide ventilation to seedlings in frames, boxes or glasshouses; do not allow water to drip on them and avoid close spacing, both of seedlings and young plants. Zineb sprays may be effective if applied fortnightly from the time when half of the seedlings have emerged. Four or five sprays should normally be adequate but must not be applied within a fortnight of harvest.

GRAPEVINE DOWNY MILDEW *Plasmopara viticola*

Originally a North American disease and first found in Europe around 1880. Now common and sometimes devastating in wine-growing areas but infrequent in Britain. It also occurs on Virginia creeper and related vines.

Symptoms. Pale yellow-green patches appear on the upper leaf surfaces, with corresponding areas of mould growth on the undersides in damp conditions. The patches later turn brownish and the mould disappears when conditions become drier. Young shoots, leaf stalks, tendrils and flowers may turn brownish and flowers and mature fruit may rot. Young fruit remain small and blacken and shrivel as the mould growth appears although in Britain, mould on grapes is more likely to be caused by the grey mould fungus (p. 352).

Biology. Resistant oospores, produced in dead leaves, survive the winter and are released into the soil when the leaves rot. They are splashed by rain onto young vines in spring to initiate new infections. Other, short-lived spores, develop on the mould growth and spread the disease during the growing season, before germinating into swarm spores which infect through stomata on the undersides of leaves.

Treatment. Collect and burn all fallen leaves in autumn. Apply Bordeaux mixture or zineb immediately before and after flowering; repeat ten days later and again when the fruit is about half-grown, but not within two weeks of picking.

LETTUCE DOWNY MILDEW *Bremia lactucae* **Pl. 11**

Common and widespread on glasshouse and outdoor lettuces but different strains also affect other members of the Compositae, including centaureas, cinerarias, gaillardias, globe artichokes, helichrysums and several weed species. On lettuce it is most serious in damp conditions and appears either as a seedling disease in spring or as a disease of mature plants late in the season. Downy mildew on lettuce is often followed by grey mould attack (p. 352).

Symptoms. Pale green or yellowish areas develop on the upper leaf surfaces, with corresponding patches of delicate, downy whitish mould beneath in damp conditions. The lesions later become brown and angular as the tissues die. Plants may eventually become generally debilitated and stunted.

Biology. The first infections are initiated by conidia, or by oospores in the soil which originated from a previous diseased crop. Spread in the growing season is by further conidia produced on the mould growth and no wounds are necessary for the infection to occur. Several races of the fungus exist and they are not all able to infect all lettuce cultivars.

Treatment. Remove and destroy diseased plant debris and maintain as long a rotation as possible between lettuce crops. Spray fortnightly with zineb from the time when half of the seedlings have emerged but do not spray within a fortnight of harvest. In glasshouses take care to keep water off the leaves, maintain good ventilation and remove infected leaves as they appear. Also see lettuce powdery mildew (p. 302) which is rare in Britain but quite common in some parts of Europe.

ONION DOWNY MILDEW   *Peronospora destructor*

Common and widespread on onions and shallots, especially in cool wet seasons and in wet areas, such as parts of the South and West of Britain.

Symptoms. Affected leaves turn yellow and die from the tip downwards but at this stage the disease can be confused with the many other problems which are discussed under onion leaf rot (p. 384). In moist conditions a fine, off-white, but later purplish mould growth develops on dead parts. Bulbs are also infected and soften and shrivel in store and develop some mould growth, but this can be confused with the commoner symptom of neck rot (p. 356).

Biology. The foliage of autumn-sown plants may be infected when young by airborne spores, the fungus then growing down into the bulb where it remains dormant over winter. In spring, it produces spores which spread the disease to newly sown plants and, in common with onion yellow dwarf virus (p. 456), this disease may therefore become more important as the growing of autumn-sown bulb onions increases in popularity. Semi-perennial onion varieties, such as Egyptian onions, commonly act as sources of the disease in gardens. Resistant oospores are sometimes produced in infected foliage and can survive in the soil for at least five years and also act as initial disease sources when they germinate.

Treatment. The disease is very difficult to control. Remove and destroy affected plants as soon as they are noticed and do not replant the same site with onions for at least five years. Avoid damp, sheltered sites and maintain good weed control to ensure air flow through the crop. Do not sow onions in spring close to autumn-sown plants, which may carry hidden bulb infection. Some protection may be given by spraying fortnightly with zineb from the time when half of the seedlings have emerged, but not within a fortnight of pulling.

PEA DOWNY MILDEW   *Peronospora viciae*

Occasional and widespread on peas and sometimes epidemic in wet seasons. It is less common on broad beans and sweet peas but no cross infection occurs between these host species. Yellowish patches develop on the upper leaf surfaces and purple-white mould grows on the undersides in damp conditions. The affected areas eventually turn brown and die. Pale green-brown patches may also occur on pods and, if plants are infected early, growth is severely checked and the pods develop without peas. Both conidia and resistant oospores form, the latter persisting in the soil and providing most of the initial infection. Fortnightly sprays with zineb from the time of occurrence of the first symptoms should check severe outbreaks but are rarely likely to be necessary.

ROSE DOWNY MILDEW   *Peronospora sparsa*
Occasional and widespread on glasshouse plants but rare in gardens, although sometimes found on briar seedlings. Irregular, more or less rounded, dull red-brown patches develop on upper leaf surfaces and these may
be confused with the symptoms of black spot (p. 419) but in humid
conditions bear grey mould growth on the undersides. Leaves may shrivel
and drop and the disease spreads to sepals and stem, where elongate dark
brown patches form. Resistant oospores sometimes form in infected leaves
but the fungus may overwinter as conidia or mycelium also. It can usually
be checked by improvement of ventilation and prevention of night-time
condensation and although it spreads very rapidly once established, weekly
sprays of zineb may then help to check it.

TOBACCO BLUE MOULD   *Peronospora tabacina*
A devastating disease in North America and elsewhere which was unknown
in Europe until 1958 when it appeared in glasshouses in England and
Holland on several *Nicotiana* spp. being grown for experimental purposes.
Subsequently it spread dramatically in tobacco crops throughout Europe
and has since caused devastation in Germany and other areas but has
virtually disappeared from England. In North America it also affects
tomatoes, egg plant and peppers and is most severe on seedlings, patches of
which turn yellowish, wilt and may be killed. Dense bluish-white mould
develops on the undersides of the leaves. The disease spreads during the
season by wind-blown conidia and persists in the soil as resistant oospores.
Suspected outbreaks in Britain should be reported to the Ministry of
Agriculture.

UMBELLIFER DOWNY MILDEW   *Plasmopara nivea*
Occasional and widespread on parsnips; less common on carrots, celery,
celeriac, chervil, parsley and other umbelliferous herbs. It is possible that
two fungi are involved and have sometimes been confused. Common on
several wild species but cross infection probably only occurs between
plants of the same genus. Yellowish spots develop on the upper leaf surfaces, and later turn black while in damp conditions corresponding white
mould patches appear on the undersides of the leaves which ultimately
shrivel and drop. Destroy diseased plant material at the end of the season
and do not overcrowd plants as this results in high humidity and favours
the infections. Take care to control umbelliferous weeds, especially wild
parsnip which is the only weed species of the same genus as the umbelliferous vegetables likely to be found in gardens. Chemical sprays are not
justified on a garden scale.

MINOR DOWNY MILDEWS   *Peronospora* spp.
This group of fungi affects many plants and although most of the species
are widespread they are only occasionally damaging. The host plants most
commonly affected are anemones, antirrhinums, campanulas, clarkias,
digitalis, geums, hebes, helleborus, laburnum, mesembryanthemum,
myosotis, poppies, primulas, rhubarb, violas and possibly a few other
related genera. Downy mildew of chrysanthemum flowers has been found
occasionally in recent years and could in time prove serious as the causal

fungus is common on related wild species in parts of Europe. It is believed that a different *Peronospora* species infects each host genus and is likely to be restricted to that genus. It is very improbable that any cross infection could occur between any of the host genera listed above.

Symptoms. Usually only serious on seedlings or young plants. Yellowish or brownish spots develop on the upper leaf surfaces with off-white, grey or purplish mould growth beneath in damp conditions. The leaves may become distorted, shrivel and drop and plants consequently become stunted or die.

Biology. Little known but probably similar to that of other *Peronospora* species (see p. 291).

Treatment. Destroy badly affected plants, paying particular attention to old diseased perennials which can act as permanent sources of infection for new stocks of the same genus. Spray severe outbreaks with zineb three or four times at ten-day intervals, taking particular care to spray the undersides of leaves.

POWDERY MILDEWS
(Plate 11)

Ascomycete fungi (Family Erysiphaceae); all are obligate plant parasites and they cause some of the most familiar and serious garden diseases. Conidia are produced on a whitish mycelial coating to leaves, stems, buds and occasionally other above ground parts of plants to which the development of the spores gives a characteristic powdery appearance. Affected parts commonly become discoloured or distorted and the effect of powdery mildews is one of a general debilitation of the host plant through the slow decline of diseased tissues. There are well over one hundred species with a world-wide distribution but they are generally most serious in warm, dry climates. Although conidia are by far the commonest spore type, most powdery mildew fungi are also capable of producing ascopores which are borne among the hyphae in very small black bodies termed perithecia. Only seven powdery mildew genera are important in northern Europe; *Erysiphe, Microsphaera, Phyllactinea, Podosphaera, Sphaerotheca, Uncinula* and *Oidium*, the genus into which are placed those species with no known ascospore-producing state. The classification and naming of powdery mildew fungi is confused and has been frequently changed. The situation is further complicated because many species attack many types of plant and because more than one powdery mildew species may attack the same host plant. To minimise confusion therefore and avoid pointless repetition, relatively few powdery mildews are described and named here individually; those that are have certain unique features either of biology or importance. The remainder are considered jointly on p. 302.

Several fungicides are effective in controlling powdery mildews but they may need applying at least fortnightly during the summer in very dry seasons. Sulphur was used in the past but damages some plants and has been replaced in most instances by dinocap and by the systemic fungicides triforine, benomyl, thiophanate-methyl and carbendazim.

APPLE POWDERY MILDEW *Podosphaera leucotricha* **Pl. 11**
Very common and widespread and one of the most serious apple diseases.
The same fungus occurs less frequently and less seriously on pear, quince
and possibly on medlar.
Symptoms. A powdery white coating begins to develop on leaves, shoots
and subsequently on flowers soon after leaf bud burst in spring. Blossoms
wither and drop and leaves fall later as the symptoms spread rapidly to
healthy foliage and shoots. A few leaves persisting at the end of an otherwise
bare shoot are commonly seen on some cultivars. Very rarely a powdery
coating may be seen on fruit but early infections which subsequently
disappear commonly cause a web-like fruit russet.
Biology. The first infections in spring arise from fungus that has lain
dormant over winter following infection in the previous season. Conidia are
produced on these primary infections and are then carried by wind to
spread the disease to secondary sites increasingly far from the original
infected shoot. Such infections take place throughout the summer except
when there is water actually lying on the leaf surfaces. Some buds may be
killed by the disease but most survive to become the primary sources of the
next year. Despite the very general development of mildew on shoots
during summer the disease dies down in winter and the fungus in the buds
is entirely responsible for the first infections in the following year. Early
symptoms and effects vary slightly depending on whether the source bud
was on a spur shoot or was terminal or lateral on a main shoot but the end
result is more or less the same.
Treatment. There are basically three main ways to treat mildew; in
winter, in spring and through the summer.
   In winter remove all buds and shoots distorted by mildew and cut back
any affected woody shoots to several buds below the limit of visible whitening. On small young trees, the cutting out of affected shoots can be continued throughout the season.
   During the early pink bud stage in spring, carefully remove as much as
possible of the affected leaves (the primary infections) but avoid showering
spores onto healthy foliage. If done carefully and thoroughly it is possible,
on young trees, virtually to eliminate mildew for the entire season in this
way.
   On large trees and as an added precaution on young trees, spraying
should also begin at the early pink bud stage and continue fortnightly to late
July. The best chemicals available to gardeners are dinocap and the systemic fungicides benomyl, thiophanate-methyl or carbendazim which also
have the merit of giving excellent control of scab (p. 335) and some control
of red spider mite (p. 249). Sulphur is a traditional remedy but the cultivars
'Beauty of Bath', 'Cox's Orange Pippin' and 'Newton Wonder' among
others, are damaged by it and it is probably better avoided.

CRUCIFER POWDERY MILDEW *Erysiphe cruciferarum*
Common and widespread, appearing as a thin powdery off-white coating to
leaves, stems and seed pods of most crucifers; especially common and
severe on swedes and turnips, sometimes on kales and increasingly so on
closely spaced commercial Brussels sprouts, although still infrequent on
this host in gardens. Powdery mildew of poppies is also probably caused by

this fungus. Badly affected leaves become yellowed and die and drop prematurely while the entire plant may wilt and roots of swedes and turnips may crack. The disease spreads rapidly by conidia in hot, dry weather and it became increasingly severe in the dry summers of the mid 1970's aided also by mild winters which it survives on crop debris. Dry light soils seem to favour this disease and early crops are very prone so sow turnips and swedes as late as possible. There is some variation in resistance among cultivars but none are sufficiently reliable to be recommended for garden use and no effective fungicides are available for small-scale application. Also see white blister (p. 304).

CUCURBIT POWDERY MILDEWS
Erysiphe cichoracearum, and *Sphaerotheca fuliginea*
Both fungi are very common and they are the most important and wide-spread of at least four powdery mildews on cucurbits, both under glass and outdoors. Both fungi also occur on several other plants. The symptoms are the typical powdery white or off-white coating which spreads from discrete patches to cover and occasionally kill leaves or entire plants. Summer spread is by conidia and these two diseases overwinter on plant debris and are abundant in all except the wettest years and are most evident late in the season. They may best be treated by the improvement of ventilation in glasshouses, by spraying with dinocap, benomyl or thiophanate-methyl at ten day intervals from the time of first appearance of the symptoms, and by cutting down and destroying debris in the autumn.

AMERICAN GOOSEBERRY MILDEW
Sphaerotheca mors-uvae
A North American disease, appearing in Europe about 1900 and spreading very rapidly. Now widespread but the intensity of attack varies con-siderably from year to year. In some recent seasons black currants have been more severely affected than gooseberries.
Symptoms. A powdery white coating appears on young shoots from April onwards; spreading rapidly to all new growth under moist, still conditions but the leaves are often least affected. The symptoms later intensify and stems and fruit develop a brown felty covering containing black dot-like bodies. The shoots are malformed and stunted and the fruit small and tasteless. On red and especially black currants, the symptoms are similar but the leaves are more severely affected. The disease occurs occasionally on other *Ribes* spp.
Biology. The mildew spreads during the season by conidia. The bodies produced later in the felt-like mycelium are perithecia within which asco-spores develop. Some perithecia fall to the ground and discharge their spores in autumn; others remain over winter and discharge in the spring but it is not known how important are the overwintering spores in initiating new infections in the following year and mycelial threads surviving in the buds are probably the most important means of persistence. On black currant it is almost certain that ascospores from perithecia on fallen leaves initiate the new infections which are usually not seen until May.
Treatment. Avoid planting in damp shady situations, leave plenty of space between bushes, and control weeds. Cut out infected shoot tips in

early autumn and maintain a free air circulation in the crop by winter pruning. Avoid excessive use of nitrogenous fertilisers. To prevent severe attacks on gooseberries, spray with sulphur starting just before the flowers open, again after fruit set and again two to three weeks later. On black-currants start the sprays as the flowers open and continue fortnightly until early July. Sulphur damages some gooseberry cultivars however and should not be used on 'Careless', 'Early Sulphur', 'Golden Drop', 'Lord Derby', 'Leveller', 'Roaring Lion' or 'Yellow Rough'. Dinocap, benomyl or thiophanate-methyl are safe alternatives for these cultivars. Spraying with a solution of washing soda is a traditional remedy for gooseberry mildew but is rarely very successful. Also see European Gooseberry mildew (below).

EUROPEAN GOOSEBERRY MILDEW
Microsphaera grossulariae
Occasional and widespread especially on old, well-shaded bushes, but rarely damaging. The disease occurs primarily on gooseberries but rarely on red currants, black currants and other *Ribes* spp. also. A faint powdery white coating develops, mainly on the upper leaf surfaces and less usually on the undersides or berries. Black perithecia also develop on the leaves and overwinter on them on the ground where they discharge their ascospores which renew the infection in spring. Control is scarcely justified but burning fallen leaves is worthwhile after severe attacks.

HAZELNUT POWDERY MILDEW *Phyllactinea guttata*
Fairly common throughout northern Europe and appears as a powdery coating to young leaves and flowers of hazelnut, cobnut, mulberry (although probably not in Britain) and also acers, alder, ash, birch, hornbeam, lilac and many other broad-leaved trees. Pale, gradually darkening perithecia also occur, mainly on the undersides of the leaves. The disease is rarely damaging but sulphur dust gives effective control in severe cases.

OAK POWDERY MILDEW *Microsphaera alphitoides*
Very common throughout Europe on most species of oak and occasionally on beech and sweet chestnut. The disease appears as a powdery white coating to leaves and shoots from mid-May onwards. It cripples young plants and spreads rapidly by wind-blown conidia during the summer. Leaf size may be reduced and shoot growth slowed or stopped. Ascospores form but apparently rarely germinate and overwintering by mycelium in the buds is probably most important. On mature trees, although mildew is often prolific on young shoots and particularly on lammas shoots in late summer, the effects are rarely serious. Spray nursery plants with sulphur as soon as symptoms appear and continue at fortnightly intervals into autumn if necessary.

PEA POWDERY MILDEW *Erysiphe pisi*
Common and widespread on peas and also on lupins as powdery white patches on the leaves and pods; rather less frequently entire plants are covered. The disease is worst on late pea cultivars in sheltered situations while lupins are sometimes devastated late in the season. Dust or spray with

sulphur, cut down and destroy plant debris in autumn, and mulch peas in dry weather to retain moisture and so reduce disease levels.

PEACH POWDERY MILDEW   *Sphaerotheca pannosa*
Widespread but commonest on plants grown under glass. The fungus is a form of that causing rose powdery mildew but cross infection from one host to the other probably does not occur naturally.

Symptoms. Powdery white patches develop on young leaves, buds and shoots of peach or, more rarely, on almonds or nectarines. The infected leaves may be elongated and distorted and the shoot growth is stunted. Infection spreads later to other leaves, shoots and fruit, especially those that are still immature, and some cracking can occur on the fruit which, in severe cases, appear completely covered by chalk-like dust.

Biology. Summer spread is by conidia and the usual method of over-wintering is as mycelial threads in dormant buds. Ascopores are produced in perithecia on leaves but are of little importance in overwintering in Europe. It is sometimes claimed that peach cultivars with leaf glands are resistant, and that mildew sometimes seems to follow outbreaks of leaf curl.

Treatment. Carefully cut out any badly mildewed shoots and begin fortnightly protective applications of sulphur sprays or dusts as soon as possible after the first mildew is seen or at the time of sepal drop, whichever is sooner.

PLUM POWDERY MILDEW   *Podosphaera tridactyla*
Widespread and particularly common on young plants and suckers but rarely serious. Also occurs on apricots, various cherries, cherry laurel, sloe and, with other powdery mildew fungi, on several other *Prunus* spp. Powdery white patches develop mainly on the undersides of the leaves and on shoot tips. Black dot-like perithecia also occur on the undersides of the leaves. Control measures are rarely justified but when attacks are severe, dust with sulphur.

ROSE POWDERY MILDEW   *Sphaerotheca pannosa*   Pl. 11
Very common and widespread. One of the most familiar garden diseases. Occurs in almost all situations but roses in dry sheltered sites such as against walls are particularly prone.

Symptoms. Not usually seen before May. Powdery off-white patches or more extensive coatings develop on leaves, stems, buds or rarely flowers. Leaves may turn yellow and/or purple, and wither and drop prematurely while buds may fail to open. On stems, flower stalks and especially sepals, there may be a denser mat-like growth, sometimes localised on stems around the prickles.

Biology. The main initial spring infection and all spread during the season is by wind-blown conidia produced on the powdery mycelium. From late summer onwards ascospores are produced in dark coloured cleistothecia within the white mats of denser mycelium. In some parts of Europe these may survive the winter to initiate some infections in spring but there is no evidence that they do so in Britain, where the disease persists as mycelium or conidia on plants in glasshouses or in other sheltered sites.

Treatment. Choose resistant cultivars where possible. There is considerable variation in susceptibility but few cultivars can be relied upon at all times and on all sites; 'Super Star' and 'Frensham' for instance were once very resistant but are now attacked in most localities. The following lists give a general indication of some commonly grown cultivars that are particularly resistant or susceptible. Members of National rose or horticultural societies may be able to obtain more extensive lists.

Resistant large-flowered bush roses (Hybrid Teas): 'Alec's Red', 'Alexander', 'Blessings', 'Champs Elysées', 'Charlie's Aunt', 'Chicago Peace', 'Colour Wonder', 'Gail Borden', 'Honey Favourite', 'King's Ransom', 'Mme Louis Laprerièrre', 'Michèlle Meilland', 'My Choice', 'National Trust', 'Peace', 'Pink Favourite', 'Pink Peace', 'Prima Ballerina', 'Rose Gaujard', 'Troika', 'Uncle Walter', 'Yellow Pages'.

Susceptible large-flowered bush roses (Hybrid Teas): 'Betty Uprichard', 'Christian Dior', 'Crimson Glory', 'Eden Rose', 'Frau Karl Druschki', 'Golden Melody', 'Hugh Dickson', 'Josephine Bruce', 'Mme. Butterfly', 'Peer Gynt', 'Picture', 'Spek's Yellow', 'Super Star'.

Resistant cluster-flowered bush roses (Floribundas): 'Allgold', 'Arthur Bell', 'City of Belfast', 'City of Leeds', 'Dickson's Flame', 'Escapade', 'Golden Slippers', 'Jan Spek', 'Korresia', 'Manx Queen', 'Marlena', 'Michelle', 'Molly McGredy', 'Moon Maiden', 'Orange Silk', 'Queen Elizabeth', 'Sea Pearl', 'Southampton'.

Susceptible cluster-flowered bush roses (Floribundas): 'Anne Cocker', 'Elizabeth of Glamis', 'Frensham', 'Iceberg', 'Lilli Marlene', 'Masquerade', 'Orange Sensation', 'Rosemary Rose', 'Woburn Abbey', 'Zambra', *Rosa gallica* 'Officinalis', *R. gallica* 'Versicolor'.

Because it is rarely possible to rely entirely on cultivar resistance, cultural and chemical methods should be practised also. Use mulches on dry sites, to retain moisture, and also restrict the use of nitrogenous fertilisers which promote susceptible soft growth. Apply chemical sprays fortnightly or three-weekly from April to September in all but the wettest seasons or most mild attacks, when less frequent sprays from May until August are adequate. Many chemicals give some control but if black spot (p. 419) and rose rust (p. 285) are also troublesome choose those proprietary products which include more than one chemical for this purpose. Dinocap is effective against mildew but may damage some cultivars; dichlofluanid is one of the few chemicals that is moderately effective against both mildew and black spot but it leaves a white deposit; benomyl and thiophanate-methyl are moderately effective against mildew only.

STRAWBERRY POWDERY MILDEW
Sphaerotheca macularis
A long-established problem on strawberries and especially serious in hot, dry weather and in the South of England. Occasional in parts of Europe on other rosaceous plants, especially filipendulas. The same fungus also occasionally causes mildew on the leaves, or more seriously, on the fruit of raspberries, blackberries, loganberries and related species.

Symptoms. Powdery grey-white patches appear on the undersides of the leaves in spring with corresponding dark reddish blotches on the upper surfaces. The leaves curl upwards at the margins, an effect that may be mistaken for drought damage. Infection spreads to the flowers, which are distorted or may fail to open, and also to the fruit, which is rendered inedible.

Biology. Summer spread is by wind-blown conidia and overwintering probably by conidia and/or mycelium on old leaves.

Treatment. The strawberry cultivars 'Cambridge Vigour', 'Crusader' and 'Templar' are very susceptible; 'Cambridge Favourite', 'Red Gauntlet' and 'Cambridge Prizewinner' much less so but even with the latter, other control measures may be necessary. Burn off old diseased foliage and spray fortnightly throughout the season, from just before flowering, with sulphur, dinocap or benomyl, although care is needed with dinocap which may damage the petals. If the disease occurs regularly on raspberries, moving the canes to less sheltered sites sometimes lessens its severity.

VINE POWDERY MILDEW   *Uncinula necator*

The first European occurrence of this disease was in Kent in the 1840's and it is now common under glass and in the open throughout Europe and sometimes very serious if routine preventatives are neglected.

Symptoms. Powdery white patches usually develop at first on the undersides of the leaves, then spread to the uppersides, and to the flowers, young fruit and shoots. In severe and early attacks flowers and fruit may drop but the later the infection, the more likely are grapes to remain on the vine, even if in a cracked or moribund state. In mild cases faint greyish discolouration of affected parts is the only symptom.

Biology. Spreads throughout the season by conidia and although in North America ascospores are produced in perithecia this is unusual in Europe, where the manner of overwintering is uncertain. Probably mycelial threads persist on young twigs or buds and establish new infections in spring.

Treatment. Good ventilation is essential in glasshouses and checks to growth during cold spells should be avoided by providing adequate supplementary heat. Sulphur is a traditionally effective remedy and four or more applications of dust should be applied on outdoor vines, particularly if they have been attacked in the previous year. Under glass it is often difficult to avoid damage by sulphur to vines or other 'sulphur-shy' plants growing nearby. Dinocap is a safe alternative; use three applications indoors, the first when the laterals are about 30 cm long, the second immediately before the opening of the flowers and the third after fruit set.

GENERAL POWDERY MILDEWS

In addition to the more individual diseases described above, there is a large and widespread group of powdery mildews, virtually indistinguishable by symptoms and all amenable to similar treatment. All occur most extensively in hot and dry seasons on closely spaced plants and may represent species of most of the common powdery mildew genera although each individual mildew species has a restricted host plant range. Garden plants on which one or more of these commonly occur are: acers (especially field maples), anemones, antirrhinums, aquilegias, asters (almost always affected in most gardens), begonias (especially serious on the cultivar 'Gloire de Lorraine'), calendulas, carnations, carrots (although probably infrequent in Britain), centaureas, chrysanthemums, cinerarias, clematis (often worst on large flowered and late flowering cultivars), cyclamens, delphiniums, digitalis, doronicums (almost invariably late in the season but usually too late to affect flowering), hawthorn, honeysuckle, hydrangeas (commonest on *H. macrophylla* in glasshouses, the mildew growth being accompanied by red-brown spots), Japanese spindle (originated in Japan and now very common in the South and West of England and elsewhere near the sea), laburnums (sometimes impairs flower formation), lawn grasses, lettuce (common throughout Europe but rare in Britain and only known there since 1961), lupins (may be devastating late in the season), marigolds, myosotis (often seriously affects flowering), parsnips, parsley and other kitchen herbs, phlox, poppies, potatoes (common in parts of Europe; rare in Britain and only found in very dry seasons), ranunculus, salsify, scabious, schizanthus, scorzonera, senecios, solidago, succulents (especially kalanchoes and sedums on which black spots occur surrounded by typical mildew growth), sweet peas, valerians, verbascums and violas.

Symptoms. Almost identical on all hosts; a powdery white coating develops on leaves, stems and sometimes on other above ground parts. It is usually seen first on the leaves as discrete off-white patches which may later spread to cover the plant extensively. The leaves may turn yellowish, die and drop prematurely and the entire plant may droop.

Biology. Summer spread is by conidia; overwintering usually in crop debris either as conidia, mycelium or ascospore-producing bodies (usually perithecia). Some species may be seed transmitted.

Treatment. There is little variation in cultivar susceptibility although some pale-flowered delphiniums may be resistant to mildew. The best chemical treatments for garden use are probably dinocap, benomyl, thiophanate-methyl or triforine sprayed fortnightly or three-weekly with the addition of a few drops of wetting agent. Care is needed with dinocap however on some cultivars of begonias, chrysanthemums or, in hot weather, cinerarias as they may be damaged by it. There is little that can be done in the way of general garden husbandry that is likely to have much impact on most mildew attacks. It is sometimes suggested that diseased parts of lightly infected shrubs such as hawthorn should be clipped out and burned but the showering of spores onto healthy leaves that this operation will cause may well do more harm. Diseased plant debris should however always be burned in autumn.

SOOTY MOULDS
(Plates 3, 4 and 11)

Black or dark brown soot-like deposits, commonly known as sooty moulds, often appear on the upper surfaces of leaves and on other parts of plants that are infested by sap-feeding pests, such as aphids (p. 161), adelgids (p. 182), whiteflies (p. 158), scale insects (p. 185) or mealybugs (p. 189). Many of these pests excrete a sticky, sugary fluid ('honeydew') which is derived from the sap on which they feed. Small droplets of honeydew accumulate on leaves, stems and fruits beneath the feeding sites of the pests and this honeydew, which can sometimes be heard pattering onto leaves, may form a varnish-like coating over exposed surfaces. Various non-pathogenic fungi, especially certain species of *Cladosporium*, then grow on the contaminated surfaces and produce the sooty deposits of fungal spores and mycelium.

Thick layers of sooty mould can build up fairly quickly and often cover all upper surfaces of affected leaves on shrubs and even on large trees. These symptoms are sometimes mistakenly thought to be the result of industrial atmospheric pollution but their true cause can usually be easily established by searching affected plants for signs of pests. Bay laurels, birches, camellias, ornamental citrus, lime trees, oaks, plums, roses, tomatoes, vines and willows are particularly susceptible but many other plants can be affected, both outdoors and under glass. Sooty moulds are especially troublesome on glasshouse and house plants since these are particularly likely to be attacked by sap-feeding pests and do not have the benefit of periodic rainfall, which washes some of the honeydew and sooty moulds off outdoor plants.

The fungi that produce sooty moulds do not cause direct damage to plant tissues but the black deposit cuts off light from the leaves, reducing photosynthetic activity and sometimes causing premature leaf-fall, with consequent weakening of the plants. The deposits also spoil the appearance of plants and contaminate ripening fruits, especially plums, grapes and tomatoes. Sooty blotch disease (p. 420) causes similar symptoms on apple, pear and plum fruits but is caused by a quite different fungus. Controlling the pests that cause the initial fouling of plants should lead to a general improvement although it may be some months before deposits disappear from older leaves. If possible, spray or sponge leaves with water to remove thick deposits. Affected fruit can be easily cleaned by washing or wiping and is usually quite edible since the moulds do not penetrate below the surface.

Honeydew and sooty moulds are sometimes a nuisance on cars, especially if they are parked regularly under limes, oaks or other large trees infested by aphids in summer. If it is impossible to avoid such situations, affected vehicles should be washed frequently to prevent damage to the paintwork.

WHITE BLISTERS
(Plate 11)

A small group of diseases caused by Phycomycete fungi of the genus *Albugo*; related to *Pythium*, *Phytophthora* and downy mildews. Only two species are of any consequence in Britain and northern Europe.

CRUCIFER WHITE BLISTER   *Albugo candida*
Widespread but varies greatly in intensity from year to year. Most usually seen on the weed, Shepherd's purse but sometimes a problem on cultivated vegetable and ornamental crucifers and commonly occurs associated with crucifer downy mildew (p. 292). Small white lesions, at first smooth but later powdery as the spores are produced, make affected plants look as if they have been splashed on the above ground parts with white paint. Distortion

of the affected organs or of the entire plant results if the fungus permeates the tissues. Spores are distributed by wind, insects or rain splash and germinate via a swimming swarm spore stage to infect through stomata. Resistant oospores are also produced and remain dormant in the soil for several months before they too germinate. Several races of this fungus exist but they are not all able to infect all crucifer species. Diseased plants should be destroyed and although further treatment is not usually necessary, severe attacks can sometimes be checked by spraying fortnightly with Bordeaux mixture up to one week before harvesting.

COMPOSITE WHITE BLISTER   *Albugo cubicus*
Occasional and widespread on salsify, scorzonera and other composites. Symptoms, biology and treatment are as for crucifer white blister (above).

NEEDLE CASTS
(Plate 11)

The premature shedding of leaves is a fairly common secondary symptom of many plant diseases but the wholesale browning and drop of conifer needles, often with little or no obvious infection of the shoots, is a difficult symptom to diagnose. In gardens, it is seen most commonly soon after planting out Christmas trees (Norway spruce) which are dying because of damage to, or lack of roots. Other types of root damage, as well as some diebacks, shoot and twig blights, leaf spots and especially winter cold damage are among other very common causes. There is additionally a group of diseases termed needle casts, often serious in forest plantations but in gardens only likely to be a problem on young trees growing close to mature plantings. All are caused by fungi (mostly of the Ascomycete order Phacidiales) and many species are common and widespread. Needle casts are most frequent on pines and larches but also occur on spruces, Douglas fir and, outside Britain, on silver firs (the commonest needle disease on silver firs in Britain is needle scorch p. 405). On trees affected by needle casts, the needles commonly turn brown and fall in scattered groups or *en masse* from entire branches or even whole trees. On close examination with a lens, tiny black dot-like fungal fruiting bodies can usually be seen on the needles but these are absent from one common form of larch needle cast and small orange bodies are sometimes present instead in early summer on affected Douglas fir needles. It should also be noted that a number of rust fungi (p. 264) infect conifer needles and produce small, usually orange or white pustules. Most of the fungi responsible for needle casts also live harmlessly on fallen needles and on those killed by other causes and it is from these that the spores carry the disease. Because they can be saprophytic however, the presence of fruiting bodies on needles need not indicate serious damage. Treatment in gardens is unlikely to be necessary but if elimination of other causes suggests a needle cast disease on small and valuable ornamentals, spraying with Bordeaux mixture in the following season may be worth a try. The timing of the sprays varies with different trees and the chemical may be damaging, especially in dry weather, so expert advice should be sought.

WILTS
(Plate 11)

This is a very common and widespread group of diseases all of which result in essentially similar symptoms. All, with the exception of Dutch elm disease (p. 307), are caused by normally soil-borne or occasionally seed-borne organisms and are therefore considered jointly. On herbaceous plants, the symptoms are often difficult to distinguish from root and foot rots and other stress factors, with which they may be associated. Host plants affected include fruit, vegetables, ornamentals and trees but the biology of the diseases is in all cases very similar. Most wilts are caused by Deuteromycete fungi, primarily of the genus *Verticillium*, especially *V. dahliae* and *V. albo-atrum*, or of the closely related genus *Phialophora*. On some plants, largely host-specific forms of *Fusarium oxysporum* cause similar symptoms but are usually less important than *Verticillium*. Only on China asters, carnations and peas is *Fusarium* wilt normally the more important disease. On antirrhinums, delphiniums, godetias and a few other plants, another Deuteromycete fungus, *Diplodina passerinii* is an important wilt-causing agent while on a few further plants, including carnations and wallflowers, bacteria may be responsible. The most important garden plant species affected by wilts are acers, antirrhinums, asters (China and Michaelmas daisies), beans, begonias, brassicas, carnations and other *Dianthus* spp., catalpas, cherries, chestnuts, chrysanthemums, cotinus, cyclamen, cucumbers, dahlias, delphiniums, elms, fuchsias, godetias, heleniums, limes, paeonies, peas (especially early in the season), potatoes, privet, quince rootstocks, raspberries, rhododendrons, rhubarb, robinias, roses, strawberries, sumach, sweet peas, tomatoes and wallflowers.

Symptoms. The only truly characteristic symptom is a brown or black discolouration seen in the conducting elements of the stem (whether herbaceous or woody) when cut through *well* above ground level. This discolouration can be traced both upwards in the plant and downwards towards the roots but if it is confined to the basal parts it is more likely to be caused by root damage than by a wilt. Accompanying effects may include epinasty (downbending) of the lower leaf stalks, wilting (often with, initially, some recovery at night), yellowing and later shrivelling of the leaves with the lower, or outer in the case of rosette plants, being affected first. Although many other factors, including root damage or cultural malpractices, such as inadequate or over-watering, can cause plants to wilt, these symptoms usually show on younger, not older leaves first. True wilt effects are sometimes confined to one side of the plant only and on raspberries, wilt disease is known as blue stripe since blue striping appears on one side of affected canes. Wilting of the leaves may sometimes occur very suddenly, especially during hot weather, and on trees it can be very dramatic when almost all the foliage droops simultaneously. Eventually plants become generally debilitated and may die.

Biology. The damage is often considered to result from the blockage of the water-conducting tissues of the stem, so starving the leaves of water. This blockage may occur because of cell collapse but is more usually because of gums and other materials causing physical obstruction. Although such mechanical blockage is a regular feature, it is now believed that the production of toxic substances by the pathogens in the conducting tissues is at least as important in causing the wilt effects. Many wilt organisms persist for several years in the soil; *V. albo-atrum* for example surviving as resistant mycelium, *V. dahliae* as small bodies termed microsclerotia and *F. oxysporum* as specialised spores known as chlamydospores. Wilt diseases are commonly associated with eelworm attack and wounds caused by these pests may provide infection sites, although on young plants at least, undamaged roots can also be affected.

Treatment. Wilt diseases are very difficult to tackle satisfactorily in gardens. Although many cultivars are resistant to wilts, few are available for garden use, although there are some notable exceptions which are usually so designated in catalogues. In glasshouses, do not plant any species, but especially not tomatoes, too early and into cold soil. Raising the temperature may help glasshouse plants to recover but the effectiveness of this treatment depends on the causal organism involved and this is impossible to determine without scientific tests. Increase the shading in glasshouses and apply no basal water but spray the foliage from above. When plants are lightly infected, pack the bases of the stems with damp peat, which may promote healthy new root growth. After removal of diseased plants, thoroughly disinfect glasshouses, all pots, boxes, benches, tools and other utensils and raise the next crop of plants in sterilised compost. In gardens, remove affected herbaceous plants together with as much as possible of the surrounding contaminated soil and avoid growing any of the above listed species on the same site for as long as possible. It should however be noted that carrots, parsnips and parsley are among the vegetables not affected while China asters and peas are both affected by specific wilt fungi and may safely be followed in a rotation by any other type of plant. On trees and shrubs it is worth trying a soil drench with spray strength benomyl or thiophanate-methyl around affected plants, repeating at fortnightly intervals until the symptoms subside or the treatment fails, when uprooting is the only solution.

DUTCH ELM DISEASE *Ceratocystis ulmi* **Pl. 11**
The most devastating plant disease that most people in Britain are ever likely to see. It also occurs less seriously throughout northern Europe, except in the most northerly areas. The disease was first recorded in Britain in 1927 and was serious for many years but then died down to occur sporadically and cause only restricted branch damage, but rarely tree death. It flared up again after the late 1960's when a new, more virulent, strain of the fungus was accidentally imported on logs from Canada and it has subsequently changed the landscape of much of England through killing many millions of trees. The disease is called 'Dutch' because of the extensive research formerly made on the problem in Holland.

Symptoms. The familiar yellowing of the foliage, usually from July onwards, is often followed very rapidly by branch and sometimes total tree death in the same season. In less severe attacks death may not occur until the following year but survival is rare once attack by the virulent strain of the fungus has occurred. Dark brown streaks may be visible when the bark and outer wood are peeled back from affected branches. Corresponding brown blotches may be seen in transverse section when branches are sawn through.

Biology. The fungus is spread to healthy trees by the elm bark beetles, *Scolytus scolytus* and *S. multistriatus* (see p. 244). The fungus in affected wood produces yeast-like cells which are transported within the water-conducting tissues. In common with other wilt diseases, the blocking of these conducting elements (so bringing about the dark streaks) and the production of a toxin, which is carried up to the leaves, result in the wilt symptoms. Under the bark of trees so weakened, the bark beetles tunnel and lay their eggs. The larvae resulting from these burrow further and so produce the characteristic fan-shaped galleries within which the spore-forming bodies of the fungus are produced. The main types of such bodies are small conidia-bearing structures and tiny, black, flask shaped peri-thecia, which produce ascospores. In warm weather during summer, the mature beetles emerge from the galleries bearing fungal spores collected either by eating or by mere physical contact. These beetles fly to healthy trees and feed on the branches, eating out small channels in the crotches of twigs, into which the spores are introduced as they do so. The fungus thus gains access to the conducting tissues of trees to repeat the disease cycle and, once killed, the tree can remain a reservoir of infective beetles for about two years. Considerable disease is spread between closely growing trees, such as those in hedgerows, through common root systems and the beetles are not involved in this process.

Treatment. Control is barely practicable. All elms are susceptible, al-though smooth-leaved elm and wych elm seem slightly less so than others. In the early years of the recent epidemic in Britain, attempts were made to limit the spread by felling affected trees but this was soon recognised as of little value except in areas where the disease was not established. Here such felling is still important. Dead trees should however always be felled as, in the early stages, they provide breeding grounds for the beetles, and in the later stages become safety hazards as they decay. Unfortunately the felling and destruction of large trees, especially in urban situations, is usually a matter for experts and can be expensive; however set against the possible damage that can result when trees fall, it is a wise investment. The timber of felled trees may safely be used but the bark must be stripped off and burned. In situations such as hedgerows where root transmission may be an important means of spread, healthy trees may be protected by digging a trench approximately 15 cm wide and not less than 60 cm deep midway between the trees and diseased neighbours. Great care is needed however, in urban situations to avoid damage to buried pipes and cables. The injection into healthy trees of sytemic fungicides related to benomyl has been used as a means of protection against Dutch elm disease. The pro-cedure has also been applied to eradicate the disease from trees in the very early stages of infection. Although it is possibly worthwhile with trees of exceptional value, it is of limited use as the procedure must be repeated annually and is by no means a complete protection. It can only be per-formed by experts with special equipment.

CANKERS, DIEBACKS, STEM and TWIG BLIGHTS
(Plate 12)

This section includes a wide variety of diseases affecting woody plants. The term canker is sometimes used to describe any disease condition in which there is sharply limited death of plant tissues resulting in a lesion. Under this definition however would be included diseases such as parsnip canker or tomato canker which in reality are forms of rot (p. 338). True cankers occur on woody plants and result from death of the cambium tissue in the inner bark, thus giving rise to an area where no new bark forms. Sometimes the lesion heals after one year's growth but commonly there is a repeated annual extension, often during the winter months while the host is dormant, and with some healing overgrowth during the summer.

Although a few viruses may give rise to canker symptoms (p. 429), their biology is very specialised and the causes of all the cankers described in this section are fungi or bacteria. Among the former, ascomycetes, and especially those of the genus *Nectria*, are important.

Cankers superficially indistinguishable from those of fungal or bacterial origin can arise through repeated spring or autumn frost damage but these are relatively unusual. More commonly, frost (p. 482), like mechanical wounding (p. 489), results in a lesion which heals after one year. Cup or dot-like fungal fruiting bodies or coloured, creamy bacterial slime on cankers are good indications of a microbial cause although not always conclusive. Several types of canker can serve as entry points for wood decay organisms (p. 371) and in such cases a toadstool or bracket-like fruiting body may appear. Small twig lesions can also be associated with certain leaf diseases or scabs.

When death of the cambium in a canker extends until the entire branch is girdled (such as commonly occurs on small twigs), the shoots beyond that point die. This is one form of dieback. In many other types of dieback, a gradual dying of stems or shoots from the tip downwards can occur in the absence of a canker lesion. Sometimes fungal fructifications occur on such branches but the causes are frequently less apparent. Wholesale dying back of entire shrubs and trees is more likely to be a secondary symptom of a root rot (p. 371) or other soil factor, such as waterlogging (p. 485) or drought (p. 482). Very sudden and complete death of entire branches can occur following wilt attack (p. 306) whereas certain environmental factors, such as frost or cold wind, can cause the death of shoot tips but usually there is no progressive killing of complete shoots. Attack by pests can also cause dying back but usually the pests themselves are present on the plants or evidence of their damage (such as boring holes) can be found at the junction of the live and dead tissues of the branches.

It is difficult to be precise over what is meant by a stem or twig blight as the word blight now has only a vague meaning scientifically. It is probably easiest to consider these conditions as diseases in which small stems or twigs are killed suddenly by the girdling action of very small cankers or by some very rapidly invading airborne organism, usually bacterial. Although the individual effects of such single infections are slight, the almost simultaneous appearance of disease over a large part of a tree or shrub can be very serious. The order cankers, diebacks, stem and twig blights is roughly one of increasing rapidity of action and of increasing difficulty of control. Slowly developing branch cankers should be pruned out and the wound treated with a sealant containing fungicide. Cankers, other than apple and pear canker or the bacterial canker of cherries and plums, on the main stems of garden trees are generally best tolerated until the tree shows obvious signs of decline or of decay from secondary infection. The cutting back of branches and twigs affected by diebacks and blights up to 30 cm beyond the visible limit of tissue death may save plants from some types of attack although in others a pruning or clipping out may need to be so drastic as to cause complete disfigurement and may compound the problem by providing more wounds through which fresh infections can occur. Chemical sprays are generally of little value against canker diseases and although they may arrest some branch and twig infections, the treatment of large trees is wholly impracticable.

APPLE and PEAR CANKER *Nectria galligena* f.sp. *mali*   **Pl. 12**
Common and widespread and by far the most serious of several types of canker on these trees. It affects a number of other trees including poplars and hawthorn, but not ash, which is attacked by a different form of the fungus.
Symptoms. Elongated sunken branch lesions develop and are often centred on a leaf scar, small side branch or bud. The bark in the centre of the canker dies and may break away. Although branches may be girdled by cankers and die back beyond the lesion, some dieback of young twigs can be caused without any true canker being formed. In spring and summer small creamy pustules may appear on cankers but more usually masses of small rounded bright red bodies are seen in autumn and winter. On some cultivars cankers may have the effect of partial ringing and may temporarily result in an increased crop while occasionally a papery bark condition may occur (see also silver leaf disease p. 324). The canker fungus may also infect the fruits and form a brown rot, usually around the eye, while they are still on the tree and cause them to remain hanging over winter.
Biology. The creamy pustules produce conidia while the red bodies are ascospore-forming perithecia. Both spore types initiate infections after transport by wind or rain splash to small bark cracks, pruning cuts or, most importantly, the leaf scars formed during autumnal leaf fall. Cracks in lesions formed by the scab fungus (p. 335) are also common infection sites while diseased fruit hanging on trees can also be a source of disease.
   Other minor but similar apple diseases include surface canker caused by *Pezicula corticola*; bark canker caused by *Phacidiella discolor*; the cankers associated with brown rot (p. 350) and bitter rot (p. 349) and a canker occasionally caused by *Leptosphaeria coniothyrium*, the raspberry cane blight fungus, when apples are grown near infected canes. Some also occur on pears and all may be treated as for *Nectria* canker. The symptoms of apple canker may sometimes be confused with those induced by bleeding canker (p. 312) or apple flat limb virus (p. 436) which require very different treatment.

Treatment. Some apple cultivars, 'Bramley's Seedling' and 'Newton Wonder' most notably, show resistance but all apples are prone to some degree of cankering. Some general cultural practices may help to lessen canker incidence; do not plant apples on wet and clayey sites, do not over-apply nitrogenous fertilisers and, if at all possible, grass down under the trees. Badly diseased trees should be cut down and burned but where attack is less severe, prune out all cankered branches, pare away the damaged tissue on large stems and treat all cuts with a wound sealant containing fungicide. Because poplars are attacked by the same canker fungus they should not be used as windbreaks around apple orchards and grey alder is a suitable alternative for this purpose. Although a spray with benomyl at the time of autumn leaf fall may lessen the number of leaf scar infections, this is only worthwhile on new trees if the garden has a history of cankering.

ASH BACTERIAL CANKER   *Pseudomonas savastanoi*
Widespread; commoner in Britain (especially Scotland) than fungal canker (see below) and also occurs throughout northern Europe. In the early stages it appears as small cracks which enlarge to become large, black, often more or less circular, deep lesions with raised and ragged edges. The mode of initial infection is not known but is believed to occur through hailstone wounds and leaf scars. The bacteria spread within the tree to initiate cankers far from the original infection site. No treatment is practicable and, since the infection is systemic, removal of individual cankers will not prevent more from forming.

ASH CANKER   *Nectria galligena* f.sp. *fraxini*
Fairly common in Britain and more frequent than bacterial canker in many parts of Europe. This is a form of the fungus that produces apple canker (p. 310) and although the ash form is able to infect apples, it is not the normal cause of cankering on that tree. It forms typical 'target cankers' with concentric rings of wound tissue within a usually elongated lesion and with small, red, ascospore-forming perithecia often occurring on the dead central bark. Unlike apple canker, it probably only invades tissues on ash trees already damaged by frost. Occasionally other fungi, particularly species of *Phomopsis*, cause large sunken lesions on ash stems but neither these nor the *Nectria* cankers can be effectively controlled.

ASH DIEBACK
A stag's head appearance of hedgerow ash trees is common in some seasons and has been observed particularly frequently in the east Midlands of England. The cause is unknown but infestations of scale insects seem to accompany the condition. Affected trees commonly recover after a few seasons and no treatment seems necessary although large dead branches should be removed before they are invaded by decay organisms.

BEECH BARK DISEASE   *Nectria* sp.
Common and widespread. This is the most serious disease of beech and is due to both fungal and insect attack. Since the 1960's it has been serious in young beech plantations on chalk land in southern England but is usually considered a problem on old trees only.

Symptoms. White woolly wax on the bark produced by beech scale insects (see p. 187) is common on beech trees and need not indicate the early stages of bark disease while dimpling on the bark is often seen as a result of earlier attack. Patches of dead bark a few centimetres across among the white woolly infestation may exude sap and become tarry; these indicate the true beginnings of the disease. Later, large areas of bark may die and the foliage turns yellow with the leaves sometimes reduced in size and number although yellowing can also be due to lime-induced chlorosis (p. 478), especially on chalky sites, and so this alone should not be used as a diagnostic criterion. Small red or black bodies may develop on the bark while bracket fungi and several species of wood-destroying insects may also appear. The trees eventually die and often snap off at four–five metres above ground level.

Biology. Beech scale, *Cryptococcus fagisuga*, causes small wounds in the bark through which spores of *Nectria* infect. The fungus then spreads in the bark, killing it in patches and so enabling a number of decaying organisms and wood-destroying insects to invade. The red bodies are *Nectria* perithecia, within which ascospores form while the black bodies are produced by other common species of fungi which subsequently develop on the dead bark.

Treatment. Trees with established fungal attack and/or decay should be felled as they will rapidly become extremely dangerous, especially if in close proximity to buildings. In the early stages, the beech scale insect can be controlled by a tar-oil winter wash on the bark (p. 189) or by spraying with dimethoate or malathion to kill adults and nymphs on the bark in May, or at other times during the summer and autumn.

BEECH CANKER   *Nectria ditissima*

Occasional in several parts of Britain but probably more frequent elsewhere in Europe. Not to be confused with the much commoner and more serious bark disease (see above). Depressed, more or less circular cankers develop, mainly on branches and twigs which may be girdled and die back while red, pinhead sized ascospore-forming perithecia are sometimes produced on the lesions. No treatment is practicable.

BLEEDING CANKER   *Phytophthora* spp.

Although known in Britain for some time on apple trees, this disease is now increasingly frequent on horse chestnuts and limes and may occur on other species also. Yellow-brown or reddish gum oozes from the dying bark of branches or stems and dries to form a hard black crust. Eventually stems may be girdled and dieback ensues. Dead central parts of the lesion also die and the bark cracks away from the underlying wood permitting decay organisms to enter through the damaged areas. The biology of the disease is little understood but the same fungi can cause root death (p. 377), some-times on the same tree although other factors including slime flux (p. 325) may result in similar symptoms and expert advice should be sought. Pro-vided decay has not set in, gouge out all dead and dying discoloured bark, sterilise cutting tools and then extend the gouged area into the healthy tissue by at least 5 cm around the periphery. Burn all removed bark and apply a slurry of Bordeaux mixture powder and a little water to all exposed parts. After this has begun to weather away, liberally apply a timber preservative such as creosote and when this has dried, seal the whole wound with a bituminous sealant.

BOX DIEBACK *Volutella buxi*
Fairly common and widespread, especially on box hedges, which develop symptoms of twig and branch death. Several fungi may be responsible for the condition but a wound-infecting species, *Volutella buxi*, is the most usual cause and produces pinkish spore masses on the leaves. There are no established control measures but dead branches should be cut out and burned; if the affected hedge or bush is small, spray three times at monthly intervals with benomyl following the first appearance of the symptoms.

BROOM DIEBACK *Pleiochaeta setosa*
Occasional and widespread. Tiny black spots develop on the leaves followed by shrivelling and blackening of the shoots and premature leaf fall. Depressed blackish cankers may subsequently develop on older shoots. Treatment is very difficult and plants severely affected are best uprooted and burned.

CLEMATIS DIEBACK *Ascochyta* sp. and other causes
A common and widespread problem but the causes are still imperfectly understood and may not always be the same.
Symptoms. Younger leaves droop suddenly, the upper parts of the leaf stalks blacken and affected leaves then wither and die, the entire process sometimes taking only a few days. Discoloured lesions may occur on the stem at or near ground level and dark patches also appear on some otherwise healthy leaves.
Biology. Very little is known about the biology of this condition. The stem lesions are caused by a fungus of the genus *Ascochyta*, which may originate from the soil, other plants or from the leaf patches. Infection possibly occurs in conditions of high humidity through small wounds caused by insects, wind damage or even stem ties but there is no extensive fungal development within the plant tissues.
Treatment. Especially difficult as the cause and biology are so incompletely known but it is important to avoid mechanical damage to clematis stems, particularly from securing ties. Sprays with dinocap are sometimes used to control powdery mildew on clematis and these probably help to suppress the *Ascochyta* on the leaves also. There appear to be some differences in cultivar susceptibility but insufficient is known to enable firm recommendations to be made. Following the death of a plant from dieback it would be wise to remove the soil to a depth of about 30 cm around the affected site and replace it with fresh soil before planting new stock.

CONIFER RESIN BLEEDING
Occasionally occurs in Britain but the distribution elsewhere is unknown. It is commonest on Douglas fir on which it appears as flat lesions or cracks on the bark of main stems and of branch bases from which there is copious resin exudation. Affected trees gradually decline in vigour and may die. The cause of the problem is unknown and no treatment is possible but it

should be noted that resin flow from stems may be associated with other more clearly defined diseases of conifers.

CORAL SPOT   *Nectria cinnabarina*   Pl. 12

Exceedingly common and widespread; the fungus occurs mainly as a saprophyte on dead branches (it is very often seen on old pea sticks) but also as a parasite causing dieback on many woody plants including acers, aesculus, apples, beech, cercis, currants, elaeagnus, elms, figs, gooseberries, limes, magnolias, mulberries, pyracanthas and walnuts. It is rare on conifers.

Symptoms. Masses of pinhead-sized vivid salmon-pink cushion-like pustules occur at all times of the year on dead and dying twigs and branches. Sometimes, usually during spring, similarly-sized dark red bodies may occur with or instead of the pink pustules and on close examination these look much like minute raspberries. Large branches may wilt in summer if they are affected near to the base while sometimes a small canker may be produced which gives rise to girdling dieback.

Biology. The fungus probably never infects living tissues directly but commonly invades through dead parts or pruning cuts, especially in early spring, and also through the base of the stems when plants are growing in very damp conditions such as among dense grass. It commonly invades frost-damaged twigs on several types of tree including walnuts. Once inside the branches, the fungus invades the water-conducting elements and so causes the wilt symptoms. Spread of the pathogen occurs by conidia produced in the pink pustules or by ascospores from the dark red perithecia. There may be several strains of the fungus differing in virulence but this is uncertain at present.

Treatment. Eradication of the pathogen is very difficult because it is so commonly present on dead wood. Rigorous hygiene can help however; do not leave piles of dead branches in the vicinity of fruit trees or bushes and never leave dead branch stubs when pruning. Diseased soft fruit bushes are best uprooted entirely and burned but valuable ornamentals should be pruned free of affected wood to at least 15 cm beyond the visible symptoms, the cuts treated with a wound sealant containing fungicide, the affected debris burned and the remaining healthy parts of the plants sprayed immediately and then twice more at three-weekly intervals with thiophanate-methyl.

CURRANT BLACK PUSTULE   *Plowrightia ribesia*

Common and widespread on red currants (on which it is often associated with coral spot) less so on black currants and gooseberries. Ovoid black pustules develop on the branches and may be followed by shoot dieback or, if on a main stem, death of the entire bush. The fungus is a wound parasite which infects dead tissue and grows from there into healthy parts in a similar way to coral spot although plant to plant spread is solely by ascospores produced in the pustules. The control procedures outlined for coral spot should be followed.

CURRANT STEM TOADSTOOL *Flammulina velutipes*

Common and widespread on dead tree stumps and occasionally on the stems of poorly growing red currant bushes. The fungus, which is probably weakly parasitic, produces toadstools, up to 7 cm tall, with yellow-brown caps and pale yellow gills beneath. The stalk is yellow at the top and brown-black and velvety below (cf. honey fungus p. 375 which may occur in similar situations but is much more serious and usually produces larger toadstools). This disease is a sign that the bushes are past their best and should be destroyed.

CYPRESS CANKER *Seiridium cardinale*

Occasional in recent years in southern England, France and possibly other parts of northern Europe on Monterey cypress. It forms resinous cankers bearing masses of pin-head sized pycnidia from which conidia are liberated and initiate new infections through tiny bark cracks. Branches or main stems may be girdled and die, the foliage turning first straw-yellow then rusty-brown, the latter symptoms being similar to those of winter cold damage. Despite fears to the contrary, this disease has not become a problem on the related Leyland and Lawson cypresses but affected branches or trees should be cut out and burned.

DOUGLAS FIR CANKER AND DIEBACK
Potebniamyces coniferarum

Common throughout Britain and northern Europe but no longer as serious a condition as it was once thought to be. It is sometimes a problem also on Japanese larch and rarely on other conifers. It may cause large stem cankers, small branch cankers with consequent girdling and dieback or, most commonly, shoot dieback in the absence of discrete lesions but with the needles rapidly turning first yellow then brown and dropping. The fungus is probably usually associated with damage originating from some other cause, such as frost. No treatment is possible; removal of the affected shoots is of no value in eradicating the fungus because it is a common saprophyte on dead twigs.

FIG CANKER *Phomopsis cinerescens*

Occasional, widespread and sometimes serious. Target-form lesions develop, usually at the bases of twigs or associated with pruning cuts and very small black dot-like pycnidia form on the cankers and in damp weather exude an off-white mass of conidia which are spread on hands, tools, and by insects, birds and rain. Cankered branches should be cut out and burned, any stubs left in pruning removed and all cuts treated with a wound sealant containing fungicide. Dieback of fig shoots is also often caused by the grey mould fungus *(Botrytis cinerea)* which usually produces fluffy growth at the base of the affected shoot but may be treated similarly.

FIR CANKERS AND DIEBACKS

Various types of canker and dieback occur widely from time to time on firs and several different fungi may be responsible. No treatment is likely to be necessary for these but the commonest type of dying back of fir shoots is that associated with needle blight (p. 405).

FIREBLIGHT   *Erwinia amylovora*   Pl. 12

This is a devastating problem in North America and was the first plant disease proved to have a bacterial cause. It was found in Kent in 1957 and has since spread northwards and westwards in Britain and now also occurs in Holland, Denmark, France, and Belgium and is likely to spread to other parts of northern Europe. It affects many rosaceous trees but is most serious on pears, apples, hawthorn, cotoneaster, pyracantha, sorbus, stranvaesia, amelanchier and chaenomeles although it has not been recorded in Britain on all the known host plants. It is a notifiable disease in England and Wales (see p. 41).

Symptoms. On pears it is most severe on vigorous trees; dead blossoms or dark-brown leaves hang from affected branches looking as if they have been scorched by fire. There may also be dark green-brown bark lesions while the fruits, if formed, are brown-black and wrinkled. All affected parts may exude a glistening white slime in warm damp conditions and if the bark is peeled from the edge of the lesions, there may be a red-brown, often mottled colouration beneath. The symptoms spread rapidly and mature trees may be killed within six months. The much less serious bacterial blossom blight (p. 347) is very similar in the early stages but infection usually does not progress beyond the spurs. On apple, the shoot tips at first wilt but remain green and exude drops of golden slime; later the leaves and shoots turn brown. On hawthorn, the leaves turn yellow, then brown and usually drop. On most other host plants the symptoms are similar to those on pear but on pyracantha, usually only the blossoms are affected.

Biology. The slime on affected parts contains the bacteria which are spread by insects or rain splash to the blossoms before moving down the blossom stalk into the fruiting spurs and branches. Host plants that produce blossom early are less severely affected because insufficient bacteria develop in time to cause much infection. Many pear cultivars which produce summer blossom are devastated however and for this reason the growing of the cultivar 'Laxton's Superb', which produces abundant summer blossom, is severely restricted by law in Britain. In host plants other than pears, infection through the shoots rather than blossoms is the normal process. Spread across country may be by bees or other insects, by transport of diseased plants or other means.

Treatment. In England and Wales, as fireblight is a notifiable disease, any known or suspected outbreak must by law be reported to the Ministry of Agriculture. Since 1968 the uprooting and burning of affected trees has usually only been required in areas where the disease is not yet established and although detailed procedures have now been worked out by which fireblight can be fairly successfully controlled they must be carried out under expert supervision. Gardeners should be alert to any suspicious symptoms on fruit trees, ornamentals and particularly hawthorn hedges and should not hesitate in calling in a Ministry adviser. The continued planting of new hawthorn hedges, especially in commercial fruit growing areas, is to be discouraged.

GARDENIA CANKER   *Phomopsis gardeniae*

Occasional and probably widespread. Brown, at first sunken, but later swollen dead lesions with rough margins appear on stems, usually close to soil level and may ultimately cause complete girdling and plant death. Black pycnidia develop on the lesions in moist conditions and from them ooze yellow masses of conidia which are spread by wind and rain to infect healthy tissues through small wounds. Destroy affected plants, obtain new cuttings from healthy stock and root them in sterilised medium and pots.

GOOSEBERRY GREY MOULD DIEBACK
Botrytis cinerea
The grey mould fungus causes dieback on many plants but is probably most serious on gooseberries, although similar symptoms occur on currants and raspberries also. The leaves first wilt then become brown and shrivelled and, as they do so, the bark cracks and in damp conditions fluffy mould growth develops. Ultimately entire bushes may die, branch by branch. If the symptoms are detected sufficiently early, it may be possible to save plants by cutting out and burning the affected parts but later, cure is impossible and the complete bush must be dug up and destroyed. Somewhat similar unthriftiness of gooseberry or currant bushes can be caused by black pustule disease (p. 314).

GRAPEVINE DEAD ARM DISEASE *Phomopsis viticola*
This disease is unknown in Britain but has occurred occasionally elsewhere in northern Europe and is very serious in other parts of the world. The branches of vines wilt and die and elongate tumours form on the stems. Any suspected outbreak should be reported to Ministry of Agriculture officials.

HEATHER DIEBACK *Marasmius androsaceus*
Occasional where extensive areas of heather occur naturally, as in Scotland, and only likely to be seen in semi-wild gardens in such areas. Clusters of tiny toadstools with pale brown caps and thin black stems up to 6 cm high, develop mainly on old shoots and are associated with branch dieback and sometimes plant death. The same fungus (and several similar species) commonly occurs among leaf litter under conifers and other trees and spreads through the soil similarly to honey fungus (which also occasionally infects heather) by rhizomorphs which in this case are thin and wiry. There is no established control for the disease but if large areas are seriously affected, it is probably worthwhile burning off the heather which will destroy some of the fungus and promote young growth, which is less likely to be attacked.

KERRIA TWIG BLIGHT
Twig death of kerrias is common in Britain and other, especially colder, parts of northern Europe. The symptom is commonest on double-flowered types but is of unknown origin, although possibly related to frost damage.

LABURNUM DIEBACK *Chondrostereum purpureum* and other fungi
Dying back of laburnum is quite common and widespread and the silver leaf fungus, *C. purpureum*, is a frequent cause, although no typical leaf silvering occurs. Masses of black crust-like perithecia developing on dead laburnum twigs are also common and are formed most usually by the

fungus *Cucurbitaria laburni* which spreads from dead tissue into living but weakened twigs and results in further death. Pruning out and burning the dead and dying branches may arrest disease development but the chances of recovery are slight and felling of affected trees is often the only solution. Another fairly common problem on laburnums is that known as fuzzy top; the terminal bud dies and the accelerated growth of lateral shoots results in a curiously misshapen tree. Although species of the fungus *Fusarium* are usually present in the affected plants they may not be the primary cause and the disease is uncontrollable at present.

LARCH CANKER AND DIEBACK

Trichoscyphella willkommii and other causes

Stem and branch cankers and various types of dieback are common, widespread and very serious on European larch (especially alpine strains) in plantations but rarer on isolated garden trees. Cankers are caused by *T. willkommii* aided by winter frost damage and dieback of small branches commonly occurs through canker girdling. Tiny yellow and white ascospore-forming apothecial cups are sometimes produced on cankers. (Similar orange and white cups on cankers and also on dead larch branches are very common but are produced by a related non-parasitic fungus). General dying back of larches, often with the production of epicormic shoots causing the trees to appear tufted is also common (again mainly on alpine strains) and often occurs in the absence of cankers but is usually associated with infestations of adelgids (p. 182) and sooty mould growth (p. 303). All these problems are very difficult to treat but it is wise not to plant larch on very frosty sites, such as frost hollows, and if there is a site history of cankering, choose east European strains or, better, Japanese larch, which is usually immune. Young trees should be sprayed for adelgid control.

LAVENDER SHAB DISEASE   *Phomopsis lavandulae*

Occasional and probably widespread but easily confused with winter cold damage which commonly affects lavender. The shoots wilt suddenly, usually in May and later, they, and sometimes the entire bush,'die. Subsequently, tiny black pycnidia develop on the affected dead stems and conidia from them are spread to infect new shoots through tiny bark cracks and pruning cuts. The disease can probably be avoided by only taking shoot cuttings of very young wood while slight attacks may be checked by spraying with thiram.

LILAC BLIGHT   *Pseudomonas syringae*

Occasional and widespread but although often described as the most serious disease of lilacs, extensive damage from it is unusual in Britain. Strains of the same bacterium affect many other host plants and cause serious problems on citrus for instance in some parts of the world. On lilac and sometimes forsythia, the most usual symptom is blackening of the buds and inflorescences and girdling of the stems and young shoots which consequently wilt. The damaged or killed parts are then commonly invaded by the grey mould fungus, *Botrytis cinerea*. Elongate canker lesions may form on large stems and small angular brownish spots may appear on the leaves but these alone are not diagnostic and may have other causes (see p. 410).

Spread from affected tissues is by wind and rain splash and infection probably occurs in winter and in spring through the stomata, the bacteria remaining dormant in the shoots until conditions are favourable for further development. The disease is very difficult to control although its development is often naturally restricted by spells of hot dry weather. Affected twigs and branches should be cut out and burned and although Bordeaux mixture sprays are sometimes recommended they can be very damaging to lilacs. Somewhat similar symptoms to lilac blight although not stem cankers, can occur on forced plants as a result of root infection by the fungus *Phytophthora syringae* (see p. 377) but this is rare in gardens.

LIME CANKER AND DIEBACK *Pyrenochaeta pubescens*
Unknown in Britain but occasional elsewhere in northern Europe causing sunken lesions on the bark of young trees. Small black conidia–forming pycnidia develop on these lesions and twigs and ultimately small branches may be girdled and die back. The problem is associated with unfavourable growing conditions and can probably be minimised by only planting limes on moist, loamy, nutrient-rich sites.

MULBERRY CANKER AND DIEBACK *Gibberella baccata*
Dying back of mulberry shoots in summer is occasional but widespread and is commonly caused by the fungus *G. baccata* which gives rise to small girdling cankers. Towards the bases of the dead shoots small reddish-brown conidial pustules form with, later, but rarely in Britain, blackish ascospore-bearing perithecia. It is difficult to treat on large trees but as far as possible diseased twigs should be cut out and burned.

OAK CANKERS AND DIEBACKS
Cankers are quite common and widespread on oaks but the causes are not known although various fungi, including species of *Nectria* and several decay fungi are sometimes thought responsible. Branch and twig dieback are also quite common and arise from several, usually fungal, causes but no treatment is practicable. The very common and widespread stag's head appearance of old oaks is thought to be induced by changes in the level of the water table.

PINE DIEBACK *Gremmeniella abietina*
Common throughout Britain and northern Europe and sometimes devastating in forest plantations but unlikely to be a problem on isolated trees. It occurs most usually on Corsican pine in Britain but other species, including Austrian and Scots pines, are also seriously affected in some parts of Europe while spruces are often attacked in Scandinavia. It is only really serious in the North and West of Britain where Corsican pine is at the limit of its range and as this tree is generally unsuited to cold wet areas at high altitude, the symptoms may be confused with exposure damage in similar situations. The older needles drop, leaving characteristic tufts of newer needles at the branch tips while the shoots, twigs and branches may die back, sometimes accompanied by the development of small cankers. On spruces, resinous main stem cankers cause girdling and death of the tops of trees. Tiny black pycnidia develop on the needles before they are shed and

conidia from them spread the disease to the buds, thus initiating the dieback. Very little can be done to minimise the effects of the disease which is best avoided by only planting Corsican pine on warm, dry and sheltered sites.

PLUM and CHERRY BACTERIAL CANKER
Pseudomonas mors-prunorum   Pl. 12

Common, widespread and may be very serious on cherries and plums but is less important on apricots, peaches and ornamental *Prunus* species.

Symptoms. These are of two distinct types. Cankers usually occur on the branches or crotches of cherries or on the main stems of plums. At first they appear as shallow depressions bearing (especially on cherries) blobs of amber-coloured gum but later enlarge and may exude quite copiously although on plum stems they may merely form elongate depressions, sometimes confined to one side only. Affected branches may die back after yellowing or stunting of the leaves. The second symptom is the development of dark brown more or less circular leaf spots in late spring which may ultimately merge and form patches of dead tissue which may drop out and leave a 'shot hole'.

Biology. Cankers are initiated in autumn or winter by bacteria which are splashed from the leaves and enter the stem tissues, usually through leaf scars. The lesions extend rapidly in the spring but then activity usually declines and the bacteria die out. No new cankers form during spring or summer when the trees are resistant to this type of infection but at this time the foliage is attacked instead to cause the leaf spots on which more bacteria form.

Treatment. No plum or cherry cultivars are resistant and a few, including 'Victoria' and 'Early Laxton' plums, are very susceptible. Control of this disease is very difficult in gardens although if care is taken when tying trees (especially plums) to support stakes this will lessen any bark damage through which initial infection could arise. The severity of the disease can be appreciably lessened by high working onto resistant rootstocks, such as the myrobalan type plum stocks or F12/1 cherry stock. Some control of the leaf spotting can be achieved by spraying in spring with Bordeaux mixture and may be worthwhile on ornamental *Prunus* spp. Spraying of cherries with Bordeaux mixture in mid-August, again two weeks later and finally in early October may be successful in reducing the number of bacteria present during the susceptible leaf-fall stage of tree growth. This is a very difficult operation to perform satisfactorily however; not only are cherry trees often large and not amenable to spraying but Bordeaux mixture can be phytotoxic and may damage the plants significantly.

Other symptoms are sometimes confused with bacterial canker. The production of gum by both stems and fruit of cherries and plums is a common phenomenon and can arise from a variety of generally unknown causes but including bark split virus (p. 461) and drought (p. 482). On plums, peaches and cherries dead branches often occur with parasitic fungi growing on them but it is generally considered that these only infect following damage or weakening from other causes. On cherries the condition known as tan disease is sometimes common on young shoots or main stems. The bark splits and peels back to reveal a rusty coloured powder composed of dead cells. The only way to minimise such conditions is to ensure that trees are not planted on waterlogged sites and that correct cultural practices, such as careful pruning, are followed.

POPLAR BACTERIAL CANKER *Aplanobacter populi*

Common and widespread throughout northern Europe but less important than formerly in Holland and Belgium. The most serious poplar disease in Britain.

Symptoms. Cracking of the wood, usually on one year old shoots, is accompanied, especially in spring, by oozing of dirty, cream coloured slime. Small shoots and sometimes large branches may be girdled and die back whereas on the biggest branches and stems, large irregular canker lesions burst through the bark. Orange staining develops in the outer wood and there may be blackening of the leaves. The oozing of slime from poplar stems may also occur as a result of slime flux (p. 325) but no canker lesions form.

Biology. Infection occurs through bud scale scars, leaf scars, small wounds, pruning cuts and similar sites. Paradoxically, bacterial slime is produced most abundantly in spring, although most leaf scar infection sites are only present in autumn, and the relationship between these two factors is not fully understood. Occasionally fungal species of the genus *Nectria* can cause cankers on poplars but the lesions are much more regular and target-like than those produced by bacteria.

Control. The only effective method of avoiding canker is to plant resistant clones, many of which are readily available; all cultivars of black poplar for instance, including Lombardy poplar, are resistant. Individual diseased trees, especially if closely associated with other susceptible types, may be felled but moderate disfigurement on ornamental plantings may be tolerable.

POPLAR CANKER AND DIEBACK

Cryptodiaporthe populea

Common; the most serious poplar disease in all parts of northern Europe except Britain where it is second only to bacterial canker (see above).

Symptoms. Lesions are usually flattened or sunken on the smooth bark of stems and occur frequently around wounds, at the bases of twigs or at nodes. Small stems are rapidly girdled and die back. Masses of tiny black pox-like bodies develop on the lesions in spring; often in concentric rings but sometimes clustered.

Biology. Conidia ooze in creamy masses from the black pycnidia in damp weather and are dispersed by wind and rain to spread the disease. Infection occurs through minute wounds, often those naturally produced, such as leaf scars, but can also take place through pruning cuts. Damage is most serious on trees growing under stress conditions such as drought or water-logging or on frosty sites. Comparable but less serious dieback can be caused by species of *Cytospora* and may be treated similarly while the apple canker fungus (p. 310) can also sometimes affect poplars.

Treatment. Some poplar species and hybrids probably have resistance to canker but no firm recommendations can be given. The best means of avoidance is to ensure vigorous growth, avoid unsuitable sites and overcrowding and not prune in winter. Affected shoots on isolated diseased trees should be cut out and burned although if all the trees in small plantations are attacked, this procedure is unlikely to lessen significantly the development of new lesions. Small single trees may be sprayed at monthly intervals during the summer with Bordeaux mixture which is sometimes a satisfactory treatment.

POPLAR LEAF and SHOOT BLIGHTS *Marssonina* spp.

Common and widely distributed. Large irregular brown blotches develop late in summer on the leaves, from the upper surfaces of which conidia exude in minute white tendrils. Such attack is often followed by premature leaf fall and dieback. *Marssonina* spp. are sometimes serious in Britain, Holland and elsewhere as causes of dieback of Lombardy poplar, typified by death of the lower shoots first, but they also affect other species. No treatment is practicable on large trees but spraying young trees with Bordeaux mixture as the leaves unfold and again when fully expanded is sometimes successful.

POPLAR TWIG DIEBACK *Venturia* spp.

Caused by two closely related fungi; one or other widespread in northern Europe on several poplar species but almost unknown in Britain. Angular black spots develop on young leaves early in the season and increase in size until the leaves are killed. The shoots turn black and wither, often with a characteristically hooked tip, and masses of dark green conidia form on such dead tissues in damp weather and subsequently spread the disease. The fungus overwinters as mycelium in dead twigs and the only effective treatment is to remove such twigs in winter although this is impracticable on large trees.

RASPBERRY CANE BLIGHT *Leptosphaeria coniothyrium*

Common in Britain but also occurs elsewhere in northern Europe. The disease does not affect black currants, loganberries or other hybrids but is particularly serious on the raspberry cultivar 'Lloyd George' although 'Norfolk Giant' and others are also susceptible.

Symptoms. Leaves on the fruiting canes shrivel and die and the canes themselves bear dark patches just above ground level on which the bark cracks and masses of small pin-head sized bodies develop. The canes also become markedly brittle and may snap at the position of the lesion.

Biology. The fungus infects the canes from the soil and both conidia and ascospores may be produced on the lesions; infection, as with spur blight, may occur through areas damaged by the raspberry cane midge (p. 214). The disease is readily spread on affected canes, spores from which can contaminate clean soil, or in soil adhering to stools.

Treatment. Cane blight can often be avoided by handling canes with care to prevent damage but when attack has occurred, cut back diseased canes to below soil level and burn them, taking care to disinfect the secateurs afterwards. Do not transplant stools from a site where the disease has occurred and do not plant strawberries near to infected raspberry canes as *L. coniothyrium* can cause a troublesome root disease of strawberries. Where cane blight attack is linked solely with cane midge damage, control can be obtained by applying an insecticide such as fenitrothion in early May and again about two weeks later.

RASPBERRY SPUR BLIGHT   *Didymella applanata*
Common and widespread on raspberries and loganberries but usually only serious in northern areas and in Britain most frequent in Scotland.

Symptoms. Although large irregular brown blotches may form on the leaves early in the summer these are seen infrequently but by August purple lesions appear on the canes and gradually turn brown-black and then whitish. Ultimately 10 cm or more lengths of cane can be affected and are sometimes girdled. As the winter progresses, the cane lesions become less distinct, turn silvery and are dotted with tiny black bodies. Canes rarely die as a result of spur blight but since buds on affected nodes are killed, the number of fruiting spurs and therefore the fruit yield are reduced in the following season. As with cane blight (above), spur blight may follow attack by cane midge (p. 214) although the insect is less significant for this disease.

Biology. The silvery sheen on the lesions is due to the shrinkage of the inner tissues and the admission of air between these and the outer bark. The black bodies are of two types: pycnidia which produce conidia, and perithecia which discharge ascospores in spring when the initial infections occur. Either type of spore can initiate such infection, normally via the leaf stalks and buds. Ascospore release and germination is favoured by wet conditions and the disease is therefore worse in seasons with a damp spring although summer drought can accentuate the problem as the fruits produced on diseased canes tend to dry and shrivel under such conditions.

Treatment. Cut out and burn affected canes and maintain normal cane thinning early in spring to prevent overcrowding. Spray established infections (or healthy canes when the disease is known to be prevalent in the area) with benomyl, dichlofluanid, thiram or thiophanate-methyl. Apply the first spray when the canes are about 15 cm high and repeat four times at fortnightly intervals. The new raspberry cultivars 'Malling Admiral' and 'Leo' have some resistance to spur blight.

ROSE CANKERS   *Leptosphaeria coniothyrium* and other fungi
Several cankers and associated diebacks are widespread on roses but stem canker *(L. coniothyrium)* is the commonest and most serious. Among others which occur occasionally are brown canker, caused by *Cryptosporella umbrina*, brand canker, caused by *Coniothyrium wernsdorffiae*, a canker and dieback caused by *Clathridium corticola*, grey mould dieback caused by *Botrytis cinerea*, and a disease called briar scab, caused by *Botryosphaeria dothidea*, which is common on wild roses but occasionally serious also on cultivated types. Some of these diseases occur also on other rosaceous plants.

Symptoms. STEM CANKER: brown, cracked and sunken lesions often with a reddish border, develop on stems. The lesions increase in size and eventually form large, irregular cankers which may girdle the stems and cause a dieback. Tiny black bodies usually develop on the canker surface. Stem canker can be a problem following budding when the disease spreads both up and down from the union.

BROWN CANKER: numerous small rounded reddish stem spots which later form off-white lesions. These may develop into light brown cankers, often with purplish margins and on which minute black pointed bodies may arise. The stems are commonly girdled and, depending on the time of year that this occurs, the upper parts of stems may die or produce knobbly gall-like outgrowths.

BRAND CANKER: reddish stem spots which enlarge, darken and develop small pale brown centres on which small black dot-like bodies may form. In spring, the damaged areas may appear black and eventually the lesions may merge, girdle the stem, and cause a dieback.

CLATHRIDIUM CANKER: small brown bark depressions with purple margins, usually around a small wound. Occasionally they girdle the stem to cause dieback but more usually remain confined to one side. The margins of the lesions characteristically remain smooth and, unlike stem cankers, never become ragged although small black bodies commonly form on damaged tissue.

GREY MOULD DIEBACK: most commonly a dying back of stems from the apex, old flower or fruit stalk, dead branch stub or similar origin. The symptoms are common on stems after the tip has been damaged by cold or frost. Fluffy grey mould growth may be present but is not invariably so.

BRIAR SCAB: masses of black crusty bodies on the stem; frequently around thorns.

Biology. Grey mould biology is described in detail elsewhere (p. 352) but this and all the other rose canker fungi infect through small wounds, dormant buds, pruning cuts or similar points of weakness. Conidia and, in some instances, ascospores are produced in the dark bodies on cankers and may bring about significant spread of disease during the growing season.

Treatment. Remove dead flower heads, buds and damaged shoot tips as a matter of routine. Taking care not to leave stumps when pruning will help eliminate potential infection sites. Cut out and burn any cankered stems and remove entire bushes if the disease is extensive but if the damage is slight, spray three times at three-weekly intervals with thiophanate-methyl after removal of the diseased parts, taking care to spray onto all cut surfaces.

SILVER LEAF   *Chondrostereum purpureum*

Very common and widespread, mainly on rosaceous trees and especially on plums, of which it is the most serious disease. Other common hosts are almonds, apples, apricots, cherries, hawthorns and roses while among plants of other families affected are currants, gooseberries, laburnums, poplars and willows, although not all develop foliage symptoms. It is infrequent on pears. This disease is caused by a close relative of many wood-rotting fungi but, unlike them, it does not induce decay.

Symptoms. Leaves develop a silvery sheen, usually at first confined to a single branch but soon spreading. Later, the leaf tissues may split and browning may occur at the margins and around the mid-rib. Affected branches have a dark brown discolouration in the wood and usually die back and, from late summer onwards after such branch death, small bracket-shaped or flat crust-like fructifications with a lilac-purple spore surface develop on the bark. The upper surface of the bracket or the under surface of flattened forms, is brown and hairy and the fructifications are soft and leathery when wet; shrivelled and crisp when dry. Spread of the disease from affected branches to the rest of the tree may occur but is not inevitable. Silver leaf is common on apples following top grafting and is often accompanied by papery peeling away of the bark, although this symptom can also accompany apple canker (p. 310) or arise as a result of poor soil texture. A common and superficially similar symptom to silver leaf is that of a nutritional disorder known as false silver leaf. It induces no staining in the wood however and may be treated by mulching the trees and giving fertiliser in the following season.

Biology. Basidiospores from the fructifications infect through wounds in the wood, normally from September to May. Toxic substances produced by the fungus are carried upwards to the leaves where they cause the leaf tissues to separate. Air entering the spaces so formed brings about the silver effect but there is no fungus present in the leaves.

Treatment. Natural recovery is quite common and it is best to wait and see for some time after the appearance of silvering. The leaves are not a source of disease and need not be destroyed but once branches begin to die, they should be cut back beyond the limit of any brown stain in the wood, flush with the parent branch or stem, and the wound treated with a sealant. If the entire tree is silvered, especially if silvering appears on the suckers, the main stem and roots will be affected and all should be uprooted and burned. Piles of logs and dead branches of any type should not be stored in orchards. Although all plum cultivars are liable to be affected, 'Victoria' is the most susceptible while damsons and greengages are usually fairly resistant.

Gardeners are commonly apprehensive regarding the legal status of silver leaf disease. In the Silver Leaf Order, 1923 it was stated that all dead wood of plums and apples should be burned on the premises before July 15th in each year and where such dead wood was in the trunk and extended to the ground, the whole tree, including the roots should also be burned. Moreover, wood of any kind bearing fructifications of the silver leaf fungus was required to be destroyed. While such actions are thoroughly to be recommended however, at the present time the legal requirements are no longer enforced.

SLIME FLUX AND WETWOOD

Common and widespread on many broad leaved trees especially elms, horse chestnuts, poplars and willows. Foul smelling fluxes ooze from cracks in the bark and may kill both bark tissues and any grasses or other plants onto which they flow. When dry, fluxes may be whitish and chalky. They are associated with soaking of the internal wood of the trees and may result in yellowing and dying back of crowns. Although bacteria are always associated with fluxes and may invade trees through small wounds, it is not known to what extent they are the causal agents. Within affected trees a very high pressure may build up and gas or fluid is ejected forcibly when holes are drilled into the trunk. Cure is probably impossible and badly affected trees should be felled.

SNOW BLIGHT *Phacidium infestans* and other fungi

Common and sometimes serious on Scots pine and other conifers in most northern parts of Europe where prolonged deep snow cover occurs in winter. Infection of the needles by ascospores occurs in the autumn but the fungus only spreads in and among the needles under snow, which provides moisture and higher temperatures. Browning in spring and subsequent death and fall of needles on young trees results. There is no effective control.

SPRUCE CANKER   *Scoleconectria cucurbitula*

Fairly common and widespread; chiefly on Sitka spruce in Britain but also on Norway spruce and other species elsewhere in northern Europe. In forest plantations the disease is usually associated with damage, such as the removal of branches. The commonest symptoms are elongate strips of dead bark on the trunk which eventually take the form of hollows or flutes and through which decay fungi may enter. Treatment is not really possible but it is worth checking for the appearance of fructifications of wood-rotting fungi, which may invade through the lesions and indicate rot within and structural instability.

A condition known as top dying occurs occasionally on Norway spruce; whole trees die back from the crown, the needles of which may appear vivid red. The cause is probably infestation by root aphids (p. 181) and the condition is associated with mild winters.

SWEET CHESTNUT DISEASE   *Endothia parasitica*

Although unknown in Britain, from where it is excluded by strict quarantine, this is a devastating disease on several *Castanea* spp. elsewhere in Europe and in other parts of the world. It is generally more severe in warmer climates and is sometimes called canker or blight.

Symptoms. Flattened, sometimes sunken or slightly swollen lesions develop on branches or stems and, on young branches, the affected pale brown areas contrast markedly with the normal olive-green bark. The bark often cracks away from the lesions and masses of small black pin-head-like bodies, sometimes exuding orange tendrils, develop on the dead tissues while beneath the bark, fan-shaped masses of fungal growth may occur. The crowns of affected trees wilt and die back when the branches are girdled but new shoots are often produced from below the lesions. The production of new shoots and death of the branches one at a time differentiate this disease from ink disease (p. 377) while the presence of the orange tendrils and fan-shaped mats distinguishes it from similar but much less serious cankers produced by other fungi such as *Cryptodiaporthe castanea* and *Diplodina castaneae*.

Biology. Conidia are present in the sticky orange tendrils that ooze from the black pycnidia and are washed down the stem by rain or carried to other trees by birds or insects where they infect through small wounds. Perithecia also form on the bark and produce wind-dispersed ascospores which provide another means of spread. In parts of Europe, lesions may heal completely and little damage results, indicating that some trees have resistance to the disease.

Treatment. Various, usually unsuccessful, methods of control have been tried in several countries. In Britain any suspicious damage to sweet chestnuts should be reported immediately to the Ministry of Agriculture or to Forestry Commission officials.

SYCAMORE SOOTY BARK DISEASE
Cryptostroma corticale

A North American disease, first found damaging trees in England in 1948 since when it has been most severe in London and the home counties but occasionally found as far north as Chester. It is also known from isolated areas in France and Germany. Outbreaks almost invariably follow hot summers and very occasionally *Acer* species other than sycamore are attacked. The foliage of affected trees wilts and the outer bark may peel away in patches to reveal a dark brown sooty mass of spores and many tiny black columnar structures beneath. In the early stages of attack, the bark, although blistered, remains attached to the tree but the disease is very difficult to detect at this time. Infection by the spores probably takes place through bark wounds and pruning cuts and therefore sycamore should not be pruned in areas where the disease is known to be prevalent. The fungus may become established in dead branches and move into healthy tissues during certain periods but its biology is imperfectly understood. No control measures are available but badly diseased trees should be felled and burned.

WALNUT GRAFT DISEASE   *Chalaropsis thielavioides*

Common when walnuts are propagated in glasshouses by grafting and produces masses of brown-black conidia on and around the graft preventing a satisfactory union. Effectively treated by immersing the stock in a 1:40 strength solution of formalin before inserting the scion.

WILLOW ANTHRACNOSE   *Marssonina salicicola*

Several fungi commonly cause small cankers and/or diebacks on willow stems, sometimes associated with leaf spots, but by far the most important in gardens is the anthracnose fungus *Marssonina salicicola*. It may seriously disfigure weeping willows and is almost epidemic in wet seasons.

Symptoms. Most obvious are small ashen or pale brown stem lesions, red-brown spotting on the leaves and a gradual defoliation and dying back of the shoot tips, sometimes to the extent that weeping trees cease to weep as their shoot length is reduced.

Biology. *M . salicicola* produces conidia on the stem and leaf lesions and although these are the means of disease spread, the biology of this condition is imperfectly known.

Treatment. Virtually impossible on large trees as the lesions and damaged twigs are usually so numerous that pruning out the affected parts is impracticable. On small trees in gardens, Bordeaux mixture sprays applied every three weeks from just before bud burst until early July (avoiding very hot spells) may prevent the disease from becoming established. Although not the best chemical treatment, it is the only one available for gardens.

WILLOW WATERMARK DISEASE   *Erwinia salicis*

An important disease especially of cricket bat willows and related forms of *Salix alba* in eastern England; similar conditions occur elsewhere in Europe. It induces progressive dying back of the crowns and as a result the trees take on a stag's head form with a colourless sticky liquid oozing from bark cracks. Commercially the most important effect is a brown staining of the timber. The causal bacteria are probably spread by insects but their biology is imperfectly understood. In Essex and parts of some neighbouring counties affected trees must by law be reported to the local authorities who may require them to be felled. Severely affected trees in any area should be felled as mere removal of diseased branches is of little effect.

GALLS, LEAF CURLS and WITCHES' BROOMS
(Plate 12)

Many malformations are caused by a disturbance of the normal growth of plant organs, usually through some disruption of the plant's hormone systems. Galls are swellings and other abnormal growths that arise from localised tissue proliferation and may occur on roots, stems, leaves or, more rarely, on flowers or fruit. Those described in this section are caused by several quite unrelated types of fungi and bacteria but they are in many respects very similar to the galls induced by some eelworms (p. 136), mites (p. 252) and insects (p. 229). The causes of two of the most serious gall diseases, clubroot of crucifers and wart of potatoes, are curious organisms, generally considered to be distantly related to fungi but lacking mycelium. Witches' brooms are abnormal proliferations of shoots on woody plants. They may be broom-shaped but are often more irregular, sometimes resembling birds' nests. Most are caused by fungi although viruses, mycoplasmas and other sub-microscopic organisms induce witches' broom symptoms on some plants while the exact causes of many are unknown.

Most stem galls and witches' brooms are of little consequence although they do disfigure ornamentals and in such cases can often be pruned out. Root and tuber galls are sometimes very serious in weakening affected plants and are caused by organisms with very long persistence in the soil. Control of them is extremely difficult and the maintenance of existing disease-free land by strict hygiene is essential.

Most leaf curls are caused by Ascomycete fungi of the genus *Taphrina* and result from unequal expansion or growth of the leaf tissues although superficially similar conditions may be caused by certain viruses, mites and insects and some other leaf diseases such as scabs can, when severe, also cause puckering of leaves. The only serious leaf curl disease is that of peach and this can usually be controlled by fungicide sprays.

CLUBROOT *Plasmodiophora brassicae*   Pl. 12
Very common and widespread. Clubroot is probably the most problematic disease to deal with in gardens. It affects all crucifers but is rarely a problem on radishes and in gardens it is most frequently seen on Brussels sprouts, cabbages, cauliflowers, swedes, turnips, stocks and wallflowers and can be particularly damaging on Chinese cabbage. Contrary to common belief, it does not cause disease symptoms on plants such as lettuce, beetroot and spinach which are sometimes confused with crucifers. It is generally worse on acid, poorly-drained soils and usually less serious in hot, dry seasons and on spring-maturing crops.
Symptoms. Sometimes the above-ground parts of affected plants wilt on hot days but may, in the early stages of the disease, then recover at night. In severe attacks plants may be stunted, take on a red-purple tint to the

foliage and occasionally topple over or die. The roots are swollen and distorted, either in the form of a single large gall (the club symptom) or, when the lateral roots are affected, as clusters of galls reminiscent of dahlia tubers (the finger and toe symptoms). Knobbly swellings occur at the bases of naturally swollen, bulbous roots such as those of swedes and turnips. In the later stages the galls are often affected by soft rot organisms and decay to an evil-smelling mass. Somewhat similar galls may be caused by the turnip gall weevil (p. 242) but these, when cut through, are seen to be hollow and may contain a weevil larva. The later stages of cabbage root fly attack (p. 218) are sometimes superficially similar but also usually contain larvae or evidence of their burrowing activity. The small hard swellings on swedes and turnips of unknown cause known as hybridisation nodules (p. 488) may be very difficult to differentiate from clubroot, while hormone herbicide injury (p. 491) to stem bases can also be confusing.

Biology. The causal organism persists in the soil as resting spores which can probably survive in the absence of host plants for twenty or more years. Ultimately, these spores germinate (possibly under chemical stimulation from the host plant roots) to produce swimming swarm spores which invade intact roots, usually through the root hairs. Further swarm spores of a different type are produced within the root hairs and, after liberation, spread the infection to other parts of the roots which then swell and become distorted. More resting spores form within the swellings and are released into the soil when the galls decay. Although this disease is less severe on alkaline, well drained sites, the reasons for this are not known. The pathogen is most commonly transferred to disease-free sites on infected transplants or in soil on plants, boots or tools.

Treatment. Extremely difficult. There is almost no resistance to clubroot in garden crucifers other than radishes and the need to keep this disease out of uncontaminated soil cannot be overstressed. Certain measures can be adopted to this end. Never obtain brassica plants from a source not known to be disease-free; in practice this means if possible raise them from seed in sterilised compost (see p. 22) rather than buy them. Scrub boots and tools thoroughly with very hot soapy water after use on contaminated land. Once a garden has been contaminated, there is no realistic means of eradicating the disease from the soil although certain procedures may help to lessen its effects. Improvement of the drainage and liming to raise the pH to about 7 may lessen the severity although care is needed not to over-lime if potatoes are to be grown on the same land as this may encourage common scab (p. 336). It is wise to maintain as long a rotation as possible between cruciferous crops although the three to four years practicable in most gardens is unlikely to have much effect on badly contaminated land. Earthing up diseased plants may stimulate new roots to form and so hold off the worst effects of the disease until after the crop matures. At the end of each crop, carefully dig up all diseased plants and as much as possible of the root system before it decays. These plants must be destroyed, ideally by drying and burning but, failing that, by burial in a deep hole in a remote corner of the garden.

Transplants may be protected by dipping the roots in a 10% aqueous suspension of pure calomel (mercurous chloride) or a 1% suspension of thiophanate-methyl or benomyl before planting out. Half a litre of suspension will be sufficient for about 200 plants which should be swirled in the liquid in bunches of about twelve at a time. Better adherence is achieved with calomel if the fungicide is made up in 1% cellulose wallpaper paste instead of water. Alternatively, if only a few plants are to be grown, raise each in an individual pot (of about 12 cm diameter) of sterilised compost or soil and plant it out in the complete pot ball; roots growing beyond the ball develop small galls but the plants are not seriously affected. There is no effective chemical treatment for direct-sown plants and although the application of calomel dust to the drill is often recommended the results are usually disappointing. In a large and badly contaminated garden the granular soil sterilant chemical dazomet may be tried in accordance with the manufacturer's directions. It will not eradicate the disease but, if applied with care, may diminish significantly the numbers of spores present near to the surface. Proprietary phenolic emulsions drenched onto the soil may have similar, localized, partial-sterilant effects. Deep digging is still likely to bring new, unaffected spores to the surface however. There are many traditional remedies for clubroot, including applications of rhubarb, manure and soot, or mothballs and egg-shells to planting holes; there is no proven scientific basis for any of these but they are probably worth trying if all else fails.

CROWN GALL   *Agrobacterium radiobacter* var. *tumefaciens*   **Pl. 12**
Very common and widespread on many woody and herbaceous host plants. This is by far the commonest gall-forming disease and is probably most serious on fruit trees. Among the plants on which symptoms are most frequently seen are apples, beetroot, begonias, cherries, chrysanthemums (marguerites), currants, dahlias, elms, euonymus, gooseberries, grape-vines, hawthorns, hollyhocks, loganberries and blackberries, lupins, marigolds, marrows, peaches, pears, pelargoniums, phlox, plums, poplars, privet, quinces, raspberries, rhododendrons, roses, runner beans, swedes, sweet peas, tomatoes, wallflowers, walnuts, willows and many wild plants.
Symptoms. These are very variable and it is almost impossible to be certain of differentiating crown gall from all other types of gall. Usually irregular but sometimes more or less spherical, knobbly swellings can arise on almost any part of affected plants but occur most typically at the junction of root and stem. They are common and often very large (1 m or more in diameter) on the trunks and branches of trees and are also frequently seen on woody stocks produced by layering at the point where the rooted layer is severed from the stool. On shrubs and trees, galls are usually very hard but on herbaceous plants they are commonly soft and fleshy and, when old, may decay through the action of soft rot bacteria. Galls may occur on leaves and in such positions are usually more discrete than those of leafy gall (p. 331). On some cane fruits, elongate masses of small round galls may burst through the cane surface. The root-forming masses known as burr knots on the branches of certain apples and other plants are superficially similar to crown gall but are not caused by a pathogen.
Biology. Imperfectly understood but several strains of this bacterium exist and they are not all able to infect all host plants. The initial infection almost certainly arises when soil-inhabiting bacteria are blown or splashed on to above-ground parts and penetrate through small wounds. The bacteria

may possibly move within affected plants to establish secondary galls some distance from the initial infection. Despite the often dramatic appearance, the effects of crown gall on the vigour of the host plant are rarely serious.
Treatment. Extremely difficult but rarely necessary. Since most damage is to fruit trees, these should not be grown on land which has supported diseased plants. Handle planting stock, especially of fruit trees, with care to avoid wounding, sever apple stocks from the stool with a clean cut and cut out and burn disfiguring galls on ornamentals. The application of sulphur to contaminated soil is sometimes claimed to be effective but the relatively minor importance of the disease in gardens probably does not justify the effort and expense.

FORSYTHIA GALL

Fairly common in some years and probably widespread. Irregular, often clustered small swellings develop, usually on the upright shoots, but also close to ground level on pendulous shoots in damp shady situations. They are probably analagous to burr knots on apple and other trees and as such, are root-producing growths. The galls are almost certainly harmless but if badly disfiguring should be pruned out.

LEAFY GALL   *Corynebacterium fascians*

Common and widespread on many ornamentals. Especially frequent on carnations, chrysanthemums (including marguerites), dahlias, geraniums, gladioli, heucheras, nasturtiums, sweet peas and strawberries (on which, together with certain eelworms (p. 135) it causes cauliflower disease). The same organism may be associated with the common distortion and blindness in certain bedding plants, especially antirrhinums, nemesias, petunias and occasionally phlox.
Symptoms. Masses of short, thickened or otherwise distorted shoots develop, usually at or near ground level, on plants that are otherwise normal, although there may be some general stunting. On bedding plants the commonest symptom is blindness (the absence of flower buds); although some plants such as petunias, may grow out of this and ultimately flower, others such as antirrhinums rarely recover.
Biology. The symptoms arise because the bacteria stimulate normally dormant basal buds into growth. The pathogen is soil-inhabiting but it is not known how infection occurs nor how long it can survive in the absence of host plants although with nasturtiums, sweet peas and possibly other species the disease can be transmitted by infected seed.
Treatment. Destroy affected plants and do not plant susceptible species in contaminated land for as long as possible. Disinfect hands and tools by thorough washing after handling diseased material. Always raise seedlings in sterilised compost in disinfected pots and boxes.

OLEANDER GALL   *Pseudomonas savastanoi*

Occasional and widespread, occurring on all above ground parts and giving rise to various symptoms; soft, spongy, gradually darkening swellings with projections on old shoots, elongate canker-like lesions on younger shoots, distortion of young leaves and seed pods and swellings accompanied by oozing slime on older leaves. Difficult to eradicate but if the damage is

limited in extent, cut out and burn the affected parts. Do not propagate
from plants showing any symptoms.

PEACH LEAF CURL *Taphrina deformans*   Pl. 12

Common and widespread on peaches, almonds and nectarines but rarely on
apricots and infrequent on trees in conservatories or other enclosed sites.
Worst following cold wet springs and particularly prevalent on trees growing in cool, damp situations such as close to expanses of water.
Symptoms. Soon after bud burst in spring, the young leaves thicken, curl
and twist, becoming yellow with reddish areas. The puckering and reddening increases and a pale bloom covers the affected leaves, which later turn
dirty brown and drop prematurely. Young shoots and more rarely flowers
and fruit may similarly become twisted and distorted and after repeated
attacks trees lose their vigour and fruit production declines. Certain aphids
also cause curling of peach leaves (p. 161) but the presence of the insects or
their cast skins on affected plants readily distinguishes this condition.
Biology. The bloom on the leaves results from the production of ascospores which are carried by wind and rain to the surfaces of the shoots where
they germinate in autumn and produce yeast-like colonies. These survive
the winter among the bud scales and in the bark, spores from them infecting
the young leaves in cold wet conditions in spring. Hot dry weather in the
summer may check disease development.
Treatment. If possible, collect and burn affected leaves early in the season
before the ascospore bloom develops. There is no advantage in destroying
the leaves later in the season nor in spraying them with fungicide. Fungicidal sprays should however be directed towards eradicating the yeast-like
colonies and their spores. One approach is to spray with Bordeaux mixture
after autumn leaf fall and again as the buds swell in late February or early
March while the alternative is to spray in early February with captan and
then repeat this during bud swelling.

MINOR LEAF CURLS *Taphrina* spp.

These are common and widespread on various host plants and all are
caused by species of *Taphrina*. *T. minor* causes a leaf curl similar to peach
leaf curl on cherries; *T. bullata*, small brown leaf blisters on pears similar to
those of the pear leaf blister mite (p. 253); *T. tosquinetii* causes severe
distortion of the leaves and young shoots of alder; *T. sadebeckii*, yellow leaf
blisters on the same host; *T. aurea*, bright yellow blisters on the undersides
of poplar leaves; *T. ulmi*, green blisters which later turn brown on elm
leaves and *T. caerulescens*, pink-purple leaf blisters with grey then brown
discolouration to the under leaf surface on oaks. The biology of all these is,
as far as is known, probably similar to that of peach leaf curl and although
none are normally serious, they may be similarly treated in severe cases if on
young trees.

POCKET PLUMS *Taphrina pruni*

Common in parts of northern Europe but infrequent in Britain, except in
the North and West. Developing fruits are elongate, twisted and one-sided,
the hollow side appearing pocket-like. At first affected fruits are smooth
and green but later they become paler and shrivel to appear prune textured.

Eventually they are covered with an off-white bloom of ascospores and then completely shrivel and drop. No stone forms. Mycelium originates in the twigs which are infected by the ascospores of the preceding year and permeates the fruit. Affected fruit should be collected and destroyed. If the disease was prevalent in the previous season, a protective spray with Bordeaux mixture about three weeks before flowering is a wise precaution.

POTATO WART *Synchytrium endobioticum*   **Pl. 12**
Widespread; commonest in Britain in northern areas but only a few cases are now recorded each year; mainly in gardens. This disease was devastating when first recorded at the beginning of the Century but it is now generally of little practical significance since legislation has limited the planting of susceptible cultivars. It is however a notifiable disease in Britain and comparable restrictive legislation exists in other parts of northern Europe (see below).
Symptoms. Often undetectable until the tubers are dug, but occasionally green-yellow cauliflower like growths may occur on stem bases or other above-ground parts close to soil level. The tubers become entirely warty and knobbly or develop whitish warty outgrowths which later blacken and decay. These may be confused with certain symptoms of gangrene (p. 358) or with the canker form of powdery scab (p. 337) although the latter causes smooth not warty outgrowths and can also affect roots, which wart never does. Symptoms are more severe in wet conditions and in dry seasons may be so slight as to be overlooked although outgrowths can arise on tubers after some time in store.
Biology. The causal organism, like that inducing clubroot, produces no mycelium but nonetheless has a complicated life history. Warted tubers contain resistant spores which are released into the soil when the tissues decay. They can persist for thirty or more years in the absence of potatoes but eventually germinate to produce swimming swarm spores which infect the tubers. This infection usually takes place through the eyes, although young tubers may be susceptible all over. More swarm spores may be produced in diseased tubers during the summer and give rise to further infections before resistant spores are produced again. The disease is commonly spread in contaminated soil adhering to boots and garden implements as well as in diseased tubers, especially if these are used as seed. Several races of the organism are known and some, particularly from outside Europe, can infect cultivars immune from the common European race. Many older cultivars are susceptible to this European race but newer ones are generally immune and in Britain all new cultivar introductions are now required by law to be so.
Treatment. Once contaminated, it is almost impossible to free land from wart disease and in Britain, under legislative orders, the discovery of wart disease must by law be reported to Ministry of Agriculture officials. A notice may then be issued prohibiting the growing of potatoes on the contaminated land while only officially approved immune cultivars may be grown in a surrounding safety zone. The Wart Disease of Potatoes (Great Britain) Order, 1973 prohibited the growing of any non-immune potato cultivars in private gardens but this restriction was lifted in 1974 and any potato, immune or not, may now be grown in gardens if the land is free from wart.

RHODODENDRON GALL *Exobasidium vaccinii*

Fairly common and widespread; usually on glasshouse azaleas. Small, often irregular swellings arise on leaves, buds or flowers; at first these are often reddish but later become chalky-white as a covering of spores develops. Flowering is reduced and general growth and vigour may suffer. The biology of the disease is little understood but the spores can probably be disseminated by insects. The problem is best treated by removing individual galls, preferably before they turn white and shower spores onto healthy flowers and foliage. Cuttings should not be taken from affected plants.

WATER-CRESS CROOK ROOT *Spongospora subterranea* var. *nasturtii*

Occasional in Britain and some other parts of northern Europe. In the past, the disease has sometimes been devastating but it is no longer a problem in commercial beds. Symptoms usually appear first in October and become progressively worse until early spring. The primary roots are swollen, stunted, kinked and often curved, crook-like while the secondary roots are swollen at the tips. Eventually the roots decay, the plants become stunted and yellowed and may lose their anchorage. Swarm spores are produced in affected roots and spread the causal organism in the water. In spring and summer resistant spores form and these probably germinate to produce more swarm spores in the autumn and so begin the first new infections. The disease is controlled commercially by the addition of zinc compounds to the water but this should never be attempted on a small scale unless under expert instruction.

WITCHES' BROOMS   Pl. 12

These descriptively named conditions are common and widespread on several woody host plants and arise from several different fungal, viral, pest or unknown causes. Any other factor such as frost, drought or browsing by animals that repeatedly damages shoot tips may also give rise to witches' broom-like symptoms. Witches' brooms are most commonly seen on birches but also fairly frequently on cherries, hornbeams, plums, ornamental *Prunus* spp., several types of conifer and, in parts of Europe, robinias.

Symptoms. Similar on all host species; large clusters of closely growing, many-branching twigs form in the crowns of trees and often bear abnormal foliage, the needles on conifer brooms for instance being markedly short. At a distance they are often mistaken for large birds' nests. Brooms increase in size and may occasionally become enormous, taking over most of the crown.

Biology. The mode of development of brooms is not fully understood. Known fungal causes are *Taphrina turgida* on birches, *T. insititiae* on plums, *T. cerasi* on cherries and ornamental *Prunus* species and *T. carpini* on hornbeams. All are present as perennial mycelium within the brooms and produce ascospores on the abnormal leaves but the initial mode of infection is uncertain. On robinia the cause is a mycoplasma (see p. 466). On conifers the causes in most cases are unknown (apart from the rust brooms on firs – see p. 276) but are probably due to some factor causing mutations in the buds. Some conifer brooms can themselves be propagated by cuttings and have been used to produce dwarf cultivars of some species. Apparently no brooms have much effect on the trees' vigour.

Treatment. Only necessary for aesthetic reasons on ornamental species although as brooms on fruit trees do not bear fruit some slight crop loss may occur. Cut out and burn the affected parts in winter.

SCABS
(Plate 13)

A small and ill-defined group of diseases which is limited in this section to those affecting fruit, swollen vegetable roots or potato tubers. Stem lesions and corm lesions, sometimes termed scabs, are treated as stem rots (p. 379) and corm rots (p. 339) respectively. Scabs are characterised by swollen, rough, usually corky lesions and although often merely disfiguring, apple scab is, for this reason, a serious problem in commercial production where the appearance of the fruit is an overriding consideration. Soil-borne scabs are caused by bacteria or fungus-like organisms and are very difficult to treat. Fruit scabs are caused by species of the ascomycete fungus *Venturia* sometimes in the form which produces only asexual spores. Sprays of benomyl and related chemicals and captan usually give satisfactory control of fruit scabs.

APPLE SCAB *Venturia inaequalis*   **Pl. 13**

Very common and widespread. Commercially, scab vies with powdery mildew as the most serious apple disease but as its effects are often only fruit blemishes, much can be tolerated in gardens although scab lesions on twigs often provide sites of infection for apple canker (p. 310).

Symptoms. LEAVES: more or less rounded greenish-brown blotches develop, especially on the leaves of the flowering spurs. The centres of the spots become blackened then grey and necrotic and some blistering of the leaf may occur. Eventually the leaves may be extensively covered with lesions and can drop prematurely. Similar symptoms can also occur on the sepals. TWIGS: blistered swellings arise which burst in spring to produce green-brown pustules. FRUIT: at first dark spots appear but these later become greenish and sometimes cracked and corky; they develop on fruit of all ages and are often very numerous. In store, small sunken, circular spots form (cf. bitter rot p. 349) but no fruit rotting develops with scab although it may occur if other organisms invade the lesions. Infection also occasionally develops on flowers. Although virus-like agents may cause somewhat similar fruit symptoms, the combination of these with leaf lesions is peculiar to scab.

Biology. The fungus overwinters primarily on fallen leaves and produces ascospores in perithecia on them in spring. These spores infect young leaves on which spots develop and conidia form. On trees severely affected with the disease, overwintering of the fungus can also occur in twig lesions which similarly produce conidia in the following year and these seem to be of increasing importance as the season progresses if no control measures are employed. The conidia are blown or splashed to leaves and to growing fruit on which further conidia form and these spread the disease from the initial point on the fruit to other parts of the same fruit, so giving the appearance of one large and many smaller lesions. It is improbable that much new infection actually takes place in store but latent infections producing pinpoint lesions may then become apparent. Wet and overcast weather, especially around blossom time, favours scab infection.

Treatment. Most of the best dessert apples are highly susceptible and 'Bramley's Seedling', once recommended as resistant, can also be seriously affected. A number of treatments are possible however. If at all practicable, fallen leaves should be collected from affected trees and burned; the overwintering stage of the fungus is thus removed and the life cycle broken. A detailed and complex technology has been developed around the timing of fungicide sprays for scab control in commercial orchards so that expensive treatments are not applied unnecessarily. In gardens, less critical standards are required and benomyl or thiophanate-methyl sprayed fortnightly from the time of bud burst onwards are usually effective. Continue the sprays until blossom drop if no wood scab infection is present. If twigs are affected however, the sprays must be continued through until harvest time. Rigorous application of this treatment should eliminate wood scab by the end of the third season when the shorter spray period can be resorted to. Because strains of the scab fungus tolerant to benomyl and thiophanate-methyl have developed, it is wise to spray in alternate seasons with captan. This fungicide should be applied first when the flower buds are still tightly closed, secondly when the opening buds are pink tinged, thirdly when most petals have fallen and fourthly three weeks later. Wood scab is indicative of neglected trees and cutting out scabbed twigs is a very valuable additional control measure, if practicable, especially as it helps to reduce the danger of canker infection. Also see: leaf spots (p. 391); viruses (p. 435).

CHERRY SCAB *Venturia cerasi*
Widespread but infrequent in Britain on cherries although rather more common on ornamental *Prunus* spp. Appears as small, dark green, velvety fruit or, less commonly, leaf spots on which conidia form. The fruit develop unripened and may drop prematurely. Treatment is unlikely to be necessary but if severe, spray as for apple scab (p. 335).

COMMON SCAB
Streptomyces scabies and other *Streptomyces* spp.   **Pl. 13**

Very common, especially after hot dry summers, and widespread throughout Britain and northern Europe. It is most severe on light, sandy, alkaline soils and serious attacks often occur after old grassland is newly broken. Scab is most serious on potatoes but also affects beetroot, radish, swedes and turnips. The remarks below refer specifically to potato tubers but are equally applicable to the swollen roots of these other host plants.

Symptoms. Scabby spots of corky tissue with more or less angular margins appear scattered over, or sometimes almost covering the tuber. The lesions may be superficial or take the form of deep pitting, when the surface of the tuber appears cratered or pock-marked. The damage is very similar to that produced by powdery scab (p. 337) which is more prevalent on heavier and wetter soils but generally common scab is the more frequently seen disease. The symptoms may also appear on roots and stolons but are rarely noticed. The yield of the plants is seldom reduced but the additional peeling necessary results in excessive waste.

Biology. Common scab is one of the few plant diseases caused by an Actinomycete, a group of organisms usually considered to be related to bacteria but, unlike them and like fungi, producing mycelium. Spores or hyphae from the soil infect through stomata and lenticels; the infection starting at the time of tuber formation or enlargement, this being dependent on the moisture content and lime status of the soil. More spores are produced in the scabs and are released into the soil where the organism persists more or less indefinitely as a normal member of the soil microflora. Since the organism is so widespread and the disease dependent on soil conditions, the planting of affected 'seed' usually presents little danger of transmitting additional infection to the resulting crop unless the eyes are infected and this does not usually occur to any extent.

Treatment. The use of cultivars which are relatively resistant to scab is the most satisfactory means of control. These include 'Arran Pilot', 'King Edward', and 'Pentland Crown'. Others including 'Desirée', 'Majestic' and 'Maris Piper' are very susceptible. Apply green organic matter, such as lawn mowings, to light soils with humus deficiency. On a small scale, lawn mowings should be applied at the rate of about three buckets-full per square metre and well dug in. Do not apply lime to scab-infested land; a common cause of increasing scab incidence on potatoes is the application of excessive lime for clubroot suppression in a preceding brassica crop. Do not throw scabbed potatoes or peelings onto a compost heap and reject any diseased tubers. On light, sandy land where scab may be expected, its occurrence may be lessened by watering potatoes just when the tubers are beginning to form and at other times when the soil is becoming dry.

PEAR SCAB *Venturia pirina*

Common and widespread. Very similar to apple scab and caused by a closely related fungus. The leaf and fruit blotches are more dark brown and velvety than on apple and the fruits are commonly distorted. Infection is also frequent on the bud scales, as well as on other organs. Treat as for apple scab (p. 335).

PRUNUS SCAB *Fusicladium carpophilum*

Occasional and widespread on plums, peaches and nectarines, causing rounded green-brown spots on the fruit which may crack, shrivel and exude gum (although this is a common response by *Prunus* spp. to several types of damage). Similarly coloured spots arise on the leaves and small gummy lesions develop on the twigs. Treatment is probably never justified in gardens but apple scab controls would be effective.

POTATO POWDERY SCAB *Sponogospora subterranea* var. *subterranea* Pl. 13

Common and widespread although generally less of a problem than common scab, unlike which, it is most serious in wet seasons or on heavy soils. The disease also affects tomato roots.

Symptoms. Although it occurs on the stolons and roots (on which the very small tissue proliferations often superficially resemble the cysts of potato cyst eelworm (p. 137)) it is only important on tubers. The tuber symptoms are often very similar to those of common scab (p. 336) although usually more rounded in outline. When small, the scabs may also very occasionally be confused with the symptoms of skin spot (p. 416) but dry brownish powder forms within each lesion. Sometimes powdery scab develops as a canker type of malformation which can often be mistaken for a symptom of the much more serious wart disease (p. 333).

Biology. Closely related to the organism causing clubroot and has a similar life history. The powder within the scab lesions comprises masses of spore balls which, when released into the soil, can probably survive for ten or more years. Ultimately the spores germinate to produce swimming swarm spores which infect host root hairs and produce further swarm spores within. These, when released, spread the infection through wounds, eyes and young lenticels, to produce new lesions. The wart-like symptoms arise when wet soil conditions promote new growth following a dry period. The increased moisture also favours the activity of the scab organism which then attacks the susceptible newly formed tuber tissue and causes the misshapen appearance of extensive scab pustules. The powdery scab organism has a secondary importance as the vector of potato mop top virus (p. 462).

Treatment. Difficult. Maintain as long a rotation as possible between potato crops; improve drainage, do not compost diseased tubers or peelings and reject scabby seed tubers.

PYRACANTHA SCAB   *Spilocaea pyracanthae*
Quite common and widespread, especially in wet seasons and probably the most frequent disease of pyracanthas. It causes small, rough lesions on twigs, olive-brown spots on the leaves (which may drop prematurely), flowers and, most seriously, the fruit, which are grossly disfigured. The fungus overwinters in diseased leaves and buds. It may be treated by spraying fortnightly or three-weekly from mid-March to July with benomyl or lime sulphur.

ROTS
(Plates 13–15)

This is an enormous group of diseases characterised by breakdown of the cells that comprise plant tissues, which are thus reduced to a more or less amorphous rotting mass. Rots may be sub-divided into a number of types, depending on the chemical component of the tissue which the pathogen attacks to effect this breakdown or upon the rapidity with which it does so. Such sub-divisions include soft rots, white rots, brown rots, dry rots and wet rot. A very wide range of fungi and bacteria can cause rotting but it is not a characteristic symptom of virus attack. Many rotting organisms, particularly those bacteria and fungi associated with soft rot, secondarily invade tissues that have been damaged by other diseases, disorders and pests. From a gardener's point of view, it is convenient to group rotting diseases according to the type of plant organ attacked as in many instances this classification both aids identification and can sometimes indicate the type of control measure to be adopted. Six such sub-divisions are used here.

BULB AND CORM ROTS
(Plate 13)

Although they are very different structures botanically, bulbs and corms are storage and perennation organs requiring similar cultural treatment and are commonly considered together. All are fleshy structures and are therefore particularly prone to rotting, either in the ground, as a result of which the plants fail to emerge or are stunted, or in store. Among the most important causal fungi are species of *Fusarium* and *Sclerotinia*, as well as certain bacteria. Similar conditions, sometimes associated with fungal or bacterial action, can be caused by eelworms (p. 132) while woodlice, slugs, mites and many insect pests may cause damage which facilitates infection by rotting organisms. Rot can sometimes be detected at planting time and elimination of diseases by rogueing at this stage is a major means of control. Routine soaking of bulbs or corms in fungicides, such as benomyl or thiophanate-methyl before planting or storage, has helped to diminish significantly the importance of many rots. The only common bulb rots not described here are bacterial soft rot, which affects many different types of plant, and onion neck rot, both of which are considered in the section on Fruit and Vegetable Rots (p. 348). Rotting of the roots of bulbous plants is also common but often does not spread into the storage organ itself and descriptions of these root rots will be found in the section on Root and Foot Rots of Herbaceous Plants (p. 361).

General Treatment. Although a few specialised treatments for certain diseases are indicated in the detailed accounts below, a number of general measures may be taken to prevent or treat most bulb and corm rots. Bulbs and corms must always be handled with great care to minimise mechanical injury which will predispose them to infection. Such careful handling is particularly important when they are lifted prior to storage while another common cause of damage at this time is allowing them exposure to direct sun. Any bulbs or corms seen to be damaged either before storage or before planting must be destroyed while storage under cool, dry, well-ventilated conditions is least likely to lead to losses by rotting. Dipping bulbs and corms for 15–30 minutes in a suspension of benomyl or thiophanate-methyl before planting is a good routine treatment and will protect them from a number of rotting conditions. Any growing plants showing symptoms should be carefully dug up, together with a spadeful of surrounding soil, and destroyed. It is wise to leave a contaminated site free of bulbs or corms for about three years before replanting with fresh stock.

BLUE MOULD ROTS   *Penicillium gladioli* and other
   Penicillium spp.   **Pl. 13**
Common and widespread on many ornamental species and one of the most distinctive types of rot.
Symptoms. Usually appear in store but the effects differ according to the

host plant. On crocus, cyclamen, freesias, gladioli, scillas and tigridias, one or a few large, irregular, red-brown and sunken lesions develop on the sides of the corms and small pinkish bodies may be present beneath the skin of the central parts of the lesions (cf. dry rot (p. 341) and hard rot (p. 342) which have smaller black bodies). White or blue-green mould growth may develop on the lesions under moist conditions and a grey-brown, firm, dry rot spreads below the lesions into the corm tissues. On bulbous irises, a soft rot spreads from a lateral lesion or, more seriously, from the base into the bulb tissues and a blue-green mould growth may also develop. Occasionally a storage rot with blue-green mould growth occurs on hyacinths, imported lily bulbs, narcissi and tulips.

Biology. The fungi probably originate from the soil and infect through small wounds before lifting although infection through undamaged tissues may also be possible and can spread from the affected organs to attached daughter bulbs or corms. The rot develops rapidly in storage especially under conditions of high humidity and low temperature and although the rot does not spread, under warm conditions, such conditions may encourage other pathogens to invade the damaged tissues.

Treatment. See p. 339.

GLADIOLUS ROTS

There are five main rotting conditions affecting gladioli and their relatives. The symptoms may be difficult to distinguish but for all practical purposes, their treatments are similar and are those outlined on p. 339.

GLADIOLUS CORE ROT   *Sclerotinia draytonii*   Pl. 13

First found in England in the late 1920's and now the most serious gladiolus disease in many parts of the world. It is found occasionally on other related species and the same fungus also causes leaf spotting on freesias.

Symptoms. Commonest on growing plants in cool wet seasons; there is yellowing and browning of the foliage, a rot spreads inwards from the outer sheathing leaf bases and the decaying tissues become covered with a greyish mould growth. Sometimes the decay appears to begin further up the stem where the sheathing leaves separate and a soft brown rot then spreads downwards. Small brownish spots with red margins may develop on the leaves and small spots also develop on the flowers as they unfold. On the corms, the commonest symptom is a soft rot of the central core sometimes so extensive that the centre drops out completely. Occasionally a dark brown spongy rot develops but this is difficult to detect until the scales are peeled away. Coral-like masses may develop on the corms while on those in store or left in ground, ovoid flat black bodies can occur.

Biology. Infection on the corms may originate from the black sclerotia which can survive in soil for several years. The disease can spread rapidly in store from one infected corm to others, especially in moist conditions. Some cultivars, especially older and possibly red flowered types are more susceptible than others but none can be recommended as truly resistant.

Treatment. See p. 339.

GLADIOLUS CORM ROT and YELLOWS
Fusarium oxysporum f.sp. *gladioli*

Common and widespread in Britain and Europe. There are two distinct symptom types, formerly believed to be caused by different pathogens but now proved to have a common origin. The same organism also causes a basal rot on bulbous irises and crocuses.

Symptoms. On corms – symptoms usually develop in storage as dark brown, concentrically ridged surface lesions, mainly at the corm base but sometimes at the nodes. The corm tissues become dry and eventually black and mummified but there are never any black bodies present (cf. the other gladiolus corm rots). On growing plants – a pale green mottling and then bright yellow striping develop on the terminal few centimetres of the outer leaves. The colouration spreads down the leaves, the tips become brown and the entire plant may die back. Small, pale brown root lesions form, gradually darken and become stripe-like while the outer root tissues may be shed. When the corms are cut into, blackened rings marking the positions of the root bases and dark streaking of the conducting tissues can be seen within. More extensive rot spreading upwards from the basal plate may follow. The symptoms on growing plants are seen quite frequently on freesias.

Biology. The causal agent is a soil-inhabiting fungus which invades the corms through the roots and corm bases. Once present in the soil, the fungus can probably persist for many years. Corms from infected plants may develop the characteristic symptoms in store but some infected corms can remain symptomless and then give rise to diseased plants when planted. Young corms may become infected from diseased parents.

Treatment. See p. 339. Drenching the soil with a spray strength suspension of benomyl may be effective in protecting freesias when diseased plants have appeared in the vicinity.

GLADIOLUS DRY ROT *Stromatinia gladioli* Pl. 13

First found in north America but became established in Europe in the early part of this century. Now a very common disease on gladioli; less so on other iridaceous plants including, in parts of Europe, acidantheras, crocuses, freesias, galanthus and montbretias.

Symptoms. On growing plants – a rapid browning and death of the leaves with a dark decay at ground level of the sheaths which characteristically bear masses of tiny black bodies. Occasionally the plants may completely topple over. On corms – many small, shallow, sunken lesions develop, especially in the rings where the scales join the corm, and these are readily visible when the scales are removed. The lesions may merge to form larger blackish areas, while narrow bands of decay extend into corm tissues and a few small black bodies occur on the affected area. The corm symptoms may be confused with those of hard rot (below). A disease that is serious in north America and elsewhere, caused by the fungus *Curvularia trifolii* f.sp. *gladioli* has recently appeared in France and Holland and may be present in other parts of Europe. The symptoms are not easily distinguished from dry rot and in cases of doubt, expert opinion should be sought.

Biology. Dry rot usually originates with small black sclerotia in the soil and they can survive there for at least five and possibly as many as twenty

years. It also survives on corms during storage and can then contaminate new soil after planting. Although it cannot spread directly from parent to daughter corms it may do so via the leaf bases or soil.
Treatment. See p. 339.

GLADIOLUS HARD ROT   *Septoria gladioli*
First found in Italy but now common throughout northern Europe on gladioli, less so on acidantheras, crocuses and freesias.
Symptoms. On leaves – small brown-purple spots bearing tiny black bodies develop on both surfaces and the central areas of the spots become ashen, then brown and finally shrivel, leaving a brown or black surrounding area. These leaf symptoms are quite common in south west England but less frequent elsewhere. On corms – small red-brown or brown-black spots form chiefly on the lower parts and are often concealed by the scales. These lesions become large, sunken and brown as they merge and small black bodies may develop on them. The symptoms are similar to but grosser than those of dry rot (p. 341). This disease is worst following wet seasons and on land low in nutrients.
Biology. In areas where the leaf spot symptom is prevalent some spread occurs during the growing season by conidia produced in black pycnidia on the spots. The origin of the disease on the corms is not known but is assumed to be from fungal mycelium persisting over winter in the soil. The fungus is unable to pass directly from parent corms to attached daughters as happens with other gladiolus rots.
Treatment. See p. 339.

GLADIOLUS SCAB and NECK ROT   *Pseudomonas gladioli*
Occasional and widespread on gladioli, freesias, crocuses and related plants, especially in wet seasons. The symptoms appear first as small, rounded, red-brown leaf spots which enlarge and darken. The lower parts of the leaves may become covered in spots and gradually rot while the leaf tips become brown and shrivelled. On the corms, rounded pale yellow spots form and enlarge to become golden yellow craters which exude sticky gum containing masses of bacteria. This gum later dries to give the impression of spots of varnish. The symptoms arise either as a result of planting diseased corms or because of the infection of healthy tissue by bacteria in the soil. Destroy affected corms and do not plant gladioli or other iridaceous plants on affected land for as long as possible.

HYACINTH BLACK SLIME   *Sclerotinia bulborum*
Occasional and widespread. Most damaging on hyacinths and scillas, less so on crocuses and fritillarias.
Symptoms. In beds, isolated plants turn yellow and topple over soon after flowering and other neighbouring plants in turn show symptoms so that bare patches develop. The leaf bases rot and the leaves are readily pulled away from the bulb, the scales of which are thin and turn dark grey. Whitish mould grows on the bulb with small, pale but darkening bodies among the scales and eventually the bulb becomes a dryish black mass.
Biology. The dark sclerotial bodies survive the winter in the soil and may germinate in spring to form stalked apothecial cups from which ascospores

are released but the role played by these in establishing infection is unknown. Probably more important is the development of mycelium from the sclerotia and its growth through the soil to infect the bulbs. Further mycelial growth then spreads the disease to nearby plants.
Treatment. See p. 339.

HYACINTH YELLOWS *Xanthomonas hyacinthi*
Occasional and widespread. One of the first plant diseases to be proved bacterial in cause.
Symptoms. The bulbs rot in the ground and fail to produce plants, or give rise to stunted individuals with brownish blotches on the leaves and yellow-brown water-soaked striping on the flower stalks. The flowers usually die prematurely and the affected bulbs, when cut, show bright yellow dots in the scales which ooze a sticky yellow slime if squeezed. These symptoms may also be accompanied by a soft rot.
Biology. Daughter bulbs may be infected from diseased parents or alternatively healthy plants become infected when bacteria are splashed from neighbouring diseased plants onto the leaves. The bacteria pass through the stomata and so downwards in the leaves to the bulb.
Treatment. Control is not really possible. Carefully remove and destroy diseased bulbs or plants, together with a few spadesful of surrounding soil, and disinfect any knife used for examining them. Slightly diseased bulbs are impossible to detect but commercially such infection can be induced to appear by storing the bulbs at very high temperatures. Healthy bulbs remain unaffected by the treatment while diseased ones can be found and destroyed.

IRIS INK DISEASE *Drechslera iridis*
Occasional on bulbous irises throughout much of northern Europe but in Britain it occurs mainly in the South and West. Very occasionally the disease also affects lachenalias and montbretias. Formerly the commonest symptom was the almost total destruction of the bulbs, leaving a black deposit encased in the scales but recognised in the early stages by black patches and streaks over the bulb exterior. More recently, leaf symptoms have been commoner, especially late in the season and also in wet seasons on Dutch irises, as yellowish streaks which become black blotches soon after emergence. The upper parts of the leaves may turn red-brown and wither. The biology of this disease is little understood but apparently the fungus persists on affected bulbs, on crop debris and as conidia in contaminated soil. It spreads rapidly among growing plants by wind-blown conidia. Any bulbs showing black, ink-like splash marks should be destroyed while sprays with dichlofluanid at fortnightly intervals may arrest the foliar symptoms.

NARCISSUS BASAL ROT *Fusarium oxysporum* f.sp.
narcissi **Pl. 13**
Common and widespread; in Britain this is one of the most serious narcissus diseases and it also causes considerable problems on crocuses and occasinally on lachenalias.

Symptoms. On NARCISSUS: symptoms appear mainly during storage and are usually apparent four to five weeks after lifting. Softening and browning of the basal ring is followed by a rich brown sometimes red-violet rot spreading through the inner bulb scales and often accompanied by pink-white mould. These symptoms are particularly common after hot summers. On CROCUS: the growing plants become yellow and the foliage dies. The corms, when lifted, show a rot spreading from the base upwards in strands through the conducting tissues and appearing at the top as rings of dark, sunken spots with pale centres. Scales often turn olive green. The rot develops further in store, sometimes associated with other pathogens or pests, and a pink-white mycelium may envelop affected corms. A closely related fungus causes similar symptoms on cyclamen.

Biology. On narcissus and probably crocus also, the fungus commonly spreads into the bulbs from the soil via the roots and is also present in the soil adhering to the bulbs when they are lifted and it may invade them later through small wounds. There are considerable differences in susceptibility between narcissus cultivars; the triandrus, jonquil and tazetta groups are highly resistant.

Treatment. See p. 339 and also the measures used to control stem and bulb eelworm (p. 132) as these can lessen the effects of basal rot also.

TULIP GREY BULB ROT   *Rhizoctonia tuliparum*   **Pl. 13**
Common, widespread and serious on tulips; less so on bulbous iris (on which similar symptoms are also caused by *Sclerotium delphinii*), ixias, colchicums, crocuses, hyacinths, lilies and narcissi.

Symptoms. In plant beds, patches of bare earth appear with few emerged plants. When they are dug up, affected bulbs will have soil adhering to them and also be seen to have a dry rot, normally from the neck downwards. The bulbs and adjacent soil may be coated with a whitish mould growth among which are very large (up to 80 mm diameter) round or flattened bodies, white at first but later darkening to nearly black.

Biology. The large bodies are sclerotia, in which form the fungus persists in soil for several years. Ultimately they germinate and the resulting mycelium infects the bulbs either shortly after planting or in the early spring. Although diseased bulbs are so severely affected that it seems unlikely that anyone would ever wish to plant them, they nonetheless remain the main means by which the disease is transmitted.

Treatment. As soon as the disease appears, remove all non-emerged bulbs and any distorted plants, together with several spadesful of the surrounding soil. Do not grow susceptible host plants on contaminated land for at least five years. Soil partial sterilisation is used commercially as a control measure but no suitable chemicals are available for garden use and although formaldehyde is sometimes recommended it is usually of little value.

FLOWER AND BUD ROTS
(Plate 13)

Flowers and buds contain some of the most delicate of plant tissues and they are therefore particularly prone to attack by rotting organisms. Damage can be serious as on both ornamental species and fruit, the end products of cultivation (flowers and fruits) are lost. These diseases almost invariably arise from fast-growing airborne fungi or bacteria, of which the grey mould fungus *Botrytis cinerea* is the commonest, and symptoms sometimes spread from the flowers and buds to the shoots and stems, so giving rise to a dieback. Routine removal of dead and diseased heads will help minimise potential sources of disease but chemical control is sometimes difficult as certain flowers may be sensitive to a number of the more traditional fungicides, although many modern products are safer. Certain other flower diseases, although not strictly rots, are described here for convenience.

APPLE BUD ROT *Gibberella baccata (=Fusarium lateritium)*

Occasional and widespread on apples or more rarely on pears. Flower buds, or less frequently leaf-buds, cease development and die soon after swelling in spring and occasionally pink pustules develop on the outer bud scales. Tiny wefts of fungal mycelium within the affected buds can be seen with the naked eye if the buds are examined carefully and the fungus probably persists in this form in dead buds over winter. The most susceptible cultivar is 'Grenadier' but 'Cox's Orange Pippin', 'Lord Derby', 'Bramley's Seedling' and 'Early Victoria', among others are prone to the disease. Dead buds and associated twigs should be pruned out and destroyed.

CARNATION BUD ROT *Fusarium poae*

Occasional and widespread in Britain and probably elsewhere in northern Europe; usually on glasshouse carnations. It is probably caused jointly by *F. poae* and the mite *Siteroptes graminum* which transmits the conidia of the fungus. Both organisms are also associated with a disease of grasses known as silvertop. The buds may develop a brown or pink mouldy internal rot and decay before the petals open or alternatively the blooms may emerge but appear distorted and there may be infestations of mites on affected buds. The disease is rarely serious but all diseased buds should be destroyed and grasses or other weeds in and around glasshouses controlled.

CHRYSANTHEMUM PETAL BLIGHT
Itersonilia perplexans

Occasional in Britain in glasshouses and in wet seasons in gardens. It is probably absent from other parts of northern Europe. The disease also affects several other species of the Compositae and also anemones. Small rust-brown spots or larger, ovoid, translucent lesions develop in the outer florets, which turn brown and soon shrivel. The entire bloom eventually shrivels but the symptoms are often rapidly masked by grey mould growth (p. 346) and can be confused with those of ray blight (p. 345). The biology of the fungus is imperfectly understood but is similar to that of the smuts (p. 287). Spray fortnightly with zineb from the time of appearance of first symptoms.

CHRYSANTHEMUM RAY BLIGHT

Didymella chrysanthemi **Pl. 13**

This was originally a North American disease but appeared simultaneously in Britain and several parts of Europe in 1961 and is now fairly widespread and increasingly serious although it is not always appreciated as such because the symptoms can be confused with grey mould damage (below) and with petal blight (p. 345). It affects all above-ground parts but is most severe on the shoots and on the flowers which develop reddish petal spots on light coloured cultivars and brownish spots on darker coloured types. The lesions spread rapidly and the flowers collapse with the ensuing decay. On shoots the disease often begins with the rotting of terminal buds while the leaves are affected by patches of brown-black decay. When large lesions form at the base of the stem, the plants may develop small, yellowed and often shrivelled leaves. Cuttings taken from affected plants may develop stem rot. Small black pycnidia develop on the affected parts and these produce conidia which rapidly spread the disease to neighbouring plants. The fungus probably persists from season to season on plant debris in the soil. Control is not easy but it is important to remove and destroy diseased plants promptly and avoid high humidity in glasshouses. Spray three times at weekly intervals with zineb or thiram following the appearance of symptoms and, where the disease is suspected or prevalent, dip unrooted cuttings in a benomyl suspension.

GREY MOULD *Botrytis cinerea* **Pl. 13**

Very common and widespread on the flowers and buds of many host plants; sometimes associated with stem and leaf rots or other damage by the same fungus. Host plants that are commonly affected by one or more flower and bud symptoms are globe artichokes, anemones (on which grey mould is sometimes associated with the fungus *Itersonilia* sp.), glasshouse chrysanthemums, cyclamen, dahlias, freesias, impatiens, helianthus, hydrangeas, lavender, rhododendrons, magnolias, myosotis, pelargoniums, poppies, roses, tomatoes, virburnums and zinnias. The symptoms are varied but are usually associated with grey mould growth (sometimes containing small black sclerotia) on the diseased tissues. There may be rotting of buds, flowers or flower stalks or petal spotting (usually off-white or brownish). The fungus spreads rapidly by means of airborne conidia produced on the mould and persists in the soil as sclerotia or saprophytically on plant debris. Grey mould control is basically synonymous with good hygiene; maintain good ventilation and raise the temperature in glasshouses; avoid overcrowded, damp and shaded situations outdoors; remove buds or flowers that have for any reason become moribund as they may easily be colonised by the grey mould which can then spread to affect more healthy organs; carefully remove affected flowers and buds (and maintain routine 'dead-heading') and spray plants with dichlofluanid, benomyl or thiophanate-methyl, repeating if the disease reappears.

Other, general, notes on grey mould biology and control will be found on pp. 352–3 in the section on Fruit and Vegetable Rots and grey mould dieback of woody plants is considered on p. 317.

PEAR BACTERIAL BLOSSOM BLIGHT
Pseudomonas syringae
Occasional in Britain during wet or cold weather at blossom time and possibly more widespread. This disease is very similar in symptoms and biology to brown rot blossom wilt (p. 350) but usually does not progress beyond the spurs. It is also easily confused with the very serious fire blight (p. 316), especially when the branches become affected. Blossom blight may be treated by pruning out and destroying any shrivelled blossom trusses.

RHODODENDRON BUD BLAST
Pycnostysanus azaleae   **Pl. 13**
Common and widespread; sometimes serious, especially in southern England on evergreen rhododendrons and azaleas.
Symptoms. Brown discolouration of the young flower buds develops in autumn and spreads during the winter until the entire bud becomes dry, sometimes silvery grey, and eventually dies but does not drop. When the lateral buds are affected entire twigs may die. Tiny, stiff, black pinhead-like structures emerge from the affected buds in spring.
Biology. The pinhead structures are coremia; conidia-bearing fungal organs which can continue to be produced from dead buds for three or more years. Bud blast attack is often associated with infestations of the rhododendron leaf-hopper insect (p. 154) but the exact inter-relationship is uncertain. Damage caused to the buds by the leaf-hoppers as they lay eggs may provide wounds for the fungus to infect as it is unable to penetrate healthy tissues. It is possible that the female leafhoppers actually carry spores to the buds.
Treatment. Some cultivars seem more resistant than others to bud blast attack but none can be firmly recommended. The best treatment is to remove and destroy all affected buds and spray with captan, zineb or Bordeaux mixture with a wetting agent just before flowering and at monthly intervals thereafter when the attack is severe. Control the leaf-hopper with malathion sprays at three weekly intervals, starting in mid-June, if feasible.

RHODODENDRON PETAL BLIGHT
Ovulinia azaleae   **Pl. 13**
Occasional in several parts of northern Europe, often appearing suddenly after rain and a very serious disease in North America. Small pale rounded spots develop on petals and enlarge and coalesce. The blooms as a result become limp and slimy and remain hanging on the plants for many months, covered with a white coating on which conidia form. These conidia are spread by wind, rain and insects to neighbouring flowers and in Europe are probably the means by which the fungus survives the winter although in North America resistant sclerotia are produced and either remain on diseased blooms or fall to the ground. In spring they germinate to produce apothecia from which ascospores are discharged and these then form the first infections. It is not known how the fungus survives in glasshouses during the summer prior to infection on winter-flowering azaleas. If only a few flowers are affected, remove and destroy them as soon as possible after

the appearance of the symptoms. If the disease was prevalent the previous year it would be wise to spray valuable plants. Spray weekly with zineb as the flowers open and while they are in bloom. Apply heavy mulches in winter following an outbreak because if sclerotia are produced, this will prevent them from germinating.

FRUIT AND VEGETABLE ROTS
(Plate 14)

Almost all fruit and vegetables are, like the ornamental bulbs and corms described on p. 339, plant storage organs and, as such contain fleshy tissues that are very prone to infection by a wide range of decay-causing fungi and bacteria. Infection and subsequent rotting can occur either in the ground or in storage and, as wounds are commonly necessary for infection to take place, careful handling at harvest time and the control of wound-forming agents, such as pests, can minimise many storage rot problems. In glasshouses, carefully control humidity and temperature to reduce damage on cucumbers, tomatoes and other fleshy produce and support low-growing fruit of such crops clear of the soil to minimise the risk of infection by soil-inhabiting pathogens. In general, the prompt removal and destruction of any rotting fruits and vegetables is desirable to lessen the chance of further disease spread although some rots in which infection takes place on the growing plants are not able to spread to neighbouring healthy produce in store. Indeed care with such cultural practices and good hygiene are often the only preventative measures that can be adopted in gardens because the fruit and vegetables themselves are destined for consumption and chemical treatments applied to them prior to storage are potentially hazardous. There are however a few instances where the application of fungicide to a growing crop can lessen the likelihood of damage on the stored produce and these are indicated in the text.

BACTERIAL SOFT ROT
Erwinia carotovora and other bacteria   **Pl. 14**
Very common and widespread everywhere in garden or store on many vegetables but especially troublesome on brassicas (particularly turnips and swedes), celery, cucumbers and other cucurbits, leeks, lettuce, onions, parsnips, potatoes and tomatoes; also on ornamentals, including cyclamen, hyacinths, irises, muscaris and zantedeschias.
Symptoms. Soft rot often follows damage from other causes such as slugs, carrot fly or other pests and diseases. The symptoms usually begin as a small water-soaked lesion around a wound but this stage of the disease is seldom noticed. The lesion enlarges rapidly and the tissues of stems, leaf bases and/or storage organs disintegrate to become an evil smelling soft and slimy mass of usually brownish putrefaction. The leaves may rot in patches and discoloured areas appear on less severely affected organs such as cauliflower curds. Storage organs, such as swedes, turnips and sometimes tomatoes and other vegetables and ornamentals, may persist as a shell of outer tissues with the centre totally disintegrated. On tomatoes, the stems may be affected and the pith becomes yellow-brown and disintegrates while the lower parts of stems turn slimy; symptoms that are often confused and

associated with bacterial canker (p. 388). Symptoms superficially similar to those of soft rot can occur from other causes on cucumbers and other cucurbits (pp. 380–1) and parsnips (p. 356) but the treatments are rather different.

Biology. *E. carotovora* is the most important of several soft-rotting bacteria. It is a motile, soil-inhabiting bacterium that is never transmitted on seed but infects through wounds and, by means of enzymes, brings about the destruction of the layer by which the cells of plant tissues are held together. Numerous different strains exist and not all are capable of infecting every host species. Several factors increase the severity of soft rot including poor drainage, especially if combined with high applications of farmyard manure; potassium deficiency or, conversely, nitrogen abundance.

Treatment. Soft rot attacks may be minimised in a number of ways: maintain as long a rotation as possible between susceptible crops; control wound-forming pests and diseases; avoid potassium-nitrogen imbalance and impeded drainage and do not over-apply manure, especially to water-logged soils, although, contrary to common belief, manure itself does not contain the soft rot bacteria. Chemical control of the disease is seldom practicable but gouging out the affected tissue on lightly diseased bulbs or corms, followed by dusting with dry powdered Bordeaux mixture, is worth trying on valuable plants. Following a soft rot attack, the affected produce should be buried deeply and never composted.

BITTER ROTS *Glomerella cingulata, Pezicula alba* and
  P. malicorticis   **Pl. 14**
One or more forms of bitter rot and the associated canker lesions are common and widespread on various fruit trees.

Symptoms. Bitter rot – common on apples, quinces and occasionally on pears. Round, brown, saucer-like depressions develop, sometimes while the fruit is still on the tree but usually more serious in store. The depressions may bear concentric rings of either off-white to yellow-brown glistening pustules or darker pustules oozing minute pink, worm-like tendrils. On cherries, only dark pustules occur and the entire fruit shrivels while on grapes the disease usually only occurs under glasshouse conditions and only pale pustules form. Perennial canker – sometimes small cankers or elliptical sunken branch lesions up to 30 cm long develop at the base of spurs, particularly on old neglected trees.

Biology. *P. malicorticis* and *P. alba* are the commonest and most serious causes of bitter rot on apples in Britain although *G. cingulata* is most frequent on pears and quinces and alone affects cherries, grapes and peaches. The fruits are usually infected when air-borne conidia germinate and invade through the lenticels; they become increasingly susceptible as the season progresses. *P. malicorticis* and *P. alba* rots are virtually indistinguishable by eye and produce conidia on pale pustules while *G. cingulata* produces them in characteristic pink slimy tendrils. Perennial canker is not a serious disease in Europe; its main cause is *P. malicorticis* although *P. alba* causes minor cankers on and death of spurs. Ascospores of both may be produced on diseased wood but are most unlikely to be a major source of the disease for attacking the fruit.

Treatment. Very little can be done to limit bitter rot attack in gardens although affected fruits should be removed and destroyed as soon as they are noticed; this can be done most effectively by examining stored apples regularly. If the disease is a problem on grapes, it can sometimes be checked by improving ventilation and removing affected fruit and the following season's blossom should not be syringed as this may encourage infection as well as pollination.

BRASSICA BLACK ROT   *Xanthomonas campestris*

Unusual in Britain although seen occasionally on seakale. Infrequent in all cool areas; common and serious in warmer areas or seasons on all crucifers and most usually seen on vegetable brassicas.

Symptoms. On seedlings the cotyledons shrivel and drop after developing blackened margins but this will be indistinguishable to gardeners from other types of seedling death. On larger plants the leaf margins darken, the veins may turn brown-black and eventually the entire leaves become yellow, wilted and may drop while the plants remain stunted. Dark specks may occur on cauliflower curds but this symptom alone is not diagnostic and can be caused by other diseases and disorders. A dark ring may be seen in cut stems and leaf stalks and also in the swollen tissues of kohlrabi and radishes and very commonly and characteristically in seakale roots.

Biology. The bacterium exists as several strains and cross infection may not take place between all types of host plant. When infected stalks of brassica seed crops are threshed, bacteria are liberated in the dust and adhere to the seeds. This seed-borne infection is the main source of disease in seedlings, although bacteria can also persist for short periods in crop debris and are spread to plants by rain splash. Infection then occurs through stomata or tiny wounds, such as those caused by insects.

Treatment. Destroy all affected plant debris by burying deeply and do not grow crucifers on contaminated land for four years. Seed-borne infection can be eliminated by treating the seeds with water at 50°C (122°F) for 18 minutes but this is unlikely to be necessary in most parts of Europe.

BROWN ROT   *Sclerotinia fructigena* and *Sclerotinia laxa*   **Pl. 14**

Very common and widespread. Probably the most serious fruit rot in gardens and often present with the associated symptoms of blossom wilt, spur blight and wither tip.

Symptoms. Brown rot – on apples, pears, plums, cherries, peaches, almonds, nectarines, apricots, quinces and rarely on medlars. Soft brown patches appear on the fruit, sometimes around a wound, either while they are on the tree or in store, and these enlarge and extend to other fruit in contact. Dirty white or yellowish cottony pustules develop on the patches, usually arranged concentrically while sometimes, particularly in store affected apples may turn completely black and develop no pustules. The fruits may ultimately shrivel and, if still attached to the tree, remain hanging there throughout the winter. Blossom wilt – on apples, pears, plums, cherries, nectarines, peaches and apricots; also on ornamental *Prunus* spp. Flowers and later spur leaves wilt, usually about a fortnight after coming into blossom. Small cankers may develop on those spurs that bear either brown rotted fruit or wilted blossoms, and extend into the branches which are sometimes girdled and killed. Spur blight – on plums and some ornamental *Prunus* spp.; causes withered leaves on the foliar spur, often with a small canker at the base. Wither tip – on plums, almonds or rarely cherries; causes withered leaves and drooping of the tip of the terminal shoot with brown lesions near the base of the lowest affected leaves. It is often accompanied by aphid infestations.

Biology. *S. fructigena* is usually the cause of brown rot on apples, pears, almonds and quince and also causes a disease known as nut drop of hazel which is associated with attack by the nut weevil (p. 244). *S. fructigena* or *S. laxa* are the causes of brown rot on cherries, plums, peaches, nectarines, apricots and medlars whereas *S. laxa* is the sole cause of blossom wilt, spur tip and wither tip. Several different strains of *S. laxa* occur and the strain infecting plums is unable to infect apples. The first infection of the fruit occurs when wind-blown or insect-borne conidia land on a wound or puncture, such as those caused by birds, earwigs, caterpillars, wasps, frost or hail. Mycelium then spreads rapidly through the tissues and into neighbouring healthy fruit, more conidia being produced on the cottony pustules. Mycelium also spreads down into the spur or branch to form the canker. Blossom wilt, spur blight and wither tip similarly arise when conidia of *S. laxa* infect flowers or leaves and then the mycelium extends down into the spurs or branches to form cankers. The fungus overwinters in cankers and also in mummied fruit which continue to produce conidia while hanging on the trees. In parts of Europe, although not in Britain, ascospore-producing apothecia develop on fallen fruit during the winter.

Treatment. There are some slight variations in cultivar susceptibility to brown rot and 'Bramley's Seedling' apple is resistant to blossom wilt. A number of measures will limit the severity of the problems: remove and destroy apple mummies and cut out and burn cankers and diseased spurs in winter, treating any cuts with a wound sealant containing fungicide. Remove diseased or damaged fruit from the trees, take care not to put any into store and avoid pulling the stalks from fruit intended for storage. A spray with benomyl or thiophanate-methyl in mid-August and again two to three weeks later will help to reduce brown rot on fruit intended for store and a similar spray when the flowers are open and again one week later will help to control the blossom wilt. Careful hygiene and handling are much more effective however in gardens than are chemical treatments.

CARROT BLACK ROT *Alternaria radicina*

A common and widespread disease equally important on seedlings and on stored carrots.

Symptoms. On seedlings – damping-off with black lesions appearing around the hypocotyl and extending up to the cotyledons. There may also be brown or black spots on the leaves and leaf stalks although the plants that survive seem to develop relatively normally. On stored carrots – a slowly developing mealy rot occurs under dry conditions with a green-grey mould and more rapid rotting when wet. The rot may be relatively localised when lateral, but if starting from the crown it extends deep into the core.

Biology. A seed-borne disease, fungal mycelium from the seeds infecting the seedlings although the fungus causing the storage rot originates from conidia and mycelium in the soil and is commonly carried by rain splash or on tools or hands.

Treatment. This is very difficult and virtually impossible on stored carrots on a garden scale. Although seed-borne infection of seedlings can be eliminated by hot water treatment of the seed at 50°C (122°F) for 24 hours, this also is not easy for gardeners to do satisfactorily.

CARROT LIQUORICE ROT and CELERY CROWN ROT *Mycocentrospora acerina*

Occasional and widespread; although *M. acerina* is the commonest cause, crown rot of carrots can arise as a result of infection by other fungi.

Symptoms. CARROTS: black, sunken lesions with brownish watersoaked margins develop on the crowns, or less frequently on the roots of plants in store or over-wintering in the ground but not during the growing season. A rather different symptom sometimes seen on carrots is that of dry, sunken, dark brown lesions at the points of lateral root emergence. This is known as crater spot and is caused by the common soil-borne fungus *Rhizoctonia solani* (see p. 363). CELERY: growing plants are stunted and yellow and on these and on stored crowns, green-black lesions develop on the roots, crown and, especially when damaged, on the leaf stalks. The margins of the leaves may appear red when cut through while severe basal rot may develop after up to two months in store and spread to the extent that the crown may fall away completely.

Biology. Commonest in moist organic soils where the fungus persists in the absence of crop plants for a few years as resting bodies (chlamydospores). The disease does not spread from plant to plant in store and infection probably arises from resistant chlamydospores and mycelium lodged between the leaf bases; some may also be present in adhering soil particles. Infection always occurs through wounds, particularly those arising from damage at lifting time and, on carrots, at the points where lateral roots have broken off. The same fungus causes one type of parsnip canker but this disease, although in some forms similar to liquorice rot, is complex and is described separately (p. 356). *M. acerina* also causes a stem and leaf blight of pansies and in other parts of the world, rots of several other types of plant.

Treatment. Very difficult. Maintain good drainage, at least three years rotation between carrot, celery or parsnip crops and avoid excessive use of potash fertilisers. Store roots when slightly immature and under conditions of low humidity and temperature if possible. No chemical treatments are available for garden use.

CUCUMBER GUMMOSIS *Cladosporium cucumerinum*

Occasional and widespread especially under cool wet conditions. Small, greyish sunken spots form on the fruit and a sticky liquid oozes from them. Velvety dark green mould then develops and eventually the fruits crack open while brownish leaf spots and stem lesions may also be formed. Diseased fruit should be removed and destroyed and the best treatment in glasshouses is to improve ventilation and if possible raise the temperature.

GREY MOULD ROT *Botrytis cinerea*   Pl. 14

A ubiquitous pathogen; the cause of many different symptoms on a very wide range of host plants. Descriptions of grey mould flower and bud rot will be found on p. 346 and of grey mould dieback of woody plants on p. 317. This description is of grey mould rot symptoms only.

Symptoms. Distinctive fluffy grey mould with a powdery surface covering appears on affected organs and a soft, usually brown rot develops beneath. A grey cloud of spores may be liberated when affected produce is disturbed and scattered black bodies may be present among the mould growth; in the later stages these bodies may occur alone, the mould itself having disappeared. It is common as a rot on apples, and less so on other orchard fruit, on grapevine (very serious under glass and in the open), strawberries (the commonest and most serious rot), figs, blackberries, raspberries, gooseberries, beans, brassicas, celery (in storage), carrot (in storage), cucumber and other cucurbits, lettuce (a very serious problem, considered in detail on p. 383), tomato, peas (sometimes causing 'chalky' peas) and potatoes (but easily confused with several other rots). On apples it may also commonly cause a symptom known as dry eye rot in which a brown sunken area appears around the calyx of the growing fruit. This may ultimately lead to the normal grey mould rot in store. Grey mould is always worse under cool, damp, overcrowded conditions.

Biology. Despite its importance and widespread occurrence, this is a weak pathogen, often attacking through some form of wound and frequent after frost damage or on stressed plants. It is common everywhere on dead or dying plant material and spreads by the numerous conidia produced on the grey mould. On beans, peas, strawberries and some other host plants, infection initially occurs through the petals and only develops later in the fruit. On other plants, such as carrots, the fungus may be present on the leaf bases, from which it spreads after roots are put into store. The small black bodies frequently formed among the older mycelium are sclerotia. These can give rise, on germination, to apothecia in which ascospores form, but in Britain they usually produce conidia once more. The sclerotia may help the fungus to overwinter although mycelium can continue to grow at low temperatures.

Treatment. Always avoid damp and overcrowded situations for both growing and stored fruit and vegetables and reduce the humidity in glasshouses at the first sign of symptoms by increasing both ventilation and heat if possible. The key to satisfactory prevention of grey mould attacks, especially in glasshouses, is good hygiene. As the fungus can exist saprophytically, it will grow on any dead or moribund plant remains which must therefore be cleared away promptly. Chemical treatments are not worthwhile or practicable on some crops in gardens but the following are sensible precautions. Spray blackberries, gooseberries and raspberries with benomyl, thiophanate-methyl or dichlofluanid just after the start of flowering and repeat at least three times at ten day intervals, but do not pick the fruit within three weeks of the last spray with dichlofluanid; spray strawberries with benomyl, thiophanate-methyl or dichlofluanid when the flowers open and repeat this treatment at ten day intervals until the fruit are at the white stage, but do not use dichlofluanid on strawberries under polythene and do not pick the fruit within two weeks of the last spray with this chemical; spray beans with benomyl or thiophanate-methyl once when the flowers are fully open; spray lettuce with benomyl or thiophanate-methyl after planting or thinning and repeat at fortnightly intervals; spray cucumbers fortnightly or three-weekly with benomyl or thiophanate-methyl beginning as the flowers open; spray tomatoes with dichlofluanid at fortnightly intervals but take care that the spray covers the lower parts of the plants and do not pick the fruit within three days of the last spray; spray outdoor grapes with dichlofluanid fortnightly starting in late May or early June but do not pick the fruit within three weeks of the last spray. Although for many of these diseases, sprays with benomyl or thiophanate-methyl are the most effective treatments, strains of grey mould tolerant to these fungicides have become established in some parts of the country and they should therefore be alternated whenever possible with dichlofluanid in any spray programme.

LEATHERY FRUIT ROTS *Phytophthora cactorum* and *Phytophthora syringae*

Common, widespread and serious on some apple cultivars and occasional on strawberries. Less frequent on pears.

Symptoms. Affected apples become leathery and crack to reveal a white colouration. On pears, either dark brown lesions with well-defined borders or lesions covered with glistening white pustules develop. Strawberry fruit become leathery, brown and shrivelled and the fruit stalks shrivel and die.

Biology. The causes are soil-borne fungi; only serious on wind-fall apples and pears and when fruits hang near to the ground, especially over bare soil. The spores are splashed upwards by rain and the disease is therefore worse in wet seasons or following severe storms at harvest time. These spores germinate to produce swimming swarm spores which then infect the fruit. The symptoms caused by both *Phytophthora* spp. are similar on apples but on pears *P. syringae* produces bordered lesions and *P. cactorum* produces glistening spore pustules. Only *P. cactorum* infects strawberries and the disease is worst when the soil is not covered with straw or other protection.

Treatment. In orchards use routine fungicide sprays to control scab (p. 335) as these help to suppress leathery rot also. Support low hanging branches, maintain grass cover under trees if possible and use straw liberally between strawberries. Three fortnightly applications of sulphur dust, beginning in early May, sometimes give some control in strawberries, especially if generous applications of potash fertilisers are also used on the crop.

MINOR ROTS OF STORED FRUIT Pl. 14

A number of rotting conditions appear commonly on apples and other fruits in store even when the produce has seemed healthy at harvest time. Although the treatment in all cases is the same, their causes are different and it is useful to be able to distinguish between them.

Symptoms. BLUE MOULD ROT: common; sometimes serious and extensive on apples and pears. The flesh is pale brown, soft and pulp-like and the affected zone is at first circular and watersoaked but later bears white pustules which ultimately turn a dusty blue-green. PINK ROT: usually associated with scab lesions (see p. 335) on apples which take on a pinkish bloom while a brown discolouration spreads from the lesions. This disease occurs commonly in garden stores as a result of apples being stored in plastic bags which give rise to warm, humid conditions. FISH-EYE ROT: occasional in Britain; commoner elsewhere in Europe on apples but rarely serious. Pale brown, depressed areas develop, often around a scab lesion, the resulting damage resembling the eye of a fish. BLACK ROTS: several types; relatively infrequent in Britain but commoner in other parts of Europe. A gradually extending brown area develops and usually the whole surface of the fruit eventually turns black and bears tiny black dots. GIBBERELLA ROT: occasional on apples and pears especially after several weeks in store but rarely serious. It begins as a small brown spot with a clearly defined margin and spreads until the whole fruit is affected and white tufts appear from the lenticels. PIN MOULD ROT: occasional, especially on apples and peaches. It appears as a soft, pale brown rot which extends rapidly and minute pin-like structures with black heads appear on affected tissues.

For other fungal rot symptoms on stored apples see bitter rot (p. 349), brown rot (p. 350), apple canker (p. 310), leathery rot (p. 354) and grey mould rot (p. 352), all of which are more common on growing fruit but can also be troublesome in store. A number of disorders of apple fruit also occur quite commonly in store and may be confused with the fungal-induced rots. Lenticel spotting occurs, particularly in dry seasons, and appears as small, dark brown sunken spots around the lenticels. Another common spotting disorder of apple fruit is bitter pit; this is related to calcium deficiency and is described on p. 476. Irregular dark blotches, usually around the eye end of the fruit but only superficial in extent are the symptoms of scald. This arises especially when large quantities of fruit are stored together and is caused by gases emitted from them. It can be avoided by wrapping each fruit in waxed paper. Glassiness or water core is a condition that seems commonest when trees have been heavily manured and when there are marked fluctuations between wet and dry spells in a season. Glassy or translucent areas appear in the fruit flesh and may extend to the fruit surface also.

Biology. BLUE MOULD ROT: caused by *Penicillium* spp. especially *P. expansum* which spreads by conidia produced on the blue-green patches and usually infects through wounds but also sometimes through the lenticels. PINK ROT: caused by *Trichothecium roseum* which can infect through the lenticels but more usually through scab lesions. The conidia give the pink bloom to the lesions. FISH-EYE ROT: caused by *Athelia epiphylla*. Infection probably occurs after basidiospores land on moribund or wounded tissue. BLACK ROTS: caused by *Botryosphaeria obtusa* and *Diaporthe perniciosa*. Both probably infect through small wounds and produce conidia in dot-like pycnidia. GIBBERELLA ROT: caused by *Gibberella* spp. especially *G. baccata* which spreads by conidia probably produced on dead wood and infects through the lenticels. PIN MOULD ROT: caused mainly by *Rhizopus stolonifer*, a very common cause of mould on fruit in general and spreads by sporangiospores produced in the pinheads.

Treatment. Only one very general recommendation may be made: remove and destroy any affected fruits as soon as they are seen. Many of these diseases will spread rapidly in store but prompt hygiene can save much fruit from decay. Always store fruit in well ventilated cool situations and avoid contact between individual fruits.

ONION NECK ROT   *Botrytis allii*   **Pl. 14**
Very common and widespread. The commonest disease of stored onions
throughout Europe.
Symptoms. Only seen after about 10–12 weeks in store as a softening of
the bulb scales followed by the development of a brown, sunken lesion
around the neck. The rot spreads from the lesion throughout the onion
tissues and fluffy masses of grey mould and small black bodies often
develop on badly affected bulbs.
Biology. It was long thought that infection arose from conidia landing on
the bulbs at harvest time but it is now known that most infection is seed-
borne, a high percentage of onion seed being contaminated in most years
although some disease may also originate from crop debris. Mycelium from
the germinated spores invades the emerging cotyledons and more spores
develop on the dying green tissues and spread the disease to other plants
but there are no obvious symptoms at this stage. The bases of the older
leaves soon become infected, then the necks, and although there is no bulb
to bulb spread in store, the symptoms do not appear simultaneously on all
onions so giving the appearance that spread is taking place.
Treatment. Control is relatively straightforward; gardeners should dust
all onion seed with benomyl before sowing as only seed for commercial use
is sold already treated. Soaking or dusting onion sets with benomyl would
probably also be effective but this is not yet proven. Maintain a three or four
year rotation between onion crops and do not dump old onions in gardens.

PARSNIP CANKER   *Itersonilia pastinacae, Mycocentrospora*
acerina and *Phoma* sp.
Widespread and fairly common, especially in wet seasons. This is a complex disease, the causes of which have only recently been elucidated.
Symptoms. Three main types. Black canker: small, ill-defined dark
lesions develop on the root, often at the bases of the lateral roots, and extend to form a brown or purple-black rot. Small black bodies may develop
on the lesion or necrotic spots with pale green halos may appear on the
leaves but these two symptoms do not occur together. Purple canker: arises
on black fen soils and is manifest as purple lesions with brownish water-
soaked margins. Orange brown canker: a brown roughening of the skin
arises around the shoulder of the root.
Biology. Black canker is caused mainly by *I. pastinacae*. This fungus also
forms the leaf spots from which spores are washed down to the soil to infect
the roots through wounds, such as those caused by carrot fly (p. 217).
Infection commonly starts where the fine rootlets join the main root,
although the fungus can also be seed-borne. *Phoma* sp., is a less common
cause of black canker and develops spores in black pycnidia on the lesions.
M. acerina causes the brighter purple lesions and is very common on highly
organic soils. The same fungus causes carrot liquorice rot and celery crown
rot (p. 352), among other diseases. The causes of orange brown canker are
not known but it is believed that several soil-inhabiting pathogens are
responsible.
Treatment. The cultivar 'Avonresister' is highly resistant to all forms of
canker and should be grown where the diseases are troublesome. Other
measures which may help are the maintenance of good drainage, rotations

of four years and the eradication of any wild parsnips in the vicinity as these are potential hosts for harbouring the disease. Attempts should also be made to control carrot fly. Parsnips with smaller roots are less prone to canker and close spacing and the delaying of sowing until the end of April or early May will produce this effect. Earthing up in summer may prevent the spores of *I. pastinacae* reaching the roots and is sometimes a successful preventative treatment.

POTATO ROTS
Potato tubers are prone to several different types of rotting of which five are described here. The identification of potato tuber damage is not easy for rots may follow attack by other diseases (particularly blight p. 385), pests or disorders. They may also be affected by more general rotting conditions, such as bacterial soft rot (p. 348) and grey mould (p. 352).

POTATO BACTERIAL RING ROT   *Corynebacterium sepedonicum*
Not present in Britain, France or Germany; common and serious in some other parts of northern Europe and devastating in North America. The oldest leaves roll and wilt, usually soon after flowering, the leaflets turn pale green with interveinal yellowing and a creamy ooze can be squeezed from cut stems. The leaf symptoms may be confused with those of late blight (p. 385) but, unlike blight, develop most in hot dry weather. The tubers may be cracked or swollen and, when cut, a creamy yellow rot appears in the ring of conducting tissues (not merely a discolouration as in *Verticillium* wilt p. 306) and later spreads to the entire tuber as soft rot organisms invade the damaged parts. The bacteria do not persist in the soil but in diseased tubers, spreading from there to the stem, foliage and, via the stolons, to new tubers. Plant to plant spread is most commonly on knives used for cutting seed potatoes before planting. If the disease is suspected in Britain, Ministry of Agriculture officials should be contacted. In areas where it is endemic, it should be treated by planting only whole tubers of healthy seed, destroying diseased material, and disinfecting contaminated tools.

POTATO DRY ROT   *Fusarium* spp.   Pl. 14
Very common and widespread but only occurs in stored tubers, never on growing plants although it is a common cause of emergence failure after planting seed thought to be sound.
Symptoms. Large tubers tend to be more susceptible than small ones. Dark brown external lesions appear on the tubers, enlarge rapidly and become sunken and shrunken, with concentric wrinkles. When cut through, the flesh around the lesions is seen to be pale brown, usually containing cavities lined with a fluffy white, blue or pale pink mould while a white mould may also appear externally. The rot may progress to affect the entire tuber or remain restricted. In moist conditions there is less wrinkling and shrinkage and bacterial soft rot may follow dry rot.
Biology. The commonest cause of dry rot is *Fusarium solani* var. *coeruleum* although *F. avenaceum*, *F. sulphureum* and other *Fusarium* spp. may sometimes be responsible. The fungi are present in the soil which adheres to harvested tubers and can survive in gardens for at least nine years in the absence of potatoes. Infection only takes place when tubers are damaged

before storing and the disease is most serious when potatoes are stored at relatively high temperatures. Tubers are less susceptible if grown with adequate fertiliser applications and in general, susceptibility is least when mature potatoes are first lifted and is highest in the spring after winter in store. The disease does not spread in store, even by contact, unless the tubers are damaged. It is worst on early cultivars but 'Home Guard', together with the maincrop cultivars 'Arran Banner', 'Pentland Crown' and, to a lesser extent, 'King Edward' are fairly resistant. 'Arran Comet', 'Arran Pilot', and sometimes 'Majestic' are very susceptible while many of the older cultivars still popular in gardens, are also susceptible.

Treatment. Buy ample seed, particularly of known susceptible cultivars, to allow for some loss. Do not use diseased tubers as seed, even if they sprout: the resulting plants are much less likely to survive than healthy stock and the progeny tubers are more likely to rot if they become damaged. Do not store immature tubers, which are highly susceptible to infection, and remove the haulm about two weeks before lifting potatoes for store as at this time the tubers are more resistant. Potatoes intended for store must be handled carefully at all times and diseased tubers should be buried deeply well away from land on which potatoes are grown. No chemical treatments are suitable for garden use.

POTATO GANGRENE *Phoma* spp. Pl. 14

Common and widespread in some European countries. Especially serious on seed potatoes in store.

Symptoms. Can be observed first several weeks after lifting. Small, but enlarging 'thumb print' lesions with well defined edges develop on the tubers and may enlarge to cover the entire tuber or remain restricted and shrivelled. The rot associated with the lesions is at first pale pinkish and watery but later becomes purple-black with cavities and black dot-like bodies may appear on the cavity walls. Crops grown from diseased seed may emerge late, show gaps and produce proportionately more small tubers but these effects are most likely to be seen when general growing conditions are poor. Gangrene may be confused with dry rot (p. 357) but there are no coloured patches of mould growth, such as are sometimes seen with that disease. Dry rot usually causes concentric wrinkling of the skin; with gangrene it is irregularly wrinkled, if at all, but expert identification may be needed as both diseases can occur together.

Biology. The usual cause in western Europe is *Phoma exiqua* var. *foveata*. The black, dot-like pycnidia on the potato haulm produce conidia which may be washed into the soil and infect either healthy tubers or those bruised or injured during lifting. Diseased seed, when planted, gives rise to diseased plants to repeat the cycle.

Treatment. In commercial practice, controlled temperature and humidity in the store, together with chemical treatment at lifting, is used to attempt to control gangrene. No chemical is suitable for garden use, however, and, although saving one's own potato seed is generally unwise because of virus problems (p. 461), gardeners doing so should take great care to avoid damage at lifting while the examination of stored tubers and removal of any suspect is essential if gangrene is not to become a problem.

POTATO PINK ROT   *Phytophthora erythroseptica*   **Pl. 14**
Fairly common and widespread but infrequently a serious problem.
Symptoms. May occasionally be seen on the haulm. These include leaf
yellowing and wilting of the plant late in the season. The principal symptoms however are obvious on the tuber, initially at the stem end, and are
usually noticed at harvest as wet rots, the largest tubers commonly being
more affected. The tissues have a rubbery consistency and internally are of
an off-white colour. This colour usually becomes rather pink in about half
an hour and finally black. Such tubers eventually decay. In the soil at
harvest time, affected tubers resemble those with blackleg symptoms but in
the latter disease they disintegrate rapidly whereas tubers with pink rot
usually maintain their consistency.
Biology. The disease is worst after warm summers on fairly heavy land and
in wet parts of gardens. The fungus can survive in the soil for several years
as resistant oospores; these ultimately germinate and give rise to swarm
spores which infect the tubers, sometimes directly but more usually via the
stolon. More oospores are produced in the roots, stolons or stem bases (but
infrequently in the tubers) and are released into the soil when the plant
decays. The disease can also be spread in storage by contact of tuber eyes
with a diseased tuber. It may also be introduced into the soil by planting
infected tubers that have few obvious symptoms.
Treatment. Destroy or bury deeply all diseased plant material away from
land on which potatoes are grown. Maintain a three or four year rotation
between potato crops, if possible, and drain waterlogged land.

POTATO WATERY WOUND ROT   *Pythium ultimum*
Occasional but rarely very serious. This fungus, in association with other
fungi, causes root rotting on many other plants.
Symptoms. Large tubers are usually more prone to attack and small dark
wet lesions develop on affected tubers, usually following damage at harvest.
The skin breaks when touched and oozes liquid. In the early stages affected
potatoes may smell of alcohol when first dug and when cut, a dark pulpy rot
with cavities is seen, sometimes with a black boundary between rotted and
healthy tissues. The affected tissues soon turn black on exposure to air and
may then have a fishy smell. Ultimately, the entire central tuber tissue may
disappear and a hollow forms but it is often difficult to differentiate these
symptoms from those of blackleg (p. 368).
Biology. Serious only when the crop is lifted early and during warm
weather. The disease arises from fungus in the soil (where it may persist as
oospores) infecting through wounds at lifting time.
Treatment. Lift and handle tubers most carefully. Destroy or deeply bury
those that are diseased away from land on which potatoes are grown and
store potatoes at temperatures below 5°C (41°F) but above freezing.

STRAWBERRY HARD ROT   *Septogloeum fragariae*
Occasional and widespread especially on the cultivar 'Paxton' but also
frequent on wild strawberries. Brown sunken patches bearing many 'seeds'
develop on green and mature fruit. There may be brown-purple patches on
the sepals and flower stalks while flowers sometimes shrivel. Large brown
leaf spots form (cf. common leaf spot p. 421) with red purple margins and

often a surrounding yellowish zone and on these, tiny black pycnidia occur and extrude white tendrils of massed conidia in moist conditions. The conidia are spread by rain splash to other plants. Control is difficult but rarely necessary. Spray with Bordeaux mixture in August and again in the following May after severe attacks.

TOMATO BUCK-EYE ROT *Phytophthora nicotianae* var. *parasitica*

Common and widespread both under glass and on outdoor plants but is usually confined to the lowest trusses. Grey-brown enlarging spots develop usually with concentric red-brown circles within ('eye-like') and a wet rot extending beneath. There may be white mould growth on the lesions and the disease may be confused in the early stages with blossom end rot (p. 477) but the latter has little or no zonation or penetrating rot. It may also be confused with blight (p. 385) but the latter causes a drier, harder fruit rot and is not most prevalent on the lowest trusses. Accompanying extensive stem and leaf rotting and the absence of root rot are most likely however to indicate blight. There is confusion over the identity of the causal fungi, which apparently vary between localities, but all can also cause foot rot, damping off and other rot symptoms. The disease originates from spores in soil splashed upwards onto the low-hanging fruit. It may be minimised by applying straw mulches, tying up the trusses well away from the soil and not splashing the soil onto the fruit when watering. Examine outdoor fruit carefully after heavy rain and destroy or use quickly any found diseased.

VIOLET ROOT ROT *Helicobasidium purpureum* Pl. 14

Common throughout Europe but less serious in cooler, northern areas; rare in Scotland. It is most important as a rot of beetroot, carrot, parsnip, potato, swede and turnip but also causes a serious root rot of asparagus, celery, seakale and occasionally many other vegetables, as well as many ornamental and weed species. The disease is sometimes serious in clamped vegetables.

Symptoms. Similar on several species; above ground, plants may appear feeble, stunted and yellowed whereas the roots, rhizomes, tubers, bulbs or corms and sometimes the crown and stem bases are covered in ramifying purplish strands containing dark spots and/or a felty mass to which soil adheres. Root tissues are affected by brownish rot, often associated with secondary bacterial soft rot, while large dark velvety bodies may occur among the mycelium and in surrounding soil.

Biology. The velvety black bodies (sclerotia) may persist for many years in the soil before they germinate and give rise to mycelium which infects plants. Persistence of the fungus on living or dead roots of the many wild host plant species is probably also important. Disease spread from plant to plant by mycelial growth during the season is slow and the symptoms, even from early infections, may not appear until autumn. Acid soils and temperatures of around 15°C (59°F) favour disease development and all host plants are uniformly susceptible although the carrot cultivar 'Chantenay Red Core' is claimed to have some tolerance.

Treatment. Burn diseased roots or bury them well away from plant beds. Remove all residual crop plants at the end of the season and suppress rampant weed growth, on which the disease might persist. Maintain high soil fertility with balanced fertiliser applications and try not to allow the soil, even in patches, to become waterlogged. Following severe attacks, do not grow root crops on affected land for three or four years. Crops that are probably safe to plant during this period are Brussels sprouts, cabbages, cauliflowers, peas and sweet corn. No chemical treatments are very effective and none is suitable for garden use.

WALNUT NUT ROTS

Rotting of walnut nuts is quite common in some seasons. After wet summers a condition known as soft shell arises in which, although the basal end is normal, the apical end remains soft and thin and may be holed. Such nuts should not be stored. A number of fungi, especially the grey mould *Botrytis cinerea*, attack the nuts, particularly when damp, and after collection they should therefore be cleaned and bleached if intended for storage.

ROOT AND FOOT ROTS (Herbaceous Plants)
(Plate 15)

This section includes many of what are probably the commonest garden disease problems. They are very frequent on seedlings when they are known collectively as damping-off, but are often troublesome on cuttings and on mature plants also. Root and foot rots are caused by a wide variety of fungi, usually soil-borne, and commonly including species of the Phyco-mycete genera *Phytophthora* and *Pythium*, although several types of fungi may act together and it is often impossible, without scientific tests, to relate particular symptoms to specific pathogens. For this reason many are here grouped together and general remarks are made on biology and treatment. Damping-off and other root and foot rots are especially serious on ornamental bedding plants although it is commonly impossible to separate such diseases from stem and leaf rots of bedding plants therefore this group of problems is also described here. On other plants any stem rot not confined to the base or foot should be sought in the section on Stem and Leaf Rots (p. 379).

Among the other plants frequently attacked by general forms of root and foot rot are asparagus, glasshouse carnations, chrysanthemums, glasshouse cucumbers, cyclamen, delphiniums, gerberas, hyacinths, hydrangeas, leeks, lettuces, lilies, lilies of the valley, lupins, narcissi, peas and beans, pot plants, primulas, radishes, saintpaulias, strawberries, sweet peas, tomatoes, tulips, violas and zantedeschias. Descriptions of these are given below whereas the root and foot rots of certain other plants are rather more individual in terms of symptoms or treatment and are listed separately. Diseases predominantly of swollen storage organs such as carrots, parsnips, bulbs and corms are described in other sections (pp. 339 & 348) while related diseases affecting a wide range of host plants but with rather more readily recognised symptoms are *Sclerotinia* Disease (p. 369); Violet Root Rot (p. 360) and White Root Rot (p. 370). Damping-off of all types of seedlings is described on p. 362.

Symptoms. Although the primary effects are on the roots or stem base, the earliest signs of damage are often indicated by the leaves which may be smaller than usual, turn yellow and wilt, sometimes with dramatic suddenness. Flowering and fruiting may be reduced and when pulled up the root systems are seen to be feeble, blackened and/or decayed. The root decay may occur in discrete areas; on tomatoes for instance the tip may rot first while a tough core commonly remains after the outer tissues have disappeared. With less severe attacks dark lesions on the roots and stem base, with little actual decay, may be all that is visible.

Biology. The causes of root and foot rots are many and varied and may differ between plant species. On peas and beans, for instance, *Aphanomyces euteiches* and specialised species of *Fusarium* are commonly responsible, while other species of *Fusarium* and of *Phytophthora*, together with *Rhizoctonia solani* and *Thielaviopsis basicola*, are probably the commonest causes on most other affected plants. All of these fungi are present in the soil as spores or other resting bodies or as mycelium and some are able to live saprophytically on dead plant remains. Many require small wounds through which to infect and are commonly more serious on plants growing under stress.

Treatment. Outdoors, affected plants should be removed and destroyed and correction of some underlying cause of stress is often the only way to minimise the likelihood of future attacks. In outdoor plant beds, improvement of drainage, nutrient status and pH, for instance, often indirectly cure root and foot rot problems. Such procedures are best combined with the introduction of cropping rotations of three or four years. Peas and beans are especially difficult to treat however and rotations of at least five years between crops may be necessary, although the digging in of quantities of brassica debris is sometimes claimed to help. In large gardens with extensive contamination the granular soil sterilant dazomet may be worth applying in accordance with the manufacturer's directions (see p. 35). In glasshouses, plants must be raised in sterilised compost (see p. 21) in disinfected pots while persistent root rot problems on tomatoes and cucumbers can often be overcome by growing the plants in ready-prepared bags. Care must be taken to ensure that water supplies such as rain-water butts are cleaned out regularly. Valuable glasshouse plants may be induced to form new roots by packing moist peat around the stem base and applying top dressing; additions of nitrogen and phosphorus-containing fertilisers are especially effective with chrysanthemums.

DAMPING OFF   Pl. 15

One of the commonest and most troublesome garden disease types, affecting the seedlings of a very wide range of plants but probably most serious on bedding plants, especially fast growing types such as alyssums, antirrhinums, callistephus, nemesias, penstemons, petunias, salvias, stocks and tagetes. Among many other plants commonly affected are brassicas of all types, cress, lettuces, tomatoes, peas and beans while very young seedlings of many trees are especially prone to the grey mould form of damping off.

Symptoms. These are exceedingly variable but all result in death of at least some of the seedlings in a box or bed. Young seedlings may die out in more or less circular patches, the larger affected plants sometimes having stem lesions at or about soil level. Seedlings may become unthrifty and have tough, shrivelled or wiry stems ('wire-stem'), usually after the first few leaves have formed. Sometimes roots may rot away completely or appear as a few discoloured stumps. On bedding plants, leaf spotting of various types may accompany the other effects while grey mould growth on the stem and leaves also commonly occurs.

Biology. The biology of the damping off organisms is as varied as the number of different types. The commonest causes of the dying out of seedlings in patches are the soil-borne fungi *Pythium* and *Rhizoctonia solani* which usually survive in the soil as spores or sclerotia respectively. The dark stem lesions are often caused by soil-inhabiting species of *Phytophthora* while the leaf spots on bedding plants are commonly formed by seed and/or soil-borne fungi and bacteria, particularly species of *Alternaria*, *Phyllosticta* and *Pseudomonas* (see also p. 391). The grey mould symptoms which often accompany damping off are caused by *Botrytis cinerea* which is described more fully on p. 352.

Treatment. Damping off is not an easy problem to deal with in gardens, largely because none of the chemicals employed successfully for soil treatment in commercial practice is available for small scale use. Control is essentially synonymous with good hygiene therefore. Affected plants and contaminated compost must be disposed of and apparently healthy seedlings from a bed or box containing some diseased individuals must not be used as they may themselves be invisibly affected and thus spread the disease to healthy land. Seed boxes, whether or not they have contained diseased plants, should be disinfected routinely before use and then filled with sterilised compost. It is not safe to assume that all commercially prepared compost is sterile and small quantities for seed box use can be treated by cooking in an oven at about 150°C (gas regulo 2) for an hour or so. Care should also be taken that water supplies (such as rain-water butts) are cleaned out regularly as they may harbour pathogenic organisms. In glasshouses and cold frames avoid water-logging, high humidity or any other factors likely to place plants under stress as this will render them more susceptible to attack. Bedding plants such as alyssums, antirrhinums, carnations, cinerarias, lobelias and zinnias should be sprayed two or three times at seven to ten day intervals with thiram to prevent the spread of seed-borne diseases to healthy plants. It is advisable also to apply a single spray of benomyl or thiophanate-methyl to all bedding plants (but especially begonias and geraniums) immediately after pricking out to prevent infection by the grey mould fungus.

BEETROOT BLACKLEG AND DRY ROT   *Pleospora betae*
Commonest and most widespread as a seedling disease in which the stem below the cotyledons blackens and shrivels. On mature plants there may be brown leaf spots with glistening black conidia and shiny black areas on the roots, the root tissues beneath these areas sometimes having a red-brown or black, relatively superficial rot. Masses of tiny black conidia-bearing pycnidia develop on affected parts. The disease is often associated with heart rot brought about by boron deficiency (p. 479) and is seed-borne, the fungus invading the flowering stalks of seed crops. Good seedling growth should be encouraged by correct applications of fertiliser, fine tilth and not sowing too early. Seed dressings with fungicide are effective in control-

ling the disease commercially but in gardens the incidence of blackleg is unlikely to be so great that insufficient seedlings are available for transplanting.

BRASSICA CANKER, BLACK LEG and DRY ROT
Leptosphaeria maculans
Common and widespread; recently it has become of increased importance on commercial oil seed rape crops and may spread from there to become more serious on vegetable brassicas.

Symptoms. Ashen grey spots which form on seedlings are usually over-looked but on mature plants two main types of symptom develop. On turnips and swedes – pale, elongated, greenish lesions develop on the exposed parts of bulbs; these gradually extend, deepen and turn brown to become a transverse crack, along the edges of which small dark bodies form. The disease is often seen to affect small groups of growing plants on which the foliage turns purplish but sometimes it is scarcely apparent until the roots are stored at which stage either grey mould (p. 352) or soft rot (p. 348) may follow the attack. On other brassicas – most commonly, elongated, sunken areas develop in the stem at ground level. These extend gradually down to the roots and around the stem until it is totally girdled by a black lesion, dotted with tiny black bodies. Small circular brown spots occur at the same time on the leaves, which may wilt and turn purplish but charac-teristically do not drop. Large-headed plants, such as cabbages, topple over while others may wilt and die.

Biology. The disease is seed-borne and when the seeds germinate, the cotyledons become infected and the seedlings may die in damping-off fashion. Small black pycnidia can be produced on these dead seedlings and conidia from them are blown or splashed to neighbouring plants. The disease can also originate from fungus in the soil or on buried crop debris or by ascospores produced in perithecia on debris at the soil surface. Spread within the crop is by conidia from the pycnidia on stem and bulb lesions while the disease may also persist on wild crucifers, such as charlock, horseradish and wild radish.

Treatment. Treatment is very difficult on a garden scale and the best plan is to remove and destroy all affected debris, maintain a rotation of at least four years between brassicas and be particularly careful in eradicating cruciferous weeds.

CELERY ROOT ROT   *Phoma apiicola*
Occasional on seedlings or mature plants; golden, self-blanching cultivars being most susceptible. Dark brown or black areas develop on the stem and root just below soil level. When pulled, the roots of mature plants break off or may appear as black stubs and although the tops are sometimes unaffected the leaves may wilt, the cotyledons of seedlings droop and the stems turn red-purple. Tiny black pycnidia containing conidia appear in diseased tissue and although the disease may be seed-borne it is not known how the seeds become infected. Destroy affected debris; do not grow celery on contaminated land for at least four years and treat seed as described for leaf spot (p. 399).

LAWN TURF ROTS

Because lawns are composed of grasses, the diseases to which they are subject are more akin to those affecting cereals and similar grass crops than they are to those on other garden plants. It is almost impossible to distinguish individual plants within turf and the five diseases described here are essentially diseases of the turf community. Although not all of them are strictly root rots they are conveniently treated here as a group. Dead patches of turf need not be due to any of these causes however and very common alternatives are leatherjackets (p. 215), soil compaction, spillages of oil or petrol from lawnmowers, spillages of garden and household chemicals, of boiling water or of urine, usually by dogs or cats (p. 264); occasionally small boys also.

FUSARIUM PATCH *Micronectriella nivalis*

Common and widespread; also known as snow mould and one of the most frequent and troublesome lawn diseases. It occurs on many grasses but is frequent on fine-leaved types, such as annual meadow grass, creeping bent and certain fescues. Although deep snow and cold wet weather in spring are often the preludes to attack, the disease can occur at any time of the year in damp cool conditions but most notably from October to March.

Symptoms. More or less circular patches develop, up to 30 cm (12″) or more in diameter, in which the grass turns brown and rots. The overall appearance is yellow-brown with a whitish-grey or pinkish colouration, especially at the perimeter, and the affected leaves may feel slimy. Several patches may merge to affect significant areas of turf.

Biology. The pinkish-white colouration is due to the abundant mycelial threads and conidia produced on the affected turf. The same fungus can be serious on some cereals, causing footrot and other symptoms, and, although ascospores are produced in perithecia on cereals in some localities, they do not occur on grasses.

Treatment. Improve drainage; thoroughly fork the surface to admit air and do not apply top dressings, especially of nitrogenous fertilisers, after the beginning of September as these encourage soft, susceptible foliage. A few fungicides are effective but the often recommended mercury-containing products should be avoided. Benomyl or thiophanate-methyl should be applied immediately after mowing as eradicant and/or preventative treatments, in accordance with the manufacturers' instructions.

DOLLAR SPOT *Sclerotinia homeocarpa*

Occasional and widespread, especially on lawns containing red fescue. The symptoms usually appear in mild wet weather in early autumn as yellowish-brown patches 5–8 cm (2–3″) in diameter. The individual leaves may have a thin dark band separating green and bleached regions. The fungus rarely produces spores and its biology is imperfectly understood. Treat as for *Fusarium* patch.

OPHIOBOLUS PATCH *Gaeumannomyces graminis*

Common and widespread especially on fine-leaved bent and similar grasses on alkaline or recently limed soils. Bronzed or bleached patches appear from late summer onwards and gradually increase in size to become shallow depressions invaded by weeds. Roots and stolons turn brown and are entangled with dark hyphal threads. The causal fungus is a serious root and foot-rot pathogen of cereals bringing about the disease known as take-all. Avoid over-liming of bent-grass turf or follow lime application with ammonium sulphate after a few months.

RED THREAD   *Corticium fuciforme*
Common and widespread especially on lawns on poor soils. It is most serious on red fescue although annual meadow grass and bent grasses are also susceptible.

Symptoms. Usually seen during late summer when irregular reddish patches of dying grass develop up to 45 cm (18″) in diameter and with pinkish strands enmeshed among the leaves. Red-pink needle-like outgrowths arise from the strands.

Biology. The pinkish strands are fungal mycelium and the needle-like structures are stromata which can survive in the soil up to two years before germinating, infection then taking place through the leaf stomata. Although basidiospores are produced on the red stromata, they are of doubtful importance and may not even germinate.

Treatment. Improve soil fertility but take care not to over-apply nitrogenous materials in autumn and so predispose the turf to *Fusarium* patch (above). Treat with benomyl or thiophanate-methyl as for *Fusarium* patch.

FAIRY RINGS
More or less circular rings of mushrooms and toadstools of several species are very common on lawns and rough grass. There are three main types: those which have no noticeable effect on the growth of the grass, those in which its growth is enhanced and those in which it is damaged. The associated basidiomycete fungi can be identified by reference to field guides, but in Britain the commonest of the damaging types is the fairy ring mushroom, *Marasmius oreades*. This produces a dryish buff cap, at first bell shaped but later flattened and up to 5 cm (2″) in diameter, sometimes with a wavy margin. The gills are thick, widely spaced, at first whitish but later turn buff. The stalk is slender and tough and the flesh off-white. The damage arises because mycelium on the inside of the ring exhausts the nutrients in the soil, dies and so produces an impervious layer which deprives the grass roots of moisture. A ring of enhanced growth of grass probably arises because of the liberation of nitrogenous compounds by fungal breakdown of organic material. The rate of radial growth of rings varies but can be 30 cm (12″) or more per year while extension may, under favourable conditions, continue for several hundred years. Treatment is difficult and highly poisonous mercurial compounds which are often recommended should be avoided. Remove a ring of turf for a radius of about 60 cm (24″) beyond the fungal ring; break up the soil and saturate it with a 1:50 solution of formalin (40% formaldehyde) at the rate of about 17 litres per square metre. If possible cover the treated area with plastic sheet for about two weeks to retain the gas. Remove the sheets, fork over the soil and re-seed after five weeks.

ONION WHITE ROT *Sclerotium cepivorum* **Pl. 14**
Very common and widespread on onions (especially salad onions), leeks
and, less seriously, chives, garlic and shallots. It is the most serious disease
affecting onions while they are still growing; neck rot (p. 356) is more
important on stored bulb onions.

Symptoms. The leaves turn yellow and die back, and the plants later
sometimes keel over as the roots rot, especially when seedlings are affected.
White cottony fungal growth develops around the bulb and bears very
small black bodies among the matted threads. The disease often appears as
patches of diseased plants which may subsequently merge to affect considerable proportions of the row or bed.

Biology. The small black bodies are sclerotia that can survive in the soil for
many years before germinating to produce mycelial threads and infect the
roots. Once one plant is infected, spread to others can occur through
contact between the roots or bulbs but the fungus is also able to grow
through the soil to a limited extent: it may for instance spread along rows
but not between them.

Treatment. Once soil is contaminated, it is virtually impossible to eradicate white rot disease although very careful removal of diseased onions and
several spadesful of soil around them will prevent its further increase by
minimising the numbers of new sclerotia that are added to the soil. Similarly, onions affected with white rot should never be composted because of
the likelihood of spreading the disease further, and land already known to
be contaminated should never be used for growing onions or related plants.
The addition of calomel, benomyl or thiophanate-methyl dusts to the drill
before sowing is a fairly satisfactory treatment for small numbers of plants
and although white rot can now be controlled very effectively in commercial crops, the fungicide used, iprodione, is not yet available to
gardeners.

POTATO BLACK DOT *Colletotrichum coccodes*
Common and widespread on many host plants but of importance in northern Europe only on potatoes and tomatoes (especially in glasshouses). It
is only a serious pathogen in dry seasons or warm areas (France particularly) and it may be that the fungus grows as a secondary invader at the
stem base of potato plants that are prematurely senescing for other reasons,
such as dry or acid conditions of the soil or infertile land. Dark brown-black
lesions develop on the stems at or just below soil level; internally, the stems
at this point often show violet-coloured mycelial growth on which many
microsclerotia are produced. Rarely the disease may completely girdle the
stems and cause yellowing and rotting of the leaves while the outer parts of
the lesions pull away from the inner tissues. Masses of tiny black sclerotia
('black dots') occur on the diseased parts, including potato tubers where
they form on silver-grey patches. Apart from the presence of the sclerotia,
which are sometimes difficult to see with the naked eye but readily visible
with a lens, the symptoms on tubers are extremely difficult to distinguish
from those of silver scurf (p. 416). Destroy diseased debris and do not grow
potatoes, tomatoes, other solanaceous plants, cucurbits, or legumes on
affected land for at least five years. Use sterilised compost for tomatoes.

POTATO BLACK LEG *Erwinia carotovora* var. *atroseptica* **Pl. 15**
Very common; often one of the earliest potato diseases to appear, especially in wet seasons. Both this bacterium and *E. carotovora* var. *carotovora* can also cause soft rots of potatoes.
Symptoms. These are generally most severe in wet seasons. The upper leaflets first curl inwards (and thus differ typically from the symptoms of secondary leaf roll (p. 462) where the lower leaves curl first and become stiff), and the colour fades to yellow-green. The stem base decays and becomes black and slimy but not all stems on a particular plant may show these symptoms. In store, affected tubers rot completely and spread the bacteria to neighbouring healthy ones.
Biology. The disease is transmitted in diseased seed tubers and the bacteria do not remain viable in the soil from one season to the next but can survive in heaps of discarded potatoes. Affected plants may produce typical blackleg soft-rotted tubers, the organism penetrating through the stolon into the stem end and rotting the tuber tissues. Such plants may also produce tubers that appear quite healthy but in which the lenticels harbour the bacterium. These bacteria have been spread in the soil from the breakdown of the infected seed tubers. Tubers may also be infected through damage at the time of lifting and these may introduce serious rotting in neighbouring previously healthy tubers in store.
Treatment. Ensure that seed tuber stores are well-ventilated, and do not use as seed any tubers that have been in contact with soft-rotted material. Discard any tubers showing browning at the heel (stem) end and do not plant potatoes on waterlogged land. No resistant cultivars are known but 'King Edward' is less susceptible than many.

POTATO STEM CANKER and BLACK SCURF
Rhizoctonia solani
Very common and often serious. The causal fungus is also responsible for a wide range of root and other diseases on many host plants.
Symptoms. On tubers – harmless, hard black patches develop to which soil adheres and which cannot be washed off, although with gentle scraping (as with a finger nail) they pull away leaving the skin unaffected beneath. Sprouts from affected tubers develop sunken brown lesions which may cause complete girdling. If conditions before emergence are dry, the tips of the sprouts are often attacked and this leads to sprout proliferation, thinner stems and retarded emergence. Growing plants – rough, brown, cracked lesions on the underground parts of stems and stolons. The entire plant may be stunted, the leaves curled and sometimes coloured yellow or purple and clusters of small, knobbly tubers can develop at or immediately below ground level. Small, green aerial tubers can also form in the leaf axils. In moist conditions, white mould grows around the stem base above soil level, while in warm conditions on early cultivars, there may be some pitting of the tubers giving symptoms rather like wireworm damage (p. 236).
Biology. Several strains of *R. solani* exist and not all are able to infect all hosts. It is a common soil inhabitant, often existing saprophytically on plant debris or as the hard black sclerotia that form the scurf on tubers. Infection can thus arise from soil as well as from affected seed tubers.
Treatment. Very difficult. Rotation of potato crops should always be

adopted and it is particularly important to avoid two successive crops on the same land. Although seed tuber disinfection is sometimes performed commercially it is scarcely practicable in gardens. The disease is often prevalent in spring in dry, cold land, particularly with high ridges, and may be reduced naturally with the onset of rain. Shallow planting is sometimes an effective control as this allows the plants to emerge and establish quickly and may reduce the chances of sprout infection. Similarly the knocking down of high ridges may help as attacks seem to cease once the shoot reaches soil level. Early lifting of the crop reduces the numbers of sclerotia on the tubers.

SCLEROTINIA DISEASE   *Sclerotinia sclerotiorum* and possibly other *Sclerotinia* spp.   **Pl. 15**

Widespread; affects many vegetables and ornamentals and the host plant list given below is not exhaustive. It is generally most serious in northern areas and is common in the north of England and Scotland but frequent in all damp, cool regions.

Symptoms. A brown, more or less wet rot develops, often associated with masses of fluffy white mould commonly containing large, usually black but sometimes paler bodies. The symptoms occur especially at the stem base of beans, celery, chicory, cucumbers, Jerusalem artichokes, lettuces, peas, potatoes, tomatoes, campanulas, glasshouse chrysanthemums, convallarias, dahlias, delphiniums, gypsophilas, helianthus, lupins and sweet peas. The effects may occur occasionally on leaves and fruit as well as on stored bulbs, corms and tubers. They are very common and serious on stored carrot and parsnip roots, the symptoms often being restricted to the crown of the root. Growing plants may wilt suddenly, show yellowing of the basal leaves and topple over at the points of infection.

Biology. The large, usually black sclerotia lie dormant over winter in the soil and germinate in spring or summer to produce apothecia which emerge just above the soil surface. Ascospores from the apothecia infect leaves (the fungus usually establishing first on wounded or dying leaves unless the conditions are very moist, when healthy tissues may be attacked) and passes from there to the stem and to other leaves. Mycelium permeates the stem but usually does so in a restricted region within which more sclerotia are produced. It is possible that mycelium from germinated sclerotia may sometimes infect directly, without spore formation.

Treatment. Collect and destroy affected material to prevent sclerotia from contaminating the soil and do not grow susceptible plants on contaminated land for at least three years. Pay particular attention to the control of weeds, on which the disease might persist. Check stored roots regularly and remove any diseased. Late planting of potatoes sometimes helps to protect this crop as fewer dying leaves are then present at the time of spore release.

STRAWBERRY CROWN ROT   *Phytophthora cactorum*

For several years this disease has been serious in many parts of Europe especially in warm seasons and on plants grown under polythene. It has recently become increasingly troublesome in Britain especially with the increased popularity of continental cultivars. The symptoms are reminiscent of drought effects; the younger central leaves wilt while the peripheral foliage, although reddening, retains its normal appearance (cf. *Verticillium* wilt (p. 306) where the outer leaves flag first). Although there may ultimately be some regrowth from the lateral buds the plants usually die. When cut through, the interior of the crown appears brown and necrotic. The causal fungus is soil inhabiting and infection arises through damaged tissues. Treatment is very difficult and the best plan is probably to destroy affected plants and replace with fresh stock on a site as far as possible from the original.

STRAWBERRY RED CORE *Phytophthora fragariae* Pl. 15
Also known as Lanarkshire disease. It was first found in Scotland in 1921 and remains commonest in the North although it now occurs also in southern England. It is a notifiable disease in Great Britain.
Symptoms. Most obvious in May or June. Patches of stunted plants appear with small and reddish central leaves and brown and stiff outer ones, the latter giving an overall brown appearance to the plants. When pulled up, the roots are dark brown or black and the outer root tissue is readily stripped off, revealing a red core; this symptom is diagnostic for the disease but is not usually visible during the summer months, when the symptoms may be confused with those of other strawberry root diseases.
Biology. Within the red tissues, resistant oospores form and these are released into the soil as the root decays. They can persist in the soil for at least twelve years but ultimately germinate, sometimes via a swimming swarm spore stage, to infect healthy roots in wet conditions. The disease is spread to uncontaminated land either by the planting of infected runners or in soil adhering to plants, roots or tools.
Treatment. It is virtually impossible to eradicate red core from contaminated land and anyone discovering strawberries believed to be affected with red core disease must, by law, inform officials of the Ministry of Agriculture.

WHITE ROOT ROT *Rosellinia necatrix*
Common and widespread on many host plants. In Britain it is only serious in the South West, especially the Isles of Scilly. It is worst in wet seasons and on poorly drained soils and is favoured by high temperatures if these are combined with moisture. The most important hosts are: almonds, apples, irises, ixias, grapevines, narcissi, pears, potatoes, privet and tulips. Diagnosis of the disease is difficult. Plants tend to be affected in groups and either die or grow slowly, with wilting and/or sparse foliage. Bulbs and corms are affected by a brown rot, often with blackening of the outer scales, and when cut through length-wise, thin white strands of mycelium may be seen between the bulb or corm tissues. On woody plants, there is often yellowing and premature fall of the leaves while the roots are at first covered with dense white or green-grey mycelium which later darkens and becomes cobweb-like. In moist conditions mould growth can occur also on the soil surface around affected plants and tiny black sclerotia may occur among the mycelium and can persist in the soil for long periods. Ascospores can be produced in subterranean perithecia but are unrecorded from Britain. It is difficult to treat white root rot in gardens but as the fungus is susceptible to

drying, repeated turning over of affected soil (preferably with a rotavator) in hot sun is sometimes very effective.

TREE AND SHRUB ROTS
(Plate 15)

These rots are caused mainly by Basidiomycete fungi; mostly of the sub-order Polyporinae which are known popularly as polypores. The basidio-spores are produced in minute vertical tubes, usually in a bracket or hoof-like fructification with pores on the undersides through which the spores are liberated. These fructifications can be very large: 60 cm (24″) or more in diameter in some species. Some genera of polypores, such as *Polyporus*, produce a new fructification annually while in others, like *Fomes*, they are perennial and persist from year to year. Some related genera however, such as *Stereum*, produce instead a flattened skin or plate-like fructification which is closely adpressed to the wood or bark. A few wood-rotting fungi are gill-bearing basidiomycetes of the order Agaricales and the most wide-spread of all, honey fungus (*Armillariella mellea*) is of this type, giving rise to annual clusters of toadstools. Many wood-rotting species are of world-wide distribution and most affect many host plants. Exceptions to Basidio-mycete causes of rots are a few species of Ascomycetes while species of the Phycomycete genus *Phytophthora* are responsible for root death without any decay in the strict sense and for ink diseases.

A feature common to all wood-rotting fungi is the ability to degrade chemically either of the two main constituents of wood: cellulose, giving rise to a brown rot, or lignin, giving rise to a white rot. The distinction between saprophytes and parasites is not always easy among decay fungi however as the heart wood of trees is actually dead tissue while the sapwood is living. Exposure of the heartwood of trees may therefore permit sap-rophytic decay fungi to enter and cause structural weakness or other damage to still-living trees.

Usually a succession of symptom types occurs in affected trees; initial staining of the wood gives way to different types of rot but the descriptions given here refer generally to late-stage rotting as this is most likely to be seen. Descriptions are also given of fungal fructifications (toadstools or brackets) as, when present, these give valuable clues to the cause of rot. It should be appreciated however that more than one decay fungus may be involved in any particular tree and also that a number of other fructification types, normally of saprophytic decay fungi, may also appear on trees in the later stages of decay and decline. Standard field guides to fungi should be referred to for the identification of these and also of the many fungi commonly found on fallen logs and timber but which never cause damage to living trees and are beyond the scope of this book.

General Treatment. Generalisations can be made concerning the pre-vention and treatment of decay in trees. A few species, such as honey fungus (p. 375) and white pocket rot (p. 378), are truly parasitic and can enter through undamaged tissues. Most wood-rotting fungi however, enter through wounds or dying tissue by the germination of air-borne basidiospores and every precaution should be taken to prevent or seal such entry sites. Remove dead or broken branches as close to the stem as possible with clean saw cuts, never leave jagged ends or stubs when pruning and treat all cuts and accidental wounds with a wound sealant containing fungicide. Once a tree is infected, the procedures are more difficult and specialised tree surgery may be necessary. Unless fungal fructifications are seen, it is often difficult to determine whether or not rot is present and even the presence of fructifications may not indicate the extent of rotting, although it usually may be considered an ominous sign. If fructifications are present on a branch, that branch should certainly be removed close to the trunk. If decay is seen to progress into the trunk, an attempt may be made to gouge it out before applying sealant but if the decay is obviously extensive, it is better to fell the tree as it is likely to be unstable and prone to windthrow. If in any doubt with trees in situations (such as gardens) where the fall may damage buildings, either fell or, especially with valuable or large trees, seek expert advice. See also p. 42 for legal aspects of tree felling.

ALDER HEART ROT   *Inonotus radiatus*
Common and widespread on alder causing a heart rot and occurring occasionally on birch, poplar and willow also. The fructifications are annual, thick, 2–6 cm (1–2½″) across, bracket-shaped, yellow-brown above but darkening with age while retaining a golden-yellow margin. The upper surface is velvety with raised radial lines and the pore surface beneath is silvery. The flesh is hard, yellow-brown and fibrous, the rot whitish or fawn and flaky. For biology and general treatment see p. 371.

ASH HEART ROT   *Inonotus hispidus*
Common and widespread; it is most frequent on ash as a serious cause of stem rot associated with wounds but also occurs on apple, elm, plane, walnut and other broadleaved trees. The fructification is annual, hoof or bracket shaped, rusty-red but later black, markedly shaggy above and with a red-yellow ragged pore surface beneath. The flesh is spongy and reddish-yellow and the rot indefinite but affected wood is lighter and softer than sound tissues. Yellowish or whitish streaks occur around the point of infection. General hygiene, as outlined on p. 371, should be followed.

The characteristic smooth, rounded, lumpy black fructifications of the Ascomycete fungus *Daldinia concentrica* are very common on dead ash wood but the organism can be parasitic and the decay may extend into living tissues. Affected branches should be cut back well into the healthy wood and the cuts sealed.

BEECH HEART ROTS   *Ganoderma applanatum* and *Pleurotus ostreatus*
Both fungi are common and widespread. They are most important in Britain and northern Europe as causes of heart rot of beech, following wounding, but also occur on other broadleaved trees. The fructifications of *G. applanatum* are flattened brackets, up to 30 cm (12″) across and often overlapping in groups; flat, zoned or lumpy and rich rusty brown above with a white or yellowish pore surface beneath. The flesh is a dirty brown and the fructifications and surrounding bark are often dusty with a covering

of rich brown spores. The rot is white and ultimately soft and spongy, bordered by a narrow dark brown band. *P. ostreatus* produces an annual, thick, fleshy, fan-shaped fructification up to 15 cm (6″) across and sometimes with a short stalk-like base. It is smooth, often moist and deep grey-blue or later, grey-fawn above with whitish gills beneath and white flesh. The rot is white and flaky and also often bordered by a dark band or zone-line.

Root and stem rot of beech and occasionally of lime and other trees can also be caused by the Ascomycete fungus *Ustulina deusta* which invades from the soil via the roots or sometimes through aerial wounds. The symptoms are not readily distinguished from the other fungi rotting beech wood and for all, the notes on p. 371 should be followed.

Other rots occurring on beech are Birch Heart Rot (below), Oak Heart Rot (p. 376) and Giant Polypore (p. 374).

BIRCH HEART ROT   *Fomes fomentarius*
Rare in England, commoner in Scotland and elsewhere in northern Europe, especially on birch and beech. The fructification is up to 60 cm (12″) across, perennial, hoof-shaped, smooth, hard, grey-brown to black with concentric zonation above and with a greyish pore surface beneath. The flesh is brown and the rot in both heart and sapwood is mottled, yellow-white with narrow dark zone-lines and sheets of yellow-white mycelium among the cracks. It is associated with wounds and usually confined to the upper parts of trees. For biology and treatment see p. 371.

BIRCH POLYPORE   *Piptoporus betulinus*   Pl. 15
Very common and widespread. It occurs almost exclusively on birch and is the major cause of heart rot of this tree in all except the more northern areas where *Fomes fomentarius* (above) is equally important although *Inonotus radiatus* (p. 372) also attacks birches. *P. betulinus* probably only attacks already weakened trees and produces an annual fructification which is sometimes up to 40 cm (16″) across and begins as a hard round grey knob and then expands to become hoof or kidney-shaped with a smooth edge. It is smooth and pale grey-brown above with a chalky white pore surface beneath and white, soft or corky flesh. The rot is reddish-brown and breaks into angular blocks before eventually becoming powdery. For biology and treatment see p. 371.

BROWN CUBICAL ROT   *Phaeolus schweinitzii*
Common and widespread in Britain and northern Europe. It is virtually confined to mature conifers and is particularly common on Douglas fir, Sitka spruce, larches and pines. The annual, usually bracket-like fructifications (but occasionally stalked) are up to 30 cm (12″) across and arise in late summer near to ground level; they are velvety and rust-brown above, with a yellow-brown pore surface often bearing liquid drops beneath. The rot is dark and red-brown and breaks into cubes which easily crumble. Sheets of pale mycelium develop between the cracks and there is usually a strong smell of turpentine. Infection occurs through the roots and control of the disease is impossible. Affected trees should be felled.

BROWN OAK DISEASE *Fistulina hepatica*

Common and widespread, usually on old oaks but rarely on sweet chestnut and other broadleaved trees. Seldom damaging but the appearance of the fructifications in autumn may cause concern. They are tongue-shaped, 30 cm (12″) or more across, purple-red or dull chocolate above, and with an off-white pore surface beneath. The flesh is marbled and reddish, exuding red juice and the fungus may stain the wood to form the brown-oak, prized by cabinet makers. No treatment is necessary.

CURRANT COLLAR ROT *Phellinus ribis*

Occasional and widespread on currants or more rarely on gooseberries. The fructifications are perennial at the base of the plant, hoof-shaped and often grouped and sometimes fusing to encircle the stem. They are yellow-brown and concentrically zoned above, at first velvety but later smooth and have a grey-brown pore surface beneath. The fungus enters through basal wounds and affected plants gradually become unthrifty and may die. Affected bushes should be uprooted and destroyed. Also see Stem Toad-stool (p. 315).

ELM BUTT ROT *Rigidoporus ulmarius*

Common in Britain; less so elsewhere in northern Europe. It occurs almost exclusively on elm and is the main cause of butt rot following wounding but is not to be confused with Elm Heart Rot (below) which causes hollowing of elms. The fructification is perennial, close to the ground, up to 30 cm (12″) across, thick and bracket-shaped with a creamy, lumpy upper surface which discolours with age and a cinnamon-white pore surface beneath with buff-white tough flesh. The rot is confined to the butt and the tissue eventually becomes dark brown and friable and splits into brick-like pieces. For biology and treatment see p. 371.

ELM HEART ROT *Polyporus squamosus*

Common and widespread as a cause of heart rot associated with branch wounds, especially in elm and sycamore, but also occurs on other acers, pears, walnut and other broad-leaved trees. The fructification is annual, rapidly growing, fan-shaped and, particularly when at the base of trunks, can be very large (60 cm (24″) or more across). It is pale fawn above with numerous dark brown scales and has a cream pore surface below and tawny flesh. The rot is white and stringy in the heartwood and eventually causes hollowing of the trunk. This disease is the major cause of heart rot in elms and renders trees very liable to windthrow. For biology and treatment see p. 371.

GIANT POLYPORE *Meripilus giganteus*

Common and widespread but it has only recently been recognised as a serious pathogen. It is especially damaging on beech and in Britain is the commonest cause of root decay on this tree. It is particularly dangerous as the root decay can become extensive and trees thus rendered unstable with few external symptoms. The fructifications are annual, appearing in late summer and autumn at the bases of trees or on the ground over roots and often in dense tufts of a metre or more in diameter; fan shaped, yellow-brown to brown above with a yellow-white pore surface beneath which a blackish stain typically shows where bruised. The flesh blackens when cut. Sparse foliage and dieback can be associated with the root decay but may have other causes so root examination is essential if no fructifications are present. The upper roots of trees are often deceptively sound and the decay may not be seen until the roots are excavated to 45 cm (18″) or more. Trees with advanced decay in the root systems should be felled but checking by an expert is advisable with any suspect large and valuable trees.

HONEY FUNGUS *Armillariella mellea* **Pl. 15**

An extremely serious pathogen of worldwide distribution recorded on almost all woody plants and several herbaceous species causing either death or decay. It is very common in gardens and is often found on sites close to old tree stumps but by no means invariably so, although a site history of broad-leaved trees is probably essential. It causes one of the most important garden plant diseases.

Symptoms. Toadstools can arise at any time between July and December but commonly disappear by October with the autumn frosts. The cap is up to 15 cm (6″) diameter, the stalk usually up to 15 cm (6″) high and they occur in clusters on stumps, roots, trunk bases and occasionally higher up on the stem. The toadstools are usually yellowish or tawny but are variable in colour and sometimes have a greenish, grey or pink tint. There are often darker specks or scales towards the centre of the cap and striations at the cap margin. The gills are whitish, becoming yellower with age and are sometimes spotted brown. The gills may merge into the top of stalk. The flesh is yellowish and the stalk similar in colour to the cap or paler, usually with a large, whitish yellow-bordered ring. The affected wood is initially stained, then a soft wet brown rot develops which eventually becomes fibrous, stringy and white, often mixed with flaky white material. The rot rarely develops for more than about 50 cm (20″) above ground level and sometimes is virtually confined to the roots. There may be dark zone-lines in the wood surrounding the most badly affected parts and often flat white sheets of fungal growth and sometimes masses of flat blackish-brown strands develop beneath the bark. Black, rounded bootlace-like strands can often be found among the soil around affected plants. Young trees are much more likely to be killed than older ones but root damage may render mature trees prone to windthrow or to death from other causes.

Despite these apparently distinct symptoms, the diagnosis of honey fungus attack is not always easy. The toadstools are only present in the autumn and do not always occur then, even on badly diseased trees. The bootlace strands are not always easy to detect in the soil and similar bodies may be formed by other fungi. If several woody plants are grown within a small area, honey fungus attack is often manifest by the gradual death of them in ones and twos over a period of years. The disease is unlikely however to kill many plants quickly in a small area or to kill trees and shrubs planted less than a year previously.

Biology. This fungus lives saprophytically in dead tree stumps and spreads from there through the soil as bootlace-like masses of aggregated mycelium termed rhizomorphs. The larger and closer the diseased stump, the more likely is honey fungus to infect neighbouring vigorous trees and although rhizomorph growth is the most important means of spread to healthy trees, infection by root contact can also occur. Basidiospores are produced on the gills of the toadstools and spread the fungus to infect new stumps. Several different strains of *A. mellea* are known but the biological importance of them is uncertain.

Treatment. This is extremely difficult, and honey fungus probably shares with clubroot more mythology than any other garden disease. There is considerable variation however in the susceptibility of plants to attack. On a site known to be infested, the following are among the common garden plants that should not be grown as they are all highly susceptible: apples, birches, cedars, cypresses and false cypresses, lilac, pines, privet, walnuts and willows. The following however are probably sufficiently resistant to make their planting on an infested site reasonably likely to succeed: ash, beech, box, clematis, Douglas fir, eleagnus, false acacia, hawthorn, holly, ivy, larch, laurel, lime, mahonia, silver firs, sumachs, tamarisk, tree of heaven and yew. Once infected a tree or shrub cannot be cured of honey fungus and the only effective procedure to limit its spread to others is the prompt removal of the diseased individual, together with its entire root system and as much as possible of the surrounding soil. The removal of large stumps is very difficult but a number of commercial contractors offer the service. Chemical methods of stump destruction with paraffin and saltpetre or proprietary products are not effective. Once removed, no woody plant, even those listed above, should be planted on the same spot for at least a year to give the fungus chance to die down. Sometimes the source of fungus for attacking plants in a particular area cannot be traced or may be on neighbouring land. In such cases, only resistant types should be grown while an attempt may be made to provide barriers to rhizomorph growth. Trenches or buried vertical sheeting might be effective but may need to penetrate to at least 60 cm (24"). Proprietary chemicals based on phenolic emulsions are sold for honey fungus control and could also be used to provide barriers if the soil is drenched with them in order to kill the rhizomorphs. Such materials may need re-applying annually however and cannot be relied upon at all times in all soil types. There seems scant evidence that they have any effect in curing already diseased plants.

OAK BUTT ROT *Inonotus dryadeus*

Widespread and fairly common on oak. The fructification is annual and usually occurs only on large, old trees, near the base. It takes the form of a thick and lumpy bracket up to 25 cm (10") across, sometimes occurring in groups and pale yellow-brown and wrinkled or smooth above with darkening drops of liquid at the edge. The pore surface beneath is white and the flesh brownish, at first soft, then corky. The rot is soft, white and pulpy and never rises to more than two metres above the base. Infection probably occurs through the roots and the disease reduces the stability of trees and renders them liable to windthrow. For biology and treatment see p. 371. Also see Brown Oak (p. 374).

OAK HEART ROT *Laetioporus sulphureus*

Common and widespread. The major cause of decay in old oaks but also occurs commonly on cherries and less frequently on apple, beech, larch, pine, sweet chestnut, yew and other trees. The fructifications are annual, up to 40 cm (16") across, thin and bracket-shaped with wavy margins, usually closely overlapping in groups. They are vivid reddish-orange above and with a pale yellow pore surface beneath although the colours fade when older. The flesh is soft, yellow and cheesy and often exudes yellow juice. The rot is deep red-brown and breaks into cubical fragments, often separated by tawny, leathery sheets of mycelium. For biology and treatment see p. 371.

OAK PIPE ROT   *Stereum gausapatum*

Common and widespread as a major cause of decay associated with wounds in oaks. It is often commoner in younger trees than *Laetioporus sulphureus* (above) which is the major cause of decay in old oaks. The fructifications occur on bark or wood and are thin and skin-like, grey-brown with a whitish margin. They 'bleed' by the exudation of red juice when scratched but are very difficult to distinguish from those of other *Stereum* spp. most of which react similarly. The rot is dark brown or yellow with streaks and bands of soft decay interspersed with sound wood. Small branches may be completely decayed with a soft yellow-white rot but generally it is very difficult to detect as the vigour of the trees is often little affected. For biology and treatment see p. 371.

PHYTOPHTHORA ROOT DEATH   *Phytophthora* spp.

Very common, widespread and serious. This disease has begun to reach epidemic proportions in parts of England where it is one of the commonest causes of the death of ornamental trees and shrubs. The species most severely affected include azaleas and rhododendrons, Lawson cypress, beech , heathers and heaths, apple, yew, limes, ornamental *Prunus* spp. and planes but many others can be attacked. In old coppiced woodland, sweet chestnut is frequently infected, the symptom being known as ink disease, while it is now thought that the condition known as 'the death' which affects fruit trees on heavy, waterlogged soils and following wet winters may have a similar cause. The fungus is solely a killer of roots and not a decay organism although other wood-rotting species often invade affected roots very rapidly.

Symptoms. Difficult to diagnose accurately. As with other root damage, the above-ground symptoms include small, yellowed and sparse foliage, partial dieback and, in severe cases, the total death of plants, sometimes only revealed when new growth fails to appear in the following season. The symptoms at the stem base are more diagnostic: the thicker parts of the roots close to the stem may be dead while the thinner younger parts are still alive but bearing scattered dead patches. Areas of dead bark may extend up the stem, either in strips or in more or less triangular patches, and if, as on cypresses, this dead patch reaches to the base of low-growing branches, a part of the tree dies. In the later stages diagnosis is difficult as the damaged tissues are commonly invaded by honey fungus (p. 375), symptoms of which mask the primary cause.

Biology. Several species of *Phytophthora* are responsible, especially *P. cambivora* and *P. cinnamomi*. Infection arises when swimming swarm-spores germinate from resistant resting spores present in the soil and penetrate the root tissues. Within the roots, mycelium and ultimately more spores form. Infection is favoured by warm, wet soils and probably also by any factor, such as farmyard manure or mulches, which increases soil water retention. Nursery sites where susceptible species are grown intensively are very prone to develop severe infestations.

Treatment. Some good chemical treatments are available but unfortunately not for garden use. The likelihood of attack can be minimised by only buying plants from reputable nurseries and avoiding any where debilitated or browned cypresses and other trees are evident. Never transplant from infested into clean land; avoid excessive watering, manuring or mulching of susceptible plants; do not allow hollows to form around stem bases – slope the soil away from trees and shrubs; do not plant into peaty or similar medium artificially introduced as pockets into heavy soils as these will rapidly become waterlogged; replace affected plants by more tolerant species – for example Leyland cypress (× *Cupressocyparis leylandii*) although less attractive, may be used instead of the susceptible Lawson cypress. In badly infested nurseries, a change to container production of plants is usually the only solution.

PLUM HEART ROT   *Phellinus pomaceus*
Common and widespread on *Prunus* spp. in Britain and northern Europe and especially frequent associated with wounds on old plum trees. The fructification is perennial, about 6 cm ($2\frac{1}{2}''$) across and either hoof-shaped and dark grey above with a grey-brown pore surface beneath or flat, grey-brown and plate-like, adhering to the bark. The rot is white and crumbly in the centre of the stem or branch with a surrounding firm dark brown-purple zone. For biology and treatment see p. 371.

RED ROT   *Stereum sanquinolentum*
Common and widespread; associated with wounds on conifers. The fructifications are thin, skin-like, cream or brownish, adhering to the bark. When scratched, they 'bleed' by the exudation of red juice but this is not unique and other *Stereum* spp. also do this. The rot is variable, white and soft in the later stages although reddish when young but never pocketed (cf. white pocket rot, below). For biology and treatment see p. 371.

WHITE POCKET ROT   *Heterobasidion annosum* (= *Fomes annosus*)
Very common throughout Britain and northern Europe. In Britain it is the most important forest tree pathogen but also occurs on broadleaved trees. Elsewhere in northern Europe it is possibly of more significance on broadleaved trees although it is still essentially a disease of conifers. It is unlikely to be of major importance in gardens. A dark red brown, perennial, bracket fructification up to 35 cm (14″) across develops, sometimes with a white margin and with a white pore surface beneath. It usually arises at soil level among leaf litter and enveloping small twigs and leaves. The rot is unlikely to be seen until well advanced and then it usually has the consistency of coconut fibre, with small white pockets within. The initial infection is of fresh stumps by spores from the fructifications and the fungus spreads from there by root contact to healthy trees where it may exceptionally extend to

three metres up the trunk. Trees of any age are likely to be killed if fresh stumps are near by and the felling of trees and replanting or thinning operations in groups of trees may therefore lead to death some years later. Protective treatment of cut stump surfaces is obligatory in most State and commercial forests and is sensible when fellings are made in small private plantings of conifers. Urea is the most generally useful material for 'painting' on to the fresh cut. Diseased trees cannot be saved and should be felled and if possible, the stumps and surrounding soil removed. A rather similar pocket rot, caused by *Phellinus pini*, is very serious in mature conifers in parts of northern Europe but not in Britain.

WILLOW HEART ROT   *Phellinus igniarius*

Common and widespread. The fungus is usually associated with branch wounds and is almost entirely confined to old willows in Britain but can affect most deciduous trees and is a serious pathogen of many species (especially birches) elsewhere in northern Europe. Related and similar species of *Phellinus* cause serious decay on aspen in more northerly and eastern parts of Europe but not in Britain. The fructification is perennial, woody, usually cracked and split, up to 25 cm (10″) across, hoof or bracket-shaped, grey-black above and with a light brown margin and a grey to yellow-brown pore surface beneath. The flesh is hard and dark red-brown and the rot soft and whitish with thin black zone-lines. For biology and treatment see p. 371.

STEM AND LEAF ROTS (Herbaceous Plants)
(Plate 15)

These are caused by a wide range of fungi and bacteria and while most arise from infection by air-borne spores, some spread upwards after the tissues are invaded by soil-inhabiting organisms. The distinction between some stem diseases and root and foot rot conditions is however often blurred and in some instances the same organism may be responsible for both. Conditions such as crown rots of delphiniums, rhubarb or strawberries for instance could be classified as either stem or foot rots. Leaf rots often spread with extreme rapidity and some are commonly known as fire diseases or blights, although the latter term is often used indiscriminately for many unrelated plant diseases and some pests. Prevention is essential with such fast-spreading problems because cure or the arresting of development is very difficult. Many of the stem and leaf rots described here are caused by species of *Sclerotinia* or *Botrytis* against which systemic fungicides related to benomyl are often very effective although it is wise to try alternative chemicals first in some instances, to minimise the likelihood of tolerant strains of the pathogens developing (p. 30).

CACTUS and SUCCULENT STEM ROTS

Several rots occur commonly on both seedlings and mature plants of most genera of the Cactaceae especially when grown as pot plants and over-watered. Among succulent and semi-succulent plants, distinct rotting conditions occur commonly on agaves, aloes, crassulas, kalanchoes and sanseveiras. The causes are imperfectly known but bacterial soft rot caused by *Erwinia carotovora* is common on some genera. There are no established chemical treatments for any of the rots but captan or benomyl sprays are worth trying for top rots, after cutting out the diseased parts. Caution with the watering of cacti and succulents can often prevent the troubles arising however.

CARNATION LEAF ROT   *Heteropatella valtellinensis*
Common on outdoor and glasshouse plants throughout much of northern Europe. Brown-yellow water-soaked spots develop on the leaves or, less frequently, on the stems. These dry to become purple or dirty white with tiny black pycnidia and may merge to affect entire leaves or occasionally to kill the plants. Adequate ventilation and light should help prevent disease establishment while, in the event of an attack, affected plants must be removed and destroyed and the remainder sprayed three times at fort-nightly intervals with thiram. Only healthy stock should be used for cuttings.

CARROT LEAF BLIGHT   *Alternaria dauci*
Occasional and widespread. This disease was first recorded in Britain in 1958 and is most usually seen in wet seasons. The fungus is seed-borne and causes dark brown-black lesions, edged with yellow, which arise first on the margins of the leaflets. These leaflets later shrivel and die while root development may be impaired because of the weakened foliage. Seed-borne infection can be controlled by hot water treatment of the seeds but this is unlikely to be necessary on a garden scale. Although unable to survive in the soil, the fungus can do so on crop debris but a three-year rotation should eliminate this source of disease.

CHRYSANTHEMUM STEM ROT   *Botrytis cinerea*
Very common and widespread in glasshouses and especially troublesome on cuttings in cool, moist conditions or sometimes following slight frost damage. The infection arises from air-borne conidia and cuttings rot, become covered with grey mould growth, and die. The stems of mature plants keel over at the point of attack, which also bears grey mould, and the disease may spread to the buds and flowers and these in turn develop grey mould growth. General principles of hygiene for the prevention of grey mould are outlined on p. 352 while it may be possible to arrest an attack by removing badly diseased plants and spraying lightly affected and healthy stock three times at fortnightly intervals with benomyl, captan or dichlofluanid.

CUCUMBER BACTERIAL CANKER   *Erwinia carotovora*
A common and widespread problem caused by the same organism re-sponsible for soft rot in many types of plant (p. 348). A soft wet rot of the stems develops at soil level and spreads both further up into stem and down into the roots, resulting eventually in plant death. It may be prevented by using only sterilised compost and avoiding excessive moisture but there is no reliable chemical treatment. A collar made from an old plant pot and

placed around the stems when the plants are young sometimes helps to keep the soil in contact with the stems relatively dry.

CUCUMBER STEM ROT *Didymella bryoniae*

Occasional and widespread on cucurbits from April onwards especially in warm areas and in glasshouses. The fungus normally only infects through wounds, such as those caused by leaf removal, and small groups of black pycnidia containing conidia and/or perithecia containing ascospores develop on the infection sites. Small pale green spots with yellow halos appear, usually at the edges of older leaves, and the centres of these spots become brown and dry and also bear pycnidia and perithecia. Occasionally the disease spreads to healthy stems, resulting in plant death, while rot in the fruits can arise either from flower infection or through the stubs where the fruit is picked. A soft wet grey-green lesion develops at the distal end of the fruit and gradually becomes black as fungal fruiting bodies are produced. The mode of persistence of the fungus between crops is unknown and may be as conidia and/or ascospores. The disease may be treated as for Grey Mould Rot (p. 352) although the fungicides suggested may be less effective against *Didymella*.

A fruit or more rarely stem rot of cucumbers is sometimes caused by soil-borne fungi, such as *Pythium* spp., when the fruit are in contact with the soil. Care to keep the fruits clear of the ground and to avoid splashing them with soil when watering should eliminate the problem. Young cucumber fruit sometimes shrivel but show little signs of rotting. This occurs most commonly under warm, wet conditions but other factors, including excessive use of manure, heavy pruning and wide fluctuations of temperature, may also be responsible.

DELPHINIUM STEM ROTS and BLIGHTS

Rotting of larkspur stems and of perennial delphinium crowns occurs occasionally and has been a problem for many years in southern England. There are several causes but fungi of the genus *Diplodina* give rise to a crown rot often associated with necrotic lesions on the stems and leaf stalks and occasionally with black leaf blotches very like those caused by *Pseudomonas delphinii* (p. 403). Abundant small black pycnidia form on the diseased tissues and affected plants often fail to survive the winter. Bacterial soft rot also affects delphinium crowns, as does grey mould *(Botrytis cinerea)* which forms a basal stem rot covered with copious grey mould growth. A crown rot with whitish mould growth and numerous yellow-brown sclerotial bodies is caused by *Sclerotium delphinii*. Plants affected with any of these diseases are best destroyed and delphiniums should not be grown on the land for as long as possible. The only rot that can be controlled effectively is grey mould which benomyl, dichlofluanid or captan sprays may arrest in the early stages.

GALANTHUS GREY MOULD BLIGHT *Botrytis galanthina*

Common and widespread on many *Galanthus* spp. but most severe on snowdrop *(G. nivalis)*. It is worst in mild winters or sometimes when frosts alternate with rain in early spring. The shoots emerge but remain small, stunted and covered with grey mould growth while the plants usually fail to flower; the leaves become brown and rotten and the disease spreads to the bulb which turns soft and pulpy. Small black sclerotia are produced on the affected bulbs and occasionally on other parts of the plants. Infection often arises from sclerotia in the soil and spread from plant to plant by conidia from the mould is probably unimportant. If applied sufficiently early, benomyl, thiophanate-methyl or dichlofluanid sprays will check attacks but badly affected plants should be destroyed and galanthus not planted on the contaminated site for as long as possible. Examine new stock carefully before planting and discard any bulbs bearing sclerotia.

HYACINTH FIRE   *Botrytis hyacinthi*

Occasional in south-west Britain and elsewhere in northern Europe but common in Holland. A brown discolouration appears at the leaf tips and spreads downwards causing the leaves to wither and shrivel. It is most severe in damp conditions when the leaves rot rapidly while the flowers are similarly destroyed and masses of grey mould develop on the diseased plants. A prompt spray with benomyl at the first sign of symptoms and repeated fortnightly should check attacks.

HYDRANGEA STEM and LEAF ROT   *Botrytis cinerea*

This is the most important disease of hydrangeas and is common and widespread especially in cool moist conditions. It affects plants at all stages of growth. The grey mould growth arises from air-borne conidia and, on the leaves, it usually begins at the extremities, sometimes associated with hydathodes. On stems, infection is often centred around leaf scars but soon spreads and the stems keel over at the point of attack while the disease often spreads to cause a flower rot. General notes on the prevention of grey mould are given on p. 352 while a spray at three-week intervals with benomyl, thiophanate-methyl or dichlofluanid should check attacks in the early stages.

LEEK WHITE TIP   *Phytophthora porri*

First found in Scotland in the late 1920's and now common in other parts of Europe, particularly in wet seasons. It occurs occasionally on other *Allium* spp.
Symptoms. These first appear around late August but are not very extensive until October when the leaf tips turn yellow, then dry and appear bleached or sometimes decay. The affected parts may hang limply or be twisted and distorted and commonly up to half the length of the leaves can die while in severe attacks they drop off at soil level; in lighter outbreaks the plants may merely be stunted.
Biology. The fungus persists as oospores in debris in the soil for three or more years. Eventually these germinate to form sporangia which are probably blown to the leaves where they start new infections, further oospores eventually being produced in the diseased tissues.
Treatment. Destroy affected plant debris and diseased plants as soon as they appear and do not grow leeks or other alliums on contaminated land for at least three years. Up to five sprays at two- to three-week intervals with captan or zineb may be successful in preventing severe attacks but should not be applied within two weeks of pulling.

LETTUCE GREY MOULD *Botrytis cinerea* **Pl. 15**
Very common and widespread in glasshouses and on outdoor plants. Often the most serious disease of lettuces.
Symptoms. On seedlings: essentially a damping-off symptom, appearing as a brown leaf rot which spreads to the stem. The plants keel over and a mass of grey mould develops on affected parts. On older plants: as the plants reach maturity, a leaf rot or more commonly a red-brown stem rot enables plants to be lifted clear of the ground by rupture of the almost non-existent stem tissues. If undisturbed, the entire plant wilts dramatically as natural severing of the stem occurs. In humid conditions, the rot may spread from the leaf margins (when it can be confused with marginal spot (p. 409), tip burn (p. 477) or, in the later stages, bacterial soft rot (p. 348)) to affect the head which becomes soft and slimy and often bears grey mould growth.
Biology. Rapid spread from plant to plant occurs by conidia produced on the grey mould whereas small black sclerotia produced on decaying tissues are possibly the means of survival in soil between crops, although the range of plants affected by grey mould is so large that there are probably always enough spores available to start new infections. The fungus is sometimes seed-borne, this being the cause of much of the damping-off symptoms.
Treatment. Prevent damage or stress to lettuce plants. Among common predisposing factors are downy mildew, soil-borne damping-off, chemical spray damage, waterlogging, mechanical damage to seedlings at transplanting, overcrowding, transplanting seedlings when too large, incorrect temperature control in glasshouses, inadequate ventilation or insufficient watering. It is nonetheless wise to spray lettuce plants routinely with benomyl or thiophanate-menthyl at pricking out time and fortnightly thereafter. Simultaneous application of zineb to control downy mildew (p. 293) is also very desirable.

LILY DISEASE *Botrytis elliptica*
Common in damp conditions and spread is often stopped by dry weather. The problem is especially frequent on madonna lily *(Lilium candidum)*. Small, red-brown, circular or elliptical leaf spots appear initially with yellowish halos. These spots enlarge and fade and others may then appear on stems, flower stalks and buds. The leaves may wither and hang down while buds are either killed or produce distorted flowers. Small black flattened sclerotia are produced on diseased tissues and these survive the winter in the soil, although the fungus may also persist as mycelium on debris. The sclerotia germinate in spring to produce mycelium and conidia by which the first infections are initiated. Further spread occurs by means of conidia during the growing season. Destroy affected plant debris in autumn and spray fortnightly with dichlofluanid or benomyl from the time of appearance of the first symptoms.

NARCISSUS FIRE *Sclerotinia polyblastis*
Occasional in wet seasons in Britain, usually in the South and West. The symptoms are small, pale brown flower spots which spread rapidly to destroy the blooms while yellowish or red-brown elongate leaf blotches later form, with glistening drops of mould growth. Conidia produced on

this mould spread the disease very rapidly and the entire foliage may be destroyed. Large black sclerotia form on plant debris and persist over the winter to germinate in spring and bring about new infections, probably through the production of ascospores. Destroy all diseased debris in autumn and spray plants with benomyl at the first signs of infection.

NARCISSUS SMOULDER *Sclerotinia narcissicola*

Common throughout Europe and generally most severe in cold, wet seasons, especially if such conditions occur early. In spring, shoots appear bearing grey mould growth and conidia produced on this mould spread the disease to other plants on which the lower parts of the leaves turn brown and later the upper regions become yellow and wilt. Eventually these plants become coated with grey mould growth. Black sclerotia may develop between the bulb scales and can be liberated into the soil where they persist and may infect plants in the following season, although not all the sclerotia commonly seen on narcissus bulbs indicate smoulder; many are produced by the grey mould fungus also. Bulbs affected with smoulder either fail to grow or give rise to diseased and distorted plants. Control is very difficult but the disease may be prevented by dipping bulbs for 15–30 minutes in suspensions of benomyl or thiophanate-methyl before planting or storage. This will protect against smoulder and certain bulb rots. Alternatively the papery scales bearing the sclerotia may be rubbed off before planting. This is usually adequate to prevent more than a few odd diseased plants from appearing.

NARCISSUS WHITE MOULD *Ramularia vallisumbrosae*

Occasional in warm damp southern parts of Britain and northern Europe. The symptoms are small ashen or yellow leaf spots or streaks which enlarge and become covered in moist conditions with white mould growth. Conidia produced on the mould rapidly spread the disease in damp weather and the entire foliage may rot while the flower stalks are attacked on late cultivars. Small black sclerotia develop on dead plants and survive the winter among debris to germinate in spring and produce spores which infect the newly emerging plants. All affected plant debris should be destroyed in autumn and narcissi should not be planted for at least one year on land on which the disease has occurred. A protective spray with benomyl is worth trying on emerging plants when the disease is known to be prevalent.

ONION LEAF ROT *Sclerotinia squamosa*

This disease is common on onions and, to a lesser extent, on leeks and other alliums, especially in conditions of high humidity. It is frequent in the West of Britain and seems especially serious on soils with high nitrogen content. Small elliptical or circular white leaf flecks develop, often with water-soaked margins and congregated towards the leaf tips. The lesions later dry out and the leaf tips shrivel, collapse and hang down while in moist conditions a grey mould growth producing masses of conidia may grow over the damaged tissues. Attacks may sometimes be checked by spraying fortnightly with benomyl or dichlofluanid. Further outbreaks may be minimised in areas where the problem is prevalent by applying potash to the soil if it is deficient, and increasing the spacing between the plants.

This is not an easy disease to diagnose and among several factors having at least some symptoms in common are hail damage (p. 485), downy mildew (p. 293), white tip (p. 382), copper deficiency (p. 480), grey mould (p. 346) natural leaf senescence, and, recently, a species of *Cladosporium*.

ONION SHANKING *Phytophthora* spp.

Occasional and widespread on onions and shallots. The leaves turn yellow and dry up, the bulbs soften and roots shrivel. Various species of *Phytophthora* may be responsible, some of which are the causes of similar diseases on tulips and other bulbous plants. Control is rarely necessary and may be difficult but four- or five-year rotations of onion crops may help.

PAEONY GREY MOULD BLIGHT *Botrytis paeoniae* and *Botrytis cinerea* Pl. 15

Common in Britain; common and sometimes serious elsewhere in northern Europe. *B. paeoniae* is the most usual and serious cause and both fungi cause a comparable leaf and stem blight of lily of the valley, which may be similarly treated. Soft brown regions develop at the bases of the leaves, which wilt and fall over, and a dark brown rot subsequently develops in the bases of both stem and leaves while grey mould growth arises on the stem above soil level. The young buds blacken and wither, the flowers and leaves bear brown patches and grey mould develops on all damaged tissues. Affected plants should be cut down to ground level in autumn and the debris destroyed. Then scrape away the top soil around the plants and replace it with fresh soil. As a further protection, spray plants with thiram, benomyl or thiophanate-methyl as the leaves expand and then twice more at fortnightly intervals.

PELARGONIUM BLACK LEG *Pythium* sp.

Very common and widespread, particularly on cuttings but occasionally on mature plants also. A blackening of the stems spreads upwards from the base and the affected tissues shrivel and take on a characteristic pinched appearance while the leaves wilt and die. Although it is so often troublesome, black leg can be controlled by taking certain precautions. All pots and trays should be disinfected before use and compost sterilised carefully. Cuttings should be taken only from healthy plants and dipped in hormone rooting powder containing fungicide. Glasshouses should be kept dry and well-ventilated and the cuttings watered only with clean water, avoiding possibly contaminated sources such as rain butts.

POTATO and TOMATO BLIGHT *Phytophthora infestans* Pl. 15

Widespread; common and serious in wet seasons, less frequent in dry ones. Potato blight is one of the classic plant diseases and the succession of severe blight epidemics in the 1840's led directly to the Irish potato famine. It is also serious on outdoor and cool glasshouse tomatoes and is sometimes known as late blight to distinguish it from the much less serious early blight (p. 387).

Symptoms. POTATOES: dark brown-black blotches sometimes with paler margins develop on the leaves and are often concentrated at the leaf tips or edges. In damp conditions, white mould grows on the undersides of the blotches which increase in both size and number and may spread to the stems. In a humid atmosphere the entire foliage may be reduced rapidly to a collapsed, rotting mass and there is usually a rapid spread of the disease to neighbouring plants in these conditions. (Fairly restricted brown or black blotches at the edges or tips of potato leaves are common in some years and follow damage resulting from walking or using implements among the plants. They are caused by the grey mould fungus and are not serious.) On tubers, brownish lesions on the surface extend gradually to envelop the whole tuber, giving a dry type of rot. Internally the affected tissue has a brownish-red speckled appearance. These tubers are often invaded by other fungi and bacteria and soft rotting may result in damp conditions. The fungus itself does not spread from tuber to tuber during storage.

TOMATOES: the foliage symptoms are similar to but less severe than those on potatoes. Dark streaks and spots develop on the stems and green fruit while more mature fruits decay rapidly with a usually dryish rot but may appear symptomless for several days after picking before the rot becomes visible. The fruit rot is usually accompanied by mould growth and the symptoms can be confused with other *Phytophthora* rots (p. 360) although these are more common on glasshouse plants. Also see blossom end rot (p. 477).

Biology. The fungus persists over winter in diseased potato tubers from the previous crop, either in the ground or more usually on dumps or in undetected seed tubers. When such tubers grow, either as seed in the ground or in large diseased heaps, they sometimes produce infected shoots. Spores may then be blown from them to infect a new crop. Tubers become infected by spores falling from the haulm onto the soil and being washed downwards by rain. Spores are produced and dispersed only in damp conditions and foggy weather in summer is therefore very favourable to the disease. Infection by blight spores is likely to take place when the temperature does not fall below 10°C (50°F) and the relative humidity does not fall below 75% during a period of 48 hours, this being known as a Beaumont period. The Ministry of Agriculture broadcasts warnings when Beaumont or the comparable Smith periods have occurred in the main potato growing regions and these indicate that commercial growers should apply protective sprays.

Treatment. There is little choice in terms of cultivar susceptibility to blight but 'Arran Comet', 'King Edward' and 'Ulster Chieftain' are particularly prone. Since rotten and unwanted potatoes often carry blighted tubers they should not be dumped but buried deeply away from land on which a new crop is to be grown. Seed tubers should be examined carefully and any showing symptoms destroyed while earthing up well may prevent spores produced on the foliage from reaching the tubers. Several chemicals are suitable for protective spraying but their application may not be generally worthwhile on potatoes in gardens. In damp seasons however, particularly if blight is known to have been present in the vicinity in the same or the previous year, a spray programme is a wise precaution. Spray potatoes with zineb or maneb in the first week of July (earlier in the South-West of Britain and other mild areas) and repeat at about 14 day intervals depending on the wetness of season (which affects both symptom severity and the extent to which the chemical is washed off the foliage). Spray or dust outdoor tomatoes as a routine in all except the very driest seasons with zineb or maneb at ten-day intervals, beginning just after the first fruit set. Dust is often most effective applied in the early morning while dew is present but the fruit must be washed before eating.

Also see: powdery mildew (p. 302) and wilt (p. 306) which are not common but have some foliage symptoms similar to those of blight.

POTATO and TOMATO EARLY BLIGHT *Alternaria solani*
Occasional and widespread but rarely causes much damage. Early blight is a North American name and is misleading as in Europe the disease usually occurs late in the season. Dark brown-black concentrically zoned, more or less angular spots develop on the leaves and are bounded by the leaf veins; the leaves later shrivel and drop. Small sunken dark spots with raised grey centres develop on the stems and a rot may develop at the sepal end of tomato fruit and bear a velvety black mould growth. A dry rot of potato tubers can occur but is not known to do so in Britain. The fungus persists on debris in the soil and is usually considered a weak parasite, often only attacking already unthrifty plants. Treatment is rarely necessary and good agronomic practices, in particular adequate fertiliser application, should render any problems from it unlikely. Two or three sprays with benomyl or one of the blight fungicides at three week intervals should check severe outbreaks.

RHUBARB CROWN ROT
Occasional in Britain. The terminal bud rots and the pith is affected by a soft, chocolate brown rot, often with cavities. Feeble lateral shoots develop but the young leaves often rot off and the older leaf bases sometimes also rot in wet conditions. Although it is often said that infection arises by the bacterium *Erwinia rhapontici* invading through wounds caused by stem eelworms, there is little actual evidence for the role of either organism in the disease and the cause in many instances is unknown. Grey mould *(Botrytis cinerea)* commonly affects rhubarb when being forced but can be checked by improved ventilation and removal of the worst affected crowns.

SAINTPAULIA STEM and LEAF ROTS
Rots of saintpaulia stems and leaves are very common and widespread. Growth of grey mould *(Botrytis cinerea)* and an associated decay of stems and leaf stalks was formerly the commonest symptom but the bacterium *Erwinia chrysanthemi* has recently been serious in some areas. The symptoms of the latter are similar to those of *Phytophthora* foot rot with black lesions at the stem base and general debilitation but no mould growth. Unlike the fungal disease, the bacterial infection causes the leaves to detach from the rosette but the precise diagnosis of this still imperfectly understood condition is a matter for expert study. *B. cinerea* is readily controlled in the early stages by spraying with benomyl, thiophanate-methyl or dichlofluanid but there is no established control measure for *E. chrysanthemi*.

SWEET PEA WHITE MOULD   *Ramularia deusta*

Common and widespread; especially important in the South and West of England and in Holland. It is easily confused with the very common powdery mildew (p. 302) but, unlike mildew, is favoured by dampness. A mealy white covering develops on the leaf surfaces and stems and sometimes water soaked spots appear which gradually turn red-brown at the leaf margins. The affected parts become shrunken and yellowed and the leaves may drop. To treat white mould, avoid damp, warm conditions in glasshouses and spray fortnightly with zineb, beginning at the first sign of symptoms. Grey mould *(Botrytis cinerea)* is sometimes the cause of a basal rot of sweet peas but is readily controlled by a spray with dichlofluanid, captan or benomyl and improvement of the ventilation, if in glasshouses.

TOMATO BACTERIAL CANKER   *Corynebacterium michiganense*

Originally a North American disease, introduced into England in the 1940's but now rarely serious because fewer outdoor tomatoes are grown. Still important in regions where tomatoes are grown routinely in the open. The symptoms are sudden wilting, usually of the upper parts of plants and often as the fruits begin to ripen. The lower leaves turn brown and die and brown streaks appear on the stems, which may split open to reveal large cavities and discoloured tissues (cf. *Sclerotinia* disease p. 369). There may be yellow streaking inside the stem while the fruit, when cut, also show yellowing of the placenta. Often a brown crescent shape is seen when the leaf stalks are cut through. The causal bacteria are seed-borne but can also persist in compost. Affected plants should be burned and contaminated soil removed. Do not save seed from diseased crops, disinfect glasshouses regularly and grow outdoor tomatoes on different land for at least one year.

TOMATO GREY MOULD STEM ROT   *Botrytis cinerea*

Common and widespread especially in cool glasshouses and when tomatoes are associated with other grey mould-prone plants, such as lettuce. Infection is normally from air-borne conidia through leaf scars or pruning cuts, which develop grey mould and rot. The symptoms may be associated with ghost spotting of the fruits (p. 422) and flower damage (p. 346) and the problem can often be avoided by adequate attention to ventilation, hygiene and other factors conducive to good plant growth. Do not leave leaf and shoot stubs when pruning and note the general comments made on grey mould biology and treatment on p. 352. Benomyl or dichlofluanid sprayed at fortnightly intervals should check any further spread of the disease.

TOMATO LEAF MOULD   *Fulvia fulva*   Pl. 15

Originally North American and now very common and widespread in glasshouses but rare on outdoor plants.

Symptoms. Sometimes appear as early as April but usually not until June or later. Patches, at first greyish but soon turning yellow, develop on the upper leaf surfaces and correspond with areas on the undersides covered with pale grey-brown mould growth. In moist, warm conditions the patches increase in size and number and darken to red-brown. Mould may then appear on the upper surface also and the leaves soon curl, wither and die but may not drop. Growth of the plant is checked, the fruit develops poorly and occasionally the flowers or fruit are themselves affected by mould.

Biology. The fungus probably survives over winter as conidia which initiate the first infections, usually on the lower leaves. More conidia are produced on the mould and spread easily by air movement and on hands and clothing. The disease is often restricted to the lower parts of plants as it is retarded by bright sunlight and is favoured by high humidity. Several races of the fungus occur and although many cultivars are resistant to some, very few are resistant to all.

Treatment. Disinfect glasshouses containing diseased plants by burning sulphur *before* the crop debris is removed (care should be taken with sulphur in non-wooden glasshouses and the manufacturer's instructions must be studied carefully). Maintain good ventilation throughout the growing season and take particular care to ensure that the lower layers of air do not stagnate. Do not allow high temperatures to coincide with high humidity; remove and destroy any lower leaves that show signs of disease and if leaf mould is a persistent problem, then apply protective sprays of benomyl, maneb or zineb at fortnightly intervals, but not within two days of picking the fruit.

TOMATO PITH NECROSIS *Pseudomonas corrugata*
Widespread and occasional especially in wet seasons and on unheated crops. The symptoms usually appear in the form of yellowing and wilting of the upper leaves as the fruit is ripening. Plants may be somewhat stunted and have elongate dry black regions on the stem, beginning about 30 cm (12″) above soil level. When split open, the stems may have blackened pith or pith cavity. The means of spread and persistence are unknown but commonly plants survive and produce fruit although any killed should be removed and destroyed. The avoidance of soft growth and prompt removal of the lower leaves will lessen the likelihood of the disease becoming established.

TOMATO STEM and FRUIT ROT *Didymella lycopersici*
Common and widespread on glasshouses and, even more so, on outdoor tomatoes, but now less of a problem than formerly. It also occurs on aubergines (egg plants).

Symptoms. Sunken, brownish lesions with small black bodies usually appear on stems at soil level with yellowing of the lower leaves and sometimes the development of roots on the stems above ground. The stem may be girdled by the lesion and secondary lesions may form anywhere on the main stem or roots. On the fruit the calyx becomes blackened and a rot spreads from it to form a black, crusted surface (but also see blossom end rot (p. 477)).

Biology. The fungus persists on crop debris or in the soil, from which the first infections arise. It can be transmitted on seed from affected fruit but this is probably of little importance. The black bodies on the lesions and fruit are pycnidia, within which conidia are produced and spread the disease very rapidly to other plants by rain splash or on hands and tools.

Treatment. Never leave diseased plants over winter in the ground but remove and destroy them in autumn and wash and disinfect hands and implements that have been in contact with diseased material. Where plants are grown on soil believed to be contaminated, protect them by drenching the bottom 10 cm (4″) of stem and the soil around each plant with captan within three days of planting. Repeat this treatment after three weeks. Similarly, treat healthy plants after removing diseased individuals from a bed, disinfect glasshouses and boxes at the end of the season and do not dip seedlings in water before planting as any infection present spreads rapidly to all other plants in this way.

TULIP FIRE *Botrytis tulipae* **Pl. 15**
Common in all parts of northern Europe and sometimes devastating.
Symptoms. Distorted leaves or shoots appear soon after emergence from the soil but are usually present only on scattered plants in a bed (cf. grey bulb rot (p. 344) where the plants die in patches). The shoots may be almost completely withered and in moist conditions are covered with grey mould growth bearing black bodies. Many small sunken yellowish leaf spots with surrounding dark green areas develop on neighbouring plants and there may be elongate dark brown patches bearing black bodies on the flower stalks. The flower buds may fail to open and be covered with mould while opened blooms may have small pale brown-white spots or blisters. On bulbs, small black bodies develop on the outer scales or on the remains of the flower stalk and the outermost fleshy scale, when exposed, may have depressed circular lesions, sometimes bearing black bodies. Occasionally the bulb may rot and have masses of black bodies around the outside.
Biology. Infection on newly planted tulip bulbs originates either from conidia on the bulbs, mycelium in the lesions on the outer scales (both usually derived from diseased parent bulbs) or from the black sclerotial bodies which can survive in the soil on the bulbs, and especially at the base of the old flower stalk, for about two years before germinating. Because of this short persistence, soil-borne infection is probably only important when tulips are grown continuously on the same land. The leaf and flower spotting is caused by conidia from the primary infections.
Treatment. Destroy affected tulips immediately as the spread to neighbouring plants can be very rapid. After removing the diseased individuals, spray the remaining healthy plants with benomyl. Carefully inspect bulbs for sclerotia and reject any affected and, as a precaution, all bulbs may be soaked before planting for 15–30 minutes in suspensions of benomyl or thiophanate-methyl, which will also give protection against certain other bulb diseases. Ideally tulips should not be grown on affected land for at least three years but in formal bedding this is impossible and in such cases, raking in quintozene dust at planting time may help, although this chemical is not normally available to gardeners.

TULIP SHANKING *Phytophthora cryptogea* and *P. erythroseptica*
Sometimes serious commercially but only on forced plants in glasshouses and rarely a problem in gardens. The commonest symptom is for the buds and leaves to turn yellow or reddish and sometimes shrivel. The roots are poorly developed or damaged and the bases of the flower stalks are shrunken and affected by a pink-brown stringy rot. The symptoms are

easily confused with those of the commoner root-attacking fungi, *Pythium* and *Rhizoctonia*. The causal fungi are soil-borne and sterilisation of the compost and growing boxes usually gives effective control.

TULIP YELLOW POCK *Corynebacterium oortii*

This disease was recently identified in Holland but had probably been present for many years. It certainly also occurs in Britain and probably other parts of northern Europe. On bulbs in storage, yellowing of the first bulb scale is seen when the brown skin is peeled back. When the bulb is cut, the conducting tissue usually appears yellow and severely affected bulbs fail to sprout, while others produce stunted plants with silvery leaf streaking (but note that silver white leaf flecking of tulips and other plants is commonly caused by hail damage (p. 385)). The leaves may eventually appear cracked and tattered and, like the bulbs, reveal yellow conducting tissues when cut, although the disease is not always easy to diagnose and the symptoms may be confused with those of frost damage or the common root-infecting fungus *Rhizoctonia solani*. Yellow pock infection probably spreads from plant to plant by wind and rain splash and can only really be treated by removing affected bulbs and plants as they are detected.

VIOLA STEM ROT *Myrothecium roridum*

This disease is occasional in England on pansies and violas but infrequent on violets. It probably occurs elsewhere in northern Europe but is most serious in warmer regions. The causal fungus may be associated with soil-borne root rot fungi and is sometimes regarded as a weak parasite affecting plants damaged by other causes. The stem just above soil level becomes dry and brittle and is covered with masses of small black pycnidia containing conidia. *M. roridum* is soil-borne and, in some host plants at least, seed-borne. It may be treated by not growing *Viola* spp., antirrhinums, dahlias, delphiniums, lupins and possibly tomatoes and potatoes (all of which can be affected to some extent) on contaminated land for three or four years.

SPOTS
(Plates 16–17)

This section is one of the largest in the book and it includes some of the commonest of all plant diseases. Thousands of species of fungi and many bacteria and viruses cause disfiguration, often of the leaves, on thousands of different host plants. This account includes only those that either lead to serious diseases or, although virtually harmless, cause symptoms that are so common or distinct that some comment upon them seems necessary, if only to allay concern. There are many instances where spots on leaves or other plant parts are only one outward expression of a disease in which some other effect is the more serious and descriptions of these will be found under other appropriate sections. Diseases, such as potato blight, in which leaves decay will be found described as leaf rots (p. 379) while leaf blistering is another special type of symptom also described separately (p. 328). Most of the lesions included in this section are characterised by death of the cells that comprise them and most arise from fungal or bacterial infection. Virus induced spots are described separately (p. 429) although a few spot diseases arising from non-pathological or unknown causes are included here. Although it is almost impossible to give any useful guidelines for deciding whether a spot symptom has a fungal, bacterial or viral origin, very small, usually black, dot-like bodies present on lesions are certain to be fungal reproductive structures. The most important spot forming organisms are fungi of the genera *Alternaria, Ascochyta, Cercospora, Colletotrichum, Gloeosporium, Phyllosticta* and *Septoria*, which together account for over 150 different spotting conditions on cultivated plants in Britain alone, although not all are described here. It should also be appreciated that other very common causes of spotting are misdirected garden chemical sprays or other airborne pollutants which are considered in detail on p. 489. Any spotting that is markedly unidirectional in its distribution on plants, or that extends across several different species is likely to fall into this category. White flecks with a similar unidirectional distribution may be caused by environmental factors, such as hail damage (p. 485) and are very common on fleshy leaved plants, including onions, tulips and other bulbous species. Extreme caution moreover should be exercised in deciding that a plant is affected by any leaf disease in late summer and autumn when normal seasonal necrosis may give rise to many misleading appearances. It is also worth noting that gardeners are often misled because the leaves and stems of some plants, of which hydrangeas are probably among the best known examples, have spots of seemingly pathological origin as part of their normal colouration.

Disfigurements may take several forms, largely reflecting the extent to which the pathogen has penetrated the plant tissues before being arrested by some resistance mechanism, which usually takes the form of specialised and toughened cells. A spot differs from a blotch in its smaller size and commonly more regular shape. Angular spots or blotches reflect inability of the pathogen to invade the leaf veins, which thus form boundaries to the dead tissue, and concentric or target spots indicate some aspect of radial growth on the part of the causal organism. Scorch is the term used to describe a symptom that gives the impression of a leaf having been held close to a flame and the name anthracnose, literally charcoal-like, is often used to describe several diseases with black sunken lesions. Dead spots or patches of tissue in leaves often drop out to produce the symptom known as 'shot-hole'. Size or number of any of the above lesions need be no criterion for judging their effect on the plant; the very dramatic symptoms of horse chestnut leaf blotch or tar spot of maples for instance are associated with very little harm to the respective trees but the much smaller and relatively insignificant lesions of black currant leaf spot, for example, can indicate considerable potential damage to the host plant. Nonetheless, even small and biologically unimportant spots can be a matter of concern when ornamental species are disfigured.

When the cause is fungal or bacterial, spots usually indicate the point of entry of the pathogen to a diseased plant organ, although virus induced spots may be of two types: local lesions, which are entry points analagous to those of fungal or bacterial origin, and systemic spots, which arise when a plant carries widespread virus infection but the siting of the actual lesions is not necessarily related to the presence of virus in those particular cells.

Many spot diseases require no treatment and those that merely cause slight disfigurement on ornamental species scarcely justify the use of chemical control measures. Because of the wide variety of fungi involved in spot causation, it is difficult to make any general recommendations for fungicide usage, although on ornamental species it is probable that two or three sprays at fortnightly intervals with either thiram or benomyl would give some control of the majority of spotting diseases. Bordeaux mixture is useful for some fungal spot diseases and is the only chemical likely to give any suppression of bacterial spots but as the fungicide itself causes spots on many plants, a general recommendation for its use would be unwise. For virus-induced necrotic spots, see the general comments on the treatment of virus diseases (p. 432).

ACER TAR SPOT *Rhytisma acerinum*   **Pl. 16**
Very common and widespread on sycamore but many species of ornamental acer can also be affected. A very similar disease on willows is caused by the related fungus *R. salicinum*.
Symptoms. Quite unmistakable large black bituminous blotches with yellow halos form on the upper surfaces of leaves from mid-summer onwards, preceded by rarely noticed yellowish patches in spring.
Several other types of spot are also common and widespread on sycamores, field maples and, to a lesser extent, on other acers. They vary from red-brown lesions to large yellowish blotches and although there is still uncertainty over the identity of the causal fungi, none of these spots normally reach serious proportions and no control measures should be necessary.
Biology. After the leaves have fallen in autumn, reproductive structures develop within the black blotches and by spring mature ascospores eject from them to bring about fresh infections as the new season's leaves begin to emerge. Different strains of *Rhytisma* may exist, each only capable of infecting particular *Acer* spp. The disease normally has no effect on the tree's vigour although this is small consolation when a plant grown as an ornamental is so conspicuously disfigured.
Treatment. Tar spot is not easily controlled. As ornamental maples are often small trees, the collecting up and burning of affected fallen leaves is usually feasible and a spray with Bordeaux mixture during the summer has been claimed as an effective treatment. Outbreaks of tar spot are less severe in industrial areas where sulphur in the atmosphere is believed to limit fungal development.

ANEMONE LEAF SPOT *Septoria anemones*
Occasional throughout northern Europe on *Anemone coronaria* and causes conspicuous sharply demarcated dark marginal spots to appear on mature leaves. The diseased tissue dries, becomes sunken and develops small dark pycnidia. There are no established control measures.

ANTIRRHINUM LEAF SPOT *Phyllosticta antirrhini* **Pl. 16**
This is the commonest of several fungal spotting diseases on antirrhinum leaves and often becomes apparent suddenly during humid weather in spring. Dirty brown, round leaf spots appear, often with dark margins, and

may fuse to form large blotches which can result in leaf death. Black dot-like pycnidia develop on the lesions and similar spots occur on the flower stalks, capsules and stems, the latter type of infection sometimes causing girdling and death of entire plants. The disease spreads from plant to plant by means of conidia liberated from the black pycnidia although the fungus, when present in dry plant debris among seeds, may also be a cause of seedling damping-off. Although seed treatment with fungicide can be used to control the damping-off symptoms, this is unlikely to be necessary in gardens although severe attacks on growing plants may be checked by one or two sprays with thiram, captan or zineb.

ANTIRRHINUM SHOT-HOLE   *Heteropatella antirrhini*
Widespread and common in wet seasons. The disease first appears as faint green leaf spots, the centres of which gradually lighten, dry and drop out producing 'shot-hole' symptoms. Occasionally the stems become distorted and spotted with water-soaked patches which also dry and become scorched. Two types of conidial spores form: one on the spots and the other in pycnidia on dead plant debris in the spring. Like *Phyllosticta antirrhini* (above), this fungus is probably transmitted with the seed. Following an outbreak, carefully collect and burn all dead material which may harbour the fungus and do not grow antirrhinums on the same soil in the following year. Spray established attacks three times at ten-day intervals with zineb. Also see downy mildew (p. 295).

APPLE LEAF SPOTS   *Asteromella mali* and *Botryosphaeria obtusa*
These are occasional and widespread but may be confused with the more serious leaf symptoms of scab (p. 335). Small, more or less rounded, light brown spots bearing black pycnidia are caused by *A. mali* although on 'Cox's Orange Pippin' similar spots (often called 'Cox spot') are common in dry seasons and have an unknown, but non-fungal cause. Large, irregularly shaped dark brown blotches on the leaves which shrivel and drop prematurely are caused by *B. obtusa*, an organism that also causes a storage rot in apples (p. 355). Neither fungus is usually serious in northern Europe and routine control measures against scab should eliminate them.

AQUILEGIA LEAF SPOTS   *Actinonema aquilegiae, Haplobasidion pavoninum*
Common and widespread. Large, irregular pale brown blotches which extend inwards from the leaf margins and often have a purplish border are caused by *H. pavoninum*. Small irregular pale brown spots which sometimes lose their centres to give a 'shot-hole' effect and are covered in a white spore-bearing mould are symptoms of *A. aquilegiae* attack. There are few reports of successful treatment of these fungi using fungicides but a benomyl spray may be worth trying. Also see: rust (p. 269); viruses (p. 437).

ARBUTUS LEAF SPOT   *Septoria unedonis*
Small irregular whitish leaf spots with purple margins are common wherever *Arbutus* is grown. They are not damaging and no control is necessary.

BEAN ANTHRACNOSE *Colletotrichum lindemuthianum* **Pl. 16**
Widespread on dwarf and less severely on runner beans but usually only apparent in cool wet summers. Brown sunken stripes develop on the stems, reddening on the undersides of the leaf veins, sometimes followed by leaf withering and death, and reddish, rounded, brown-bordered spots on the pods. In damp conditions pink-brown slimy masses of conidia may form on the spots. The pod spots result in infection of the seed which may be detected as dark spots on light coloured seeds or indefinite pale marks on dark seeds. Such infected seeds result in infected seedlings. These are usually killed and conidia from them splash onto healthy plants. The disease is rarely severe enough to warrant treatment in Britain where many dwarf bean cultivars are resistant to some races of the fungus. Also see: broad bean rust (p. 272).

BEAN CHOCOLATE SPOT *Botrytis fabae*
Widespread and common on broad and field beans but varies greatly in intensity from year to year and may appear at any time from mid-winter onwards although most serious when late frosts precede wet mild weather in early summer.
Symptoms. Dark, chocolate brown, more or less rounded spots develop on the leaves (mainly the upper surfaces), on the stems (where the lesions are more elongate), and on the leaf stalks, pods, seed coats and flowers. The spots may merge and result in blackening and death of organs or entire plants and even slight attacks can result in reduced numbers of pods.

Other types of spot can also occur quite commonly on broad bean leaves and may be confused with chocolate spot. Sunken, ashen spots with a raised brown perimeter on leaves are caused by *Cercospora zonata*. Similar leaf lesions, but more usually with black pycnidia present and sometimes occurring also on the pods, are caused by *Ascochyta fabae*. Greyish blotches with veined, net-like margins and covered by a dark brown mould growth, with occasional leaf death, are the symptoms of net blotch caused by *Pleospora herbarum*. None are usually serious and treatments are not normally justified.
Biology. Since this disease is favoured by damp, shaded, overcrowded conditions, it is generally not nearly as common in gardens as in commercial field beans. Although *Botrytis fabae* is responsible for most chocolate spot seen in Britain, the grey mould fungus *(Botrytis cinerea)* can sometimes cause at least some of the symptoms. Most infections originate from conidia over-wintering on plant debris but the seeds may also sometimes carry the disease. Plants infected in early spring seem much more likely to recover than plants infected later in the season.
Treatment. Apply potash fertilisers, improve drainage, increase spacing between plants and avoid sheltered sites. If all else fails, fortnightly sprays of benomyl may check outbreaks.

BEAN HALO BLIGHT *Pseudomonas phaseolicola* **Pl. 16**
Seed-borne bacterial diseases of dwarf and runner beans are quite common and widespread but there has been confusion in the past over the identity of the organisms responsible. It is now clear that in Britain *P. phaseolicola* is the most important.

Symptoms. Small water-soaked spots are sometimes visible on the cotyledons of newly emerged seedlings but are more likely on the first leaves.
Leaf lesions gradually darken and dry and become surrounded by characteristic yellowish zones or halos. Later, inter-veinal yellowing appears on
the leaves, which sometimes become entirely yellow, and lesions may also
appear on the stems. Seedlings wither and die in severe attacks but less
badly affected plants remain stunted. The pods may also become infected
and develop greasy spots.

Biology. There are usually only a few infected seeds in any one batch but
rain splash rapidly spreads bacteria from the resulting diseased plants to
neighbouring healthy leaves, the bacteria gaining entry through stomata.
Infected pods give rise to the infected seed. The search for resistance to this
disease is complicated by the occurrence of two races of the pathogen.
Although sources of resistance to both races are known, resistant cultivars
are not yet available.

Treatment. Never soak bean seeds in water before planting as this will
transfer the bacteria from contaminated to clean individuals. If it is
necessary to pre-germinate seeds, place them for a time in damp peat which
does not facilitate such transfer. One or two sprays with Bordeaux mixture
will restrict the spread of the disease on growing plants but should not be
applied after the first small pods are formed. Also see: broad bean rust (p.
272).

BEECH LEAF SPOT *Discula quercina*

Fairly common, probably more so in Britain than elsewhere in northern
Europe and said to be worst under damp conditions. Brownish irregular
blotches and/or spots develop, especially along the mid-rib. Pycnidia develop on fallen diseased leaves in autumn and ascospore-producing perithecia occur on them in the spring, the spores being liberated to infect the
new foliage. Control is unlikely to be necessary and is probably impossible.

BEET LEAF SPOTS *Ramularia beticola* and *Cercospora beticola*

Common and widespread on red beet and spinach beet, *C. beticola* also
being a serious pathogen on sugar beet. The spots are small, more or less
circular, with pale ashen centres and brown purple margins. Neither pathogen need cause concern as little damage ensues. Also see: blackleg (p. 363).

BLACKBERRY CANE SPOT and PURPLE BLOTCH
Septoria rubi and *Septocyta ramealis*   **Pl. 16**

Both are occasional, widespread and very similar and also easily confused
with raspberry cane and leaf spot (p. 418) which sometimes affects blackberries. Both cause grey-white elliptical spots with purple borders on the
canes, each bearing black dot-like pycnidia. Cane spot also causes similar
but circular spots on the leaves and such symptoms are sometimes severe on
the cultivar 'Himalaya Giant'. No control is normally needed although
three sprays with thiram at fortnightly intervals after the first appearance of
symptoms should check disease spread, but must not be applied less than
one week before picking the fruit.

BOX LEAF SPOTS

Leaf lesions arising from various causes are very common and widespread. Pale brown spots with distinct purple margins are caused by *Phyllosticta* spp.; large yellow blotches, especially on the undersides of the leaves, are caused by *Hyponectria buxi* but there may well be others. Benomyl or Bordeaux mixture sprays are worth trying.

BRASSICA LEAF SPOTS

Leaf spotting is exceedingly common on brassica leaves and heads. A number of different types occur, some being associated with other, more serious symptoms. Fortunately the various types of spot are fairly distinct, the commonest confusion arising between virus spots and dark leaf spot and between white blister, white spot and light leaf spot:

BRASSICA BACTERIAL LEAF SPOT
Pseudomonas maculicola

Occasional and widespread on the leaves of most brassicas and sometimes a problem on cauliflower curds. Tiny, usually angular water-soaked spots appear and soon dry and become dark brown to black. The spots often merge to damage large areas of the leaves which may turn yellow and/or drop. The disease is almost certainly seed-borne and it seems unlikely that it would persist for very long in the absence of a crop. The removal of debris is however important in limiting its impact although the avoidance of the overlapping of brassica crops, which would be even more effective, is scarcely practicable in gardens.

BRASSICA DARK LEAF SPOT   *Alternaria brassicicola,*
A. brassicae

Very common and widespread on several types of brassicas, including Brussels sprouts, cabbage and cauliflower. Black/purple spots develop on old leaves and often later become covered with a dark sooty mould. The disease is only of practical importance in commercial seed crops however where the seed pods are infected and no control should be necessary in gardens where dark spots on cabbages are more likely to be caused by viruses (p. 440).

BRASSICA LIGHT LEAF SPOT   *Cylindrosporium concentricum*

Common, particularly so in northern Britain, on Brussels sprouts, cabbage, cauliflower and less commonly on other brassicas. Tiny, fluffy white lesions form, particularly on the under surfaces of older leaves, but are found also on other parts, including cauliflower curds. They occur in groups in characteristic ring formations, the individual rings occasionally fusing and the fluffiness being caused by masses of tiny white conidia which are squeezed out from beneath the leaf surface. Infection arises from fungus persisting in plant debris in the soil. Tiny black apothecia, in which ascospores are produced, can also occur but are rare. The disease may be troublesome in gardens where the two-year break between brassica crops which limits its impact is not always practicable.

BRASSICA RING SPOT   *Mycosphaerella brassicicola*   **Pl. 16**

Frequent throughout northern Europe but in Britain usually only serious in the South West in cool wet seasons. It is most frequent on Brussels sprouts, cabbages and cauliflowers but is found also on other brassicas and is commonest on intensively cropped and heavily manured land.

Symptoms. These are most usually seen on the older leaves. The Rounded brown, sometimes faintly purple spots develop, often bearing more or less concentric target-like rings of abundant black dot-like bodies and often with a paler yellowish margin. The entire leaves may ultimately become yellow and shrivel. Lesions can also occur on other parts of the plants and on Brussels sprouts the buttons are often severely affected.

Biology. The tiny black bodies may be either pycnidia of unknown function or perithecia, within which ascospores develop, these spores, released from perithecia on plant debris in cool damp conditions being the main sources of disease. Transmission of the fungus in seeds produced in infected pods has been suggested but not proved definitely.

Treatment. Remove and burn all affected plant remains. No chemical treatment is suitable for gardens but 'Roscoff' types of cauliflower have some resistance.

BRASSICA WHITE SPOT *Pseudocercosporella capsellae*
Common and widespread on turnips and even more so on swedes but infrequent on other brassicas. Especially common in wetter areas in the West of Britain. Tiny more or less circular dirty white spots develop with a faint purple border on both leaf surfaces. These spots gradually enlarge and merge to result in yellowing and sometimes death of the leaves and defoliation of the plant. Conidia produced on fine mould growth over the surface of the spots bring about rapid disease spread and small resistant sclerotia survive over winter. Fortnightly sprays with Bordeaux mixture usually check outbreaks. Also see: white blister (p. 304); downy mildew (p. 292).

CAMELLIA LEAF SPOTS *Pestalotiopsis guepini* and
   Phyllosticta camelliae
Both species are common throughout northern Europe but are usually only serious under glass. *P. guepini* causes large yellow-brown spots with dark borders which spread inwards from the leaf edges. They are easily confused with the symptoms of severe iron deficiency (p. 478). *P. camelliae* forms smaller, brown, often circular spots with reddish borders. The fungi often colonise spots brought about by over-watering or nutrient imbalance and care with general growing conditions is therefore important. A spray with Bordeaux mixture may be effective in checking severe outbreaks.

CAMPANULA LEAF SPOTS *Ramularia macrospora* and
   Ascochyta bohemica
Several types of leaf spot occur on various *Campanula* spp., many of them common throughout northern Europe. *R. macrospora* causes brown irregular spots and *A. bohemica* produces similar lesions with purple margins and bearing dot-like pycnidia, but there are others. Bordeaux mixture or benomyl sprays may be effective.

CELERY LEAF SPOT *Septoria apiicola*   **Pl. 16**
Common throughout northern Europe and serious on celery, less so on
celeriac.
Symptoms. Brown spots with either a lighter or darker border develop
initially on the older leaves but they soon spread to other foliage and
occasionally to the leaf stalks. The spots are sometimes very numerous and
may fuse to form much larger blotches which cause leaf death. On and
around the spots tiny black dots appear and in severe attacks the crowns
may develop feebly as a result of the leaf damage.
Biology. The initial source of the disease is fungus persisting on the coats
of seeds from affected plants, the seedlings which emerge from such seed in
turn becoming infected. Tiny black pycnidia develop on them and produce
conidia which are spread by rain splash onto healthy plants on which the
leaf symptoms then develop. It is possible that diseased celery debris may
also serve, for a limited period, as an initial source of the disease.
Treatment. Although attacks of the disease may be contained by one or
two sprays with benomyl, eradication of the fungus from the seed is the
only certain method of avoidance. Hot water treatment of the seed for 25
minutes at 50°C (122°F) gives good disease suppression although soaking
seed in a suspension of thiram is even more effective but both methods are
difficult on a garden scale.

CELERY PALE LEAF SPOT *Cercospora apii*
Probably does not occur in Britain but frequent and sometimes serious
elsewhere in northern Europe. Tiny yellow spots develop on either or both
leaf surfaces, extend rapidly and turn ashen. In damp conditions a grey
mould appears on the spots and produces conidia which quickly spread the
disease throughout a bed. Destroy affected crop debris at the end of the
season and spray growing plants with Bordeaux mixture to check attacks.

CHERRY LEAF SCORCH *Gnomonia erythrostoma*
A common and widespread problem; one of the most frequent cherry
diseases.
Symptoms. Large yellow patches develop on the leaves early in the season
and gradually enlarge and turn brown. The leaves eventually wither but
remain hanging conspicuously on the trees throughout the following winter. Fruit may also be infected but this is rarely serious, although hard dark
spots can develop within the flesh. Repeated and severe annual attacks can
seriously weaken the trees.
Biology. From the leaf lesions mycelium grows into the leaf stalks and
prevents development of the abscission layer by which the leaves should
separate from the tree in autumn. During the winter ascospores develop
within the dead leaves hanging on the branches and these are discharged in
spring and are the sole means of infection. Wild cherries are often badly
affected and can serve as disease sources when growing close to orchards
and gardens.
Treatment. There are marked differences in susceptibility between
cherry cultivars; 'Early Amber', 'Frogmore Bigarreau' and 'Waterloo' are
badly affected while 'Napoleon' and 'Turk', among others, are highly
resistant. Remove and burn all hanging dead leaves from trees in winter to

eliminate the source of disease and, for further protection, spray the trees with zineb as the leaves are unfolding and preferably once again after petal drop. Also see: *Prunus* leaf spots (p. 417).

CHRYSANTHEMUM LEAF BLOTCH *Septoria* spp.

Widespread but rarely serious. The commonest cause of this disease is *Septoria chrysanthemella* although other *Septoria* species can also be associated with it. Dark brown or black, more or less circular spots form, sometimes with purple tints and brown margins and may bear concentric markings. At first they appear on the lower leaves but soon spread to others and spots may merge, especially under moist conditions, to affect large areas of the leaves. Dark pycnidial dots appear on the spots and conidia from them bring about rapid disease spread. There are some differences in cultivar susceptibility but none can be recommended as resistant. Treat by stripping all diseased or suspect leaves before taking cuttings; do not replant in soil that has carried diseased plants and dust with sulphur or spray with benomyl, captan, thiram or zineb at fortnightly intervals. Large dark brown spots with a central white area and often with concentric markings are common on marguerite *(C. leucanthemum)*. They are caused by *S. leucanthemi* and can be controlled by fortnightly sprays with benomyl or thiophanate-methyl.

CINERARIA LEAF SPOTS *Ascochyta cinerariae* and *Alternaria cinerariae*

These are very common and widespread seed-borne diseases, especially serious on bedding plants. More or less circular, gradually enlarging dark brown or black leaf spots are caused by *Ascochyta cinerariae*. The entire leaves may rot and shrivel and, in Britain at least, the same fungus also causes a rot (p. 363). *Alternaria cinerariae* causes similar symptoms and, when the leaves are held up to the light, rot can sometimes be seen within the veins. Seedling damping-off caused by *Alternaria cinerariae* may result from using contaminated seed. It is not really feasible to eliminate such seed-borne infection but growing plants may be treated by removing affected leaves and spraying with Bordeaux mixture, zineb or maneb.

CLIVIA LEAF SPOT *Colletotrichum cliviae*

Red or red-brown spots are common and widespread on clivias. The causes are often obscure but *C. cliviae* is one possibility and can result in leaf death, although excess moisture may be a contributory factor. Spray treatments are not worthwhile.

CONVALLARIA LEAF SPOTS

Red brown, darkening and extending leaf spots bearing black pycnidia are caused by *Dendrophoma convallariae* and are common and very damaging in parts of northern Europe, although unknown in Britain. They are often associated with attack by the grey mould fungus and may be controlled by two or three sprays of thiram during the summer. In Britain and elsewhere attack by *Septoria* spp., and possibly other fungi, occasionally causes concern; *S. convallariae-majalis* in particular being the cause of angular, sometimes merging dark green spots with dark brown borders on all green parts of lily of the valley in some areas. There are no established control measures but thiram or zineb sprays at fortnightly intervals are worth trying. *Gloeosporium convallariae* forms small round or oval reddish-bordered leaf spots, and sometimes causes premature leaf drop in parts of northern Europe. Fortnightly sprays with thiram are an effective control.

CRATAEGUS LEAF SPOTS and BLOTCH
Phleospora oxyacanthae, Septoria crataegi and *Sclerotinia crataegi*

P. oxyacanthae is common in Britain and elsewhere in northern Europe and causes masses of tiny yellow spots which often spread to scorch the entire leaves. Tiny brown pycnidia develop on the undersides. Blotch, caused by *Sclerotinia crataegi*, is also sometimes severe and produces large brown lesions with a grey sweet-smelling mouldy growth. It infects the flowers and overwinters in fruit mummies. *Septoria crataegi*, unknown in Britain but occasional elsewhere in northern Europe, forms tiny red-brown leaf spots which become paler in the centre and bear black pycnidia. Treatment for all these diseases is largely speculative as there are no established control measures but benomyl sprays would probably be effective.

CUCUMBER ANTHRACNOSE   *Colletotrichum lagenarium*
Common and important throughout northern Europe on cucumbers and less commonly on vegetable marrows and melons.

Symptoms. Very pale green, often transparent, spots develop on the leaves and later enlarge, become dry and red-brown in the centres. The spots may coalesce and the entire leaf may be killed and depressed, dry, powdery dark patches appear on the leaf stalks and stems. Similar sunken pale areas may develop on the fruits, which may eventually crack and die. When seedlings are attacked a damping-off symptom may result.

Biology. Masses of conidia are produced on the powdery patches in humid conditions. These adhere readily to hands and clothing and spread the disease to other plants and can also lead to fungal development on old woodwork in glasshouses and frames, on straw and on other plant debris and similar material.

Treatment. Provide adequate ventilation and routinely disinfect the glasshouse fabric, avoid wide fluctuations in temperature and remove all plant debris. At the first sign of symptoms destroy affected parts, spray the plants with a sulphur suspension and repeat at intervals of a week or ten days.

CUCUMBER BACTERIAL LEAF SPOT   *Pseudomonas lachrymans*
Occasional, primarily in glasshouses, in parts of northern Europe, including Britain. Brown angular leaf spots form and exude small, clear drops of liquid containing bacteria. The centres of these spots may later fall out and similar small lesions, on which the ooze dries to a white crust, can form on the stem and leaf stalks. Small spots, at first water-soaked but later whitish, also form on the fruits and exude an orangey liquid, this symptom often being followed by rotting of the fruit. The bacteria are probably spread by water splash or insects but initial infection is present on the seed. No chemicals are available for seed treatment in gardens but spraying plants

with Bordeaux mixture is claimed to arrest disease development if applied sufficiently early.

CUCUMBER LEAF BLOTCH *Corynespora cassiicola*
Much less common since the introduction of resistant cucumber cultivars, such as 'Butcher's Disease Resister' but still a problem in some areas of Europe on both cucumbers and melons.

Symptoms. Small, pale green, more or less water-soaked spots form on the upper leaf surfaces, although in the early stages they are often overlooked. They swiftly increase in size, multiply and turn ashen before they dry and usually fall out to give a 'shot-hole' symptom. The entire leaves subsequently shrivel and rot while the symptoms may also occur on the fruit as dark sunken patches which can spread to occupy large areas of tissue. A somewhat similar symptom on outdoor cucumbers in parts of Europe, although not in Britain, is browning, caused by infection from conidia of *Alternaria pluriseptata*.

Biology. Infection arises from conidia produced by fungus overwintering in and on the fabric of glasshouses. Disease spread can be very rapid in moist humid conditions and is brought about by such means as physical contact between plants or on hands and clothing.

Treatment. The best solution, where leaf blotch proves a consistent problem, is to grow resistant cultivars, although rigorous glasshouse hygiene and good ventilation will also keep outbreaks to a minimum. Remove and destroy affected leaves as they appear, reduce the glasshouse temperature and dust the plants with sulphur. Browning disease is difficult to control but a long rotation between cucumber and melon crops will lessen the likelihood of attacks developing.

CURRANT ANGULAR LEAF SPOT *Mycosphaerella ribis*
A relatively uncommon disease on black currants, red currants and gooseberries. It is easily distinguished from the commoner and much more severe leaf spot (below) by the more angular and generally larger spots with paler centres and black pycnidia on the undersides. Perithecia develop on fallen leaves from which ascospores are discharged in the spring to initiate new infections. Treat as for common leaf spot (see below).

CURRANT LEAF SPOT *Pseudopeziza ribis*   Pl. 16
Common and widespread. It is most important on black currants but also affects red currants, white currants and gooseberries.

Symptoms. Small irregular brownish spots or blotches develop on the leaves from May onwards although they are often not noticed until well into June. The spots appear first on the older leaves, which may turn yellow if the spots are numerous, but they often merge and the disease may then spread up the plant, the leaves falling progressively as it does so. Spots occasionally occur on stems, leaf stalks, and also on unripe fruit, which then shrivels. Small sparkling droplets may appear on the spots in wet or damp conditions.

Biology. The tiny droplets on the spots are masses of conidia which spread the disease during the summer. During the winter apothecia develop on fallen leaves and from these ascospores are discharged in the spring and are

carried by the wind to initiate new infections on the lower leaves. Severe attacks may result in a weakening of the bushes and a reduced crop in the following year.
Treatment. No resistant black currant cultivars are known but 'Baldwin' is probably the most susceptible and the red currant cultivars 'Fertility' and 'Fays's Prolific' are also particularly severely affected. Collect and burn affected leaves and spray the bushes with zineb, benomyl or thiophanate-methyl after flowering and repeat once or twice at fortnightly intervals, but no later than a month before the fruits are picked. Spray again immediately after the fruits have been picked. Also see: currant angular leaf spot (p. 402).

CYCLAMEN LEAF SPOTS *Cryptocline cyclaminis,*
  Phyllosticta cyclaminis and *Septoria cyclaminis*
It is possible that none of these leaf spots occur in Britain but they are common in the warmer parts of northern Europe. *C. cyclaminis* causes distinctly zonate, watery, pale green circular spots bearing pink spore pustules on the flower buds, flowers and flower stalks and on leaves in contact with the flowers. It is only important in glasshouses. *P. cyclaminis* causes yellow-brown leaf spots, often with a reddish border, which later shrivel and drop out to form 'shot-holes'. *S. cyclaminis*, probably the most common of the three, causes red, concentrically zonate and later grey circular leaf lesions. Lowering the temperature in glasshouses and spraying with Bordeaux mixture have proved successful treatments in Germany where the diseases have been troublesome.

DAPHNE LEAF SPOT *Marssonina daphnes*
Widespread and the commonest of at least two leaf spots on daphnes. Green, gradually darkening spots develop on the leaves (especially towards the base) and on the leaf stalks. This sometimes results in defoliation and, after repeated attacks, death of the plants. Remove and destroy affected leaves and spray fortnightly with Bordeaux mixture.

DELPHINIUM BLACK BLOTCH *Pseudomonas delphinii*
Common and widespread. The same organism probably causes tomato speck (small irregular leaf spots) which is known only from Denmark in northern Europe but is serious in North America and elsewhere. Large bituminous-black blotches develop on the leaves, stems, leaf stalks and flower buds. The spots may coalesce and affect entire leaves but if these symptoms are associated with crown rot, see also p. 381. Black blotch attacks arise from bacteria in the soil, which are spread by rain splash and infect through the stomata. Destroy affected plants and protect others by spraying with Bordeaux mixture. Drenching the chemical onto the soil around the plants is also sometimes suggested but is rarely successful.

DIANTHUS LEAF SPOTS *Ascochyta dianthi* and *Septoria dianthi*
Two common, widespread and similar leaf spots occur although there may be others (see also ringspot (p. 404) and leaf rot (p. 380)). *A. dianthi* is mainly a problem on sweet williams but sometimes also on carnations under mist propagation. *S. dianthi* occurs only on carnations. *A. dianthi* spots are

pale brown with darker margins and bear clustered pycnidia. *S. dianthi* causes light brown or ashen leaf or stem spots with purple margins and also frequently causes extensive pale straw-coloured discolouration from the point of attack along the tip of the leaf. The leaves may curl longitudinally and black pycnidia develop on both leaf surfaces. It is more severe on the older leaves and is especially troublesome in southern parts of northern Europe. Both diseases may be controlled by fortnightly sprays with thiram.

DIANTHUS RING SPOT   *Mycosphaerella dianthi*   **Pl. 16**
Two common and widespread strains exist, one affecting carnations and the other sweet williams. Cross infection does not occur between the two host plants and the disease is probably more serious outdoors than in glasshouses. More or less circular greyish spots with very marked red-purple margins develop on the leaves, stems or rarely on the sepals. Individual spots may fuse and whole leaves shrivel under damp conditions. Irregular patches and later concentric rings of dark grey mould appear in the centres of the spots and conidia, produced on this mould, rapidly spread the disease to other plants. The fungus over-winters as a perennial mycelium in the leaves. The chances of attack may be lessened in glasshouses by avoiding the development of high humidity. Outbreaks may sometimes be checked by spraying with captan or with Bordeaux mixture at three-week intervals during the summer although the latter chemical disfigures many cultivars.

DIGITALIS LEAF SPOTS
At least three types of spots occur on *Digitalis*; two are relatively common, widespread and sometimes serious on *D. lanata* and/or *D. purpurea*. Tiny brown-purple spots, which gradually become lighter and colourless, developing on first lower, and then on other leaves are caused by *Colletotrichum fuscum*. Badly affected leaves become yellowish, then brown and may die. Seed transmission of the fungus seems possible but in gardens a spray with Bordeaux mixture and a wetting agent at the first sign of symptoms should keep this disease in check. Variously sized irregular brown spots with a purple border and bearing minute black pycnidia are caused by *Septoria digitalis*. There is general spotting on *D. lanata* with this disease and it is a serious problem on commercial crops in some areas, but on *D. purpurea* the symptoms are usually confined to the leaves and stems. The spots ultimately turn necrotic and develop small black pycnidia. Like *C. fuscum, S. digitalis* may also be seed-borne but can be controlled by fortnightly sprays of captan on growing plants.

DRACAENA LEAF SPOTS
Among a number of causes of the various common and widespread spots and blotches on dracaena leaves are *Coniothyrium concentricum* (see p. 424) and *Phyllosticta draconis*. Benomyl or captan sprays are worth trying in severe cases. Bordeaux mixture is sometimes recommended for these diseases but is likely to cause as many spots as it cures.

ELM LEAF SPOTS
At least five different types of leaf spot occur irregularly on elms throughout Britain and northern Europe. Among the commonest are small brown

spots, predominantly on the veins and leaf margins, caused by *Gloeo-sporiun* spp.; conspicuous black tar-like spots caused by *Platychora ulmi* and red-brown spots with whitish undersides which are associated with some premature leaf fall and are caused by *Mycosphaerella ulmi*. None of these are serious and no treatment is either necessary or practicable.

FIR NEEDLE SCORCH   *Sydowia polyspora*
Widespread in northern Europe on a number of fir species although probably most common on silver fir, Grecian fir, pindrow fir, Spanish fir and noble fir. Newly formed needles become red-brown and gradually darken and shrivel but usually remain hanging on the trees until the following season. In severe attacks the entire shoot may become distorted and die back and occasionally the entire tree may be killed. Conidia form in black dot-like pycnidia on the upper surfaces of the dead needles and ascospores develop in similarly black perithecia. Little control can be exercised other than cutting out the diseased branches. Infestations by adelgids (p. 182) often accompany the disease.

FLY SPECK   *Schizothyrium pomi*
Occasional and widespread on apples and plums and often associated with sooty blotch disease (p. 420). Small groups of tiny black circular dots of fungal growth arise on the fruit. These dots are easily rubbed off and not serious.

FORSYTHIA LEAF SPOTS   *Marssonina forsythiae* and *Phyllosticta forsythiae*
These fungi are relatively common in northern Europe and cause large brown or grey-brown leaf spots but are rarely reported in Britain. They are occasionally serious enough to warrant treatment and a single spray with benomyl or zineb should be sufficient.

GERBERA LEAF SPOT   *Ascochyta gerberae*
Common in the warmer parts of northern Europe and in the South and West of Britain. Large, rounded brown leaf spots form with a very distinct purple margin and masses of dark dot-like pycnidia. The spots eventually coalesce and the leaves shrivel. Two or three sprays with thiram at three-week intervals are usually effective in checking the disease.

GRAPE ANTHRACNOSE   *Elsinoë ampelina*
Occasional in Britain but common and serious in many other parts of northern Europe. Irregular grey leaf spots with dark margins develop and may later dry and drop out, leaving holes. On the fruit, similarly bordered but more or less circular red-brown spots appear and may eventually merge and bear a greyish mould growth producing conidia. Control of this disease is difficult. Cut out and burn affected shoots and spray vines with lime sulphur during the winter when they are dormant. Then spray with Bordeaux mixture when the shoots are approximately 20 cm (8"). long, again just after flowering, again ten days later and finally when the grapes are half grown. In Britain, grey mould on fruits and the absence of leaf spots is a much commoner occurrence and is caused by the grey mould fungus (p. 352). Also see: *Parthenocissus* leaf spot.

GRASS LEAF SPOTS

Spotting, usually brownish and taking various forms, is common and widespread on lawn grasses. Several fungi may be responsible but species of *Drechslera* are probably the most significant. Smooth-stalked meadow grass and perennial rye grass are the species most likely to be attacked and sometimes death of entire leaves can occur and result in small bare patches of turf. Spraying with captan immediately after the appearance of symptoms and at three-weekly intervals during the summer would probably be effective but is rarely likely to be necessary.

HEBE LEAF SPOT   *Septoria exotica*

Common on shrubby veronicas in parts of northern Europe. It has been particularly serious at times in the Isles of Scilly but now frequently causes problems on some cultivars in many parts of southern England. More or less circular brown to grey-brown leaf spots form with distinct purple margins and bear dark brown dot-like pycnidia. There may be serious dying back of the shoots and plants gradually become unthrifty. There are no established control measures but spraying at two- or three-week intervals with benomyl or thiram are likely to be effective.

HELLEBORUS LEAF SPOT   *Coniothyrium hellebori*   Pl. 16

Very common and widespread on most *Helleborus* spp. but usually less damaging on tougher leaved species, such as *H. corsicus*.

Symptoms. Large, often irregular but sometimes round or elliptical dark brown and/or black, gradually coalescing lesions develop on both sides of the leaves and are most frequent on soft, immature foliage. The blotches appear scorched with concentric rings and may have minute black dots clustered towards the centre on the undersides. The leaves may turn prematurely yellow and die, when the symptoms spread to the rest of the plant, which suffers from the shortage of foliage. Lesions may also develop on the stem which in turn shrivels; the immature flower buds also shrink and wilt and the entire plant may topple over on its weakened stem. In less severe cases, the flowers may be spoiled by black spots on the petals.

Biology. Ascospores, released from the dot-like perithecia, spread the disease but require constant moisture for the initial infection. Under appropriately damp conditions in autumn or spring an entire bed can be affected in a few days and no initial damage is needed for the infection to occur.

Treatment. Remove and destroy all infected foliage; spray at monthly intervals with Bordeaux mixture or captan, beginning in October, and continue until the new growth ceases, except for the period when the flower buds are expanding. Avoid excess moisture on Christmas roses lifted in autumn for forcing.

HONEYSUCKLE LEAF SPOTS

A number of common and widespread leaf spots occur on honeysuckles. Large brown spots of varying shape, which gradually dry in the centre to leave a dark border, are very frequent and are caused by *Ascochyta vulgaris*. Black pycnidia may be present on these spots. Smaller rounded spots, also with pycnidia, are probably caused by *Colletotrichella periclymeni*. Both can look serious but damage seems slight and treatment unnecessary.

HORSE CHESTNUT LEAF BLOTCH
Guignardia aesculi **Pl. 16**
Common throughout northern Europe; increasingly so in Britain on horse chestnut and other *Aesculus* spp. Irregular brown blotches with yellow halos appear on leaves from July onwards, although similar symptoms on leaf margins can arise from a variety of causes, including normal autumnal necrosis. Conidia are produced on the blotches and ascospores, which give rise to the first spring infections, originate on fallen leaves. There may be some premature leaf fall but usually there is little harm to the tree and no treatment is necessary.

HORSERADISH PALE LEAF SPOT *Ramularia armoraciae*
This is the commonest of a few species of fungi causing spots on horse-radish leaves throughout Europe (see also p. 397). Numerous very pale, often almost white spots form with a more or less brown margin. The centres commonly drop out leaving a 'shot hole' but there is no evidence that the quality of the roots is impaired and no treatment is necessary. A very similar spotting is caused by the closely related fungus *Ramularia rhei* on rhubarb leaves but likewise it causes no damage.

HYDRANGEA LEAF SPOTS
Leaf spots of hydrangea are common, widespread and varied. Their causes are largely unknown but several fungi and/or bacteria may be involved, although damage seems slight and control is not necessary. Note however that abundant spotting on hydrangea stems is normal and not pathological.

ILEX LEAF SPOTS *Coniothyrium ilicis* and *Phyllosticta aquifolina*
Greyish spots with darker brown borders are caused by *P. aquifolina*; white spots are caused by *C. ilicis*. Both are occasional throughout northern Europe but neither is very common in Britain, although the latter in particular is potentially damaging. There are no established control measures but note also that holly leaf miner (p. 223) produces similar symptoms.

IMPATIENS LEAF SPOT *Ascochyta impatientis*
Occasional on cultivated and wild *Impatiens* spp. throughout northern Europe but apparently occurs only on wild species in Britain. The symptoms are large pale brown blotches with black or brown pycnidia and paler margins. There are no established control measures and care is needed with fungicides as the foliage is very susceptible to phytotoxic damage.

IRIS LEAF SPOT *Mycosphaerella macrospora* **Pl. 17**
This disease has been known in England since the end of the last century and is common throughout northern Europe but varies considerably in its seriousness from season to season. It is worst in wet years and on wet sites and lime deficiency apparently aids disease development in some irises. It also affects freesias, hemerocallis and gladioli, among other iridaceous plants.

Symptoms. Tiny brown spots with yellowish margins develop on the leaves of rhizomatous irises in spring but are often not noticed until later. The spots enlarge to become elliptical with finally a grey mould in the centre and deep red-brown margins. On bulbous irises the symptoms arise earlier in the season, the spots lack borders and are ashen in colour. Parts of or entire leaves may die, occasionally early in the season, but more usually after flowering, with the result that the plants become gradually weakened. **Biology.** Conidia produced on the spots spread the disease rapidly in the summer by rain splash or physical contact with other leaves. The fungus overwinters as mycelium or (in England) possibly as conidia and in Germany at least, perithecia are known to develop within the leaf tissues and survive the winter to produce ascospores in spring. **Treatment.** There is some variation in cultivar susceptibility while among *Iris* species, *Iris sibirica* and *I. versicolor* are said to be highly resistant. Destroy all dead foliage in autumn and apply lime to the soil, although this must be done in moderation or the plants may become prone to attack by soft-rot organisms. Bordeaux mixture or sulphur sprays have been reasonably successful in the past; zineb or maneb more recently but the addition of a wetting agent is essential for fungicide adherence to the waxy foliage. Begin the sprays with the first spring leaf growth and continue at about monthly intervals until the onset of cold weather in the autumn.

IVY LEAF SPOTS

Ivy is very prone to a number of leaf spots; the commonest and most widespread being large and pale grey areas, sometimes transluscent and with brown margins and caused either by *Mycosphaerella hedericola* or by *Phoma hedericola*. Little treatment is possible for large plants but clipping, or spraying with benomyl might be worthwhile in severe cases. Also see p. 470.

LABURNUM LEAF SPOTS

Leaf spots are common and widespread on laburnums; usually they are large and grey-brown with a very narrow brown border. These are probably caused by *Phyllosticta cytisi*. Young plants are often damaged by small but enlarging brown blotches causing premature leaf fall and occasionally death. The cause of these is *Pleiochaeta setosa* and regular spraying with Bordeaux mixture is said to be a satisfactory control.

LAVATERA LEAF SPOT   *Colletotrichum malvarum*

This disease is devastating on hollyhocks in North America. It has long been known in several parts of northern Europe and was found in Britain around the turn of the century. Small yellow-brown spots appear on the upper surfaces of the leaves and on the leaf stalks and stems. They often appear first at the stem base but can affect all parts up to and including the flowers, although the spots on the stems are generally paler than on the leaves. They enlarge rapidly and merge, become oval, and 5 cm (2″) or more in length. The stems are sometimes girdled by large lesions on which dark brown stiff spore-bearing bristles appear. Spray fortnightly with Bordeaux mixture to check established attacks and do not grow *Lavatera* spp. on the same land for as long as possible. Although the disease may be seed transmitted, it is difficult to eradicate from this source. Also see hollyhock rust (p. 278).

LAVENDER LEAF SPOT   *Septoria lavandulae*
Small brown circular spots are very common and widespread but often overlooked. They are particularly frequent on plants affected by shab disease (p. 318) but no treatment is necessary.

LETTUCE LEAF SPOT   *Pleospora herbarum* and *Septoria lactucae*
These are the commonest of lettuce leaf spots, apart from ringspot (below) and are frequent through northern Europe, less so in Britain. *P. herbarum* causes circular, pale brown, more or less rounded leaf spots which enlarge, coalesce and form a dry necrotic patch. The centres may fall out giving a 'shot-hole' and although rarely serious enough to warrant treatment, ringspot controls should be effective. *S. lactucae* causes angular pale brown leaf or stem spots bearing small black pycnidia. This fungus is probably seed-borne and may be eliminated by hot water treatment at 48°C (118°F) for 30 minutes, although there may be some suppression of germination. Treat established infections as for ringspot. Also see: rust (p. 280).

LETTUCE MARGINAL LEAF SPOT   *Pseudomonas marginalis*
Shrivelling of lettuce leaf edges is a common and widespread problem in gardens. Infection by the soil-borne bacterium *P. marginalis* splashed onto the leaves is one cause. The infection occurs through stomata, either in wet seasons out of doors or on carelessly watered glasshouse plants, and narrow black lesions develop at the leaf margins. The entire plants may become dull green and collapse. Any measures to reduce soil moisture and avoid wetting the leaves will help to suppress this disease. The symptoms are often confused with those of grey mould (p. 383) and certain mineral deficiency effects, including tip burn (p. 477).

LETTUCE RINGSPOT   *Marssonina panattoniana*
This disease only arises occasionally in gardens but can be very damaging when attack does occur. It is most frequent in cold wet weather on winter lettuce outdoors or under cold glass but also attacks chicory and endive. Sunken areas develop on the stems and these may easily be confused with slug damage (p. 139). Small brown-yellow spots form on the undersides of the leaves and their centres gradually dry and fall out, giving the leaves an irregularly perforated appearance. In severe attacks the entire leaves may become ragged and rotten and the disease spreads to the hearts while the entire plant takes on a yellow and stunted appearance. Conidia produced on pinkish mould on the spots under moist conditions spread the fungus to other plants and it may persist in plant debris in the soil and probably also on seeds. Plant lettuces on well-drained land, do not overwinter plants in the open, destroy plant debris and use as long a rotation as possible between lettuce crops. Fortnightly sprays with thiram or captan up to two weeks before cutting may be effective but care is needed to cover the undersides of the leaves with fungicide.

LILAC LEAF SPOTS

Large, variously shaped grey-brown blotches caused by *Heterosporium syringae* are common and widespread. As the lesions dry, a blackish conidia-bearing mould growth develops on them. Damage is normally slight and control rarely necessary but a fungicide other than Bordeaux mixture should be used as this often damages lilacs. Another leaf spot fungus common in Britain is *Phyllosticta syringae* which produces pale areas with a brown border and bearing small black pycnidia. Elsewhere in northern Europe, spots are caused by several other fungi but the damage is not serious. Also see: lilac blight (p. 318).

LILY LEAF SPOT   *Cercosporella inconspicua*

This disease is unknown in Britain but quite common in other parts of northern Europe on several *Lilium* spp. It appears as rounded chestnut-brown spots with black margins, gradually becoming covered with a grey mould on which conidia are produced. In damp conditions defoliation may result but treatment is difficult and there are no established control measures.

LIME LEAF SPOTS

Brown, dark-bordered leaf spots are common on lime trees in Britain from early summer onwards and occasionally spread to the leaf stalks and shoots and rarely cause leaf death or defoliation as the damaged stalks break. The cause is *Apiognomonia errabunda*. If the affected trees are small, collect the fallen leaves on which the ascospore-forming perithecia develop to lessen the likelihood of a repeat attack. Tiny brown spots with paler centres are very common in other parts of northern Europe but occur also from time to time in Britain. The cause is *Cercospora exitiosa* and, as with *A. errabunda*, defoliation can result from an attack. Other spots also occur occasionally on lime leaves but are not serious enough to warrant treatment.

LUPIN LEAF SPOT   *Pleiochaeta setosa*

Occasional and widespread on ornamental *Lupinus* species but sometimes important in parts of northern Europe on commercial seed crops. Occurs also on *Cytisus* in some areas, although not in Britain. Small purplish-black spots initially, develop on both surfaces of the basal leaves but rapidly spread upwards. They increase in size and turn grey-brown, more or less circular and with the outer edge merging into a light yellow-green zone. The centre is very dark and outside it light and dark green rings alternate in roughly concentric circles. The leaves shrivel and die and the stems become bared and further growth is spindly. Spray fortnightly with Bordeaux mixture plus wetting agent.

MAGNOLIA LEAF SPOT   *Phyllosticta magnoliae*

Common and widespread causing irregular pale spots on *M. grandiflora* but not damaging and treatment is not necessary.

MARGUERITE LEAF SPOT *Ramularia bellunensis*

Rare in Britain, common in other parts of northern Europe but not a problem on outdoor plants. Brown-grey, gradually darkening spots form at the leaf margins and may enlarge and merge to affect entire leaves. The disease is only serious under damp shady conditions, such as occur in winter, and the best treatment is to improve the lighting and ventilation in glasshouses.

MEDLAR LEAF BLOTCH *Sclerotinia mespili*

Widespread; the most common disease of medlars. Very conspicuous, large, dark brown, sweet-smelling blotches develop and may almost cover the leaves. The fungus can spread from the leaf blotches into the shoot and sometimes cause a serious wilt. Conidia develop on mould-like growth on the blotches and are dispersed by insects attracted by the characteristic aroma. Infection occurs when insects deposit these conidia on flowers which then fail to set fruit but become 'mummies' packed with mycelium. Ascospore-bearing apothecia develop on the fallen 'mummies' in parts of Europe but this is rare in Britain where the fungus overwinters by forming sclerotia on the leaves. Collect and burn affected leaves, preferably early in the season before the mould and conidia form, and spray diseased trees three or four times at fortnightly intervals with thiophanate-methyl.

MULBERRY BACTERIAL LEAF SPOT *Pseudomonas mori*

Widespread. Numerous small angular black leaf spots form, often with a yellow halo, and entire leaves may turn yellow and the vigour of the tree be reduced after severe attacks. Elongated ragged lesions, within which the bacteria overwinter, may arise on shoots and as a result dieback can ensue. Such affected twigs should be cut out and burned in the autumn and the remainder of the tree sprayed in spring with Bordeaux mixture to protect the new foliage. Repeat the spraying two weeks later and at monthly intervals throughout the summer but do not use fruit within a fortnight of spraying.

MULBERRY LEAF SPOT *Phloeosporella maculans*

Very common and widespread. Small, dark brown, irregular leaf spots with pale green–yellow halos appear in the spring. The spots later increase in size, the centres become whitish and the halos become brown necrotic margins. Small dark brown–black pycnidia arise on the lesions and defoliation can occur after severe attacks but this is rare in Europe. Spray with Bordeaux mixture as the leaves emerge in spring, repeat two weeks later and again at monthly intervals throughout the summer but do not use the fruit within a fortnight of spraying.

MYRTUS LEAF SPOTS *Pestalotia decolorata* and
Pseudocercospora myrticola

Very tiny but rapidly enlarging and sometimes coalescing red–purple spots with gradually lightening and ultimately creamy centres are caused by *P. decolorata*. Withering of the leaves can ensue. The disease is always common and widespread, and while the leaves commonly wither, occasionally attacks may be devastating with wholesale defoliation. Spots with pale yellow edges and dark reddish centres and often with grey mouldy conidia-producing growth on the undersides are caused by the less serious pathogen *P. myrticola*. The two may occur together and can be treated by collecting and destroying affected leaves and spraying the plants fortnightly with Bordeaux mixture. Adequate ventilation is essential in glasshouses.

NARCISSUS LEAF SCORCH *Stagonospora curtisii* Pl. 17

This disease is common throughout northern Europe but was not recognised as serious until the later 1920's. It is commonest on narcissi but also affects *Amaryllis* spp., including *A. belladonna*, as well as sternbergias, crinums and galanthus.

Symptoms. The tips of young shoots take on a red-brown, typically scorched appearance soon after they emerge, to be followed, under damp conditions, by a gradual spreading of this symptom down the leaves and the appearance of brown spots. The leaves turn yellow, shrivel and die and similar symptoms develop on the spathes and leaf stalks while disfiguring brown blotches can occur on the flowers. Masses of tiny black dots appear on damaged areas.

Biology. Conidia produced in the dot-like pycnidia are dispersed by rain splash and bring about rapid spread of the disease onto healthy tissues and neighbouring plants. The fungus probably overwinters between the papery and fleshy leaf scales of the dormant bulb, from whence it infects new leaves as they emerge.

Treatment. The poeticus, polyanthus and poetaz groups of narcissi are most susceptible and should not be grown where the disease is prevalent. In gardens, prompt removal of scorched leaf tips will restrict the disease and, if this is not possible, a fortnightly spray with Bordeaux mixture or zineb may limit further spread but will not eradicate the primary infection. Bordeaux mixture may cause unsightly deposits if applied during flowering. Soaking the bulbs for two hours in 0.05% formalin before planting is also effective although care should be taken as any solution stronger than this is likely to be damaging if not lethal to the bulbs. Storage of the bulbs at low temperatures, especially if combined with late planting, sometimes increases the likelihood of damage.

NYMPHAEA LEAF SPOT *Ramularia nymphaearum*

Occasional and widespread. Spots on both leaf surfaces are at first rounded with concentric zoning but later enlarge irregularly and change from reddish to grey-brown or black and finally rupture in the centre. The mycelium is superficial and forms compact white spots on the lesions. Remove affected leaves but never apply fungicides to these or other aquatic plants as serious water contamination may result.

OAK LEAF SPOT *Sclerotinia candolleana*

This is the only leaf spot disease of any consequence on oak and is common and widespread. Yellowish leaf spots appear in mid-summer and are at first small and rounded but enlarge to become brownish blotches, the centres of which may later drop out. It is probably potentially damaging but is rarely so recorded.

ONION SMUDGE *Colletotrichum circinans*
Widespread and formerly common on onions, shallots and leeks but now infrequent. Masses of small dark bodies arise beneath the outer bulb scales and are sometimes scattered but usually concentrated in concentric rings in smudge-like spots. In moist conditions in garden or store, spores form on the black bodies and in severe attacks the fungus grows from the spots into living tissues which are affected by a yellow stain. Protect harvested onions from the damp and if persistent and severe attacks develop on particular sites, grow only coloured cultivars which are resistant.

ORCHID LEAF SPOTS
Spots are very common and widespread on many orchid genera and arise from a wide variety of causes. Commonest are small pale straw-coloured spots, of irregular shapes on the upper leaf surfaces. These rapidly enlarge, become sunken, brown-purple and show through to the undersides. The diseased areas shrivel and die but characteristically do not fall out. It is believed that local chilling is the cause but several other types of spot, usually brown and jointly known as 'Orchid spot' are largely of unknown origin, although *Colletotrichum* spp. commonly occur on orchid leaves and may be at least partially responsible. Avoid draughts, dampness and, particularly, the exposure of leaves to dripping water. No other treatments are really effective.

PAEONY BLOTCH *Septoria paeoniae*
Common and widely distributed. Grey-brown spots with reddish margins and bearing black pycnidia develop on the leaves or stems. This disease is not usually serious and sprays with thiram or zineb are effective in severe cases.

PALM FALSE SMUT *Graphiola phoenicis*
Fairly common and probably widespread on date palm and other palms in glasshouses. Small yellow spore masses appear on dark warty scabs on the leaves, large areas of which may eventually become discoloured and die. Treatment is not easy but remove and destroy affected leaves, improve ventilation and reduce humidity, if possible. Fortnightly sprays with Bordeaux mixture may be effective.

PARSLEY LEAF SPOT *Septoria petroseleni*
Very common throughout Europe. Small angular brown leaf spots arise bearing tiny black dot-like pycnidia within which conidia are produced. The causal fungus is closely related to that causing leaf spot of celery (p. 399) and the biology of the two diseases is similar. The treatment is as for celery leaf spot but is probably never worthwhile in gardens.

PARSNIP LEAF SPOT *Ramularia pastinacae*
Pale leaf spots with brownish borders are common and widespread on parsnip and more than one fungus may be responsible. The leaf spots are not usually serious but if a chemical spray seems called for, then Bordeaux mixture may be worth trying. The spots may be confused with certain symptoms of parsnip canker (p. 356).

PARTHENOCISSUS LEAF SPOT *Elsinoë ampelina*

Greyish leaf spots with a dark border are common and widespread in Britain and northern Europe on parthenocissus and other garden vines. The spots may fall out leaving a 'shot-hole' but the effects generally seem less serious in Britain than elsewhere. The cause may be *E. ampelina*, an organism that causes a serious disease of grapevines in some areas (p. 405), but the fungi causing leaf spots on vines are very difficult to identify. Treat by spraying at three weekly intervals with benomyl or captan during the growing season and the leafless shoots once with Bordeaux mixture in the winter.

PEA LEAF and POD SPOT *Ascochyta pisi* and other
fungi **Pl. 17**

Common throughout Europe and although known since the mid-nineteenth century only recently have all the causal fungi been identified.
Symptoms. The commonest symptoms are brown-yellow, often somewhat sunken spots which bear dark brown pycnidia and have darker perimeters. These arise on the leaves, flower stalks and pods. Less frequent are purple-brown spots sometimes with concentric markings on the leaves, pods and, in more elongate form, on the stems. The stem spots may also occur on very young seedlings which die as a result of the attack.
Biology. The commonest symptoms are caused by *A. pisi*, the less frequent ones by *Phoma medicaginis* var. *pinodella* or *Mycosphaerella pinodes*; all are more severe in wet seasons. Diseased pods result in infected seeds which are the primary origin of the disease, although some carry-over of conidia in crop debris also occurs.
Treatment. Reject any seeds showing the sunken brown-purple lesions, burn plant debris after severe attacks and do not grow peas on the same land for as long as possible.

PEACH SHOT HOLE *Stigmina carpophila* and other causes

'Shot-hole' effects are common on the leaves of peaches and to a lesser extent on cherries, nectarines and almonds. The causes are often obscure and may be related to unsuitable soil conditions, such as waterlogging, although sometimes in Britain and probably more frequently elsewhere in Europe they are due to infection by *S. carpophila*. The symptoms are often confused with those of scab (p. 337) as the disease can also infect the twigs and fruit, on which scab-like lesions form. If there is no other obvious environmental cause for shot-holes, apply spray treatments as for apple scab (p. 335). Also see: plum and cherry bacterial canker (p. 320).

PEAR LEAF FLECK *Mycosphaerella pyri*

Occasional and widespread. Masses of small spots form each with a dark margin and pale grey centre, and may cover the leaves of pears, particularly where routine fungicide sprays for scab control (p. 337) have not been made. Conidia are produced on the spots and overwintering ascospores develop on fallen leaves. The disease is rarely severe in Britain and sprays used against scab should eliminate it, but remove fallen fruit and leaves as a precaution. The spots of pear fleck are larger and of a different colour from those of quince leaf blight (p. 417) which can also affect this host plant.

PELARGONIUM LEAF SPOTS
Leaf spots are very common and widespread on pelargoniums. There may be several causes but *Xanthomonas pelargonii* is known to cause such spots in parts of northern Europe and probably in Britain also. Small, brown, more or less rounded spots, which may merge to form larger lesions, develop on the leaves. Blackening of the stems may also be associated with leaf spots (but see p. 385) and the problem is most serious in overcrowded glasshouses. It is important therefore to avoid warm, damp conditions and over-watering. Affected leaves should be destroyed, and spraying fortnightly with Bordeaux mixture from the first appearance of symptoms may be effective. Brown water-soaked areas on the leaves and blossoms, bearing grey mould growth, are commonly caused by *Botrytis cinerea*, which may also be associated with larger but similar lesions and rotting on the stems. Spray with dichlofluanid or benomyl at the time of appearance of the first symptoms and improve ventilation.

PHLOX LEAF SPOTS   *Septoria* spp.
These are very common and widespread and although the identity of all the causal fungi is uncertain, *Septoria drummondii* is almost certainly one cause on *Phlox drummondii*. More or less circular, often merging spots form with pale centres and dark brownish-red rims and often result in premature death of the leaves. Black pycnidia or more rarely perithecia may occur on the lesions. Spray with benomyl immediately before pricking out and as soon as the first spots are seen.

PLANE ANTHRACNOSE   *Apiognomonia errabunda*
Common; this is the most serious disease wherever planes are grown and affects several species of *Platanus*.
Symptoms. Some young leaves and shoots become brown and die early in the season. Later, angular brown patches, clearly defined by the veins, appear on the other leaves and eventually result in their death while very small, creamy pinhead-like structures develop on the undersides of affected parts. Small twigs can be killed and cankers form on larger ones but although trees may be weakened and disfigured they are rarely killed.
Biology. Young leaves are infected from two sources: ascospores produced in perithecia on fallen leaves and conidia from pycnidia on the bark. The pinhead structures on affected leaves are also conidia-bearing and bring about rapid disease spread during the summer. Damp spring weather favours initial infection and the same fungus causes similar blotches on lime and oak leaves but apparently these are not serious.
Treatment. Collect and burn fallen leaves, if practicable; prune off affected twigs and branches and spray small trees at three-weekly intervals from early in the season with Bordeaux mixture.

PLUM LEAF BLOTCH   *Polystigma rubrum*
The causal fungus of this disease probably never infects plums in Britain although it does so elsewhere in northern Europe and can cause defoliation. It is common in Britain on blackthorn and bullace and there is a possibility that it could spread from there to plums. Attacks are readily recognisable by the orange or red leaf blotches. After autumnal leaf fall and over-wintering,

ascospores are produced on these blotches and initiate new infections in spring. Collect and burn affected leaves in autumn. Spray diseased plums twice with benomyl at fortnightly intervals starting at the first sign of symptoms but the spraying of blackthorn in Britain is probably unnecessary.

POPLAR LEAF SPOTS
Many fungi cause leaf spots on poplars; several are rarely if ever found in Britain but are serious or potentially so elsewhere in northern Europe, particularly in the low countries where the host tree is more important. Irregular brown blotches, especially on Lombardy poplar, are often associated with dieback disease (p. 322). Small, black or brown spots which sometimes merge and often have paler centres are said to be more common on the lower leaves, although this may be because the tops of poplars are rarely examined. The overall damage is slight and the causes are *Septoria* spp. Small brown leaf spots in spring are common in Holland and other parts of northern Europe but unknown in Britain. They enlarge and become whitened as conidial masses of the causal fungus *Septotinia populiperda* develop on both surfaces. This disease is always associated with leaf damage from other causes, especially insects, but although it can be very common on young plants, no control seems necessary.

POTATO SILVER SCURF   *Helminthosporium solani*   **Pl. 17**
Common and widespread but merely a disfiguring skin blemish and only likely to be of consequence in gardens if the tubers are required for show purposes. The disease originates from fungus on the seed tubers and symptoms appear as brown or greyish patches, not usually obvious at lifting but extending in storage under humid conditions and developing a silvery sheen as air penetrates between the cell layers. A sooty appearance may develop as dark conidia are produced but the symptoms are entirely confined to the tubers, unlike black dot (p. 367) with which this disease is sometimes confused. Treatment is not worthwhile in gardens.

POTATO SKIN SPOT   *Polyscytalum pustulans*   **Pl. 17**
Common and widespread but usually only a disfiguring condition although it is one of the major causes of sprouting failure on seed stocks. It also causes wastage of tubers through the extra peeling necessary when cooking. The disease is worst in years when lifting conditions are wet and subsequent storage is cold.
Symptoms. Small brown-black pimples arise on the tubers or occasionally more penetrating brown lesions form, often in small groups in and around the eyes but sometimes virtually covering the entire organ. These tuber symptoms do not usually become evident until December–January. There may also be brownish discolouration of the surfaces of roots and stolons, sometimes with faint white mould growth on the upper part of the main root. Eyes and young sprouts may also be damaged and sprout tips may be killed although in itself this is not diagnostic.
Biology. Affected seed potatoes are the commonest sources of disease and superficial contamination develops on stem bases, stolons and roots, from where it spreads to new healthy tubers. Infection takes place through

lenticels, skin or small wounds on growing tubers or on mature tubers before lifting but develops slowly until after harvest, when cold damp conditions are very favourable for further development. The lesions form when the tubers produce corky tissue to prevent penetration of the fungus. *P. pustulans* can also persist in soil on plant debris but in Britain at least, this is probably unimportant. Sprout infection may result in total tuber failure and a gappy crop.

Treatment. No fully resistant cultivars are known although 'Home Guard' is seldom affected in Britain and could be chosen where the disease is troublesome. 'King Edward' is very prone to eye infection and consequent tuber loss. Examine seed potatoes carefully, reject any showing appreciable amounts of disease and then sprout them before planting to reveal any with damaged or killed sprouts.

PRIMULA LEAF SPOTS
Several types of spot are very common on primulas, polyanthus and more rarely on auriculas. In Britain they are most frequently caused by *Ramularia primulina* or *Phyllosticta primulicola* although other fungi may be involved. All are commonest in wet seasons but none is usually serious. The commonest symptoms are bright yellow-brown irregular spots which gradually fade but retain a yellow margin and eventually bear conidia-producing mould, especially on the undersides. Sometimes the centres may drop out to give a 'shot-hole'. Among other types, spots with dry centres and bearing black pycnidia are also frequent, as are chocolate brown lesions with no surrounding halo. In severe attacks, spray with benomyl or zineb at three-weekly intervals.

PRIVET LEAF SPOTS   *Mycosphaerella ligustri*
This is by far the commonest and most widespread of privet leaf spots. Brown spots form and are often circular but may be irregular and gradually lighten in the centres to leave a darker border. They are rarely damaging and no treatment is necessary.

PRUNUS LEAF SPOTS
Among several types of leaf spot on ornamental cherries, the most serious is probably that caused by *Blumeriella jaapii* which is common in parts of northern Europe but not found in Britain. It is most severe in wet seasons and can result in extensive defoliation. Faint yellow-brown spots with dark margins are however common in Britain on cherry laurel. The lesions darken as brown conidial pustules develop and finally they fall out leaving a 'shot-hole'. The cause may be either *Trochila laurocerasi* or *Stigmina carpophila* and summer pruning has been blamed for aiding infection by providing wounds on young leaves. It is sensible therefore to confine clipping to the autumn.

QUINCE LEAF BLIGHT   *Diplocarpon maculatum*
The most important and widespread disease of quinces; commonest in wet seasons.

Symptoms. Irregular spots develop on both surfaces of the leaves; these are at first red but gradually darken to become almost black. Minute dot-like pustules appear on the spots and the leaves may fall prematurely after severe attacks. Spots are also sometimes found on the tips of the shoots and less frequently on the fruit while pears are also sometimes attacked, especially when in the vicinity of diseased quince trees.

Biology. In Britain the early infections in spring probably arise from conidia produced in pustules on dead twigs and spread to the leaves by rain splash. Once infection is established, more conidia form in pustules on the leaf spots and spread the disease to other plants.

Treatment. Cut off and burn dead and diseased twigs (and if possible, leaves also) during the winter. Spray with thiophanate-methyl at the first sign of symptoms and again two weeks later to limit severe attacks.

QUINCE LEAF BLOTCH   *Monilinia linhartiana*

Common and identical in virtually all respects to medlar leaf blotch (p. 411) which is caused by a very similar fungus, although infection cannot be transferred from one species to the other. Treat as for the medlar disease.

RASPBERRY CANE and LEAF SPOT
Elsinoë veneta   **Pl. 17**

A common and widespread disease on raspberries but also affects blackberries although the similar but less serious blackberry cane spot or purple blotch diseases are more common on this plant. It is often a problem on loganberries.

Symptoms. Purple spots which later develop grey-white centres arise on the stems, leaves, flower stalks and, especially with loganberries, on the fruits which commonly develop one-sided. Leaves may drop when severely infected. On the canes, the lesions become elongated and the bark may split to form small cankers. Spots produced later in the summer and autumn are smaller and bear black dot-like bodies. The fruit yield is reduced and in severe attacks the canes may be distorted or killed.

Biology. Conidia from black fungal fruiting bodies infect young canes in May or June and more conidia form on the cane lesions to spread the disease to the leaves and fruit. The lesions on the fruiting canes develop from infections in the previous year.

Treatment. Do not grow susceptible cultivars such as 'Norfolk Giant' and 'Lloyd George' where this disease is a problem. Cut out and burn affected canes. Spray with benomyl, thiram or dichlofluanid at the time of bud burst and repeat fortnightly until the end of flowering.

RHODODENDRON LEAF SPOTS   **Pl. 17**

Dark red-brown leaf blotches bearing black pycnidia are increasingly common, widespread and most serious on azaleas and often result in defoliation. The cause is *Septoria azaleae* and the disease is particularly severe on plants newly imported for forcing in glasshouses. The best preventative treatment is by spraying at fortnightly intervals from mid-summer onwards with Bordeaux mixture, although some azaleas are susceptible to damage and trial sprays should be made first on a few plants. Among other common leaf spots are rusty brown lesions caused by *Phyllosticta rhododendri*; large irregular brown blotches on several *Rhododendron* spp. caused by *Gloeosporium rhododendri* and frequent on the continent but unknown in Britain are the nearly black lesions produced by *Pestalotia rhododendri*. The closely related *Pestalotiopsis guepini*, the cause of camellia leaf blotch (p. 398), occurs in Britain on rhododendrons and is probably of increasing significance. However, most of these minor spots and blotches need not normally cause concern although the destruction of affected leaves and a spray with benomyl or thiram in the spring following an attack may be advisable.

ROSE BLACK SPOT   *Diplocarpon rosae*   Pl. 17
Extremely common and widespread on most rose cultivars as well as on wild roses.

Symptoms. Dark brown or black, occasionally coalescing blotches with irregular margins appear from late spring onwards on both leaf surfaces. The leaves frequently turn yellow and drop prematurely; repeated and severe attacks resulting in much weakened shoots and bushes.

Biology. Severe attacks can result in potentially dormant buds giving rise to feeble shoots in autumn which are then killed by frosts and so render the plant weak when growth should restart in spring. Conidia produced on the spots are released under moist conditions and are spread by rain splash or on hands, clothing and tools. Warm moist conditions favour further infection and the disease is worst in warm, wet seasons. In Europe most initial infections in spring are from conidia produced on over-wintering spots that develop on shoots in the autumn, although some may originate from fallen leaves. Recently the ascospore producing stage of the fungus has been found for the first time in Europe on over-wintering leaves. This development is potentially serious as it gives scope for new races of the fungus to arise and so infect cultivars that have some resistance at present.

Treatment. Some cultivars are partially resistant but none is totally so. Susceptibility to powdery mildew and rust are also factors to be borne in mind in choosing black spot resistant cultivars while there may be marked variations between sites in terms of disease susceptibility. Generalisations can be misleading but certainly many of the older cultivars and proportionately more yellow than red or pink cultivars are susceptible to black spot. The following lists are not by any means exhaustive or to be relied upon in all areas but give a general indication of some of the most commonly grown cultivars:

Resistant large-flowered bush roses (Hybrid Teas): 'Alec's Red', 'Alexander', 'Blessings', 'Champs Elysées', 'Charlie's Aunt', 'Chicago Peace', 'Colour Wonder', 'Gail Borden', 'Honey Favourite', 'Jiminy Cricket', 'King's Ransom', 'Mme. Louis Laperrière', 'Michèle Meilland', 'My Choice', 'National Trust', 'Peace', 'Perfecta', 'Picture', 'Pink Favourite', 'Pink Peace', 'Prima Ballerina', 'Rose Gaujard', 'Shot Silk', 'Super Star', 'Sutter's Gold', 'Troika', 'Uncle Walter', 'Yellow Pages'.

Susceptible large-flowered bush roses (Hybrid Teas): 'Ballet', 'Belle Blonde', 'Blue Moon', 'Femina', 'Fragrant Cloud', 'Gay Gordons', 'Harry Wheatcroft', 'Lydia', 'McGredy's Yellow', 'Memoriam', 'Mrs. Sam McGredy', 'Piccadilly', 'Wendy Cussons'.

Resistant cluster-flowered bush roses (Floribundas): 'Allgold', 'Arthur Bell', 'City of Belfast', 'City of Leeds', 'Dickson's Flame', 'Escapade',

'Golden Slippers', 'Jan Spek', 'Korresia', 'Manx Queen', 'Marlena', 'Michelle', 'Molly McGredy', 'Moon Maiden', 'Orange Silk', 'Queen Elizabeth', 'Sea Pearl', 'Southampton'.

Susceptible cluster-flowered bush roses (Floribundas): 'Anna Wheatcroft', 'Charleston', 'Copper Delight', 'Elizabeth of Glamis', 'Evelyn Fison', 'Iceberg', 'Korona', 'Orangeade', 'Orange Sensation', 'Pineapple Poll', 'Sarabande', 'Shepherd's Delight', 'Sir Lancelot', 'Tip Top', 'Zambra'.

If some susceptible cultivars must be grown then mixed planting with more resistant types sometimes decreases disease incidence while there are several cultural practices which may help to lessen black spot. Removing the first infected leaves and shoots early in the season may delay disease spread, although the collection of fallen leaves in autumn is probably of little value because most over-winter survival is on the shoots. Severe spring pruning of these shoots does therefore significantly reduce the potential infection while the improvement of drainage, avoidance of shading and of excess nitrate in fertilisers may also be beneficial. There are several possible chemical spray treatments, of which sulphur is traditionally effective, and in industrial areas with a high atmospheric sulphur content the disease is rarely serious whereas the general significance of black spot has undoubtedly increased with the passing of anti-air pollution legislation. Coppercontaining materials should be avoided as they can cause disfiguring blotches. Captan has proved reliable in the past, if applied immediately after pruning and at least fortnightly through the season, but is ineffective against mildew and should therefore be applied as a proprietary mixture with another chemical. Systemic fungicides such as benomyl apparently vary in their effectiveness. Many gardeners have found the very similar material thiophanate-methyl more reliable but all normally need applying at least fortnightly throughout the season.

Several leaf spots in addition to black spot are also common and widespread but rarely serious. *Elsinoë rosarum* causes small rounded dark brownpurple spots with paler centres and occasionally a discolouration of the leaf. Similarly coloured scab-like spots occur on stems. *Sphaerulina rehmiana* causes somewhat similar circular grey-brown spots with black pycnidia and reddish margins. Both may be treated as for black spot.

SANSEVEIRA LEAF SPOT *Gibberella fujikuroi* (= *Fusarium moniliforme*)

Occasional and probably widespread. Round or oval, sunken red-brown spots with yellowish margins appear on either or both leaf surfaces. The spots may coalesce and bring about death or partial death of the leaves. Avoid wetting leaves when watering (an important precaution with all plants with succulent foliage) and destroy badly affected leaves. Apply a spray of zineb, captan or thiram at the time of the first appearance of symptoms.

SOOTY BLOTCH *Gloeodes pomigena*   Pl. 17

Common and widespread on apples, pears and plums, especially in cold wet seasons and on shaded trees. On apples and plums it is often associated with fly speck (p. 405). In Britain it is most prevalent in the South-West. The fruit bear dirty green-brown circular smoky blotches of fungal growth which can continue to spread in storage. In severe cases the fruit remain small and plums may shrivel after picking. Many cultivars are susceptible but it is particularly serious on pale coloured plums and gages. Routine scab control sprays (p. 335) should however prevent the disease from developing on apples. Also see: Sooty mould (p. 303).

SPINACH LEAF SPOT   *Cladosporium variabile*
Common and widespread but not often damaging. Usually small, well rounded, almost white spots with a narrow, gradually drying margin appear scattered over the leaves but may be more concentrated towards the apex. The spots may fuse and take on a dark mouldy appearance as conidia are formed on them. Remove such affected leaves as they appear.

STRAWBERRY LEAF BLOTCH   *Gnomonia fragariae* **Pl. 17**
First recorded in Britain in 1941 and sometimes proves a serious problem in gardens although there is dispute over the identity of the causal fungus.
Symptoms. Black sunken patches form on the leaf stalks and stolons and gradually enlarging brown and purple, yellow-bordered blotches arise on the leaves. A fruit rot may also be caused by the same fungus.
Biology. On the black areas on the leaf stalks and stolons, ascospores are formed but on the leaf spots and also on the fruit, only conidia develop. Abundant moisture and wet seasons favour the disease. This fungus has also been found on *Geum* × *borisii* and may possibly transfer from geums to strawberries.
Treatment. Do not plant strawberries in damp and shady places. The cultivars 'Cambridge Favourite' and 'Redgauntlet' are particularly prone to the fruit rotting condition and should be avoided where the disease is known to be prevalent. Remove and burn affected leaves, spray with zineb when growth begins in spring and again a fortnight later or alternatively, apply dichlofluanid three weeks before flowering and again a fortnight later.

STRAWBERRY LEAF SCORCH   *Diplocarpon earlianum*
Less of a problem than other strawberry leaf diseases but sometimes serious, especially in the north of Britain. Purplish spots form, similar to those of strawberry leaf spot (below) but unlike them, they do not develop a grey centre and eventually may turn dark brown. Sometimes the spots merge together to give larger scorched patches on the leaves, which usually turn yellow. Conidia develop in shiny black swellings on the spots. Burn off old foliage as described for leaf spot.

STRAWBERRY LEAF SPOT   *Mycosphaerella fragariae* **Pl. 17**
Very common and widespread but rarely serious although it often causes concern to gardeners.
Symptoms. Appear as distinct, usually more or less circular deep reddish-purple leaf spots with a quickly developing grey centre. This centre frequently drops out and leaves a shot-hole. Rarely, spots may also cover the leaves and occasionally occur on leaf stalks and flower stalks also. A white

mould and, later, tiny black dots may develop on the lesions. Rather similar symptoms, if associated with a hard rot of the fruit, may be caused by *Septogloeum fragariae* (p. 359).

Biology. Conidia develop on the mould and spread the disease during the summer whereas the infrequently found black dots are ascospore-producing perithecia. Small hard black resistant sclerotia are often formed and these survive over winter and germinate in the spring to produce the conidia responsible for the first infections. Probably there are several fungal strains or races, not all of which are able to infect all cultivars.

Treatment. Burn off dead foliage at end of the season; if straw has been used between the plants this makes an ideal tinder. There are no effective chemical control measures for use in gardens.

THUJA NEEDLE SCORCH   *Didymascella thujina*

Common throughout northern Europe and now the major disease of western red cedar and related species. It is particularly serious on young plants. Individual scattered needles on the shoots turn straw-coloured and die while less frequently there is total shoot death. On the needles small brown cushions, within which the ascospores are produced, gradually darken and then open in a flap-like manner. After spore discharge the entire fructification falls out leaving a characteristic pit. Avoid close planting of young susceptible species as this gives rise to high humidity which favours the disease. No chemical is readily available for garden use and the often recommended Bordeaux mixture spray is of little value.

TOMATO GHOST SPOT   *Botrytis cinerea*   **Pl. 17**

Common under conditions where very high moisture persists for short periods and therefore often seen in cool glasshouses towards the end of the growing season; it is frequently associated with other grey mould symptoms. Tiny pale rings with raised brown dots in the centres, form on the fruit. When dampness persists for long periods, grey mould fruit rot may set in. For treatment see the general notes on grey mould on p. 352 but note that benomyl and related chemicals are ineffective against ghost spot and only dichlofluanid should be used.

TOMATO LEAF SPOT   *Septoria lycopersici*

Occasional on outdoor tomatoes in warm, moist parts of northern Europe, especially in Germany, but rare in Britain. Circular or angular spots develop which are at first water-soaked then become brown or grey with dark margins and bear black pycnidia. They are mainly on the leaves but can occur on other green parts and rarely on the fruit also. The affected leaves turn yellow, shrivel and fall, so weakening the plants. Such defoliation also exposes the fruit to sunscald damage (p. 484). Conidia are spread from the leaves by rain splash or on hands, tools and clothing and the fungus overwinters mainly in debris and on solanaceous weeds but may also be seed-borne. Treat by rotations of at least four years between tomato crops and destroy plant debris. Fortnightly or weekly zineb sprays may check established attacks but do not pick fruit within a fortnight of spraying. Also see: tomato speck, discussed under delphinium black blotch (p. 403).

TROPAEOLUM LEAF SPOT *Acroconidiella tropaeoli*
First found in Britain in 1969, unknown elsewhere in northern Europe but serious in North America. It is commonest late in the season. Tiny red leaf spots enlarge to become irregular or rounded blotches with a surrounding yellowish area. When many such spots are present, the leaves may turn brown and wither. Green-brown mould growth develops on the lesions, especially on the undersides of the leaves in moist conditions. This disease is seed-borne and hot water treatment of the seed is used effectively to control it in North America. Otherwise, spray with captan, zineb or Bordeaux mixture to limit spread on growing plants.

VIBURNUM LEAF SPOTS
Various types of spot are common and widespread on most species of viburnum. Most frequent are rounded necrotic leaf spots with purplish margins, often associated with elongate lesions varying from pale brown to purple-black on the stems and branches. Occasionally the branches are girdled and dieback ensues. The tips of the shoots are also sometimes affected and turn black and curl downwards. Similar symptoms can occur on cuttings, which turn black from the base upwards and die. The commonest cause is *Phoma viburni* with other *Phoma* spp. of lesser importance but there are no established control measures.

VIOLA LEAF SPOTS
Many different types of spot form on violas and/or pansies and most are common and widespread. Brown, white or pale buff spots (the so-called white spot) with dark brown borders are frequent on *Viola odorata* and *V. canina*. In damp conditions tiny white conidial clusters develop and eventually the spots may drop out giving shot-hole effect. The cause is *Ramularia lactea*; other *Ramularia* spp. occur similarly outside Britain. There are no established control measures but a spray with benomyl or captan is worth trying.

WALNUT LEAF BLOTCH *Gnomonia leptostyla* **Pl. 17**
Widespread; the commonest disease of walnuts. Tiny dark brown-black spots form and enlarge to become dirty yellow blotches on the surface of leaves with a corresponding greyish colouration beneath. Dark brown or black, often somewhat sunken lesions can also occur on young green nuts (which are thus impaired), and occasionally on the shoots. Premature defoliation may result. On the blotches, conidia develop and are spread by rain splash to other leaves and ascospores are formed in perithecia on the blotches (primarily those on fallen leaves) and these bring about the first new infections in spring about a fortnight after discharge. This disease is not easy to control, the oft-recommended procedures of collecting up infected leaves or spraying with Bordeaux mixture or zineb being fairly satisfactory on young plants but of little value to a gardener with a 15 m high tree. In the latter cases there is probably nothing that can be done.
    Much smaller angular black spots on walnut leaves and leaf stalks are symptoms of bacterial leaf blight *(Xanthomonas juglandis)* a potentially devastating disease on nursery stock but rarely significant on full-grown trees in Britain. In other parts of northern Europe severe damage may

result from black spotting on fruits. It can be controlled on young plants with Bordeaux mixture sprays at fortnightly intervals between bud break and the end of June.

WATERCRESS LEAF SPOT *Septoria sisymbrii*

Occasional in Britain since 1967 and sometimes serious although it probably does not occur elsewhere in northern Europe. Pale yellowish circular spots form on both leaf surfaces; these gradually enlarge and merge and bear dark brown-black pycnidia from which, under moist conditions, conidia ooze in pale gelatinous tendrils. It differs from the somewhat similar spots caused by downy mildew in the presence of the pycnidia and absence of the mould growth that often occurs with mildew. Severely affected leaves become yellow, collapse and shrivel. Probably no spread of the disease occurs under water but cut down aerial leaves very frequently, remove uncultivated cress plants in the vicinity and destroy or bury diseased and waste plant debris. Never apply fungicides to cress beds unless under expert instruction.

WISTARIA LEAF SPOT *Phyllosticta wistariae*

Sometimes very common in Britain and elsewhere in northern Europe. Spotting has been attributed to *P. wistariae* although other fungi, such as *Septoria wistariae* may also be involved. The spots are highly disfiguring but damage seems slight and controls have never apparently been attempted although a spray with benomyl might be worth trying in very bad cases.

YUCCA LEAF SPOT *Coniothyrium concentricum*

Very common and widespread. Large greyish spots form with a brown-purple border and concentric rings of tiny black pycnidia appear on the lesions. The causal fungus is one form of a species which (probably together with other fungi) gives rise to similar lesions on other more or less succulent plants, including agaves and dracaenas. Remove badly affected leaves and spray plants with Bordeaux mixture.

ZANTEDESCHIA LEAF SPOT *Phyllosticta richardiae*

Common and probably widespread. Variously sized blotches arise on the leaves, flower stalks, spathes and particularly on the leaf stalks. The lesions often appear as small, depressed grey-black thumb marks which increase in size and ultimately encircle the stalk, causing leaf drop. The affected areas are then black with greyish margins and bear black pycnidia. The disease spreads rapidly and results in reduction in vigour and numbers of flowers. The likelihood of attacks can be minimised by not planting zantedeschias in excessively cool, damp situations but after an outbreak destroy the affected leaves and spray the remaining plants two or three times with Bordeaux mixture.

SLIME MOULDS
(Plate 18)

Slime moulds (Myxomycetes) are non-parasitic organisms, probably related to fungi. They sometimes cause concern when found on lawns and garden plants. They are characterised by a mobile protoplasmic mass, the plasmodium, which may be likened to a giant amoeba that wanders freely among vegetation feeding on living bacteria, fungal spores and particles of dead organic matter before settling down to produce a spore-bearing body, the sporocarp. Several species occur commonly in gardens and these may be sub-divided on the basis of their favoured habitats:

Lawns. The commonest species is *Physarum cinereum* with globose, stalkless, off-white, clustered sporocarps and off-white or yellow plasmodia. Another common species is *Badhamia foliicola* with globose, stalkless or short-stalked sporocarps in small clusters and yellow-orange plasmodia. On large lawns or on rough grass in orchards and shrubberies, especially on limey soil, *Mucilago crustacea* is fairly frequent; it has large, chalky, sponge-like sporocarps and creamy white to pale yellow plasmodia.

Surrounding or engulfing herbaceous plants in borders. Two species sometimes cause concern: *Brefeldia maxima*, with a purple-black sporocarp on a silvery plate-like structure up to 30 cm. (12″) in diameter and white plasmodia, and *Fuligo septica*, also with a large but cushion-like sporocarp of either white, yellowish, green, pink, red-brown or violet appearance and yellow or creamy plasmodia. Both usually occur in the vicinity of old tree stumps.

Cool glasshouses or frames. *Physarum gyrosum*, with coral-like masses of off-white to brown-red sporocarps and white or yellow plasmodia, may infest plants so heavily as to reduce their effective leaf area and *Fuligo septica* var. *candida* sometimes smothers young plants, especially cucumbers, with its white sporocarps.

Various habitats. *Didymium difforme*, with smooth, white, stalkless sporocarps and colourless or yellow plasmodia, is common on a wide variety of plants and seems particularly frequent on hyacinth bulbs grown in water culture.

All slime moulds are relatively harmless although some damage may result from the smothering of young plants, the blocking of stomata and the exclusion of light. No chemical control is necessary and washing them off with water is the easiest means of dispersal.

LICHENS
(Plate 18)

Very many species of lichen occur on all types of garden habitat, including lawns, but they are especially common on old walls, roofs and tree trunks. Although often attractive in such situations, they can be very unsightly when they grow on ornamental shrubs, as often happens on azaleas, heathers and rhododendrons, for example, especially on shaded sites and poor soil. They usually form either a grey-green crust-like or somewhat leafy covering to bark or scaly or woolly growths on twigs and branches. Lichens are biologically unique as they are dual organisms, comprising a fungus and an alga growing in close association. They are slow growing and this is why they are only likely to be seen on such plants as woody perennials and on lawns. Although enormous cloaks of lichens cover trees in the wetter, western parts of Britain, they are not generally believed to be harmful, although their development is probably greater on trees that are growing poorly, for various reasons. Lichens are very sensitive indicators of environmental pollution – generally speaking the more woolly or frondose the lichen species, the less polluted is the prevailing atmosphere. They can often be brushed off sturdy plants but tar oil winter washes (see p. 40) may also be effective and are probably necessary on the inaccessible parts of trees.

ALGAE
(Plate 18)

Many algal species, especially those of green algae, such as *Pleurococcus*, form green coatings on plants growing in damp or shaded sites. They are very common but harmless on the north-facing bark of tree trunks and, unlike lichens, can grow in very polluted air. Among other garden habitats where algae occur frequently are on the glass panes of glasshouses and frames, from which they should be cleaned off regularly to prevent light transmission being impaired. They can also cause blockages in glasshouse watering systems, contribute to the blocking of garden ponds, cause paths to become slippery and form small jelly-like masses on neglected lawns. Usually physical removal is the most satisfactory treatment but if serious problems are caused a proprietary algicide should be applied in accordance with manufacturer's directions.

HARMFUL FLOWERING PLANTS
(Plate 18)

Although only a few genera of flowering plants are definitely parasitic, several can be physically harmful in gardens. The latter are of much greater consequence than the former in Britain and northern Europe and in this category are clematis, honeysuckle and ivy. Parasitic or partially parasitic plants are represented by broomrapes, dodders, mistletoes and toothworts.

BROOMRAPES and TOOTHWORTS *Orobanche* spp. and
Lathraea spp.   **Pl. 18**
There are several species of these closely related plants which lack chlorophyll and are parasitic on the roots of wild and cultivated plant species. Many are widespread but only locally common in Britain and northern Europe. They produce erect, scaly, often brownish or purple shoots, bearing similarly coloured flowers. Toothworts are parasitic on trees and shrubs, especially elms, hazel, poplars and willows, and grow beneath them. Broomrapes occur on a wide variety of plants, including potatoes and tomatoes in some parts of Europe, but none are likely to cause much damage and are primarily of interest as curiosities.

CLEMATIS *Clematis vitalba*
Common, especially on alkaline soils in the southern parts of northern Europe. It is closely related to the large flowered, mainly Asiatic garden

species and scrambles over hedgerows, shrubs and sometimes quite large trees, forming a blanket of vegetation. The shoots are covered from July onwards with greenish white flowers and from late summer with large, conspicuous hairy silver seed heads, commonly known as 'old man's beard'. Despite its very attractive appearance and valuable cover for wild-life, clematis can smother young plants, largely by the exclusion of light from them, and some cutting back may be essential.

DODDERS   *Cuscuta* spp.   Pl. 18

There are two native species, *C. europaea* and *C. epithymum*. Only the latter is of importance in Britain although the former can also be troublesome in parts of Europe. They are widely distributed in southern Britain and Europe south of central Scandinavia and are infrequent in Scotland. Both species are parasitic; the seeds germinate on the ground but leafless, reddish shoots twine counter-clockwise around the host plants and soon become wholly aerial. The shoot derives nutrients from the plant by inserting feeding structures into the stem and bears clusters of pinkish-white flowers. They may form a completely matted blanket over shrubs and herbaceous plants. Dodders are commonest on gorse and heather but are sometimes found on other plants and can cause stunting, but only rarely death. The only treatment is the total destruction of the infested plant or branches, preferably before the seed is formed and shed.

HONEYSUCKLE   *Lonicera periclymenum*   Pl. 18

Common throughout Europe except the extreme North; the same species as garden honeysuckle. Grows on moist soil types and is often frequent on well-shaded sites, but it only flowers in good light and growth is generally more vigorous in the open. It commonly scrambles around the bases of trees but is most damaging when the shoots twine around young trees and strangle them. It is essential therefore to keep young trees free of honey-suckle shoots and to cut back and untwine any already established. Once trees have developed sturdy trunks, damage is very unlikely to occur and honeysuckle may be allowed more or less free rein.

IVY   *Hedera helix*   Pl. 18

Very common throughout Britain and northern Europe, either trailing along the ground, over walls or climbing to 30 m up tree trunks. The flowers and black berries form only when a shoot can climb no further. Many hedgerow trees, especially old elms and ash, almost invariably sup-port some ivy growth. It is generally only harmful on already unthrifty trees which may be partly constricted by large ivy stems around the trunk or may have their top growth suppressed by a mass of flowering shoots in the crown. The latter development can also render trees unstable and liable to windthrow. Ivy should only be removed if absolutely necessary because of the abundant cover that it gives to wildlife and then by severing twice close to the ground and removing a length of stem.

MISTLETOES   *Viscum album* and other species   Pl. 18

These members of the family Loranthaceae are very serious on trees in some parts of the world but there is only one species, *V. album*, in Britain and it

occurs roughly from Yorkshire southwards. It is common elsewhere in
Europe south of southern Scandinavia, growing on a wide range of deciduous trees and especially common on apples, limes, hybrid black poplars
and sycamore. It is semi-parasitic; it has yellow-green leaves for photo-
synthesis but obtains minerals and water from the conducting tissues of the
host tree, with which it is intimately connected by a special organ which
replaces a conventional root. Seeds produced in the sticky white berries are
transferred by birds which lodge them in the bark. It generally causes only
slight damage in Britain by weakening trees and even this is often offset by
the value of the mistletoe itself. Elsewhere in northern Europe, forms,
probably of the same species, infest conifers and can cause considerable
damage, especially to firs. *Loranthus europaeus*, a non-British species with
short, dark green leaves and yellow berries, sometimes damages oaks in
parts of Europe. Where they are seriously disfiguring small ornamentals,
mistletoes can be controlled by pruning off the affected branches, although
such treatment usually results in considerable mutilation of the tree. In
gardens mistletoes are better tolerated, or possibly even actively encouraged. Attempts to infest trees artificially with mistletoe usually fail for no
apparent reason.

SPECIFIC REPLANT EFFECTS

These are widespread and well known conditions which include some
forms of declines and soil sickness. They most commonly affect apples,
cherries and peaches; less significantly pears, plums, rootstocks of *Rosa
canina* and strawberries. The symptoms are poor growth of plants in the first
year of planting on land that has previously (up to seven years in some
instances) borne plants of the same, or usually closely related species. The
root systems are feeble and blackened but the plants may subsequently
recover if they are not attacked by pests and diseases while in the unthrifty
state. These effects differ significantly and importantly from general crop
deteriorations due to low soil fertility, most soil-borne pathogens or other
factors in being fairly specific for certain plant species. Although intensively studied, the causes remain unknown but a soil-inhabiting pathogen such as *Pythium* is now suspected. Partial soil sterilization is used in
commercial nurseries and orchards but in gardens the problem usually
arises when old orchards are replanted and can be cured simply by transfer-
ring the trees to a new site.

ROOT NODULES
(Plate 18)

Small nodular swellings occur on the roots of many plants. They may be
symptoms of pest or disease attack, such as crown gall (p. 330), root knot
eelworm (p. 136) or clubroot (p. 328), and occasionally are of unknown
origin. On leguminous plants they are formed by bacteria of the genus
Rhizobium and are essential in maintaining not only the health of the plants
concerned, but also the general soil fertility. *Rhizobium* bacteria are able to

fix atmospheric nitrogen, that is, converting it to an organically combined form which is available for plant growth. In this they play an essential role in the garden and for this reason it is wise, in rotations, to follow leguminous crops, such as peas and beans, with those, such as brassicas, that derive particular benefit from the increased soil nitrogen.

MYCORRHIZA
(Plate 18)

Mycorrhiza occur regularly on many trees and other woody plants and sometimes cause unwarranted concern to gardeners when root symptoms are seen or toadstools appear. Repeated branching of fine roots and thickening of the root tips, sometimes with sheathing whitish growths, is commonly visible on close examination of the young root systems of beech, birch, cedars, chestnut, firs, oaks, pines, poplars, spruces and other trees. Many of the toadstools which emerge from the soil under trees in autumn or spring are the fructifications of the soil-inhabiting fungi, the enveloping mycelial threads of which give rise to the root symptoms. The fungal mycelium forms an intimate contact with the root tissues and mycorrhiza are often mistaken for root hairs, whose function they virtually usurp in acting as a medium for the transfer of nutrient materials from soil to roots. In return, the mycorrhizal fungi take certain nutrients from the roots. Plants that are always naturally infected with mycorrhiza have been shown experimentally to be quite capable of sustained healthy growth when the fungus has been removed and why in nature mycorrhiza should be so essential therefore is unknown. Nonetheless on many infertile soils, they have been deliberately introduced to aid plant growth.

VIRUSES AND VIRUS-LIKE ORGANISMS
(Plates 19–20)

Viruses and other sub-microscopic organisms are important and widespread plant pathogens. The symptoms they cause have long been known to be transmissible by grafting from diseased to healthy plants but although such 'graft-transmissible agents' were proved many years ago to be smaller than bacteria, not until the advent of the electron microscope could they be 'seen'. Viruses have been described as absolute parasites in making use of the cellular organelles of parasitised cells, but structurally they are simple, comprising a nucleic acid and one or more proteins. They possess the characteristic of usually occurring more or less systemically in diseased plants and are not confined to the parts in which symptoms appear. This results in the transmission of virus from generation to generation of vegetatively propagated plants and is the reason why such plants, which include potatoes, carnations, chrysanthemums and bulbous ornamentals, are so commonly subject to serious virus problems. A further important feature of viruses is the ability of many to exist in host plants as latent infections; although present in the tissues, no symptoms result, but they can be induced to appear when the virus is transmitted by grafting to a different cultivar or variety. Related to this phenomenon is the fact that only now that many plants can be artificially freed of virus (see below) may it be seen that those normally and naturally carrying latent infection are of considerably reduced thriftiness and that their full growth potential is far from being realised. For instance, the flowers of the finest garden pelargoniums seem miserable now that virus-free plants can be raised and compared with them. Latent virus infection is not, of course, the only reason that plants may be unthrifty!

Recently a number of so called 'virus-like' organisms have been recognised, of which mycoplasmas, first shown in 1967 to be the causes of a number of what were previously thought to be virus-induced symptoms, are the best known. Mycoplasmas may be thought of as similar to bacteria, although lacking a cell wall, but it is as yet impossible to differentiate between different types of mycoplasma. There remain a large number of diseases, characterised by symptoms typical of viral or mycoplasmal infection and transmissible by grafting, but in which no organism has yet been recognised. They are described here as caused by graft-transmissible agents.

The naming of viruses and the diseases they cause can be confusing. The description of a virus by a Latin name comparable with those given to other organisms is not recommended and most viruses have been named after the host plant in which they were first found and the symptoms that they caused on it. Thus names like cucumber mosaic virus or passionfruit woodiness virus were self explanatory and the diseases that they caused were known as cucumber mosaic and passionfruit woodiness respectively. Indeed in many instances the name of the disease was given before the causal virus was found. However, it soon became apparent that most viruses affect more than one host plant and may cause quite differing symptoms in them. Thus among the symptoms caused by cucumber mosaic virus are mild stunting of dahlias and, indeed one form of woodiness in passion-fruits. With a number of viruses, such as arabis mosaic virus, the original host plant subsequently turned out to be much less significant in terms of disease symptoms than others identified later. In an attempt to minimise gardeners' confusion therefore, this section is arranged alphabetically under host plants and for each, disease symptoms are described under the name of the causal virus, irrespective of whether the name of the disease is the same as the causal agent or, indeed if the disease itself has a name at all. The ubiquitous cucumber mosaic virus is described in detail under its original host.

Most plants are known to be, or are probably, the hosts for at least one virus but conifers are, with dubious exceptions, not known to be affected by them, although witches' brooms and other virus-like symptoms do occur and are believed in some instances to be induced by other sub-microscopic organisms. Mycoplasmas, in as far as they have been identified, infect a wide range of ornamental and fruit plants but are not known to cause any symptoms in vegetables, apart from potatoes and tomatoes.

Symptoms. A wide range of symptom types occurs, including many similar to those of diseases induced by fungi and bacteria. They include cankers, dwarfing and stunting, leaf fall, rosetting, wilting and witches' brooms and similar proliferations. A few symptom types, most notably rotting, are however never induced by viruses although tissues may be damaged by virus attack so that they are pre-disposed to them. A further range of symptoms, typical of and, in several instances, peculiar to virus attack are illustrated in Plates 19–20 and the principal types are listed below. It should be remembered that only part (commonly, with trees, a very small part) of the plant may display the effects and that symptoms may disappear entirely especially during hot weather. Cool, dull days are generally best for observing the effects but it can be difficult to diagnose virus-induced leaf symptoms in autumn when normal seasonal colour changes and necrosis make identification uncertain. Although many other factors can induce symptoms in certain plants similar to those caused by viruses, the commonest confusion probably arises with environmental effects, such as late frost (p. 482) or drought (p. 482), and with mineral deficiency symptoms (p. 473). In general, however, such effects usually appear on all plants in a particular crop while viruses normally leave at least a few healthy individuals or, with large plants, a few healthy branches. The following, in addition to those listed above, are the more common and characteristic symptom types and this list should be read in conjunction with examination of the plates:

Leaf symptoms

YELLOWING: a loss to some extent of the normal green colour of the leaves. It may affect the entire leaf, leaf edges or leaf veins. When localised in patches, yellowing is known as *variegation* and takes four main forms:

MOSAICS: yellowed areas of varying size but usually angular in form through restriction by the veins. On monocotyledonous plants with parallel-veined leaves, mosaics take the form of *streaking* while, when bands of either dark or light cells occur along the main veins, the effect is known as *vein banding*. Related to the latter symptom is that of *vein clearing*, in which the veins become clear and translucent rather than yellowed.

MOTTLES: yellowed areas of more rounded or diffuse shape, commonly distributed in the same way as mosaics. Sometimes intermediate symptoms occur between the two. Small, irregular mottles may be known as blotches, dots, flecks, spots or similar descriptive terms.

RING SPOTS: rings of yellow or pale green tissue with normal green centres, or concentric rings of light and dark tissue.

LINE PATTERNS: irregular single or multiple yellow or pale green lines and bands taking many forms, one of the most characteristic resembling the outline of an oak leaf and known as *oak leaf pattern*.

MALFORMATION: may take many forms, the most common being known descriptively as *crinkling, crumpling, curling, narrowing* and *rolling*. Very characteristic are the outgrowths from leaves known as *enations* while sometimes leaves of virus-affected plants display *epinasty*, the more rapid growth of the surface of the leaf or its stalk, resulting in pronounced down-bending.

NECROSIS: death of selected tissues resulting in patches in the form of spots, ring spots or other shapes.

GRASSINESS: the excessive proliferation of leaves on monocotyledonous plants; almost invariably mycoplasma-induced.

Flower symptoms

FLOWER-BREAKING: a very characteristic symptom and the first virus-induced effect ever to be described in plants. Dark-flowered plants show either darker or, more usually, white or yellowish streaks and patches on the petals. No comparable effects occur in plants with white or yellow flowers.

GREEN-FLOWERS: a greater or lesser degree of green colour to the petals, usually with other flower malformations, such as dwarfing or *phyllody*; the development of leafy outgrowths. Almost invariably mycoplasma-induced.

Biology. The intra-cellular biology and behaviour of viruses is complex and will not be considered here. Of very considerable importance to gardeners however is the means by which viruses are transmitted, for the recognition of this may aid control measures. Transmission may be by vectors or by other means. Among aerial vectors, beetles, mites, thrips, weevils, whiteflies and other arthropods transmit a few viruses but by far the most important agents in temperate regions are aphids, of which the potato-peach aphid, *Myzus persicae*, probably outweighs all others in its significance. Although the relationship between viruses and aphids is complex, the facet of greatest significance to the gardener is probably the length of the retention time – the period for which a virus remains infective within or on an aphid – as this dictates whether insect control measures are likely to be of much importance in controlling the disease (see below). Among soil-borne virus vectors, eelworms are the most important, although a few viruses are transmitted by soil fungi with swimming swarm spores. Mycoplasmas seem to be almost invariably transmitted by leaf-hoppers. Although generalisations over the relationship between symptom type and means of transmission can be misleading, it is true that many mosaic and mottle type effects are commonly caused by aphid-transmitted viruses while ringspots are commonly induced by those that are soil-borne. While aphid-transmitted viruses are often of importance because of their very rapid spread among crop plants, soil-borne viruses are significant because soil, once contaminated, is very difficult to free of the infestation, which thus persists from crop to crop.

Non-vector transmission may be for example via pollen or seeds or even by physical contact, sometimes by the slightest touch. Many viruses have no known means of natural transmission and in some, such as certain fruit tree viruses, this seems to be extremely slow with very little plant to plant spread.

Treatment. There are no viricidal chemicals comparable with insecticides and fungicides although a few substances may suppress certain symptoms. Some measures can be taken to minimise the effects of viruses however. The first requirement, where possible, is to ensure that planting stock is virus free. Such stocks have been available for a few crops for some time – certified potato 'seed' for instance, raised in Scotland or other northern areas where aphid vectors are few. With seed-transmitted viruses, such as lettuce mosaic virus, contamination-free seed should of course be used if available but it should be noted that the designation 'virus tested' or 'mosaic tested' applied to seed is for practical purposes meaningless and a very high proportion of the seed may still carry virus. Only if the percentage of contaminated seed present is quoted and less than 0.01% or so should any reliability be placed on the description. (The term 'virus-tested' is however also applied to planting stock of some fruits, such as raspberries, and in these particular instances does imply that the material is of the highest quality).

More recently, methods have become available by which viruses are not merely avoided but can actually be removed from plant material. This is usually achieved by heat treatment of the diseased plants and from the plants initially treated, virus-free stocks can be propagated. The most conspicuous success in this field has been with the EMLA rootstocks and scions of apples, pears, plums and cherries developed in Britain at the East Malling and Long Ashton Research Stations and the certified raspberry stocks from the Scottish Horticultural Research Institute (now the Scottish Crop Research Institute).

An alternative method of raising virus-free stock is by meristem culture; the initiating of plants in the laboratory by culturing a few cells from the meristem (the fastest growing tissue of the plant) where the plant's growth may just exceed the rate at which the virus can spread into the cells. A continually extending range of virus-free stocks of plants, particularly of ornamentals, is becoming available, initially for professional growers but commonly available to gardeners on enquiry at the most reputable suppliers. Even if such stocks are not available, the buying of carefully selected plants from reputable suppliers will go far towards ensuring good quality planting material. Once planted, virus-free fruit stocks, apart possibly from raspberries, usually remain so as with a few exceptions, the natural means of spread of fruit tree viruses, although unknown, seems very slow and the plants are not re-infected.

Even when vegetable and ornamental plants are freed of viruses, which include those readily and rapidly transmitted by aphids, reinfection of the stock is commonly not as fast as might be expected and they can be used for five or six years at least. The importance of rogueing of plant stocks for any showing virus symptoms cannot be overstressed however. With many virus-induced diseases, the prompt removal and destruction by burning of such plants is the only means of protecting further healthy plants from becoming diseased.

Viruses transmitted by insects can commonly be controlled by the application of insecticides to eliminate the vector. This is not invariably so however as aphids transmit many viruses in what is termed a non-persistent manner; the virus remains viable on the insects for a very short period only, often a matter of hours, and control of the insects does not therefore reduce significantly the amount of virus available for infection of further plants. Control of the vectors of soil-borne viruses is exceedingly difficult but where a crop has been infected by a soil-borne virus, fresh stock should be planted on a different site away from contaminated soil. In some circumstances, such as non-availability of alternative land, the use of the granular soil sterilant chemical dazomet in accordance with the manufacturer's instructions may be worthwhile. Where control of the vector is indicated in this account as a means of virus disease control, reference should be made to the description of the insects in the Pests section of the book. In relatively few instances are cultivars of plants available with resistance to virus-induced diseases; notable exceptions being the tomatoes recently introduced as resistant to tobacco mosaic virus.

Weed control is as important for eradicating potential virus hosts as it is for other reasons. In particular, the presence of weeds botanically related to the crop among which they grow is undesirable. The prompt removal and destruction of old plants remaining at the end of the season is also very important in reducing the population of viruses available to infect new stock, although among some vegetables, such as brassicas, it may of course be very difficult to avoid having crops of different ages in the garden at the same time.

Absence of a recommended treatment from the following descriptions indicates that one is either impracticable or not necessary.

ABUTILON

ABUTILON MOSAIC VIRUS: causes the familiar and attractive variegation of *A. thompsoni* and other species which takes the form of a vivid yellow and green angular leaf mosaic. The symptoms occasionally disappear and plants 'recover' from the disease, while if they are kept in darkness for short periods, leaves initiated during that phase also lack the variegation. Natural transmission in native localities of South and Central America and India is by a species of whitefly. This does not occur in Europe where the horticulturally desirable symptoms are perpetuated by propagation as cuttings.

AESCULUS

Overall yellowing of leaves and sometimes line patterns or vein yellowing are locally common in south-west England, on *Aesculus* species, especially *A. hippocastanum*. The symptom is sometimes selected for propagation because of its attractiveness but the effect on plant vigour is not known. *Prunus* necrotic ringspot and strawberry latent ringspot viruses may be at least partially responsible.

ALMOND

PLUM LINE PATTERN VIRUS: quite commonly causes yellowish line patterns and/or rings on leaves. See p. 460.

AMARYLLIS

HIPPEASTRUM MOSAIC VIRUS: very common and widespread; causes irregularly shaped, dark and light green mottle on the leaves, leaf stalks and occasionally stems but generally causes little reduction in flowering or vigour. Transmitted by aphids and, unless symptoms are very severe, can probably be tolerated.

ANEMONE

ANEMONE MOSAIC VIRUS and CUCUMBER MOSAIC VIRUS: common in some seasons (in Britain particularly in the south-west) and the most frequent associates of mottling and distortion of the leaves, sometimes accompanied by a flower break symptom and flower stalk necrosis. Plants degenerate very quickly, largely as a result of grey mould and bacteria invading the damaged leaves and crown. Anemone mosaic virus is a strain of turnip mosaic virus and both it and cucumber mosaic virus are aphid transmitted, although they may also be carried in the corms. Good control of aphids early in the season is important in protecting anemones from the insects, from the viruses they carry and from the fungal and bacterial problems that may follow.

APPLE

Many unknown graft-transmissible agents and also some identified viruses and mycoplasmas affect apples although a number are latent in cultivars grown for fruit production.

Symptoms and biology

Organisms causing predominantly fruit symptoms

APPLE CHAT FRUIT MYCOPLASMA: occasional throughout much of northern Europe and, in England at least, almost confined to the cultivars 'Lord Lambourne' and 'Early Worcester'. Although the development of the fruits begins normally, growth virtually ceases around mid-season and they may remain small and green until harvest. Affected trees sometimes stand out from others by a vigorous upright growth habit. Means of natural transmission unknown.

APPLE GREEN CRINKLE (graft-transmissible agent): occasional through-out much of northern Europe affecting many cultivars. After about one month's development, the fruits become dimpled and this symptom be-comes progressively more severe as the season progresses. Cracks or warty outgrowths may develop in severe attacks. Means of natural transmission unknown.

APPLE ROUGH SKIN (graft-transmissible agent): common throughout much of northern Europe affecting several cultivars. Small round or more irregular patches of rough skin develop on the fruits and the agent is possibly one of the commonest causes of this often observed symptom. Sometimes tiny cracks occur (cf. apple star crack disease (below) which may possibly be caused by the same agent). It can be confused with the russetting that is normal on many cultivars, with capsid bug damage (p. 152), and also with the fruit symptom of scab (p. 335) but unlike scab, there are no effects on leaves or stem. Means of natural transmission unknown.

APPLE STAR CRACK (graft-transmissible agent): common throughout northern Europe but in England most frequent on 'Cox's Orange Pippin'. The most typical symptom is the reduced yield and small star-like cracks on the fruit, sometimes concentrated towards the calyx. Very small cankers develop on one-year-old shoots in January and the shoot tips may be killed while leaf and blossom burst may be delayed by up to three weeks. These symptoms can arise from a variety of other causes however. Means of natural transmission unknown.

Organisms causing leaf and fruit symptoms

APPLE LEAF PUCKER and related diseases (graft-transmissible agents): similar symptoms are quite common and widespread throughout most of northern Europe. Leaves may be irregularly flecked with yellow or pale green and the leaf blade may be puckered or crumpled and appear as if the main vein has been pulled short. The leaves formed during hot sunny weather are not affected. The fruit symptoms are variable and only very apparent when early summer is cool; most characteristic are rings of russet-ting. Means of natural transmission unknown.

Organisms causing predominantly leaf symptoms.

APPLE MOSAIC VIRUS: common and widespread throughout northern Europe. Many cultivars are affected but in Britain 'Lord Lambourne' is probably the most sensitive (as it is to many diseases caused by virus-like organisms). The obviously visible symptoms are very variable and almost entirely confined to the leaves, although fruit yield is reduced by up to one fifth. The leaves develop yellow or whitish spots, blotches, line patterns or vein banding, most noticeably in early summer. The latter is an important diagnostic character as very similar patterns can arise during normal autumnal necrosis. Means of natural spread unknown but very slow and most dispersal of virus probably occurs through propagation of infected stocks. Also see: leaf pucker (p. 435).

Organisms causing predominantly stem symptoms

APPLE FLAT LIMB (graft-transmissible agent): occasional and widespread throughout northern Europe causing reduction of yield of many cultivars of apple and pear but 'Gravenstein' is traditionally one of the most severely affected. The symptoms are elongate flattenings, depressions, furrowing and twisting of shoots or branches, most commonly the older ones. In severe outbreaks, trees can be seriously reduced in vigour and possibly killed. Means of natural transmission unknown.

APPLE PROLIFERATION MYCOPLASMA: unknown in Britain but occasional and very important throughout much of continental northern Europe, with the possible exception of Scandinavia. Causes probably the most serious virus-like disease of apples. Import of apple trees carrying the disease is forbidden by law in Britain although related ornamental species are exempt from this restriction and could therefore introduce the problem here. The most characteristic symptom is the development of witches' brooms, especially near the main shoot apex. The leaves on diseased trees may be abnormally small, with fine irregular serrations; the stipules may be abnormally long. The fruits are often very small and poorly flavoured. General vigour of the tree may be diminished but death is unusual and recovery may take place after a few years, especially if soil fertility is increased. Diseased trees seem especially prone to powdery mildew. Unlike other virus-like conditions affecting apples, proliferation spreads rapidly and transmission is by leafhoppers.

APPLE RUBBERY WOOD MYCOPLASMA: occasional and widespread throughout northern Europe causing reduction of yield on many cultivars of apple and pear but on 'Lord Lambourne' the symptoms are unique and unmistakable; shoots and younger branches become flexible and 'rubbery'. Trees in fruit appear pendulous and the stems bend downwards. Shoots arise commonly from the lower parts of the trunk on affected trees. Means of natural transmission unknown.

Organisms causing latent infection in apples

Several viruses and virus-like organisms are very common and widespread in many apple cultivars grown for fruit production. The infections are latent however and no symptoms are produced in apple cultivars, although on certain rootstocks, the organisms are damaging or lethal and cause conditions including those described as chlorotic leaf spot, platycarpa dwarf, Spy 227 decline, stem grooving and stem pitting.

Treatment. As with other trees, the virus and virus-like diseases of apple cannot be cured. Some measures are possible to alleviate the problems however. Trees that produce consistently few or poor quality fruit that can be traced to a virus or virus-like cause should be grubbed up and burnt. Top grafting an infected tree with a scion of a cultivar unaffected by the particular condition is valuable in some instances but insufficient information is available to make any general recommendations in this respect; one notable instance however is the top grafting of 'Lord Lambourne' when infected by apple chat fruit as the condition is almost unique to that cultivar; 'Cox's Orange Pippin' or 'Worcester Pearmain' may be used as scions. Of particular importance when planting a new orchard or replacing diseased trees is to ensure that the new stock is free of disease. Fortunately a number of virus-free rootstocks and cultivars are now readily available and have been included in the EMLA scheme (see p. 433).

APRICOT, NECTARINE and PEACH

There are four virus-induced diseases of apricots, nectarines and peaches in northern Europe of which by far the most serious is plum pox. This is the only one producing marked fruit effects and the identification of its symptoms and differentiation of them from those of the remaining less serious viruses is given on p. 460.

PLUM LINE PATTERN VIRUS: various types of yellowish line patterns and/or diffuse yellow rings, sometimes accompanied by stunting, are common and widespread. The causal agent or agents are known as plum line pattern virus and are described more fully on p. 460.

PRUNUS NECROTIC RINGSPOT VIRUS: probably occurs in peaches throughout northern Europe but passes unremarked in many areas as peaches are not an important fruit crop. Occurs also in almonds and, often symptomless, in several other *Prunus* spp. The symptoms can vary considerably between cultivars; necrotic and/or yellowish ringspots are characteristic and may be accompanied by leaf shot-holes, death of buds and entire shoots and small twig cankers. In combination with prune dwarf virus, *Prunus* necrotic ringspot virus induces, in peaches, the condition known as peach stunt. Symptoms are delayed leaf burst, wavy leaf margins, upward leaf rolling and a stiff, upright leaf habit. There may be shortening of some shoots and, early in the season, yellowish mottles, line patterns or rings on leaves. Also see plum (p. 460) and cherries (p. 442). The natural transmission of prune dwarf and *Prunus* necrotic ringspot viruses is almost certainly by pollen but most outbreaks result from the use of diseased planting material. Severely affected trees are best grubbed up and destroyed.

AQUILEGIA

CUCUMBER MOSAIC VIRUS: occasionally affects columbines and induces stunting, vein clearing and a mottled mosaic. Older leaves may develop necrosis and few flowers form. See p. 444.

ARABIS

Infection by the well known arabis mosaic virus, which has a very wide and important host range, is insignificant in *Arabis* itself. A strain of turnip mosaic virus however commonly infects *Arabis* causing pale yellowish spots and ring patterns.

ASCLEPIAS

Asclepias and other members of the family Asclepiadaceae, such as hoyas, stapelias and stephanotis, are prone to infection by a number of viruses, of which the commonest is probably cucumber mosaic virus (see p. 444). The symptoms are varied but include mosaics, mottles, stunting and occasionally ring spots. The leaves may be abnormally long and narrow.

BEAN

Viruses affecting broad, dwarf and runner beans

BEAN YELLOW MOSAIC VIRUS: a common and widespread virus, existing as several strains, some of which affect peas and also non-leguminous plants, such as freesias and gladioli. On dwarf and runner beans, the leaflets curl and bend downwards with a gradually extending yellow mosaic. The plants become stunted and bushy in habit. Pod production is reduced and the pods become distorted. On broad beans there is a slight yellowing of the leaves which retain small irregular dark green patches. Transmitted by several species of aphid, including *Acyrthosiphon pisum* and *Aphis fabae*. The commonest sources of virus are red clovers and gladioli, therefore beans should not be planted close to these crops. Maintain strict aphid control throughout the season.

Viruses affecting dwarf and runner beans only

BEAN COMMON MOSAIC VIRUS: common and widespread. Symptoms variable and similar in some respects to those induced by bean yellow mosaic virus but there is no downward rolling of the leaf margins and no regular leaf mosaic patterns, merely the development of irregular light and dark green patches. Pods are littled affected. Transmitted to some extent in the seed and also by a number of aphid species, of which the most important are *Aphis rumicis* and *Macrosiphum pisi*. Maintain good aphid control, buy seeds from reputable suppliers and, if saving own seeds, never do so from plants showing any suspicious symptoms.

Viruses affecting broad beans only

PEA LEAF ROLL VIRUS: this and bean yellow mosaic are the two commonest viruses in broad beans. It causes a characteristic yellowing and rolling of the top-most leaves which spreads until the entire upper part of the plant has bright yellow, thickened and brittle foliage. Plants are stunted and the leaves often drop. Transmitted by aphids, including *Myzus persicae* and *Acyrthosiphon pisum*. Good aphid control should restrict the problem.

BROAD BEAN STAIN VIRUS and BROAD BEAN TRUE MOSAIC VIRUS: common and widespread in field beans and to a lesser extent in broad beans. Symptoms almost identical; some leaves show mild green mosaic or are yellowish with isolated dark green patches. Eventually the leaves become puckered, with irregular outlines, and the seeds are discoloured or stained. Both viruses are transmitted by weevils of the genera *Apion* and *Sitona* and also to some extent in the seeds. Routine insecticidal treatments on broad beans in gardens should keep the weevils and the virus diseases in check.

BEETROOT

BEET YELLOWS VIRUS and/or BEET MILD YELLOWING VIRUS: common, causing the disease known as beet yellows, the former virus usually being more widespread. Very serious on sugar beet. The leaves, especially (and unusually for virus-induced diseases) the outer leaves, are thickened, dry and brittle with a red-yellow colouration spreading downwards from the tips between veins. There may be necrotic spots in the inter-veinal region or a bright orange-red colour of the older leaves which stand upwards stiffly. Symptoms vary depending on the virus involved but by mid-summer plants may become so brittle as to rattle when shaken. Transmitted by aphids, predominantly *Myzus persicae*, and control of these, preferably with a systemic insecticide, will go far towards keeping the disease in check. Sow seed early as this usually enables plants to become established before aphid infestation begins and if possible, avoid gappiness in the crop as this is more likely to encourage aphid build up.

BEET MOSAIC VIRUS: fairly common and widespread, producing variable symptoms. Most common is the appearance of small yellow flecks, followed by larger and vivid green or yellow mottling, sometimes with depressed areas on the leaves. There is usually vein clearing and often leaf distortion and curling back of the tip and margins. The entire plant is stunted in severe attacks. Transmitted by aphids, especially *Myzus persicae* and *Aphis fabae*. Can probably be avoided by regular attention to weeding as several weed species can harbour the virus. Especially important in this respect are chenopodiaceous weeds such as docks and goosefoots. Aphid control is of little value as the virus is non-persistent in the insect.

BEGONIA

Bedding begonias in particular commonly show mottling as a result of infection by cucumber mosaic virus (p. 444) and maintenance of good aphid control is important in restricting spread of the problem in spring. Begonias also develop mottling, stunting and ringspots caused by the thrip-transmitted tomato spotted wilt virus and, together with calceolarias, dahlias and zantedeschias, are among the few plants now severely affected by this once common pathogen.

BETULA

Leaf mosaics, mottling and line patterns are sometimes found on birches, symptoms commonly being apparent only in cooler autumn weather. Some check to growth may occur also. Apple mosaic virus and the pollen-transmitted cherry leaf roll virus are possible causes.

BLACKBERRY

A complex mosaic occurs commonly on blackberries but this is described under raspberry (p. 464), the more important host plant.

RUBUS STUNT MYCOPLASMA: common and widespread, sometimes reaching epidemic proportions; affects blackberries, loganberries, raspberries and related fruits, such as boysenberries. In Britain and northern Europe it is often serious on loganberry and parsley-leaved blackberry. Symptoms vary somewhat between host plants but most characteristic is the production of new canes which are short and feeble. This may be accompanied by phyllody and flower proliferation. In the following season few or no flowers or fruits develop and eventually the entire plant becomes a stunted bush with thin short canes. Plants already affected by mosaic disease are likely to be killed by subsequent stunt infections. Transmitted by leafhoppers, usually *Macropsis fuscula*, and control of these insects, together with removal and burning of diseased plants, should help eliminate the problem. Certified virus-tested stocks should always be used for new plantings.

BRASSICAS

The four commonest and most widespread of several viruses affecting brassicas are cauliflower mosaic virus, radish mosaic virus, turnip crinkle virus and turnip mosaic virus. They may cause symptoms either separately or in combination with each other and with other viruses and it is often impossible to be certain of the cause of particular effects. All may affect all brassicas but on swedes and turnips, turnip mosaic virus is easily the most important. Some F_1 hybrid Brussels sprouts have proved very susceptible to virus problems. Radishes usually escape serious attack, probably because they grow and mature so rapidly.

CAULIFLOWER MOSAIC VIRUS: very common in gardens, affecting Brussels sprouts, cabbage, cauliflower, Chinese cabbage, other brassicas and ornamental crucifers but most serious on winter cauliflower. The symptoms are not usually seen on very young plants and are commonly indistinct in summer. In winter or autumn cauliflowers, it appears as a general stunting, especially of the central leaves, and a poor quality curd. Cold hardiness may be reduced and plants die as a result. The leaves are mottled with alternate light and dark green and commonly show a narrow dark green band on both sides of the main veins. On winter or spring cabbage this vein-banding is usually less pronounced and the veins tend to be yellower. Transmitted by aphids, especially *Brevicoryne brassicae* and *Myzus persicae*, but non-persistent and the use of insecticides is therefore of limited value. Chemical control should always be combined with destruction or burying of all old brassica plant debris at the end of season to eliminate sources of virus for new plants. If one brassica crop is still growing while new ones are being raised, the older plants should be treated with insecticide to prevent aphids flying from them with virus. Spraying young plants is unlikely in itself to prevent disease reaching them but may prevent plant to plant spread within a bed.

RADISH MOSAIC VIRUS: commonly produces mosaics, ringspots and leaf crinkling on turnips and radishes; indefinite yellowish patches on cauliflower leaves and yellowed or necrotic areas on cabbages. Transmission is by flea beetles (*Phyllotreta* spp.) and control of these should eliminate the problem (p. 238).

TURNIP CRINKLE VIRUS: occasional but probably only in England and Scotland. Causes leaf crinkling, mottling and some stunting in turnips but only mild mottling in other brassicas. Like radish mosaic virus (above), flea beetle transmitted.

TURNIP MOSAIC VIRUS: common; on swedes and turnips produces yellowish inter-veinal mottling and vein clearing with crinkling and stunting of younger leaves. As the leaves age, dark green raised areas may take the place of yellow mottling; growth may be seriously retarded. On cauliflowers, small, pale green, raised, more or less circular spots or rings form but vein-clearing and vein-banding usually do not (cf. cauliflower mosaic virus above). On Brussels sprouts and more especially on cabbage, black necrotic spots and rings are formed (the so-called cabbage black ringspot symptom) and this virus is probably the commonest cause of black spotting on cabbage heads. Transmitted by the aphids *Brevicoryne brassicae* and *Myzus persicae*. Control is difficult and procedures outlined for cauliflower mosaic virus should be followed.

BUDDLEIA

CUCUMBER MOSAIC VIRUS: very common and widespread; buddleias are among the many sources of this virus for infecting other garden plants (p. 444). Leaves may show irregular mottling, large angular mosaic, irregular edges crumpling or other distortion, and are commonly narrowed. Inflorescences may be dwarfed and branched. Trees showing severe signs of these symptoms should be grubbed and burnt or many other garden plants may become diseased year after year.

CALCEOLARIA

TOMATO SPOTTED WILT VIRUS: commonly produces a leaf mottling. See begonia (p. 439).

CALLISTEPHUS

CUCUMBER MOSAIC VIRUS: quite common and probably widespread (see p. 444). Produces yellowing and mottling of the leaves with small, deformed and often abnormally numerous flower heads. The symptoms are somewhat similar to a serious and very much studied mycoplasma-induced disease known as aster yellows which is familiar in North America and elsewhere but is probably rare in Britain.

CAMELLIA

CAMELLIA YELLOW MOTTLE VIRUS: causes the common and familiar variegation, sometimes propagated for aesthetic appeal. The symptoms are irregular yellowish-white blotches or speckling on the leaves but the same virus may be responsible for a flower-break symptom. No known means of natural spread in Europe.

CAMPANULA

CUCUMBER MOSAIC VIRUS: commonly causes mosaic and distortion of the leaves with a crumpled flower symptom in which the blooms fail to open properly. See p. 444.

CARROT

CARROT MOTTLE VIRUS with CARROT RED LEAF VIRUS (carrot motley dwarf disease): common, originating in Australia but now widespread and caused by two viruses acting together. Plants are stunted and the leaves usually show reddening and a fine yellowish mottling. There is often twisting of the leaf stalk to the extent that the leaves may become totally inverted. Transmitted by the aphid *Cavariella aegopodii*; carrot mottle virus only being capable of aphid transmission when accompanied by carrot red leaf virus. Persistent in the aphid and therefore insecticide treatment is often effective, especially at the time of sowing and with systemic material.

CELERY

Mosaics, mottles, line patterns and ringspots are quite common and wide-spread on celery and eight or more different viruses may be responsible, although cucumber mosaic virus is probably the most important (see p. 444). Some are aphid-transmitted and some soil-borne and although little can be done to control them in gardens, new plants should always be raised well away from land which has carried a diseased crop.

CHEIRANTHUS

TURNIP MOSAIC VIRUS: common, giving rise to small, stunted plants usually with yellowish and dark green mottling on the leaves. A flower break symptom is very common, especially on dark red cultivars, such as 'Blood Red', where the petals develop yellow stripes and flecks. Diseased wallflowers are an important source of virus for infection of brassicas (p. 440) and should be destroyed.

CHERRY

Although well over thirty different viruses infect cherry naturally and cause several important diseases in commerce, few are likely to be of very wide-spread concern in gardens, especially as sweet cherries are rarely grown successfully on a small scale. Many of the viruses and the diseases they cause are imperfectly understood.

A complex of viruses, normally PRUNE DWARF VIRUS with either ARABIS MOSAIC VIRUS or RASPBERRY RINGSPOT VIRUS, causes a serious condition known as European rasp leaf or Pfeffinger disease. The symptoms are a reduction in the size of the leaves which become narrow and toughened with coarse serrations along the edge and enations on the underside. During the early part of the season, clearly marked yellowish areas appear on the young leaves. Situations in which disease may especially tend to increase are where prune dwarf virus spreads into orchards where latent infections with the nematode-transmitted raspberry ringspot virus have occurred previously. Alternatively, rasp leaf may increase if raspberry ringspot virus spreads into trees already affected with prune dwarf virus. Little treatment can or need be done in gardens although badly diseased trees are better grubbed up and new stock planted on a different site.

CHERRY EUROPEAN RUSTY MOTTLE VIRUS: common throughout much of northern Europe on sweet cherries. Although the virus is latent in many cultivars, symptoms do develop on some; they are normally first seen in July as clearing and yellowing of groups of fine leaf veins. The leaves gradually darken and rusty red leaf spots develop, usually during August. Natural means of transmission is unknown.

CHERRY LEAF ROLL VIRUS: occasional and widespread throughout north ern Europe except Scandinavia. Probably the most severe cherry virus in its effects. Leaves and flowers are abnormally late appearing; leaf margins roll upwards in summer and may turn purplish; trees become less vigorous, die back and, as with many afflictions of *Prunus* spp., may exude gum. Total death may ultimately result, although the latter symptoms can be confused with root damage, bacterial canker and other causes. Transmission is via the pollen. Grub out diseased trees and replant new virus-free stock as far as possible from the orignal site.

CHERRY LITTLE CHERRY MYCOPLASMA: occasional and widespread throughout much of northern Europe on ornamental and sweet cherries. Symptoms vary greatly between cultivars but most usual is the production of half-sized fruit that are late and sometimes uneven to ripen and bitter to taste. Natural transmission is by leafhoppers.

PRUNUS NECROTIC RINGSPOT VIRUS: this common and widespread virus, which affects several *Prunus* spp., is associated with three separate conditions on cherries. The virus acting alone on sour or sweet cherries can cause pale green or yellowish rings or more irregular areas on leaves with necrotic spots and/or patches which may fall away leaving a 'tatter-leaf' symptom. A somewhat similar symptom of *Prunus* necrotic ringspot virus infection on sour cherries and also widespread except in Scandinavia is known as Stocklenberg disease. This is caused by a distinct strain of the virus. (See also plum (p. 460) and peaches (p. 437).) In association with prune dwarf virus, a disease known as sour cherry yellows can occur. Yellowish patches and sometimes ringspots develop on leaves. By the second or third season, considerable defoliation can occur and the crop is reduced in total weight, although individual fruits may be abnormally large. No insect vector of *Prunus* necrotic ringspot virus is known but in cherries the virus is pollen-borne and healthy trees become diseased when pollinated with virus-carrying pollen. The virus can also be transmitted in seed. Trees affected with sour cherry yellows disease are best grubbed up and destroyed but the other conditions can probably be tolerated in gardens.

PRUNE DWARF VIRUS: although one of the joint causes of European rasp leaf and sour cherry yellows diseases, can cause symptoms when present alone but usually these consist merely of rings, line or oakleaf patterns on leaves, all of which are commonly masked in subsequent seasons.

CHRYSANTHEMUMS

Chrysanthemums are prone to infection by a large number of viruses and virus-like organisms and five different diseases are important and widespread in Britain and northern Europe. Although some viruses may be largely eliminated by rogueing, virus-free stocks of chrysanthemums are available and should be used if possible when establishing fresh plantings, especially if the site has a history of virus attack.

TOMATO ASPERMY VIRUS: common and widespread. Blooms on affected plants are abnormally small, with deformed petals (commonly confined to one side only), this giving an overall ragged appearance. Red, pink and bronze cultivars usually develop a flower-breaking with pale streaks on the florets. Aphid-transmitted, predominantly by *Myzus persicae*. Can often be eliminated from garden stock by careful and persistent rogueing out of any plants showing symptoms.

CHRYSANTHEMUM GREEN FLOWER (probably mycoplasma): occasional, especially in 'Balcombe'-type cultivars which develop small pale green flowers. The condition is self-destroying as the root system is irreparably damaged and the affected plant dies. Spread between plants is almost negligible and removal of diseased individuals should nullify the problem. This condition may be caused by the agent that gives rise to a similar symptom on heleniums (see p. 450).

CHRYSANTHEMUM STUNT VIROID: widespread but only really common in stock not bought from reputable nurseries and especially if derived from imported American plants. The causal agent is termed a viroid and is structurally different from and much smaller than conventional viruses. The symptoms are not easy to diagnose but entire plants, including flowers, tend to be midget, i.e. totally scaled down. Flowers also are of weak colour and open 7–10 days early. Not usually any definite abnormal leaf colouration. Natural means of spread is unknown but the disease is very contagious and spreads readily by physical contact or handling. Affected chrysanthemums should be destroyed and new plants obtained from a reputable supplier.

CHRYSANTHEMUM LEAF MOTTLE VIRUSES: at least five different viruses are probably widespread and related to these conditions in chrysanthemums but their effects are usually only apparent when diseased plants are compared with virus-free stock, which are seen to be larger, more vigorous and of superior flower quality.

SOIL-BORNE VIRUSES: arabis mosaic virus and strawberry latent ringspot virus; both eelworm-transmitted, are among the commonest causes of the ring patterns and sometimes serious leaf distortion occasionally seen on outdoor plants. Glasshouse plants raised in sterilised compost are not affected but outdoors new stock should be obtained and the same site not replanted.

CORYLUS
Leaf yellowing is quite common on hazels and may take several forms which seem to vary from area to area. In England, yellow ring and line patterns predominate while in Wales bright yellow patches are more usually seen. Other diffuse line patterns and mosaics have been seen elsewhere. One virus associated is apple mosaic virus but there are certain to be others.

CUCUMBER and OTHER CUCURBITS
Only two viruses need be considered in connection with cucurbits; cucumber green mottle virus and the ubiquitous cucumber mosaic virus, which has an importance far beyond its original association with cucumbers.

CUCUMBER MOSAIC VIRUS: exceedingly common and widespread. Garden plants commonly affected, in addition to cucurbits, include anemones, aquilegias, asclepias, begonias, buddleias, callistephus, campanulas, celery, cyclamen, dahlias, daphnes, delphiniums, gladioli, hoyas, lettuce, lilies, loniceras, magnolias, matthiolas, narcissi, nicotianas, passifloras, petunias, primulas, spinach, stapelia, stephanotis, tomatoes, tulips, violas, zantedeschias and many common weeds.

Symptoms. For non-cucurbit hosts, see individual entries under hosts. The virus is present in most cucurbit stocks and symptoms comprise small yellow flecks and spots, at first on the young leaves, but later spreading until the entire plant has a very obvious mosaic. The leaves are crinkled, about half normal size and the entire plant is stunted. Mottling may also appear on the fruit, especially of cucumbers, where it begins at the stem end. The fruits are also reduced in number.

Biology. The importance of seed transmission of cucumber mosaic virus in cucurbits is uncertain but it certainly occurs in some weeds, such as chickweed. A little spread can also occur by physical contact when for instance knives or secateurs contaminated with sap from diseased plants are used on others. The most frequent and important means of transmission however is by aphids; usually *Aphis gossypii* in glasshouses and *A. fabae* and *Myzus persicae*, among many others, outdoors.

Treatment. Always maintain scrupulous aphid control in glasshouses but especially where cucurbits are grown. Such rigorous control is usually not practicable outdoors but care should be taken to minimise potential sources of the virus. Do not plant the most susceptible host plants, such as cucurbits and tomatoes, close to any of the perennial hosts listed above and take special care to avoid dahlias, delphiniums and primulas. For the same reason, any measures to reduce weed populations, especially of perennial weeds such as common bryony, common mallow and teasel, will also restrict the sources of virus.

CUCUMBER GREEN MOTTLE VIRUS: relatively infrequent. Causes a light and dark green mottling (not yellow as with cucumber mosaic virus) with associated leaf distortion but usually not fruit symptoms. Natural means of transmission not known but it is spread very easily by knives or other tools and on hands.

CURRANTS

BLACK CURRANT REVERSION (graft-transmissible agent): widespread and probably almost ubiquitous in Britain in all garden black currants not grown from selected stock. Most black currant bushes in gardens probably only achieve a small proportion of their potential yield because of reversion disease. The symptoms are an overall reduction in number of the flowers and of the size of the main veins on the primary leaves with some yellowing of those leaves that are produced during the flowering period. Although an expert eye can distinguish a number of other visible symptoms, gardeners are generally unaware of the presence of reversion because they do not appreciate that their plants are yielding so poorly, this being the major effect. No specific virus has been identified but it is known that big-bud mites *(Cecidophyopsis ribis)* are associated with the disease and it is assumed that they transmit the causal agent. Control is difficult, especially when the presence of the disease is not appreciated, but always obtain currant bushes from a reputable supplier, plant as far as possible from older plantings, maintain good mite control and renew stock every 7–8 years.

GOOSEBERRY VEIN BANDING VIRUS: causes pale yellow vein banding on currants on the first leaves to expand in spring. See under gooseberry (p. 449).

CYCLAMEN

CUCUMBER MOSAIC VIRUS: an occasional mosaic pattern on the flowers and some leaf striping can arise following infection by this virus (see p. 444). Cyclamens sometimes seem to become infected from nearby buddleias.

DAHLIA

There are three common virus diseases of dahlias, of which dahlia mosaic is the most important.

DAHLIA MOSAIC VIRUS: very common and almost universal wherever dahlias are grown. Yellowish or pale green vein-banding on the leaves is the most usual symptom but on some cultivars it may be accompanied or replaced by some crumpling and twisting of the leaves and/or overall stunting of the plant. It is transmitted by aphids, most notably *Myzus persicae*, but with a fairly short retention time so insecticidal control is of little value once plants are in the bed, although it is important while cuttings are being raised in an enclosed environment, such as a glasshouse. Although there is no overall solution to the problem, careful selection and multiplication from symptomless plants is essential.

CUCUMBER MOSAIC VIRUS: faint vein-clearing and stunting can occur as a result of infection but the leaves are generally more mottled and crumpled than with dahlia mosaic virus (see p. 444).

TOMATO SPOTTED WILT VIRUS: occasional, causing a classic ring pattern symptom on the leaves. See under begonia (p. 439).

DAPHNE

CUCUMBER MOSAIC VIRUS: the commonest so far identified of many viruses infecting daphnes and gives rise to mottling and crumpling of the leaves as well as rendering the plants very susceptible to frost damage. Daphnes commonly display similar leaf mosaic and mottles, as well as an inexplicable unthriftiness, which is probably due to other virus infections, while the difficulty in propagating *D. mezereum*, for example, may well be due to the viruses present. Cuttings should never be taken from plants showing any abnormal symptoms or growth habit.

DATURA

Daturas are very susceptible to almost all the viruses affecting solanaceous plants and will almost invariably show symptoms of disease if growing near virus-infected potatoes. Full descriptions are given on pp. 461–4 but in general, because of the potential two-way traffic in viruses, the two plants are best not grown in close proximity. Similarly daturas are commonly the source of tobacco mosaic virus infection for tomatoes with consequent very serious effects (see p. 468). It should also be noted that daturas are susceptible to the tobacco-infecting strain of tobacco mosaic virus, which can be spread on hands that have handled cigarettes.

DELPHINIUM

At least seven different viruses infect delphiniums in Britain but the commonly seen symptoms are most usually caused by cucumber mosaic virus (see p. 444). Perennial delphinium stocks show a general yellowing, with vein-banding and mottling, while larkspurs are also often stunted and flowerless. Because some of the other viruses infecting delphiniums are soil-borne, it is wise to plant new stock on a different site.

DIANTHUS

Glasshouse carnations are very prone to virus diseases but while twelve different viruses occur in Britain, only two are of major importance, carnation mottle virus and carnation ringspot virus, although a number of aphid-borne viruses, such as carnation etched ring virus, can be significant in warm areas and on imported plants. Outdoor carnations are much less commonly diseased and then by carnation vein mottle virus, which is also important in sweet williams. Virus-free carnation stocks are available and should always be used, if possible.

CARNATION MOTTLE VIRUS: very common and widespread; probably present in almost all non-selected stocks but, although it markedly reduces vigour compared with virus-free plants, other symptoms are often indistinct. Pale, yellowish mottling or diffuse rings are all that are likely to be seen. It is readily spread through the handling of diseased plants; natural means of spread is unknown but a beetle or similar insect may be responsible.

CARNATION RINGSPOT VIRUS: common and widespread but probably now only in more neglected stocks. Although it commonly occurs jointly with carnation mottle virus, the symptoms are much more definite: lateral buckling and leaf-bending, often with pale brown tip necrosis; yellowish and partially necrotic rings, concentric rings and ringspots; overall plant stunting, with sometimes purplish colour at the bases of older leaves. The flowers are small, distorted, pale-coloured and reduced in number but the symptoms may be masked in hot summer conditions. Vigour is reduced so that fewer cuttings are available and those that are produced often root late or poorly. Readily spread through the handling of diseased plants; natural means of spread unknown.

CARNATION ETCHED RING VIRUS: occasional and widespread but usually associated with other aphid-transmitted viruses. Symptoms are superficially similar to those induced by carnation ringspot virus but the absence of leaf-buckling and the presence of rings etched into flowering stems are diagnostic. Usually seen in Britain on imported plants.

CARNATION VEIN MOTTLE VIRUS: locally common on sweet williams and sometimes on outdoor carnations, causing slight, diffuse yellowish spots and mottling on younger leaves. The number and quality of the flowers is reduced and flower-breaking can occur. All symptoms are more pronounced when carnation mottle virus is present also. Transmitted by the aphid *Myzus persicae*.

EUONYMUS

Mosaics and mottles occur occasionally on *Euonymus* spp., sometimes apparently caused by strawberry latent ringspot virus, and a type of virus-associated fasciation is also seen sometimes.

FICUS

FIG MOSAIC VIRUS: the only virus of any significance infecting figs. Common throughout northern Europe. The leaves develop irregular yellow-green blotches or bands, sometimes associated with the veins and with red-brown margins. Transmitted by the mite *Aceria ficus*. No effective treatment known and normally unnecessary.

FORSYTHIA

The commonest virus-induced symptom on forsythia is a net-like vein-yellowing, for which the eelworm-transmitted arabis mosaic virus may be one cause but a number of other viruses occur also.

FRAXINUS

Several species of ash and most other members of the family Oleaceae are prone to infection by a number of viruses. In Britain and northern Europe, oak leaf and other ring patterns are the commonest symptoms and the eelworm-transmitted arabis mosaic virus is a common cause. It is possible that ash and other hedgerow trees may serve as important sources of this virus for infecting other plants. In North America a virus seems to be associated with ash dieback but in Britain this seems to have a different cause (p. 311).

FREESIA

BEAN YELLOW MOSAIC VIRUS: common, especially in Dutch stocks; causes yellowish leaf-mottling and failure of the flowers to open fully. Transmitted by several aphids, including *Myzus persicae*, but non-persistent in the insects and chemical control therefore of little value. For further details see p. 438.

FREESIA STREAK VIRUS: a little-understood cause of an unmistakable white leaf-flecking which appears as rows of tiny white dashes and, later, often leaf senescence. It is probably soil-borne and it is therefore wise not to replant an infested site with freesias.

GLADIOLUS

There are four main viruses infecting gladioli in Britain and northern Europe; two are very widespread and aphid-transmitted; two less common and soil-borne but gladioli are important sources of virus infection for several other plants.

BEAN YELLOW MOSAIC VIRUS: very common and widespread, few cultivars being free of infection by this virus. It causes pale yellow mottling of the leaves and commonly thin dark streaks on the flowers of all except white, cream or yellow types. A reduction occurs in the number of corms produced by diseased plants and the life of the stock is probably also reduced. Transmitted by aphids, including *Aphis fabae, Myzus persicae* and *Acyrthosiphum pisum*. The virus is spread very rapidly to and from broad and dwarf beans, peas and sweet-peas, and gladioli should not be planted near these crops.

CUCUMBER MOSAIC VIRUS: causes white blotches on the flowers and a greyish leaf-mottling. The effects on growth are similar to those of bean yellow mosaic virus. See also p. 444. Gladioli are common sources of cucumber mosaic virus for infecting cucurbits, pansies, primulas and other garden plants.

Tobacco ringspot virus, tomato ringspot virus and other eelworm-transmitted viruses sometimes occur in Britain, especially in imported gladioli. The former may cause yellow ringspots on the leaves and white streaked flower-break while tomato ringspot virus induces the formation of abnormally short flower stalks and general stunting. If either are suspected in Britain, Ministry of Agriculture advisers should be consulted.

A so-called grassiness condition in which the leaves are narrowed, bunched and grass-like, occasionally occurs in gladioli and is presumably mycoplasma-induced.

GLEDITSCHIA
In common with other leguminous woody (as well as non-woody) plants, such as laburnums, robinias and wistarias, gleditschias are prone to infection by many viruses. Most commonly these are aphid-borne and induce mosaic type symptoms. Such plants should always be considered as possible sources of virus for the infection of other leguminous garden plants.

GOOSEBERRY
GOOSEBERRY VEIN-BANDING VIRUS: common and widespread on both gooseberries and currants. Pale yellow vein-banding occurs on the leaves, sometimes accompanied by some leaf distortion. A somewhat similar symptom can result from infestation by the aphid *Hyperomyzus lactucae* but the virus-induced symptom may be seen early in the season on the young unfolding leaves before aphids are present. Diseased bushes may be somewhat stunted and generally unthrifty. The virus is transmitted by the aphid *Nasonovia ribisnigri*. Remove and destroy badly affected bushes and replace with virus-free stock. Aphid control is probably of little value as the virus is non-persistent.

GRAPEVINE
Several virus and virus-like organisms infect grapevines but their significance in Britain is generally unknown. It is possible that the following four, all important elsewhere in northern Europe, are widespread in Britain also. Virus-free stocks are available in some areas.

GRAPEVINE YELLOW MOSAIC VIRUS: common and widespread throughout Europe although not definitely identified in Britain. It causes deep yellow mottling or more complete yellowing of the leaves in spring. The yellow areas may later turn necrotic and the leaves drop prematurely. Vines gradually lose vigour and degenerate. Soil-borne; transmitted by the eelworm *Xiphinema index*, a non-British species. Destroy affected vines and plant fresh stock on a new site, if possible. If planted on a fixed site under glass however, destroy the vines, remove as much soil as possible and preferably replace with fresh soil that has been sterilised.

GRAPEVINE FANLEAF VIRUS: common, widespread and very serious in many areas but not definitely identified in Britain. Symptoms are several and varied and none are absolutely diagnostic but most characteristic is a malformation of the leaves which take the form of a fan, although this is less pronounced on some cultivars. Ring and line patterns and green or yellow mosaic may precede the fanleaf development. Shortening of the shoots, fasciation, the abundant development of lateral shoots and fruit-dropping may all also occur. Vines gradually become unthrifty and in very warm areas with shallow soils, they may be killed. Like grapevine yellow mosaic virus, this disease is transmitted by the eelworm *Xiphinema index* and should be treated similarly.

GRAPEVINE LEAF ROLL VIRUS: occasional and widespread but is latent in rootstocks and also in many cultivars, which are distinguished by their general unthriftiness. From mid-season onwards, an inward rolling of leaves may develop, beginning at the base of the plants. The leaves feel thick and leathery and are of a brittle nature. Red-fruited (but not black-fruited) cultivars may bear grapes lacking in colour. No natural means of spread is known. Any vines showing signs of gradual decline should be replaced.

GRAPEVINE FLAVESCENCE DORÉE MYCOPLASMA: important but apparently largely restricted to certain old wine growing regions of France and Germany, in some of which it can be almost epidemic. Symptoms are most apparent after periods of sunshine. Growth is stunted and sometimes vines assume a weeping habit. The leaves become hardened and downward rolling and overlie one another in scale or tile fashion. Small black pustules may develop along the shoots and creamy-yellow spots appear on the leaf veins. The grapes shrivel, dry and drop. It is transmitted by the leafhopper *Scaphoideus littoralis* but in some vineyards a restricted form of the disease occurs in which there appears to be almost no spread from diseased to healthy plants. Destroy affected vines, replace with certified planting stock and maintain routine insecticide treatment to control leafhoppers.

HELENIUM

A green flower condition occurs occasionally, associated with phyllody, general flower degeneration and yellowing of the leaves. It is caused by a mycoplasma, presumably leafhopper-transmitted and affected plants should be removed and destroyed.

HORSE RADISH

TURNIP MOSAIC VIRUS: common and widespread and the most important virus affecting horse radish. Produces large yellow blotches on the older leaves in spring, giving a diffuse mosaic-like appearance. Plants may be quite severely stunted and later necrotic ringspots and flecks on the leaves, leaf stalks and veins may follow. Transmitted by the aphids *Myzus persicae* and *Brevicoryne brassicae*. Affected plants should be uprooted as they will provide sources of disease for turnips and other crucifers.

HOYA

See Asclepias, p. 438.

HYACINTH

Hyacinths and related liliaceous plants, such as galtonias, lachenalias and ornithogalums, very commonly have fine leaf mottling of white and dark green which may become more conspicuous and yellow or greyish as the leaves mature. On ornithogalums there may be light and dark green blotches on the flower stalks and streaking on the flowers. The cause may be either hyacinth mosaic virus or ornithogalum mosaic virus. Aphid-transmitted, several species probably being involved. Severely affected plants should be destroyed and replaced by fresh stock. Any individuals showing slight symptoms should be rogued out as quickly as possible.

HYDRANGEA

HYDRANGEA RINGSPOT VIRUS: very common and of worldwide distribution. Symptoms are very varied and include yellowish ringspots, pitting on the stems and leaf-mottling and distortion. Natural means of spread not known but spreads very easily by handling. As vigour can be depressed markedly, diseased plants are best destroyed. Among other virus-like diseases affecting hydrangeas, the one most likely to be seen by gardeners is a green flower condition, presumably caused by a mycoplasma. Affected plants should be destroyed.

IRIS

The only viruses of importance affecting irises are aphid-borne, although eelworm-transmitted viruses occasionally affect both bulbous and rhizomatous types, particularly in Holland. The aphid-borne viruses are all non-persistent and insecticides are therefore of little value in controlling the diseases they cause.

Viruses affecting rhizomatous irises

BEARDED IRIS MOSAIC VIRUS: very common and widespread; produces yellowing of the leaves of most cultivars, most markedly early in the season, the effects often fading in warmer temperatures. Flower-break commonly occurs on deeply coloured types. Probably does not affect any other host plants in Europe and is transmitted by the aphid *Macrosiphum euphorbiae*. Cucumber mosaic virus (p. 444) is the only other virus found commonly in rhizomatous irises but its effects seem slight.

Viruses affecting bulbous irises

IRIS MILD MOSAIC VIRUS and IRIS SEVERE MOSAIC VIRUS: symptoms are similar but of differing intensity: yellowing of leaves and, with severe mosaic, there may be stunting and flower-breaking. Transmitted by aphids, most commonly *Myzus persicae*. Mild mosaic is now the more prevalent of the two viruses and is almost universal in bulbous iris stocks.

JASMINUM

In common with other oleaceous plants, jasmine may be infected by a number of viruses, especially arabis mosaic virus. It is probable that this, or some other virus, is the cause of the deliberately-propagated yellow variegation. One of the earliest demonstrations of the graft-transmission of yellowing was in a species of jasmine around the end of the seventeenth century.

JUGLANS

A witches' broom symptom occurs occasionally on walnuts in some areas and is caused by a mycoplasma. Yellow ring patterns are also sometimes seen and are associated with infection by cherry leaf roll virus.

LACHENALIA

See Hyacinth, p. 450.

LATHYRUS

Many viruses infect sweet peas but only three or four are of importance in gardens in Britain and northern Europe. As may be expected, peas and beans have virus diseases in common with sweet peas and the likelihood of cross infection between the host plants should always be borne in mind when sources of virus are considered.

PEA ENATION MOSAIC VIRUS: varies enormously from year to year in incidence. The symptoms generally arise from mid-season onwards and are so characteristic as to be almost unmistakable; translucent windows develop on the leaves (note however that sawfly attack may be superficially similar although the small larvae can usually be found); flowers (especially red and blue flowered cultivars) show colour breaking and general loss of quality. Transmitted primarily by the aphid *Acyrthosiphon pisum* and since clover is another host plant for the virus, a source of the disease in gardens may be infected clover fields or presumably lawn clovers nearby. As the virus is non-persistent in aphids, however, insecticides are of limited value in preventing the disease from developing.

BEAN YELLOW MOSAIC VIRUS: symptoms are vein-clearing of the youngest leaves, followed by slight leaf-mottling and brown necrotic streaking of the stem and leaf stalks. There may be some flower-breaking. The virus is aphid-transmitted, but, like pea enation mosaic virus, is non-persistent. The source of the virus is uncertain as several strains of bean yellow mosaic exist, not all infecting all host plants and some capable of causing disease in plants outside the family Leguminosae.

PEA MOSAIC VIRUS: common; produces mottling and leaf yellowing, with some flower-breaking. Aphid-transmitted. See also under Pea (p. 458).

WHITE CLOVER MOSAIC VIRUS: causes leaf-mottling and flower distortion, accompanied by greatly reduced vigour; can be very severe in cool seasons but with few symptoms apparent in hot years. Natural means of spread unknown but highly contagious and transmitted by handling and other contact.

LETTUCE

There are four common and widespread lettuce viruses, each producing distinct symptom types.

BEET WESTERN YELLOWS VIRUS: first found in North America but known in Britain since the late 1960's and now common and probably widespread there and elsewhere. The symptoms are usually found on summer crops and are infrequent in autumn. The outer leaves of otherwise normal plants develop intense inter-veinal yellowing, usually two to three weeks before they are ready for cutting. Eventually the entire plant may become yellow or whitish, with dead brown margins to the leaves. The symptoms are very similar to, and may often have been confused with, magnesium or iron deficiencies but with virus infection, a few plants normally remain green and healthy, even in severe attacks. It commonly infects a number of weeds and probably survives the winter in them. Transmitted by aphids, especially *Myzus persicae*, and control of these on lettuce crops is essential throughout the life of the plants. Destroy old lettuce debris before planting a new crop and keep beds weed-free.

LETTUCE BIG-VEIN (transmissable agent): first found in California and confirmed in Britain and other parts of Europe since the mid-1950's. Now common and widespread. The main symptom is pale yellow or whitish vein-banding, usually most marked near the bases of the outer leaves and often accompanied by a bubbling or other distortion of the foliage. If infection takes place early, plants may be stunted but commonly there is little loss of yield. The causal agent is transmitted through the soil and into plant roots by swimming swarm spores of the fungus *Olpidium brassicae*. Control is very difficult and the agent can persist in the fungus in soil for fifteen or more years and is also maintained in related wild host plants. In gardens the best treatment is to raise plants in sterile soil in small peat pots and plant them out as large transplants. This delays the onset of the disease and minimises the damage it causes. If odd diseased plants are found, remove them carefully and also a spadeful of the surrounding soil.

CUCUMBER MOSAIC VIRUS: common in some seasons; produces symptoms similar to those of lettuce mosaic. See p. 444.

LETTUCE MOSAIC VIRUS: first found in North America but now common and widespread in Europe on outdoor plants. It is most serious on over-wintered lettuce, producing symptoms in early spring when the plants are stunted, pale green or mottled and often blistered or crinkly. Little heart forms and dramatic vein-clearing sometimes occurs. Later in the season, and especially on cos types, all plants start to bolt, and many small, often coalescing, dead spots appear on the leaves, predominantly at the edges. Vein necrosis may occur also but is unusual on cos types. Transmitted in seed and by aphids, most notably *Myzus persicae* but also *Aphis gossypii*, *Macrosiphum euphorbiae* and other species. The best means of avoidance is by the use of certified seed containing less than 0.01% infection: this can readily be obtained from large seed merchants and is essential if many plants are to be grown. Note however that the simple designation 'mosaic tested' is meaningless unless the level of infection is stated also. In gardens it is difficult to prevent aphid-transmitted virus infecting at least some plants however because infected plants in neighbouring gardens or allotments are usually close at hand. Control of aphids is however of little value for disease control because the virus is non-persistent. Some, mainly Dutch, cultivars are resistant, most notably 'Hilde' and 'Suzan' types, and are well worth trying if mosaic is a persistent problem.

LIGUSTRUM

Yellowing, sometimes vivid oakleaf and ringspot patterns as well as mosaics and crinkling all occur quite commonly on privets. Several viruses, including tomato bushy stunt, raspberry ringspot and arabis mosaic, may be the causal agents and the latter, although normally eelworm-transmitted, may occur in the absence of the vector, perpetuated by the vegetative propagation of the bushes.

LILIUM

Lilies, particularly *L. auratum*, are extremely susceptible to infection by viruses, especially those transmitted by aphids. Gardeners often find lily stocks have a much shorter working life than for instance irises and narcissi before they succumb to the build-up of viruses. The reason is that, in Britain at least, aphids colonise lilies and thus introduce enormous quantities of virus whereas on irises and narcissi they make only short probing feeds before moving on. Fortunately lilies can be propagated very rapidly and are easily replaced while virus-free stock will gradually become available. Three main viruses occur in garden lilies in Britain and northern Europe and all generally cause some shortening of the flower stem and rosette development.

TULIP BREAKING VIRUS: produces severe leaf-mottling, distortion of the flowers (sometimes giving rise to notched petals) and a general reduction in growth. Transmitted by the aphids *Myzus persicae, Macrosiphum euphorbiae* and *Aphis fabae*; diseased Darwin tulips commonly being the source of virus.

LILY SYMPTOMLESS VIRUS: a misnomer for although symptoms are not obvious, a general stunting does occur.

CUCUMBER MOSAIC VIRUS: similarly produces little effect, other than growth reduction, but when both this and lily symptomless virus occur together, characteristic elongate flecks develop on the leaves; at first yellowish but later grey or necrotic. The flowers may appear abnormal but this symptom is not invariable. Lily symptomless virus is transmitted by the aphid *Aphis gossypii*; for further notes on cucumber mosaic virus see p. 444.

LOBELIA
Slight leaf-mottling or distortion can occur on lobelias as a result of infection by the aphid-transmitted cucumber mosaic virus (p. 444) or eelworm-transmitted arabis mosaic virus, but the effects are rarely serious.

LUPINUS
Two aphid-transmitted viruses commonly infect garden lupins producing symptoms that are indistinguishable. The two are lupin mottle virus, probably a strain of the bean yellow mosaic virus/pea mosaic virus group of viruses, and alfalfa mosaic virus. The commonest symptom is a dark green vein-banding, with pale green or yellowish inter-veinal areas but various ringspots, oakleaf and other patterns can occur. Necrotic streaking on the flower stalks and colour-breaking can also be found. When the effects are serious, plants should be replaced with fresh stock.

MAGNOLIA
Mottles, line patterns, rings and oakleaf patterns occur occasionally on magnolias and are believed generally to be caused by cucumber mosaic virus (see p. 444).

MALUS
Almost certainly most, if not all of the virus and virus-like organisms causing symptoms in apples also infect ornamental malus species (see p. 435). Scions of most ornamental *Malus* spp. either fail completely to take or decline very rapidly after grafting onto apple rootstocks that have not been propagated from virus-free source material and many of the latent viruses of apple cause severe symptoms on certain ornamental cultivars.

MATTHIOLA
Stocks, like anemones, are commonly infected by anemone mosaic virus and cucumber mosaic virus (see p. 444). The predominant symptom is that

of flower-breaking, sometimes accompanied by leaf distortion and mottling.

NARCISSUS

Narcissi of all types are commonly affected by a number of viruses and, while most stocks contain, and tolerate, at least one, when two or three are present they degenerate rapidly. Although some viruses do relatively little apparent harm, the virus-free stocks now available are remarkably more vigorous and showy. Susceptibility to the viruses differs between *N. pseudonarcissus*, *N. tazetta* and *N. jonquila* while nerines are also susceptible to at least some of the narcissus-infecting viruses. The narcissus viruses fall into two groups, aphid-transmitted and eelworm-transmitted, with one additional important mechanically-transmitted virus. Only the aphid-transmitted viruses are likely to induce recognisable symptoms, however. Gardeners rarely see aphids on daffodils in Britain as they are infrequently colonised by wingless forms and, although visited by winged aphids, these are not usually noticed. It is probable that the aphid-transmitted viruses are carried mainly by migrating aphids. In general, control is not easy but, if possible, virus-free stock should be obtained and any plants with symptoms should be rogued out early in the season. Because eelworm-transmitted viruses may be associated with losses in vigour, new stocks should if possible be planted on a fresh site.

NARCISSUS YELLOW STRIPE VIRUS: ubiquitous on daffodils and jonquils and probably the most important of the narcissus viruses. It causes yellowish streaking and curling of the leaves and a distortion of the flower stalks. There may also be flower-breaking and general yield reduction. Spread slowly by *Macrosiphum euphorbiae* and many other aphids.

NARCISSUS WHITE STREAK VIRUS: common and widespread on *N. tazetta* types but, in Britain, rarely seen before the end of May. Induces longitudinal streaks, at first purple but later white and coalescing, on the leaves and flower stalks. The leaves later die but as this is commonly mistaken for natural senescence at the end of the flowering season, the symptoms go unremarked and virus-infected stocks are allowed to perpetuate. Transmitted by *Myzus persicae*, *Aphis fabae* and other aphids.

Other aphid-transmitted viruses, including cucumber mosaic virus (see p. 444), are usually more or less symptomless but bring about severe reductions in vigour. 'Grand Soleil D'Or' virus causes yellow leaf striping and general deterioration of that cultivar in the Isles of Scilly. Eelworm-transmitted viruses include arabis mosaic virus and several ringspot viruses but, in common with the mechanically-transmitted narcissus mosaic virus, all are virtually symptomless and associated with a general decline in thriftiness.

NERINE
See Narcissus, (above).

NICOTIANA
In common with many other solanaceous plants, such as potatoes, tomatoes, capsicums, daturas, eggplants, petunias and ornamental solanums,

all nicotiana species, including tobacco, are extremely susceptible to infection by a wide range of viruses. Many of the common virus-induced symptom types are likely to occur, most notably leaf mosaics and mottles and associated leaf distortion and flower-breaking. It is not possible to mention all the symptoms or common causal agents but the ubiquitous tobacco mosaic virus, cucumber mosaic virus, and, especially if planted near to potatoes or tomatoes, potato viruses X and Y are among the most frequent. If tobacco or other nicotianas are grown, a few general rules should be followed. Always maintain scrupulous aphid control; do not plant near to cucurbits, potatoes or tomatoes, to minimise the likelihood of two-way transfer of aphid-transmitted viruses; do not replant a site with potatoes or tomatoes if it has recently carried nicotianas, because of the possibility of a residue of soil-borne viruses; always wash hands and tools thoroughly after contact with nicotianas and before handling tomato or potato plants because a number of the nicotiana-infecting viruses are readily transmitted by handling.

ODONTOGLOSSUM
See Orchids, (below).

ONION and RELATED PLANTS
ONION YELLOW DWARF VIRUS: the only important virus affecting onions. Originated in North America but is now widespread in Europe. It is common in Britain on shallots (and causes a disease often known as shallot yellows) but is more serious elsewhere in Europe on leeks; less frequent on onions. If the virus is present in sets, these, when planted, produce leaves with short yellow streaks at the base. Eventually the leaves may turn entirely yellow, become crinkled, flattened, twisted and keel over. Bulbs from such plants are small but generally not misshapen. The virus is transmitted by many species of aphid notably *Aphis maidis*, *A. rumicis* and *Rhopalosiphum prunifoliae*. Important sources of virus are overwintered *Allium* spp. and in Holland the growing of leeks all year round is one reason for the importance there of this disease. Similarly, the increased growing of overwintered bulb onions may result in the disease becoming serious in Britain. It may be possible to buy sets certified free of disease but if they are kept from existing garden stock, it is essential to retain only those from plants with wholly green leaves. Removal and destruction of diseased plants is often very effective in limiting spread in gardens.

ORCHIDS
Virus diseases of orchids are very imperfectly understood. Most of the common virus-induced symptoms can occur in orchids but in relatively few instances is it possible to relate particular effects to individual viruses. In general, cattleyas are probably the most severely affected, followed by cymbidiums, dendrobiums and odontoglossoms. General methods of avoidance of virus problems are the use of virus-free stock, the rigorous control of aphids and sterilisation of cutting tools before use. Only the more fully understood orchid viruses are described below.
CYMBIDIUM MOSAIC VIRUS: commonest and most widespread orchid virus. Causes a mosaic taking the form of gradually enlarging, elongate and

usually sharply defined yellowish streaks with some black streaking on cymbidium leaves and irregular, elongate leaf necrosis in cattleyas. There is no flower-breaking although blooms may be few and small. Very easily transmitted on cutting tools but natural means of spread is unknown.

ODONTOGLOSSOM RINGSPOT VIRUS: common and widespread in odontoglossoms and other orchids. Causes necrotic spots or rings up to 2 cm ($1\frac{3}{4}''$) in diameter on older leaves or yellowish, circular, oval or diamond-shaped areas on younger leaves. There may be some premature leaf drop but there is no flower malformation. Means of transmission is unknown but probably by physical contact; affected plants should be destroyed.

CATTLEYA MOSAIC VIRUS: occasional and widespread; characterised by flower-breaking and sometimes by malformation in cattleyas, with an irregular streaked leaf mosaic, occasionally with a bumpy appearance. Diseased cymbidiums develop yellowish rectangular areas on the leaves. Transmitted by the aphid *Myzus persicae*, control of which is essential.

CATTLEYA BLOSSOM BROWN NECROTIC STREAK: widespread and disfiguring necrotic streaking on petals and sepals about one week after opening. May be long, yellowish irregular streaks on leaves. Caused by a combination of viruses, probably cymbidium mosaic virus and odontoglossom ringspot virus.

ORNITHOGALUM
See Hyacinth, p. 450.

PAEONY
PAEONY RINGSPOT VIRUS: common and widespread; causes distinctive irregular yellowish rings or sometimes an irreguar yellow mosaic on the leaves. Probably soil-borne but generally does little harm and can usually be tolerated in gardens.

PARSLEY
A widespread stunting (often to the extent that plants take on a rosette form), yellowing and sometimes, later, reddening of parsley are caused by the carrot motley dwarf complex (carrot mottle virus with carrot red leaf virus) (see p. 441). It is transmitted by the aphid *Cavariella aegopodii* and cross infection to carrots may occur. Control is scarcely justified in gardens but re-sow parsley as far as possible from the site of previously diseased plants.

PARSNIP
PARSNIP YELLOW FLECK VIRUS is the commonest cause, in Britain at least, of the frequent and widespread net-like veinal necrosis or yellowing which is sometimes followed by a faint yellowish mottling with yellow flecks. In common with other mosaic and mottle–inducing parsnip viruses, it is probably normally transmitted by the aphid *Cavariella aegopodii*. No treatment is justified in gardens, beyond routine aphid control. Diseased plants among healthy ones should however be uprooted and destroyed.

PASSIFLORA
The commonest garden species, *P. coerulea*, commonly shows leaf crumpling, puckering and/or a diffuse mosaic. These symptoms may be due, at

least in part, to infection by cucumber mosaic virus (p. 444) as this and another aphid-borne virus, passionfruit woodiness virus, cause serious disease in edible passion-fruits *(P. edulis)* in parts of the world.

PAULOWNIA
A witches' broom symptom occurs occasionally on paulownias and, although of no great pathological importance, is of interest as it was one of the first conditions in a tree shown to be associated with a mycoplasma.

PEA
Several viruses affect peas but the two commonest, most widespread and serious are pea mosaic and pea enation mosaic, although neither usually causes much damage in gardens.

PEA MOSAIC VIRUS: produces very variable symptoms but most commonly a marked leaf mottle with pale green, depressed areas, usually between the veins but not affecting the pods. The plants are stunted. Transmitted by aphids including *Myzus persicae* and *Aphis fabae*. Preventative measures are not justified in gardens but removal of badly affected plants is sensible.

PEA ENATION MOSAIC VIRUS: produces very severe mottling and crinkling of both leaves and stipules. May be pale yellow, white or sometimes necrotic leaf spots and on some cultivars diagnostic enations may occur on the undersides of the leaves. The pods develop poorly and become rough and crinkly. Transmitted similarly to pea mosaic virus and although the more serious of the two, similar treatment should usually be adequate.

PEAR
Several virus and virus-like conditions affect pears of which those described below are the most prevalent and important. General notes on biology and treatment, described under apple (p. 435) are in many respects applicable to pears also.

Organisms causing predominantly fruit symptoms
PEAR STONY PIT VIRUS: common and widespread throughout northern Europe. Serious on many cultivars, causing pits or dimples in the fruit. The fruit tissue at the bottom of the pits is usually necrotic and severely affected fruits are woody and difficult to cut. Means of natural transmission unknown although some spread can occur within an orchard and affected trees should therefore be grubbed out and replaced if possible with virus-free stock.

Organisms causing predominantly leaf symptoms
PEAR RING PATTERN MOSAIC VIRUS (apple chlorotic leaf spot virus): common and widespread throughout northern Europe. Symptoms when present are obvious and distinct pale green to yellow rings and line patterns on the leaves but many cultivars show little response to infection. No known effect on yield and little on fruit quality and therefore the disease can usually be ignored in gardens. When it occurs jointly with pear vein yellows virus (below) however, the effects can be more severe. Means of natural transmission not known.

PEAR VEIN YELLOWS VIRUS: common and widespread throughout northern Europe. Produces yellow banding on the finer leaf veins, most usually on young leaves only, and often accompanied by red flecking. It may result in significant reduction in yield as diseased trees are smaller than healthy ones. As the individual fruits are not impaired however and the yield per unit volume of tree is not reduced, it is doubtful whether replacement of existing mature trees is worthwhile in gardens. Can possibly be seed-borne but other natural means of spread unknown.

Organisms causing predominantly stem symptoms

PEAR DECLINE MYCOPLASMA: known only from France and Germany in northern Europe but widespread and serious in other parts of the world. Somewhat similar symptoms are known in Britain and elsewhere and can arise from a number of climatic, site, nutritional and other factors and accurate diagnosis of the condition is almost impossible without micro-scopical examination of the tissues. Affected trees show poor shoot growth, dying back of lateral and terminal shoots, upward rolling of leaves, which are reduced in size, a bright red autumn leaf colour and premature leaf fall. Trees may be killed or may survive to produce few small fruit. Natural means of transmission is by the pear sucker, *Psylla pyricola* (p. 156), but the problem can be avoided by the use of virus-free stocks.

PEAR BLISTER CANKER VIRUS: common and widespread throughout much of northern Europe. Small blisters and, later, cracking develop in early spring on one to two year old stems. Trees or branches are sometimes killed but may survive to produce less fruit than normally. Although, when slight, the symptom is little more than a bark roughening, when severe it can be confused with fungal canker (p. 310). Means of natural transmission is unknown. Severely diseased trees should be grubbed up and replaced with virus-free stock.

PELARGONIUM

A number of viruses induce a range of symptoms on pelargoniums but virus-free stocks are available and are remarkably more vigorous and showy than even the best of normal plants.

PELARGONIUM LEAF CURL VIRUS: widespread but inappropriately named as it does not induce leaf curling but small, pale yellowish spots, which may later become intensely yellow and star-shaped. Eventually the spots may become necrotic and the leaves puckered. These symptoms generally appear only in spring on cuttings taken from the previous year's plants and are not seen as the new foliage develops. Cuttings taken in autumn, however, again show symptoms in the spring. Unusually for viruses, the symptoms may be more intense in hot dry conditions. The problem is soil-borne but the vector is unknown. Among other symptoms sometimes seen are yellowish ringspots, caused by several soil-borne viruses, including pelargonium ringspot virus and pelargonium ring pattern virus, and a flower-breaking which appears as pale streaks on the petals and is caused by pelargonium flower-break virus. Gardeners should also be aware that some pelargonium stocks, especially of American origin, are commonly infected with tomato ringspot virus, a pathogen not well estab-lished in Europe. While causing relatively little harm to pelargoniums, it could spread to and be serious in a wide range of other crops as, although its natural eelworm vector does not occur here, closely related species do.

PETUNIA

In common with most solanaceous plants, petunias are susceptible to infection by a wide range of viruses and can be considered for practical purposes in the same way as nicotianas (p. 455).

PLUM and DAMSON

In common with other *Prunus* spp., plums are infected by many viruses but in several instances the diseases they cause are imperfectly understood. By far the most important is plum pox virus, which poses a serious threat to plum growing in several areas. The only general treatments that can be recommended to combat plum viruses are the use of virus-free stocks when planting new trees and the grubbing up and destruction of diseased stock. All of the viruses described below, except bark split virus, can be serious on other *Prunus* spp.

Organisms causing predominantly fruit symptoms

PLUM POX VIRUS: causes the disease known as sharka; common and widespread throughout much of northern Europe, except Scandinavia. It is devastating in southern Europe and has been known in Britain since 1966 in sporadic outbreaks. It is a very serious threat to commercial plum crops and also affects apricots, nectarines, peaches and several ornamental *Prunus* spp., including cherry plums *(P. cerasifera atropurpurea)*, dwarf flowering cherry *(P. japonica)*, Japanese plum *(P. salicina)* and flowering almond *(P. triloba)*, as well as related wild types. It does not occur on almonds or edible cherries. Symptoms vary between cultivars but generally the leaves develop pale green, diffuse spots and rings, not easily differentiated from other virus effects or from aphid and other damage. Fruit symptoms are more distinct and diagnostic, especially on red coloured types, such as 'Victoria', which show dark bands and rings, often with one sharp and one diffuse edge. On yellow and dark fruited cultivars, grooving and pitting is apparent on the fruit surface with a browning underneath and often premature fruit drop. The fruit is useless and acid tasting. Transmitted by aphids (in Britain at least by *Myzus persicae*) and this is the most important aphid-transmitted European fruit tree virus. Non-persistent in the aphids which are therefore unable to spread the disease across the English Channel, but this has proved no protection to British plums as many rootstocks carrying undetectable disease are imported. Any suspicious symptoms on plums or other host species in Britain should be reported immediately to Ministry of Agriculture officials. Prompt destruction of the trees and of any subsequently emerging suckers is essential.

Organisms causing predominantly leaf symptoms

PLUM LINE PATTERN VIRUS: common throughout northern Europe on plums and also on other *Prunus* spp. It is now known that at least two different viruses can cause the symptoms of line pattern disease, which are very variable and include yellowish line patterns, oakleaf patterns, ringspots, banding, striping and mosaics. There may also be some stunting and/or killing of shoots but this is rarely serious enough to warrant destruction of trees. The natural means of transmission is unknown.

PRUNUS NECROTIC RINGSPOT VIRUS: widespread on plums, except in the most northern areas but produces few or mild ringspot symptoms on the leaves. It reduces yield significantly however in many cultivars and this justifies the use of virus-free stocks. The natural means of transmission is unknown. See also cherries (p. 442) and peaches (p. 437).

Organisms causing leaf and shoot symptoms
PRUNE DWARF VIRUS: common and widespread throughout northern Europe on plums and also on other *Prunus* spp. The symptoms are usually well-marked; the leaves are small and narrowed with very irregular margins and thickened blades. Shoot growth is retarded severely and, despite the often abundant blossom, total fruit production is reduced drastically, although individual fruits are normal. The natural means of spread is unknown. Badly diseased trees should be grubbed out and replaced with virus-free stock.

Organisms causing predominantly stem symptoms
PLUM BARK SPLIT VIRUS: occasional through much of northern Europe. The symptom is a bark canker; brown-red areas develop on the bark and become sunken and often split. Dead areas of bark form around the splits and may result in the death of branches or trees when severe. Natural means of transmission is unknown. Badly affected trees are best uprooted and burned and replaced with virus-free stock. Also see: bacterial canker (p. 320).

POPULUS
POPLAR MOSAIC VIRUS: common in most poplar clones. Symptoms are variable but typically are diffuse irregular yellowish spots along and between the leaf veins. Some clones, such as 'Gelrica' and 'Robusta', produce a much more distinct mosaic pattern. Although serious reduction in timber quantity and quality can result in some clones, the disease is unimportant on garden trees.

POTATO
Potatoes are among the most important plant virus hosts. At least fifteen viruses and mycoplasmas affect potatoes in northern Europe although many are localised on particular types of site or in certain cultivars. Potato leaf roll virus and potato virus Y, which causes severe mosaic disease, are by far the most important. Most of the virus-induced symptoms in potato can fairly readily be distinguished from each other but not always from certain non-viral causes. The gradual deterioration in potato stocks, known as 'running out', which occurs when seed is saved from year to year, is due to the gradual build-up of viruses within the stock. The regular use of certified new 'seed' tubers from Scotland or other northern areas where aphids are less frequent will go far towards avoiding many virus problems. The symptoms produced by some of the viruses described below are divided into first-year symptoms, which arise when plants raised from healthy 'seed' are infected while growing, and second-year symptoms, which arise when plants are grown from diseased 'seed'. Most, except tobacco rattle virus, produce little effect in most cultivars on the tubers or yield in the year of infection and the second-year symptoms are the most important. As with many viruses, the symptoms are variable and those described below are the ones most commonly encountered.

Symptoms and Biology

POTATO LEAF ROLL VIRUS: still the commonest and most widespread potato virus, occurring even in the best stocks in Scotland and Northern Ireland. First year symptoms: slight upward rolling of the upper leaflets, often with a faint pinkish colouration but this is difficult to distinguish from similar effects with other causes and is often overlooked. Second year symptoms: upward rolling of the lower leaves a few weeks after emergence, spreading later to the upper leaves also. The rolled leaves are stiff and brittle (unlike plants infected for instance with black leg (p. 368), which may be superficially similar), and the plant may rattle if shaken. Transmitted by *Myzus persicae* and other aphids.

POTATO VIRUS Y: common and widespread; causes severe mosaic disease, sometimes in association with other viruses. The first year symptoms vary between cultivars. Most common are dark streaks or spots along the veins with the leaves then turning yellow and shrivelling before hanging by characteristic, threadlike stalks. On 'Arran Pilot', 'Ulster Chieftain' and a few other cultivars, only a mild mottling develops. Second year symptoms: plants are stunted with roughened or puckered leaves and indefinite yellow-green mottling with the tubers small and few in number. Transmitted by *Myzus persicae* and other aphids.

POTATO VIRUS X and/or POTATO VIRUS A: occasional and widespread, especially in low-lying relatively warm areas and, in the case of virus A, where large populations of aphids occur early in the year, such as in much of England, parts of Ireland and the Scottish Lothians. The symptoms are indefinite and ephemeral; faint leaf mottling, as with many virus diseases, is best seen on dull and overcast days. The yield of tubers may be reduced greatly. Transmission of potato virus X is probably by contact between plants, by animals such as rabbits and on clothing or tools; the wart disease fungus has been claimed to be implicated also. Potato virus A is transmitted by aphids, especially *Myzus persicae*.

POTATO PARACRINKLE VIRUS (potato virus M): an aphid-transmitted virus which, although it produces virtually no symptoms, reduces yield severely. For a long time 'King Edward' stocks were contaminated with it and often with the similar potato virus S also. 'King Edward' stocks in Britain must now, by law, be free of paracrinkle virus.

TOBACCO RATTLE VIRUS: common and important locally on light sandy soils, such as occur in north Norfolk and Lincolnshire, parts of the west Midlands of England and the east coast of Scotland. It causes one form of spraing disease (see also potato mop top virus, below). The symptoms may take the form of small yellowish rings on the leaves (but not on the stem) although most characteristic are brown, corky crescent-shaped marks in the tuber flesh when cut. It is a soil-borne virus transmitted by eelworms of the genera *Trichodorus* and *Paratrichodorus*.

POTATO MOP TOP VIRUS: like tobacco rattle virus (above), causes a disease known as spraing but is common in wetter areas, such as much of the 'seed' producing region of Scotland, most of Ireland and the Groningen region of Holland. Symptoms are external and internal. In some instances blistery rings are obvious on the tubers after washing, together with internal crescent-shaped marks and brown, bright yellow or most often pale yellowish 'V'-marks on the foliage. The name is derived from the mop-like appearance of the haulm. A soil-borne virus transmitted by the powdery scab fungus (p. 337).

MINOR POTATO VIRUSES: several viruses and mycoplasmas are of very restricted importance and the following are the only ones likely to be encountered by gardeners. The aphid-transmitted potato aucuba mosaic virus causes bright yellow leaf spots or mottles and internal tuber blotches in 'Majestic' and a few other cultivars in Britain but produces no symptoms at all in other areas, such as Holland. Tomato black ring virus is eelworm-transmitted and occurs most notably in Scotland and Germany. It most often produces necrotic leaf spots. Tobacco necrosis virus, transmitted by the soil-borne fungus *Olpidium brassicae*, in Britain affects a few cultivars, such as 'Duke of York', 'Dunbar Standard' and 'Maris Peer', producing tuber cracks or blisters which are very difficult to distinguish from those arising from other causes, such as gangrene (p. 358). Mycoplasma-induced witches' brooms occur on potatoes but only infrequently in Britain and are more likely to be seen in warmer parts of Europe, where they are commonly called stolbur. Like most mycoplasmas, the causal agents are probably transmitted by leafhoppers.

Treatment. There are several ways to minimise potato viruses and their effects. The regular use of certified 'seed' grown in northern areas will ensure that the tubers are virtually free of aphid-transmitted viruses at planting times. Several cultivars possess resistance to one or more virus-induced diseases and these should be selected for growing in areas where those diseases are known or are likely to occur. Conversely, particularly susceptible cultivars should not be used in such regions: those susceptible to tobacco rattle virus, for instance, should not be planted on sandy soils. The following table gives the reactions of several commonly grown cultivars to the most important potato virus diseases.

Cultivar	Potato Leaf Roll	Potato Mop Top	Potato Virus X	Potato Virus Y	Tobacco Rattle
Arran Comet	M	U	M	H	H
Arran Pilot	M	H	M	H	U
Craigs Royal	M	U	U	M	H
Desirée	M	U	M	U	H
Home Guard	M	U	M	M	U
King Edward	M	U	U	H	M
Majestic	M	U	H	H	H
Maris Page	M	U	H	M	M
Maris Peer	M	U	M	M	H
Maris Piper	M	U	U	M	M
Pentland Crown	U	H	M	U	H
Pentland Dell	M	U	U	M	H
Pentland Ivory	M	U	M	U	H
Record	H	U	M	H	U
Stormont Enterprise	M	H	M	M	U
Ulster Chieftain	H	U	M	U	U
Ulster Prince	M	H	M	U	H
Ulster Sceptre	M	H	M	H	H

H = highly likely to be affected   M = moderately likely to be affected
U = unlikely to be affected

The control of aphids is important to prevent viruses being brought into the crop from outside sources or spreading from odd diseased plants within. The use of insecticides early in the season, when aphid populations are at their height, is therefore particularly critical. The prompt removal and destruction of odd diseased plants within an otherwise healthy crop is also important in limiting disease spread. In gardens where large numbers of potatoes are grown and the soil-borne viruses, such as tobacco rattle, mop top or tobacco necrosis are persistent and damaging, the use of dazomet granular soil-sterilant in accordance with the manufacturer's instructions may be worthwhile.

PRIMULA
Yellowish leaf mosaic and often an accompanying severe stunting, usually caused by cucumber mosaic virus, as well as a number of ringspot symptoms induced by soil-borne viruses occur occasionally on primulas. A mycoplasma-induced green petal condition is also sometimes seen but none of these problems is likely to be serious in gardens.

PYRUS
It is probable that ornamental *Pyrus* species are prone to infection by most if not all of the viruses and virus-like organisms occurring in pears (see p. 458).

QUERCUS
Although a wide range of virus-like symptoms is commonly seen on oaks, especially *Quercus robur* and *Q. petraea*, there is as yet little evidence of the nature of the causal agents. Symptoms to be expected include green and yellow mosaics, necrotic spots, shot holes, vein clearing, general yellowing, interveinal yellowing, witches' brooms, leaf rosetting and some general growth stunting.

RASPBERRY and LOGANBERRY
Many viruses affect raspberries and in some instances several combine to give rise to particular disease symptoms. Four main groups of virus and virus-like diseases are important in Britain and northern Europe and of these *Rubus* stunt is described under blackberry (p. 439), its more significant host plant. Virus-free stocks of raspberries are available and should be used whenever new plantings are established.

RASPBERRY MOSAIC: a very wide range of mosaic and mottling symptoms occurs on the leaves of raspberries and loganberries and also on blackberries and related plants. The symptoms vary greatly in intensity and may be accompanied by stunting and rosetting of the plants, poor fruit formation and occasionally plant death. On blackberries a brilliant yellow vein-clearing sometimes occurs although on raspberries this symptom is most likely to arise because of infection by arabis mosaic virus (below). At least five different viruses may be implicated and it is not practicable or necessary to be able to relate specific symptoms to each. In Britain and northern Europe all are transmitted by aphids (usually *Amphorophora rubi* or *Aphis idaei*) in a non-persistent manner, therefore control of the insects does little in minimising the effects of the viruses. Obviously diseased plants should be removed and burned, as should any showing general unthriftiness without mosaic symptoms as some cultivars are symptomless carriers of some of the mosaic viruses. Wild raspberries and blackberries are also important sources of virus and should be destroyed when near to cultivated plants. The cultivars 'Delight', 'Leo' and 'Malling Orion' have resistance to *A. rubi* and therefore escape most aphid-borne viruses. The cultivar 'Malling Admiral', although lacking this resistance to aphids, does not seem susceptible to virus infection.

RASPBERRY BUSHY DWARF: an occasional and widespread disease seen almost exclusively on the cultivar 'Lloyd George' and characterised by severe stunting and a greasy appearance to the leaves, which curl downwards. Although affected plants contain raspberry bushy dwarf virus, this, at least on its own, does not cause the symptoms. It is seed- and pollen-transmitted and, in common with the viruses described below, which can also be seed-transmitted, a crumbling of the berries may occur, but this is not diagnostic. If 'Lloyd George' plants are severely affected, change to a different cultivar.

ARABIS MOSAIC VIRUS: causes the disease known as raspberry yellow dwarf, which is locally important in raspberries in England but infrequent elsewhere. Affected plants usually occur in patches and are characterised by small yellow leaf spots (reminiscent of a mild mosaic symptom) or, usually on the lower leaves, yellow vein clearing. The plants are stunted and produce few fruits. The virus is transmitted by the eelworm *Xiphinema diversicaudatum* (which also transmits the commonly associated strawberry latent ringspot virus) and is also capable of seed transmission. Remove and destroy affected plants and do not replant the same site with fresh stock. A few other soil-borne viruses (transmitted by longidorid eelworms) are also important, especially in Scotland, and include raspberry ringspot virus, which causes raspberry Scottish leaf curl disease, and tomato black ring virus. They induce yellowish ring spots on the leaves, which may show downward curling and brittleness while the canes may also be brittle and stunted. These symptoms also occur on plants in patches and may be treated similarly to yellow dwarf disease. In common with raspberry bushy dwarf virus these soil-borne viruses can induce the formation of crumbly berries but this symptom is not diagnostic and among other causes of it are environmental effects, damage by certain insect pests and genetic abnormalities.

RHUBARB

Several symptom types are common and widespread and usually most severe in spring and early summer. Among those most frequently seen are small ringspots, mottling, and yellowish-light green irregularly-shaped leaf patches. The commonest of several viruses responsible are turnip mosaic, cherry leaf roll and arabis mosaic, but not all occur together in all cultivars. Turnip mosaic virus is transmitted by aphids, of which *Myzus persicae* is probably among the most important, but spread in this way seems slow. Arabis mosaic virus and other less common rhubarb-infecting viruses are transmitted by eelworms, especially *Xiphinema diversicaudatum*. Virus-free stocks are available and re-infection of these by viruses is fairly slow, so they may be effective for several years. Once disease is present however, rhubarb will serve as a serious source of virus for other garden plants and should then be destroyed and, because of the danger from the soil-borne viruses, fresh stock should be planted elsewhere.

ROBINIA

In common with other leguminous plants, robinias are prone to infection by many viruses and among the symptoms most commonly seen are mosaics of several types, with associated leaf deformation. The aphid-transmitted robinia mosaic virus is one cause of these symptoms while a mycoplasma-induced witches' broom is also seen occasionally.

ROSE

Although virus-like symptoms are common and widespread in roses and a number of viruses have been identified, the causes are still imperfectly understood. The following are those generally considered most important, at least in Britain. Plants showing suspicious symptoms should be destroyed and replaced with fresh stock obtained from a reputable grower.

ROSE MOSAIC: the commonest virus disease of roses with very variable symptoms, including various types of yellow mosaics, bands, oakleaf and line patterns, sometimes associated with dwarfing, leaf puckering and other effects. A number of viruses have been shown to be associated with the condition and of these the most important seem to be *Prunus* necrotic ringspot virus and apple mosaic virus. Natural means of spread unknown.

STRAWBERRY LATENT RINGSPOT VIRUS: probably the most important virus infecting roses. The symptoms are characteristic yellow leaf flecks or a faint yellow vein mottle with a gradual decline in plant vigour and sometimes ultimately death. Transmitted by the eelworm *Xiphinema diversicaudatum*, in common with arabis mosaic virus which also sometimes occurs in roses, usually as latent infection. New stock should be planted on a fresh site.

ROSE WILT (graft-transmissable agent): causes a very serious rose disease in Australia and New Zealand and, although not proved to be present in Europe, symptoms similar to those it induces have been seen in recent years in England and described as rose stunt, rose proliferation or dieback. Masses of small stunted shoots arise from around the scion bud of maiden plants or one or more long tapering shoots develop and eventually cease growth. Natural means of transmission unknown. The problem is only likely to be seen in nurseries where roses are propagated and suspicious symptoms on such sites should be referred to Ministry of Agriculture officials.

SAMBUCUS

Yellow vein-banding and sometimes ringspots and mosaics, crinkling and leaf rolling occur commonly, not only on elder *(S. nigra)* but also on *S. canadensis, S. ebulis, S. racemosa* and other cultivated species. Many different viruses, including arabis mosaic virus, tomato black ring virus and cherry leaf roll virus may be responsible.

SOLANUM

All of the ornamental species of *Solanum* are liable to be infected by any of the viruses infecting other solanaceous plants, such as potatoes and tomatoes. General notes are given under Nicotiana (p. 455).

SPINACH

Cucumber mosaic virus is common and widespread on spinach, causing the disease known as spinach blight. The symptoms are yellowing (sometimes inter-veinal) of the younger inner leaves and later, more seriously, on the outer leaves. The inner leaves distort, pucker and may show inrolling of the margin and dwarfing. The outer leaves turn yellow, go limp and lie flat on the ground. Eventually growth ceases and the leaves become infected by rotting organisms, resulting in total death of the plant or reducing it to a tiny rosette. Some cultivars have resistance and those that combine this with resistance to downy mildew are the most useful.

STAPELIA

See Asclepias, p. 438.

STEPHANOTIS

See Asclepias, p. 438.

STRAWBERRY

Several viruses and virus-like organisms infect strawberries and symptoms of some diseases are invariably induced by a combination of viruses acting together.

Symptoms and Biology

STRAWBERRY MILD YELLOW EDGE VIRUS with STRAWBERRY MOTTLE VIRUS or, less frequently STRAWBERRY CRINKLE VIRUS: common and probably widespread throughout northern Europe, producing the disease known as yellow edge. The symptoms are a general stunting of the plants and the development of yellow margins to the younger leaves, apparent in late summer. There is considerable variation in cultivar susceptibility and some newer types develop few symptoms.

STRAWBERRY MOTTLE VIRUS, in combination with either STRAWBERRY CRINKLE VIRUS or STRAWBERRY VEIN CHLOROSIS VIRUS: produces a condition known as crinkle. Symptoms develop in early summer and/or autumn and take the form of yellowish leaf spots which may become necrotic and accompany crumpling of the leaves. All the above viruses are transmitted by the aphid *Pentatrichopus fragaefolii*.

ARABIS MOSAIC VIRUS, RASPBERRY RINGSPOT VIRUS and TOMATO BLACK RING VIRUS: common and probably widespread, the former most frequent in England and Wales, the latter two in Scotland. A wide range of symptoms develops, including general stunting, yellow mosaic, spots or blotches (which may later turn red) on the leaves and leaf crinkling, but the foliage symptoms may fade during the height of summer. If infestation is present in the soil, then symptoms may occur in patches; if carried on planting stock then it is more likely to be seen along the rows. Arabis mosaic virus is transmitted predominantly by the eelworm *Xiphinema diversicaudatum* while the other two viruses are transmitted by eelworms of the genus *Longidorus*.

STRAWBERRY GREEN PETAL MYCOPLASMA: occasional and widespread. Young leaves are stunted, yellowed and may be irregularly shaped. After flowering, the older leaves become reddish. The flowers are usually small and distinctly green. The fruit are either absent or remain small, unswollen and worthless. Transmitted by the leafhopper *Aphrodes bicinctus* and, to a lesser extent, *Euscelis lineolatus*, the pathogen usually coming from clover plants, on which it gives rise to phyllody disease.

Treatment. Always use certified stock for establishing new beds and maintain its health by controlling aphids and leafhoppers with a systemic insecticide. Runner beds should be sprayed more frequently than fruiting plants, to protect them from infection by virus carried by migrant winged aphids during the summer. Do not replant strawberries on a site which has been affected by eelworm-transmitted virus symptoms. If a large area is involved and no alternative site is available, partial soil sterilisation with the granular material dazomet, according to the manufacturer's instructions, is worthwhile.

SYRINGA

Virus-like symptoms occur quite commonly on lilacs and most frequently seen are leaf malformations of varying intensity, rings and line patterns of numerous types. The most usual form of this condition is known as lilac ringspot but no specific virus has been shown to be the cause, although lilacs quite commonly contain a number of viruses, including arabis mosaic virus, elm mottle virus, lilac ring mottle virus and tobacco rattle virus. Care should be taken to select for propagation only plants completely free of symptoms and *Ligustrum* should not be used as a rootstock as it too is commonly virus-infected.

TOMATO

There are three very widespread viruses affecting tomatoes, with tobacco mosaic virus outweighing all others in importance. A common symptom, especially on outdoor tomatoes, that is often thought to be due to virus attack is upward rolling of the leaves, especially in hot weather. It arises usually however from overfeeding or overstripping the foliage and is quite harmless.

TOBACCO MOSAIC VIRUS: exceedingly common and widespread, especially in glasshouses, causing the disease known as tomato common mosaic.

Symptoms. Vary depending on the strain of virus concerned and environmental conditions.

Foliage: most familiar is a leaf mottling of very variable colours, ranging from an indefinite pale green patchiness to a vivid mosaic of bright yellow and green. This is often accompanied, or sometimes preceded, by wilting of the youngest leaves on sunny days, differentiated from similar symptoms arising from other causes by the absence of root damage or stress factors, such as waterlogging or drought. Groups of leaves, (but not usually all, cf. herbicide damage p. 491) may be puckered or dwarfed and, especially in winter, show progressive narrowing and indentation (the 'fern-leaf' symptom) until in severe instances they may be almost tendril-like. The entire plant is commonly reduced in size and becomes pale green and spindly with few fruit.

Fruit: tobacco mosaic virus may be one cause of 'dry set' which affects the two or three trusses above the one flowering at the time when infection occurs. This localised effect may help differentiate the cause from other, more general ones (see p. 482). It is still uncertain which of many abnormal ripening effects of tomatoes are due to the virus but the most important are bronzing or internal browning and pitting. With the former, brown patches develop beneath the surface, especially near the calyx. They are most obvious on older but still green fruit and are commonly masked as the fruit reddens, although in severe instances they may remain discoloured. Pitting is more extreme and takes the form of grey-green areas localised around pits or streaks radiating from the stem end. The fruit symptoms are often confused with blotchy ripening (p. 476) but in the latter the fruit remains completely normal while still green. Occasionally long, thin dark brown streaks may occur on the stem.

Biology and Treatment. This virus is transmitted mechanically and in the seed and is exceedingly infectious, the merest physical contact with contaminated material being sufficient to initiate infection. It is not transmitted by aphids. Treatment depends on prevention by scrupulous hygiene and the importance of this cannot be overstressed. Seed-borne infection may be treated most easily by heat and it is easiest if seed can be bought ready-treated as the procedure requires 70°C (158°F) (dry heat) for four days. Not all virus is eliminated by this process. Gardeners saving their own seed should always select from healthy plants and preferably from fruit of the first or second trusses. Raise seedlings well away from any established plants and do not grow any tomatoes in the immediate vicinity of antirrhinums, peppers, petunias, potatoes or tobacco, all of which are very common sources of the virus. Sometimes the streak symptom described above can be caused by the combined effects of tobacco mosaic virus and potato virus X and for this reason particular attention should be paid to the removal of 'stray' potato plants from tomato beds. Remove and destroy diseased seedlings immediately and thoroughly wash hands and any implements used with hot soapy water after this and any other operation involving the handling of infected material. Transplant seedlings as soon as possible, since susceptibility to disease increases with age, but take care not to place very young plants into cold soil as this may render them prone to infection by wilt (p. 306) or root rot (p. 361). Destroy all plant debris at the end of the season and, if possible, use fresh medium, such as peat bags, for raising plants. If soil is used in glasshouses, maintain an interval of three years between successive tomato crops; outdoors rotate tomatoes with crops other than potatoes and tobacco. Chemical soil partial sterilants will not eradicate virus from plant debris. Do not smoke in glasshouses where tomatoes are grown as the virus may be present in the tobacco, although infection from this source is not common. Wash hands after smoking and before handling tomatoes. Some tomato cultivars have resistance to tobacco mosaic virus and among the best of those available to gardeners are 'Kirdford Cross' and 'Pagham Cross'. Such cultivars should not be grown in mixture with susceptible types however as this may increase the likelihood of the resistance breaking down.

CUCUMBER MOSAIC VIRUS: occasional, especially on outdoor plants; produces leaf mottling similar to that of tobacco mosaic virus and also a fern-leaf symptom at all times of the year. (See p. 444).

TOMATO ASPERMY VIRUS: occasional, especially on outdoor plants in the vicinity of chrysanthemums, from which the disease is spread by *Myzus persicae* and other aphids. Plants develop a bushy appearance with distorted and mottled leaves. Commonly the fruit are small and often seedless. Do not plant tomatoes close to chrysanthemums and maintain good aphid control on both crops at all times. See also p. 443.

TREES and SHRUBS

Because tree and shrub virus diseases have been generally less well studied than those of other plants, a number of quite commonly occurring symptoms have unknown but presumed virus or virus-like causes. The following (among of course, many other) plants are frequently found to show some of the possible symptom types listed on p. 431 but very little is known of their cause, biology or treatment: berberis, chaenomeles, clematis, cornus, crataegus, escallonias, hederas, kerrias, laburnums, medlars, mulberries and philadelphus.

TROPAEOLUM

Flower-breaking, mosaic mottling, leaf crinkling and distortion and general stunting or pale greenish leaf-spots, sometimes ring-shaped, are seen occasionally on garden nasturtiums. Among the most usual causes are the aphid-transmitted turnip mosaic and broad bean wilt viruses.

TULIP

Several viruses commonly cause symptoms in tulips but tulip breaking virus is easily the best known and most important. The effects it causes can be seen in tulips depicted by Dutch flower painters of the 17th century.

TULIP BREAKING VIRUS: occasional and widespread. Causes colour breaking on the petals of pink, purple and red flowered cultivars although white and yellow flowered types are not affected. Breaking may take the form of white or yellow streaks, of streaking of a darker shade than the normal petal colour or of a mixture of the two. In some cultivars a mottling or striping of the leaves also occurs. Transmitted in glasshouses and gardens by the aphids *Myzus persicae, Macrosiphum euphorbiae* and *Aphis fabae* and, because of the time of appearance of these insects, the symptoms are seen most commonly in late flowering tulips. The retention time is very short in the aphids and insecticides are therefore of little value as control measures. The best means of treatment is by very careful rogueing and removal of affected plants. White and yellow flowered cultivars should be examined carefully for leaf symptoms but tulips should not be planted near to lilies which can also be infected and are thus a potential source of virus.

TOBACCO NECROSIS VIRUS: causes Augusta disease which appears mainly on early flowering types as brown necrotic streaks on the leaves, stems and/or insides of young bulbs. In severe attacks plants may be killed. Transmitted by swarm spores of the soil-inhabiting fungus *Olpidium brassicae*. Remove and destroy affected plants as soon as they are detected. The eelworm-transmitted tobacco rattle virus sometimes causes elongate yellowish or transparent leaf flecks on the leaf veins when tulips are grown on sandy soil.

CUCUMBER MOSAIC VIRUS: sometimes serious after mild winters, causing corky fleck, which appears in late flowering cultivars as grey-brown sunken spots, sometimes in the form of arcs and rings on the outer and inner bulb scales. Symptoms appear during storage and may predispose bulbs to attacks by fungi and mites. Affected bulbs give rise to deformed plants, sometimes with aborted or broken flowers and yellowish or necrotic leaf streaks. See also p. 444.

ULMUS
Virus-like symptoms occur commonly on elms and one, probably frequent, cause is elm mottle virus. The effects usually seen are ringspots and line patterns on the leaves. A very serious condition caused by a leafhopper-transmitted mycoplasma is elm phloem necrosis which, although common in North America, is not known to occur in Europe, although it may do so. The gross symptoms are not easily distinguished from Dutch elm disease but in its late stages, before tree death, the inner bark of the trunk base and large roots become first golden yellow and then dark brown.

VACCINIUM
Virus-like symptoms occur quite commonly on *Vaccinium* and on *V. myrtilus*, a mycoplasma-induced witches' broom is sometimes seen.

VIOLA
CUCUMBER MOSAIC VIRUS: sometimes infects violas and causes a curling and yellowing of the leaves with a rather hazy and indefinite appearance to the petal colouration (see p. 444).

VITIS
See Grapevine, p. 449.

WATERCRESS
Turnip mosaic virus is quite common, especially on green cress. It causes yellow-green mottling or yellowish leaf spots and is transmitted by the aphids *Myzus persicae* and *Brevicoryne brassicae*. If plants are propagated from cuttings, use only material with fresh green leaves and reject those with any trace of yellowish discolouration. If possible, raise plants from seed as this never carries the virus.

WISTARIA
Several viruses are known to infect wistarias but the fairly frequently seen bright vein yellowing or diffuse spotting and mottling on the leaves is commonly caused by the aphid-transmitted wistaria vein mosaic virus.

ZANTEDESCHIA
See Begonia, p. 439.

Disorders

A healthy growing plant is in harmony not only with the biological environment but also with its physical components. These can at times operate to the plant's detriment however and in many instances, their effects produce distinct and recognisable symptoms. These symptoms, produced without the intervention of pest or pathogen, are termed disorders. For convenience, in this book, disorders are divided into those that arise purely from nutrient deficiency and those that arise from non-nutritional causes.

MINERAL NUTRIENT DEFICIENCIES
(Plates 21–22)

It is common knowledge that all garden plants require chemical nutrients from the soil and that, from time to time, supplementation of the natural supply is necessary in the form of fertilisers; either 'organic' (composts, farmyard manure, dried blood, bone meal and so on) or 'artificial' (sulphate of ammonia and super-phosphate for example). The soil chemical elements essential to plants are nitrogen, phosphorus and potassium (all required in large amounts) and calcium, magnesium and sulphur (required in lesser amounts), these six being known as major nutrients. Additionally plants require smaller amounts of the minor or trace elements, iron, manganese, boron, copper, molybdenum, zinc and sometimes, sodium and chlorine. A deficiency of any of these elements gives rise to specific symptoms and in this section an attempt is made, not to give general guidance on fertiliser practice, but to indicate the principal symptoms caused by the various deficiencies.

In general, deficiencies of nitrogen, and to a lesser extent of phosphorus and potassium, are readily recognised by gardeners as overall unthrifty growth. Certain of the other deficiency symptoms are however sometimes difficult to distinguish from each other, or from other quite different causes, and reference should be made to Plates 21–22 to give supporting information to the written descriptions. It is also very important to note whether symptoms appear first on the youngest or the oldest leaves as this can give a valuable clue to the element concerned. As with most leaf symptoms however, attempts to diagnose mineral deficiencies should not be made in the autumn when normal seasonal necrosis produces confusing effects.

There is chemical interaction between the elements in the soil and an excess of one element can give rise to deficiency symptoms of another. In such instances the second element may be present in adequate amounts but be chemically unavailable to the plants. Such an effect is known as induced

473

deficiency and although the same phenomenon can also arise under certain physical conditions, such as waterlogging, the most important predisposing factor is incorrect soil pH. All the trace elements except molybdenum are more or less unavailable to plants in alkaline soils, and even a pH of 7 (i.e. neutrality) can result in symptoms of nutrient deficiency. In gardens, soil alkalinity is probably the commonest cause of trace element deficiency symptoms. In addition to confusion with each other, the foliar symptoms of the various nutrient deficiencies are most commonly confused with those of virus-induced diseases or with environmental effects, such as air-borne pollutants, drought, frost or wind. Deficiency symptoms have been fairly intensively studied on fruit and vegetable crops but little on ornamentals, although they may be expected to show the same general features.

Deficiencies of sulphur are virtually unknown in British soils while chlorine, sodium and zinc deficiencies are also uncommon and restricted to very few crops. These four elements are therefore not considered here in detail. Excessive amounts of some mineral elements, may have damaging toxic effects on plants and these are referred to on p. 491.

MAJOR NUTRIENTS

NITROGEN   Pl. 21
Nitrogen can be, and commonly is deficient in almost any type of soil (although those of low organic matter content are most prone) and it is much influenced by the way in which the land has been cropped. In gardens, when little fertiliser is used, and especially when peas and beans are not grown (see p. 428) and brassicas repeatedly are, nitrogen deficiency symptoms may be expected. Sometimes soils, such as peats containing ample organic matter which is insufficiently decomposed, can also become deficient in nitrogen. Many different types of plant are affected but brassicas are particularly prone and the symptoms are generally readily recognisable.

Symptoms. The leaves are reduced in size, pale coloured and sometimes, as on brassicas, with yellow, red or purplish tints. The symptoms appear first on the older leaves and then spread upwards. All growth is restricted and weak, and flowering and fruiting may sometimes be reduced and delayed. Potatoes produce few tubers; apples tend to be smaller and red (not always undesirably so) and the many large, leafy vegetables yield little usable produce.

Physiology. Nitrogen is essential for many aspects of plant growth and is a major constituent of protein, of protoplasm and other plant components and is incorporated into a wide range of organic chemical compounds. Many nitrogenous materials are highly mobile and readily transported to young from old leaves, which explains why the deficiency symptoms therefore appear there first. Atmospheric nitrogen gas is 'fixed' or converted to forms usable by plants by the nodule bacteria on the roots of leguminous crops (see p. 428).

Treatment. Apply top dressings of nitrogen-containing fertilisers, such as ammonium sulphate, at or before planting and periodically thereafter in accordance with manufacturer's directions. Adequate nitrogen is commonly supplied with phosphorus and potassium in balanced NPK fertilisers but it should be remembered that nitrogen, unlike phosphorus and potassium, is readily leached from the soil and fresh applications must be made for each crop. In gardens however the regular application of compost provides a steady supply of nitrogen that is not as readily leached as that in 'artificial' fertilisers and many gardeners find they are able to satisfy all the soil's needs in this way.

PHOSPHORUS   Pl. 21

This is most usually deficient in acid soils, from which it is readily leached, and also in those in areas of high rainfall. Deficiencies occur additionally in many clay soils and in the poor soils of chalk downs. Phosphorus deficiency effects are generally different from those of nitrogen although there may be confusion between the two on brassicas. A wide variety of other plants may also commonly be affected.

Symptoms. As with nitrogen deficiency, all growth is restricted and weak and flowering and fruiting are reduced and delayed. The leaves are reduced in size and, beginning with the oldest, may drop prematurely. Unlike nitrogen deficiency, it rarely induces yellow or red colours in leaves, which instead become dull blue-green with dull purple tints. The latter symptoms can also however arise from a number of other causes, including various types of root damage. Currants typically show dull bronzing with brown-purple spots and although a few plants, such as potatoes, may have scorching at the leaf margins, this is not a general response.

Physiology. Like nitrogen, phosphorus is involved in many aspects of plant growth, including the chemistry of fats, proteins and carbohydrates, and it is particularly important in the ripening of fruits and the ripening and germination of seeds. Because phosphorus is closely associated with the functioning of nitrogen in the plant, it is surprising that there are not more similarities between the deficiency symptoms they each induce.

Treatment. Fertilisers containing phosphorus should be applied according to manufacturer's directions and this is done most readily in combination with nitrogen and potassium in balanced NPK fertilisers. Phosphorus alone is most readily available as super-phosphate, which should be placed as close to the seeds or plants as possible; in commercial practice it is commonly applied with the seed. The reasons for this are complex but germinating seeds and very young plants seem to require a particularly high phosphorus concentration.

POTASSIUM   Pl. 21

Potassium is commonly deficient in light sandy soils and also on chalky and peaty soils low in clay content. Where potassium demanding crops, such as very leafy vegetables, potatoes, tomatoes, beans and many fruits, are grown intensively, deficiencies can occur even on clay soils and these crops most commonly show symptoms.

Symptoms. Variable, but most typical is browning or scorching of the tips and margins of the leaves, often accompanied by brown spotting on the undersides. These effects appear first on the older leaves and the affected leaf edges are often curled. Somewhat similar symptoms of marginal scorching can occur as a result of exposure to drying winds, drought or air-borne pollutants, including salt spray. If these seem improbable causes, potassium deficiency should be strongly suspected, although sometimes deficiencies of phosphorus or calcium can also bring about similar effects. Additional symptoms may include a blue-green leaf colouration, some inter-veinal yellowing or spotting, general plant stunting and dying back of shoots. On tomatoes, potassium deficiency may be a cause of blotchy ripening; the fruit appear normal while still green (cf. tobacco mosaic virus p. 468) but unripened areas persist as it reddens.

Physiology. This is very imperfectly understood but potassium is known to be highly mobile in plant tissues and accumulates particularly in young leaves and at growing points. It is believed that most of the element is present in the cell sap rather than the cell walls. Among possible roles suggested for potassium are those of controlling water loss from plants and in part of the photosynthetic process.

Treatment. Where potassium alone is deficient, apply sulphate of potash before sowing and rake in thoroughly. Where nitrogen and/or phosphorus are also deficient, NPK fertilisers containing appropriate proportions of the three elements should be applied in accordance with manufacturer's instructions.

CALCIUM   Pl. 22

Calcium deficiency symptoms occur most commonly in plants growing on acid peats (i.e. not fen peats) and in acid soils originating from rocks of low calcium content, such as many granitic types and silica sandstones. They occur also on light, free-draining and easily leached sandy soils; less usually so on poorly draining clays, but rarely on alkaline soils derived from limestone or chalk. Although external leaf symptoms of calcium deficiency are rarely seen, a number of disorders arise from inadequate calcium levels in fruits, roots and also in the heart leaves of leafy vegetables. Plants commonly affected by clearly recognisable and distinct symptoms are apples, Brussels sprouts, cabbages, carrots, celery, chicory, lettuces, peppers, potatoes and tomatoes. It is probably extremely rare for there to be an actual soil calcium deficiency, the effects almost always being induced by an imbalance of calcium with other elements (see below) or a restriction on the availability of calcium to the plant because of some interfering environmental factor. Many plants, commonly termed calcicoles, require the particular iron-calcium relationship that occurs in calcium-rich soils while conversely those known as calcifuges require a different relationship between these elements and do not thrive under such conditions. Consideration of these falls outside the scope of this book however and horticultural and gardening texts should be consulted.

Symptoms. GENERAL: when foliar symptoms occur they are most common at young leaf or shoot tips which may be curled inwards or ragged, scorched and killed. APPLE: fruit; the commonest symptom is bitter pit – dark spots or pits on the surface with brownish spots beneath and also scattered throughout the flesh. Bitter pit is most frequent on young, vigorously growing trees and other symptoms on apples include glassiness (translucency of the tissues), lenticel spotting and low temperature breakdown (the death of fruit tissues when stored under too cold conditions). BRUSSELS SPROUTS (and sometimes cabbages): internal browning; a very common browning and death of the tips of some of the internal leaves of the buttons or head. CARROTS: cavity spot; elongate transverse spots on the roots which crack open to reveal small craters, is sometimes thought to be due to calcium deficiency but the condition is imperfectly understood and bacteria have also been claimed to be associated with it. CELERY AND CHICORY: black heart; usually only seen on fen soils, the central leaves of the crown becoming blackened and stunted. LETTUCES: tip burn; marginal leaf scorching is common in lettuce and calcium deficiency may be one cause; attack by bacteria, grey mould or other fungi may be others (see p. 409). POTATOES: the commonest symptoms are the production of 'leggy shoots' with rolled leaves and many very small tubers. TOMATOES (and to some extent peppers): blossom end rot; a more or less rounded dark brown gradually shrinking and toughening area of skin at the blossom end of the fruit, often induced during periods of water shortage. Calcium deficiency can also be a contributory cause towards splitting in tomatoes (see p. 488).

Physiology. Calcium is relatively immobile in tissues, not moving readily from old to young leaves, which can therefore be 'starved' of the element and display the deficiency symptoms. It is a vital component of plant cell walls and is also essential for the correct functioning of growing points and for several aspects of root growth. Within plants, it is generally in greatest concentration in the older leaves with less in young leaves and fruit. There is a complex interrelationship between calcium availabiity and that of other elements. For various reasons excess potassium, and to a lesser extent magnesium and ammonium-nitrogen, in the soil decrease the availability of calcium to plants. Other factors limiting calcium availability include water stress and high humidity. Boron deficiency can also lead to calcium deficiency symptoms, probably because boron is needed for the transport of calcium within plant tissues.

Treatment. This is not easy because of the interrelationship with other factors. In acid soils, apply lime to raise the pH to about 6.5, although this will not help on sites where calcium is not deficient, merely unavailable to the plants. It is important to appreciate that the effects of liming are not quick to appear. In gardens the best plan is to arrange for a pH test to be conducted professionally (or even by using one of the kits sold for garden use) and then apply lime in accordance with the recommendations issued for different soil types by the Ministry of Agriculture. It is important however not to apply lime to areas where calcifuge plants or soft fruit are to be grown and, except for brassicas on clubroot-infested soil, a pH of about 6.5 is probably sufficiently high for most vegetables. The time of application of lime is not very critical but hydrated lime should not be applied immediately before planting as it can be harmful to plants. Calcium deficiency is also minimised by reducing applications of potash and ammonium, while symptoms such as bitter pit in apples or blossom end rot in tomatoes can sometimes be cured by spraying at fortnightly intervals with solutions of calcium nitrate or calcium chloride containing approximately 2 g/l.

MAGNESIUM   Pl. 22
Commonly deficient on light, acid, sandy soils and others where calcium deficiency also occurs. Magnesium is readily leached and so the effects are worst in wet seasons and also after excessive applications of potassium fertilisers. Symptoms are commonly seen on many plants but are particu-

larly noticeable on apples, tomatoes, some brassicas, annual bedding plants, lettuce and potatoes.

Symptoms. Symptoms vary greatly between plant types but are most common late in the season. In general the most characteristic symptom is that of yellowing between the leaf veins, giving a marbled effect, the veins and areas immediately adjacent remaining green. Red-pigmented plants, such as beetroot, may turn purplish instead of yellow and similar colours can arise on normal green plants as the symptoms progress. The symptoms always arise first on the older leaves (cf. iron deficiency below) and sometimes, especially on peas and beans, the marbling may be confined to the central parts of the leaf, the margins remaining relatively normal. On apples severe premature defoliation can result.

Physiology. Magnesium is a constituent of the green colouring matter chlorophyll and also has several other metabolic functions in plants. It is very mobile in plant tissues and this explains why deficiency symptoms arise first in the older leaves, the magnesium having been transported to the young tissues during periods of shortage.

Treatment. Because magnesium does not usually affect yield (except perhaps on apples), treatments are usually applied only to improve appearance. As calcium is also commonly deficient with magnesium, the use of magnesian limestone when liming should correct the problem.

MINOR NUTRIENTS

IRON   Pl. 21

Actual deficiency of iron in soils is uncommon and the presence of iron compounds generally gives the familiar red and brown colouration to many soils. It is often rendered unavailable to plants however, especially on alkaline sites. Deficiency symptoms are thus most commonly seen on soils overlying chalk or limestone while high organic matter content tends to increase its availability to plants and lessen the severity of the symptoms. Many plants can be affected by iron deficiency but in gardens symptoms are most likely to be seen on azaleas and rhododendrons, camellias, ceanothus, chaenomeles, hydrangeas, roses and some fruit trees and soft fruits especially raspberries and strawberries. There may be wide variations in susceptibility between cultivars of some plants.

Symptoms. Familiar as 'lime-induced chlorosis': bleaching or yellowing, usually of the youngest leaves but not affecting the veins, which stand out as conspicuously dark green. Plants may become generally unthrifty and in severe conditions fail to flower or fruit and may die.

Physiology. Iron is associated with several functions in plants but most importantly with the formation of chlorophyll, and plants therefore become pale when it is absent. It is relatively immobile in plants and the bleaching symptoms can sometimes arise because of an increase in this immobility. Although the presence of abundant calcium in the soil induces iron deficiency symptoms, the mechanism of this process is not fully understood. High concentrations of a number of other elements, notably phosphorus, can also bring about iron deficiency.

Treatment. Mere application of inorganic iron compounds to soil on which deficiency symptoms occur rarely improves matters as the added iron will itself become unavailable. To eliminate the causes, liming must be severely restricted, although of course on naturally alkaline soils the lime content cannot readily be reduced. Superphosphate fertilisers should be used sparingly on iron deficient plants or on those susceptible to such deficiency. The most satisfactory treatment is to apply iron in a form in which it is available to plants; such compounds are on sale to gardeners and are known as chelates or sequestrenes. In these compounds, the iron is bound to organic molecules which largely prevent its reaction with other soil chemicals and thus its unavailability. Sequestrenes should be applied to the soil or sometimes as foliar sprays in accordance with manufacturer's directions.

MANGANESE Pl. 22

This element is deficient on a wide range of soil types but mainly on poorly draining sands, highly organic soils, such as fen peats and marshy land, and other wet areas of low acidity. A wide range of plants is affected, including beetroot, brassicas, parsnips, peas and beans, potatoes, spinach and several types of fruit trees and bushes.

Symptoms. Variable but there is commonly some inter-veinal yellowing which usually appears first on the older leaves. Patches of dead tissue may appear among the yellowed areas and on some plants, such as potatoes, there may be a general paleness to the younger foliage and also an upward rolling of the leaves, although this alone can arise from a variety of other causes. Peas, and to a lesser extent beans, develop a symptom known as marsh spot – more or less circular brown areas seen within the seed when the two cotyledons are pulled apart. (But also see: broad bean stain viruses p. 438).

Physiology. This is imperfectly understood but one role is associated with the formation of chlorophyll. Manganese deficiency may be induced by an excess of soil iron and vice versa, although deficiencies of both elements do frequently occur together.

Treatment. Affected plants can be cured by spraying with a solution of manganese sulphate at the rate of $1.5g/l/2m^2$. It is difficult to correct soil manganese deficiency permanently but over-liming of susceptible soils, such as those listed above, should be avoided.

BORON Pl. 22

Boron is sometimes deficient in soils derived from parent rocks, such as granites, which are low in boron content, or in freely-draining sands from which it is readily leached. More commonly, boron is present but unavailable either because it occurs in minerals, such as tourmaline, which plants are unable to utilise or because the soils have a high lime content. Deficiency symptoms commonly arise therefore after liming or in dry summers following wet winters or springs. Adequate boron supplies occur in soils derived from marine sediments and when the soil is enriched with the remains of boron-containing plants. The importance of boron is linked with the availability of calcium (see p. 476) and deficiency symptoms of the two elements are sometimes similar. Plants in which boron deficiency symptoms are most marked include apples, beetroot, cauliflowers and other brassicas, celery, lettuce and sweet corn.

Symptoms. GENERAL: as with calcium deficiency, the symptoms commonly appear as distortion, blackening or death of young growing points of leaves and shoots while leaves flag and become yellow and scorched. Root growth may be seriously retarded. APPLE AND PEAR: there may be some dying back of shoots in spring giving a tufted or rosette appearance but this can also arise from other causes. The leaves are narrowed and thickened and may have smooth margins while cracking and corky patches may develop on the surfaces and within the fruit (corky pit) (but also see viruses p. 435, and scab p. 335). BEETROOT, SWEDES AND TURNIPS: rough patches form on the skin surface and watersoaked brown or sometimes black patches and rings develop within the heart of the root tissues (a symptom known in some areas as 'raan'). The growing point of the shoot is shrivelled or otherwise distorted. Frequently there is a corky development on the surface of shoots, leaf stalks and other areas. CAULIFLOWER: water-soaked and later brown patches develop on the curd and some leaves may be stunted and brittle. When cut lengthwise, brownish patches and hollows are seen within the curd and pith and the curd taste is bitter. The pith of Brussels sprout and cabbage stems may similarly be brown and hollowed. CELERY: general unthriftiness with yellowing of the leaves and brown corky mottling and cracking of the leaf stalks, usually on the inner side. LETTUCE: indistinguishable from the symptoms of calcium deficiency (p. 476). SWEETCORN: cobs usually fail to develop, the growing point dies and transparent or whitish stripes appear on the leaves. Similar leaf symptoms are however also caused by frit fly (see p. 216).

Physiology. Imperfectly understood but boron is relatively immobile in plant tissues and performs many functions, including roles in the regulation of water uptake and of the germination of pollen; it may be concerned in some way with hormone control systems and probably with the movement of calcium, this explaining the close inter-relationship between symptoms caused by deficiencies of the two elements.

Treatment. Soil deficiencies can be corrected by raking in borax at the rate of 3 g/m^2 at planting time but although the effects are rapid, the chemical will not persist long in the soil and repeated applications may be necessary from year to year.

COPPER

Deficiencies can occur on highly organic soils, such as peats, and are relatively unusual on mineral soils. Heathland and chalky soils in Dorset, Wiltshire and S. Devon often give rise to deficiency symptoms. A wide range of plants may be affected but onions, peas and beans, tomatoes and some fruit are most likely to show symptoms, which arise first on the younger leaves. The foliage may turn blue-green and later yellow and be partly wilted. On fruit trees, the shoot tips may die back but all symptoms are particularly difficult to diagnose and in serious suspected cases expert opinion should be sought. Although much has been suggested, there is little evidence for the precise role of copper in plant nutrition but deficiency effects in gardens can be corrected by spraying growing plants with 5 litres of 0.05% copper oxychloride solution per 10 m^2.

MOLYBDENUM
Pl. 22

Rarely deficient in soils but sometimes rendered unavailable to plants on acid sites. Many annual plant species can be affected but in Britain the symptoms are only likely to be seen on broccoli and cauliflowers.

Symptoms. There is yellowish mottling of the leaves and suppression of growth and/or death of the leaf blade, giving rise in cauliflower to the 'whip-tail' effect. In many other plants the growing tips distort and die but this symptom is far from diagnostic.

Physiology. Molybdenum is needed in plants for the reduction of nitrate to ammonium, and also for nitrogen fixation by the root nodule bacteria of leguminous plants (see p. 428).

Treatment. On acid sites, increase pH by lime application (see p. 476) or water affected plant beds after seedling emergence with a solution of ammonium or sodium molybdate at the rate of 2.5 g/0.5l/m^2.

NON-NUTRITIONAL DISORDERS
(Plates 23–24)

This section includes the many physical factors of the environment, other than nutrients, that can affect plants adversely. Many are climatic such as wind or frost; others are man-made, such as pollutants or mechanical injuries, and some arise from combinations of these factors. Greatest emphasis is given in this account to those agencies which are, like pollutants and frost, very commonly harmful to plants. Other factors, such as soil moisture, the regulation of which falls more under the heading of general plant husbandry, are considered only in as far as they can give rise to well defined symptoms of disorder. For convenience, two special categories are also included here: one covers the genetic disturbances that arise through some inherent aberration in the plants while the other is described as disturbed growth phenomena and includes those problems that arise through particular combinations of factors which are commonly manifest as unseasonal growth patterns.

Particular mention must also be made of glasshouse and, more especially, of house plants since these are grown in an artificial environment, free from the likelihood of lightning, precipitation or wind injury but still affected by the temperature, moisture, light and atmospheric purity of their surroundings. Although it is often possible to pinpoint the causes of poor growth or abnormal symptoms on house plants, they frequently arise through combinations of factors. The most common cause is waterlogging but fluctuating temperatures, shortage of water, too much or too little light, incorrect nutrient status of the growing medium and any of a range of household chemicals from gas to hair lacquer can give rise to particular effects. It must also be stressed that house plants embrace a very wide range of plant types (far more than can be described in detail in this book), each with its own environmental requirements. Reference should be made therefore to specialist books which describe in detail the growing needs of each type.

DROUGHT   Pl. 23

Water is necessary for the growth of all plants, both as an essential chemical in its own right and also as the solvent within which other nutrients are dissolved. Its shortage in the soil or sometimes in the air can result in conspicuous damage or ultimately death and plants with large, smooth leaves, such as many vegetables, are particularly prone to damage from such shortage. Shortage of soil water may result from insufficient rainfall or poor water retention by the soil while a dry atmosphere can be caused by low humidity, wind or high temperature. This then causes dessication in the plant because it loses through its leaves more water than it is able to replace from its roots. Hence wilting occurs. The related phenomenon of physiological drought arises when there is plenty of water in the soil but the plant is unable to extract and transport it sufficiently fast to prevent damage occurring to the above-ground parts. Such conditions can arise when strong or warm drying winds are prevalent or when the soil is frozen.

Symptoms. SOIL DROUGHT: a general dullness of the leaves followed by wilting is the first and most obvious symptom. As water shortage continues, leaves and roots become toughened and leaves gradually turn brownish. In severe conditions they may drop and plants run to seed and/or die. A common symptom on tomatoes and capsicums is blossom end rot, a problem caused by an imbalance of calcium within the plant, which is very often brought about by water shortage; a dark-coloured sunken area arises at the blossom end of the fruit. DRY AIR: in many respects the symptoms are similar to those induced by soil drought but the browning or scorching of the leaves is the prevalent symptom (see also under wind damage p. 486). Other common effects resulting from a dry atmosphere are the dry set conditions, especially frequent on tomatoes, in which the fruit fails to develop beyond a few millimetres in diameter, and also the shedding of flowers and buds which occurs on many house and glasshouse plants.

Treatment. Maintain adequate but not excessive watering of plants at all times. Take particular care at the time of flowering and when the fruit are forming and pay special attention to tomatoes or other plants in growing bags as the compost can easily dry out unseen. Mulches should be used on sites, such as those with light or shallow well-drained soils, where drying out of the land is likely to occur. Dry air and its effects cannot be avoided outdoors but in glasshouses spray plants such as tomatoes, cucumbers and vines at flowering time with water in morning or evening to aid pollination.

FROST   Pl. 23

Several phenomena associated with low temperatures can cause damage to garden plants and each results in a particular type of injury. By far the most important is frost, although its symptoms and effects are complex and can be considered very briefly. Other related and damaging phenomena are winter-cold, glazed frost, snow and hail.

Symptoms. Damage due to frost is usually self-evident and plants not normally harmed by the severities usually experienced in particular areas during the winter are said to be hardy under such conditions. The above-ground parts of non-hardy types rapidly blacken, wilt and shrivel on exposure to the first autumn air frost. Especially early autumn frosts can cause blackening or dying back of still-growing shoots of hardy species but this is relatively unusual. Of much greater importance is damage caused to hardy and non-hardy plants alike by late spring frosts. Even in southern England, frosts are not infrequently experienced in May and they damage young growing leaves and especially flowers of a wide range of plants while if such frosts occur immediately before bud burst the buds may be killed. Leaves may be merely scorched at the edges or killed but do not usually drop. Shoots may be scorched at the tips or die back to the old wood and flowers are usually killed outright. Repeated spring frost damage to some trees can render them permanently dwarfed, with an appearance reminiscent of browsing damage, while sometimes, although relatively unusually, stem and branch cankers or splitting can arise. An enormous range of plants is prone to spring frost damage but of outstanding importance among hardy trees and shrubs are blossoms of both top and soft fruit (but also see apple bud rot p. 345), and shoots, leaves or flowers of beech, camellias, ceanothus, eucalyptus, larches, magnolias, skimmias, sumac, viburnums, and walnuts. Severe virus infections very commonly render plants more liable to be injured or killed by frost.

A rather different type of frost injury is frost lift or frost heaving; the forcible ejection of plants from frozen soil. This commonly occurs with small or seedling plants, few of which are likely to be in the open ground during the winter, but may be a problem in nurseries where young woody plants are being raised. It can also sometimes be a problem, especially on highly organic soils, with over-wintered vegetable crops which can suffer severe damage through the roots being broken.

Prolonged exposure of potato tubers, particularly during storage, to temperatures slightly above freezing commonly gives rise to chilling injury which appears as small diffuse, blackish patches in the flesh. The causes of all such discolourations of tuber flesh are not easy to determine however and viruses (p. 461) are another possibility.

Biology. The meteorology of frost causation is complex and beyond the scope of this book. The mechanism of frost hardiness in plants is also complex but depends on the amount of water that can be drawn from the protoplasm of the plant's cells. Frost damage occurs largely through the formation of ice crystals in the plant's tissues; this can only happen if water can be drawn from the cells to be frozen. In hardy plants very little such water can be drawn out during certain times of the year. There is no relationship between winter and spring hardiness however; a plant with little 'available' water in winter may be physiologically very different in spring and so quite liable to be damaged at that time.

Prevention. A number of measures can be taken to minimise freezing injury to plants. Species hardy for local conditions should always be chosen; nurserymen or experienced local gardeners can advise on the selection. Certain sites are especially prone to frosts; often only experience will indicate where frost-hollows (which may be as large as a valley or small as part of a garden) occur. Such areas should naturally be avoided for more sensitive plants. Notoriously frosty areas in England include the Breckland of East Anglia, the Cheshire plain and the Vale of Evesham. Many methods of physical protection outdoors can be employed, including cold-frames, earthing-up or covering with straw or hessian; note however that plastic covered cloches offer little protection and indeed can become colder inside than out. Large heaters may be used to protect blossom in extensive and valuable orchards but few gardeners would be likely to undertake such precautions. Shelter belts of hardy species are of limited value; they will protect plants from wind or air frosts when cold air is transported over land but not from radiation or ground frosts (such as most spring frosts) when freezing occurs because of the rapid loss of heat from the soil. Relatively few plants possess resistance to freezing injury but indirect benefit is achieved when, for instance, later flowering types can be used to avoid spring frost damage; a number of fruit trees and bushes are now being selectively bred for this purpose.

HIGH TEMPERATURE   Pl. 24

Can induce a variety of disorders in a wide range of plants. The two most common conditions are both known as sun scald. One form arises when hot sun strikes the bark of thin barked trees, such as beech, cherries, maples or poplars and results in its death either in patches or down the entire south or south-west face of the trunk. The other form of sun scald occurs when fruit are similarly struck by hot sun; apples, gooseberries, grapes, pears and tomatoes are most commonly affected. On tomatoes also, a condition known as greenback is common; the fruit develop a hard green or sometimes yellow area on the side where it is struck by high sunlight. Occasionally leaves, especially of glasshouse or house plants, can be scorched in a similar way and if flower bulbs are exposed to hot sun after lifting they may be damaged and fail to flower as a result. Most commonly the scorch symptoms result in papery, pale brownish patches on the affected parts. The problem can be minimised on glasshouse plants by adequate ventilation and particularly by summer shading. Greenback incidence on tomatoes can also be lessened by not over-stripping the foliage and thus exposing the fruit to too much sunlight. Additionally cultivars bred with greenback resistance may be chosen; those derived from 'Moneymaker' and most F_1 hybrids have this attribute.

A second type of high temperature injury is caused to the stem bases of plants by hot soil. On woody plants the symptom is usually known as heat canker and commonly heals satisfactorily. On other plants, especially on root vegetables, the effect can be very damaging and is usually termed strangles.

LIGHT   Pl. 24

Adequate light is essential for the satisfactory growth of all green plants and its absence, through shading, is the main reason for the death of the lower and inner branches of trees and shrubs and for the yellowing and subsequent general debilitation of sun-loving plants when overhung or overgrown by larger species. There are also occasions when the effects of high light have been thought damaging. In most of these instances, such as greenback development in tomatoes, the damage is probably caused by the high temperatures associated with the high light level. The development of green chlorophyll in potato tubers (which are stem, not root tissues and so potentially photosynthetic), so rendering them inedible is however an almost exclusively light-induced effect. Potato tubers should always be stored in the dark therefore. A curious and unique effect associated with high light intensity (and a certain measure of warmth) occurs when cold water is splashed onto the leaves of saintpaulias and other gesneriaceous plants. This results in ringspot and line pattern symptoms indistinguishable from those induced on many plants by viruses. The problem is remedied by watering such plants only from below and avoiding their exposure to direct sunlight.

LIGHTNING
Pl. 24

Although lightning strikes can damage low growing plants, gardeners are only likely to see the effects on trees. Isolated trees are more prone to damage than are groups and it is sometimes thought that certain species, such as beech, are damaged less frequently. Symptoms are normally either complete shattering of the trunk or a long jagged scar down one side. Trees need not be killed by a lightning strike nor need they fall, although in a garden they should normally be removed as the damage is usually extensive and predisposes them to attack by decay organisms. Although some sites are more prone than others to lightning strikes, only in the most exceptional situations, such as very valuable trees close to buildings, does the risk justify the fitting of lightning conductors.

PRECIPITATION

Hail. Injury caused by hail is usual in most years; soft fleshy plants are particularly prone and symptoms are commonly seen on the leaves of onions, tulips and similar plants as small white flecks. More severe hail storms can cause serious damage to fruit and to young shoots and twigs, on which the bark may be pitted or torn. Damage can be differentiated from that arising from other causes by its presence on one side only of affected plants. On larger woody stems and on fruit, even small hailstone wounds can provide entry sites for decay organisms.

Rain and glazed frost. Although rain normally causes only temporary damage through beating down flowers, it is a very important agent in the dispersal of spores of parasitic fungi; many fruit diseases for example can be minimised by supporting the produce well clear of the soil and out of reach of rain splash and contaminated earth. Glazed frosts, the result of periods of freezing rain, are relatively uncommon in Britain but when they do occur can be very damaging because of the considerable weight that layers of ice can impart to both herbaceous and woody plants. This weight can be sufficient to break branches and may even uproot large shrubs and trees.

Snow. A covering of snow for long periods of the winter is normal in many parts of northern Europe and is an important factor in protecting plants from winter cold. Heavy falls of snow can cause damage similar to that of glazed frost.

WATERLOGGING

Although drought injury occurs commonly on both outdoor and house plants, damage through waterlogging is relatively infrequent outdoors. It is probably the commonest cause of the decline of house plants however and although its effects arise in a number of ways, the most important are through the limiting of oxygen supplies to the roots and the inability of carbon dioxide to diffuse away. The commonest symptom is the yellowing of leaves, sometimes with the development of dry angular blotches and a general growth stunting, but the secondary effect of encouraging the development of root rotting organisms is often more serious. The treatment for house plants is usually simple; check carefully the water requirements for each type of plant – they differ widely but many do not flourish when standing permanently in dishes of water. Outdoors, take appropriate measures to lighten heavy soils and improve drainage. Actual installing of drainage systems is scarcely practicable in most gardens but thorough and deep digging in autumn and working in organic matter will help.

WIND   Pl. 24
Wind may induce damage in a number of indirect ways, such as by the transport of pollutants and by its influence on temperature and humidity. These are described under the appropriate headings but wind may also have a more direct or purely mechanical effect on plants. Wind damage of this type is naturally most common on exposed sites, such as high land and the sea coast, and, not surprisingly, large and/or tender plants are those most affected. The wind effects of most importance to gardeners are:

WINDTHROW: the uprooting of plants, especially trees, by gales. Gardeners will probably consider this beyond their control although it is possible to predict those situations in which it is most likely to occur. Tall trees with large crowns (or, as often happens with elms for example, crowns overgrown with ivy), standing on shallow soils and with defective or small root systems are very prone to windthrow. Trees that have always grown isolated from others will generally be inherently more stable than those either in plantations or from around which neighbouring trees have been removed.

WINDROCK: very commonly occurs on young trees, shrubs and herbaceous plants with a large top in relation to the root system. Plants are moved to and fro in the ground but are not uprooted. The consequent effect can be that the stem base is chaffed and pathogenic organisms invade. If rainwater fills the soil hollow produced at the stem base and then freezes, further serious damage or death can occur. Any plant rocked by strong wind should always be firmed in again promptly.

WINDBREAK: the snapping off of tree trunks or branches or stems of other plants at some distance from the ground commonly occurs after winter gales and if branches are broken off in this way from garden trees, the broken stub should subsequently be sawn off close to the trunk and the cuts treated with a wound sealant containing fungicide. No particular types of tree are especially prone to windbreak although it is very common on beech affected by beech bark disease.

WIND-SCORCHING and WIND-PRUNING: on very exposed sites trees often appear permanently leaning over, as if blown by the wind. This effect is due to the death of buds on the windward side of the trees by the direct drying effect of the wind. The buds are killed by induced localised drought. On other exposed sites, scorching or wind-blast damage caused by similar dry and/or cold wind effects can affect foliage and thus tree growth. It is considered the major factor limiting the growth of forests at high altitude in Britain. On the sea coast it is commonly accompanied by the damaging effect of windborne salt spray. The symptom may appear as an overall browning of the foliage on the side of the plant facing the wind, while on individual leaves the margins may be very markedly more browned than the centres. Sensitive plants can be protected by shelter belts of resistant tree species such as beech, Monterey cypress, Austrian pine or sycamore. Even in relatively sheltered gardens and orchards cold wind is probably the commonest cause (apart from pesticides) of russetting on apples and other fruit (but also see scab p. 335, viruses p. 435, and boron deficiency p. 479). It is scarcely practicable or necessary to attempt to control this in gardens and commercially russetting is only a problem because it affects the appearance and thus the marketability of the produce.

NON-CLIMATIC EFFECTS

DISTURBED GROWTH PHENOMENA   Pl. 23

Several forms of disturbed plant growth can give rise to concern and perplexity in gardens. They are quite different in their causes and effects but are conveniently grouped together although galls, which are a special form of disturbed growth, are considered in detail on p. 328.

Blindness. Arises when growth stops through absence or malfunction of the growing point. Among vegetables it can be particularly troublesome in broccoli, cauliflowers and other brassicas while tulips and other bulbous plants are the ornamentals most commonly affected. The causes of blindness may be varied and damage arising from pest or pathogen attack, nutrient deficiency, waterlogging or dryness are common causes. The exposure of bulbs to high temperatures after lifting or during storage is another frequent cause.

Bolting. The premature flowering and seeding of vegetable plants, which often causes frustration to gardeners when crops are lost in this way before maturity. Vegetables that are very prone to bolting include beetroot, brassicas, celery, lettuce, onions and spinach but the factors controlling it are very complex and vary from crop to crop and even between cultivars of the same crop. In general however, in order to flower, such plants require exposure to cold temperatures at a certain critical stage of their growth and for certain lengths of time. Once they have been exposed in this way, nothing, not even prolonged periods of high temperature, can prevent flowering. There is no real way to prevent bolting although some cultivars have been selected for resistance to it and may be so indicated in seed catalogues. In general however, early cultivars will bolt much more readily because they are more likely to experience the appropriate cold temperatures. Bolting will be less of a problem in a season following a mild and early spring and more serious after a late and cold one. Brassicas for spring planting, if raised under glass, will be more prone to bolting than plants raised outdoors if low temperatures occur in late spring as they will be further advanced and will have already reached the critical growth stage. It must be added, however, that day-length also influences the onset of flowering and can interact with, or modify, the temperature-induced response.

A rather different type of bolting is that commonly arising during periods of dry weather when plants are under stress. A natural response of most plants under such conditions is to flower and produce seed but the likelihood of this can be lessened by giving them additional water at such times. Conversely, ornamental plants that fail to flower often do so because they are too well looked after, i.e. not placed under sufficient stress, and the withholding of water from them can often induce flower formation.

Oedema. Sometimes known as dropsy or intumescence; a rough and warty condition in which small outgrowths develop on the undersides of the leaves and on the stems of a wide range of plants, including begonias, brassicas, cacti, camellias, capsicums, pelargoniums, peperomias, solanums, tomatoes and vines. It occurs most commonly in glasshouses as a result of overwatering and too high humidity or other factors leading to an excess of water in the plant. This water excess induces extra growth of discrete patches of cells and so the formation of the warts. Oedema can be prevented by a reduction in watering and improved ventilation.

Splitting. Very common on many plants but especially so on cabbages, carrots, celery, cherries, onions, parsnips, plums, potatoes, swedes and tomatoes and is most usually brought about when rapid growth is induced by the onset of wet conditions following a long dry spell. Related symptoms include the development of hollow fruit on apples. The only way that these conditions can be minimised is by improvement of the water retention of the soil by incorporation of organic matter, and by regular and careful watering (if permitted) during periods of drought, although cultivars of some plants have been bred with resistance to splitting and may be so indicated in seed catalogues.

Woody shoot malformations. A number of sometimes bizarre shoot malformations on woody plants may cause concern although none is harmful. Flattened stems, sometimes known as fasciation, occur occasionally on a wide variety of trees. Deeply ridged bark on young shoots, also known as winged cork, is common on field maples and on elms, especially as hedgerow plants. Spiral growth is also fairly common on many trees and shrubs and some species when so affected are deliberately propagated as 'contorta' cultivars. A symptom superficially like spiralling also occurs when stems are entwined with honeysuckle (p. 427). Bursting into life of dormant buds occurs in the form of lammas shoot production on oaks and other trees late in the season, and also as epicormic shoot growth, particularly frequent on larches affected by canker and/or dieback.

GENETIC DISTURBANCES Pl. 23

A few types of plant malformation are genetic in origin; they are natural in the sense that they are produced under the plant's normal genetic control but may only appear under certain environmental conditions, following a particular breeding history or a particular genetic mutation. Among examples commonly seen in gardens are the hard swellings on swede and occasionally turnip roots, often mistaken for clubroot and known as hybridisation nodules; very pale coloured or albino seedlings that only survive until the food reserve in the seed is exhausted – lacking chlorophyll they are unable to produce their own nutrient; and silvering in tomatoes which appears as silver-green patches on leaves and stem with sterility of trusses on affected plants.

MECHANICAL INJURIES   Pl. 23

Most of the damage in this category could be described as originating from maltreatment as it is usually induced by avoidable actions by unthinking or uncaring persons. The question of subsequent treatment or reparation usually only arises with trees and shrubs and although it is inappropriate to consider in detail all features of the incorrect husbandry of such plants, the following commonly cause trouble in gardens.

Covering of bases of established plant stems. Compacted soil, asphalt or worse, concrete (such as that laid for paths) will upset the water-air balance in which a plant and its root system have become established and may well give rise to asphyxiation and death of the tissues. Symptoms usually appear as a wholesale dying back of the plant and although removal of the asphyxiating surface, if done sufficiently early, may result in recovery, it frequently does not.

Incorrect planting. Constriction of roots can result in the development of a very shallow root system and consequent instability while roots badly arranged in the planting hole can subsequently encircle the stem base and cause a girdling dieback. Soil should be sloped away from newly planted trees to prevent rainwater from collecting and either giving rise to injury when it freezes or predisposing the stem base to attack by decay organisms.

Pruning damage. Incorrect pruning can be very harmful, especially if branch stumps are left unprotected; these will die and provide entry points for decay organisms. Inadequate pruning can result in a mass of twigs which leads to general unthriftiness and provides conditions in which many pathogenic organisms can flourish. Pruning should be performed carefully, at the appropriate season for each plant and large wounds should be dressed with a sealant containing fungicide.

Wounding. From any cause, ranging from penknives to over-tight support wires, can injure trees, and shrubs. Any agency, such as lawnmowers or gates, that physically strikes plants is likely to remove bark and/or other tissues and so enable decay organisms to enter. Support wires, stem ties or fencing nails can cause damage which results in a gradual dieback. Any constricting agencies should be removed and stem wounds should be cut clean and treated with a wound sealant. Such wounds may be mistaken for cankers but unlike them do not usually display any annual extension of damage. Careless handling of fruit and vegetable produce, especially at harvest time, will almost always lead to injuries and subsequent losses through decay if they are later stored.

POLLUTANTS   Pl. 24

Many chemical substances can damage plants. While it is commonly impossible even for experts to determine the specific causes of chemical damage solely on the basis of symptoms, it is useful for gardeners to be able to distinguish such damage from other non-chemical causes. Although it is also usually impossible to avoid the effects of air-borne pollutants, resistant plants may sometimes be selected. Many chemicals harmful to plants are commonly used in the home and garden however and knowledge of their effects may not only explain otherwise puzzling damage but may enable its recurrence to be prevented. It is convenient therefore to divide pollutant chemicals into those that are air-borne or widespread and those that are more localised but because of the frequent similarity of symptoms caused by widely different chemicals, a considerable measure of commonsense is necessary in considering possible causes. On exposed coastal gardens, subject to seawater spray but far from towns for instance, marginal scorching of leaves is more likely to be caused by airborne salt or cold wind than by fluorides produced in a chemical factory.

Widespread pollutants

The commonest of the widespread pollutants in Britain and northern Europe are probably carbon monoxide, sulphur dioxide, nitrogen oxides, chlorides (including common salt, sodium chloride), fluorides, ethylene, various hydro-carbons, ammonia, ozone, aldehydes, arsine, phosphine, pesticides and herbicides (especially those containing copper, polysulphides or insecticidal oils or emulsions), various heavy metal salts and dusts such as smoke and cement. Many of these materials can be produced by a wide range of industrial processes and a general indication of their presence (especially that of sulphur dioxide) can be gained by an examination of the type of lichen growing in the area (see p. 425). Automobile exhausts are another common source of several pollutants. Airborne pesticides and herbicides commonly arrive in gardens as drift from agricultural spraying but their effects are described more fully under localised pollutants (below). As a widespread pollutant, common salt is usually only a problem when it is blown by wind as seaspray onto coastal plants but it can have a more local importance in gardens (below). Sulphur in the atmosphere can actually have a beneficial effect through its fungicidal action on certain plant diseases, such as the well known instance of its suppression of blackspot of roses in industrial areas.

Symptoms and Physiology. Symptoms are commonly divided into chronic, where tissue is injured but not killed, usually by exposure to low levels of pollutants for long periods, and acute, where tissues are killed, often by a brief exposure to a high level of pollutant; pesticide damage for instance usually falls into the latter category. Despite the wide variation in symptom types, the commonest are necrotic or bleached flecks or stipples on leaves, especially on the upper surfaces and between the veins, or broad bands of marginal necrosis or scorching. Premature leaf fall is common. These symptoms, induced by pollutants, can sometimes occur uniformly over plants, but wherever any type of leaf injury is markedly unidirectional in its disposition (especially on the side facing the prevailing wind) and/or extends over several species, an airborne pollutant should be immediately suspected. Young plants are generally more susceptible than older ones and evergreen more so than deciduous types while damage is greatest close to the source of the pollution and usually in spring. The mechanism of the injury is as varied as the chemicals causing it.

Prevention. Real prevention depends on removing the source of the pollutants but with widespread materials this is impossible and the use of plants resistant to damage either for their own sake or as wind-breaks is the best that the gardener can hope to achieve. There is considerable argument over the relative susceptibilities of plants to different chemical pollutants but the following list includes some generally thought to have well-defined responses to chronic pollution injury and may be useful in badly polluted areas. Few plants are likely to resist acute injury. Unlikely to develop pollution damage: apples, aquilegias, cabbages and Brussels sprouts, ericas, gladioli, hornbeams, laburnums, lilacs, limes, marguerites, planes, privets, prunus, roses, sycamores and sugar maples, thujas, and tsugas. Likely to develop pollution damage: asters, barberries, beans, beech, beet-root, begonias, birches, cauliflowers, chrysanthemums, cinerarias, crocus, dahlias, dianthi, firs, freesias, gooseberries, hyacinths, irises, larches, let-tuces, lupins, narcissi, nerines, some types of oak, onions, passifloras, petunias, many types of pine, potatoes, primulas, salvias, scillas, spruces and tobacco. In addition, the following trees and shrubs are usually re-sistant to salt-spray damage and may be useful in coastal gardens: Mon-terey cypress, escallonias, hebes, holm oaks, hydrangeas, olearias, Austrian and Monterey pines and tamarisks.

Localised pollutants

Although the effects of these are commonly similar to those of widespread pollutants, the symptoms are often more uniformly distributed on vege-tation. Among the agencies commonly causing such damage in gardens are tar, hot water, paint (quite common in newly painted glasshouses), zinc (damage can occur on strawberries for example if a cage of new galvanised wire is erected over them), creosote (near to fences), oil and petrol (very often spillages from lawn mowers cause bare patches on lawns), urine (especially on lawns), fertilisers used in excess, salt used on roads or garden paths in winter (a common cause of injury to or death of roadside trees and hedges or plants in borders adjoining paths), or gas (from leaking mains; sections of hedges sometimes die when a leak from an underground main kills part of the root system).

FUNGICIDES AND INSECTICIDES. Although most agricultural and garden pesticides are intended for application onto growing plants, damage can ensue if they are used on plants for which they were not intended, at too high a concentration, under inappropriate climatic conditions or too fre-quently. Most pesticides are therefore potentially phytotoxic or damaging to plants. Damage in gardens from such pesticides may be due to spray drift from nearby agricultural land but frequently misuse of pesticides by gar-deners themselves can cause damage. Rules that should be observed when using pesticides are described on p. 32 and the necessity of adhering to them cannot be overstressed. Symptoms of pesticide damage are generally very similar to those of other airborne pollutants (see p. 490).

HERBICIDES. Herbicides (weed-killers) differ from insecticides and fungi-cides in that they are intended to kill plants. Problems arise in gardens when weed-killers are inadvertently blown or sprayed onto crop plants or are misused in some way. Where the effects are not fatal to the plants, the symptoms are similar to those of other airborne pollutants, although one group of herbicides produces quite different effects and may cause damage much more insidiously. These are the hormone herbicides which are ab-sorbed into plant tissues and cause abnormal growth; leaves may become narrowed (sometimes extremely so), straplike and thickened; stems twis-ted, callused and contorted and fruits set poorly and become irregularly shaped. Such herbicides may occur in gardens as spray drift (from farms or from within the garden itself if they are being used on lawns), or as residue in inadequately washed watering cans. A very common additional source of damage is from contaminated straw, derived from cereal crops on which the herbicides have been sprayed. The symptoms then develop on plants on which a straw mulch has been used or if farmyard manure containing such contaminated straw has been applied to the soil. Tomatoes and cucumbers are especially sensitive to this type of damage. In such circumstances all soil contaminated with the straw should be removed to at least a spade's depth and if the problem persists in glasshouses it may be worthwhile growing the next one or two crops in containers to give the chemical time to disperse from the ground.

Bibliography

Books

The following selected list of books provides a basis for further reading on various specialised aspects of the recognition, biology and treatment of pests, diseases and disorders. Several of the older works are now out of print and/or out of date, especially with respect to control measures, but they are virtually the only sources of more detailed information and may therefore still be worth consulting through public libraries.

General

The natural history of the garden by M. Chinery, 1977, Collins.
A field guide to the insects of Britain and northern Europe by M. Chinery, 1976, Collins.
Beneficial insects and mites by B. D. Moreton, 1970, MAFF Bulletin 20, H.M.S.O.
Horticultural pests. Detection and control by G. Fox Wilson (3rd edition revised by P. Becker, 1960), Crosby Lockwood.
The pests of protected cultivation by N. W. Hussey, W. H. Read and J. J. Hesling, 1969, Edward Arnold.
The pocket encyclopaedia of plant galls in colour by A. Darlington, 1968, Blandford Press.
Pathology of trees and shrubs with special reference to Britain by T. R. Peace, 1962, Clarendon Press.
Decay of timber and its prevention by K. St. G. Cartwright and W. P. K. Findley, 2nd edition, 1958, H.M.S.O.
Collins guide to mushrooms and toadstools by M. Lange and F. B. Hora, 2nd edition, 1965, Collins.
Introduction to fungi by J. Webster, 2nd edition, 1980, Cambridge University Press.
A textbook of plant virus diseases by K. M. Smith, 3rd edition, 1972, Longman.
Plant viruses by K. M. Smith, 1977, Chapman and Hall.
The diagnosis of mineral deficiencies in plants by visual symptoms by T. Wallace, 3rd edition, 1961, H.M.S.O.

Fruits and vegetables

The pests of fruits and hops by A. M. Massee, 3rd edition, 1954, Crosby Lockwood.
Pests and diseases of fruit and vegetables in the garden by A. M. Toms and M. H. Dahl, 1976, Blandford Press.

Diseases of fruit and hops by H. Wormald, 3rd edition, 1955, Crosby Lockwood.
Diseases of vegetables by L. Ogilvie, 5th edition, 1961, MAFF Bulletin 123, H.M.S.O.

Ornamentals

Pests of ornamental plants by P. Becker, 1974, MAFF Bulletin 97, H.M.S.O.
Garden pests and diseases of flowers and shrubs by M. H. Dahl and T. B. Thygesen, 1974, Blandford Press.
Pests, diseases and nutritional disorders of chrysanthemums by N. E. A. Scopes, 2nd edition, 1975, National Chrysanthemum Society.
Fungal diseases of turf grasses by J. D. Smith, 2nd edition, 1965, Sports Turf Research Institute.
Diseases of bulbs by W. C. Moore, 2nd edition, 1979 revised by A. A. Brunt, D. Price and A. R. Reeves, MAFF Book HPD1, H.M.S.O.

Ministry of Agriculture and Agricultural Research Council publications

The Ministry of Agriculture produces a comprehensive series of Bulletins and Advisory Leaflets. They are intended for use by farmers and commercial horticulturists but contain much information that is of interest to keen gardeners. Some of the relevant Bulletins are noted in the book list above and a full list of current publications can usually be consulted at Government Bookshops or may be obtained direct from MAFF Publications, Lion House, Willowburn Estate, Alnwick, Northumberland, NE66 2PF.

Several of the research institutes supported by the Agricultural Research Council, especially East Malling Research Station, the Glasshouse Crops Research Institute and the National Vegetable Research Station also produce occasional booklets and leaflets that contain useful information for gardeners.

Forestry Commission publications

The Forestry Commission is mainly concerned with commercial forestry but useful information may be found in the Leaflets, Forest Records and Bulletins that it produces. A list of current publications can usually be consulted at Government Bookshops or may be obtained direct from Forestry Commission Publications, Forest Research Station, Alice Holt Lodge, Wrecclesham, Farnham, Surrey, GU10 4LH.

Magazines and Journals

Frequent comment and occasional articles on pests, diseases and disorders appear in the following publications and some of them also run advisory services to answer queries from readers.

Amateur Gardening (weekly)

Garden News (weekly)
Popular Gardening (weekly)
The Garden (monthly Journal of the Royal Horticultural Society)
Gardening World (monthly)
Greenhouse (monthly)
Practical Gardening (monthly)
Scottish Gardening (monthly)
Southern Gardening (monthly)

Promotional literature of commercial firms

The major firms engaged in marketing chemicals for pest and disease control and for the treatment of disorders in garden plants produce various leaflets, booklets and wall-charts on these subjects. They are up-dated frequently so are useful sources of information. 'Be your own garden doctor' and 'Be your own vegetable doctor' by D. G. Hessayon are useful illustrated summaries that include both chemical and non-chemical recommendations for treatment.

The British Agrochemicals Association, Alembic House, 93 Albert Embankment, London SE1 7TU publishes an annual 'Directory of garden chemicals' and all BAA member companies listed in this directory will supply charts, booklets or other information.

Societies

Many national and local societies exist to promote general and specialised interest in horticulture and most of these are involved to some extent with problems posed by pests, diseases and disorders. The Royal Horticultural Society, Vincent Square, Westminster, London SW1P 2PE, is the major national society with interests in all aspects of horticulture. Membership is open to anyone and members are entitled to many facilities, including a free advisory service on the diagnosis and treatment of pests, diseases and disorders that is provided by professional scientists working in the Entomology and Plant Pathology Departments at the R.H.S. Garden, Wisley. Much useful information is also provided by articles published in the Society's monthly journal 'The Garden' and in various booklets, of which the following are especially relevant.

Fruit pests diseases and disorders Wisley Handbook 27 (1980)
Vegetable pests diseases and disorders Wisley Handbook 28 (1979)
The fruit garden displayed (revised 1977)
The vegetable garden displayed (revised 1975)
Garden pests and diseases (in the *RHS Encyclopaedia of practical gardening*, 1980)

Other national societies with specialised interests are too numerous to list here but a list, with addresses, is published annually in the R.H.S. gardener's diary.

Glossary

Aeciospore. A type of spore produced by a rust fungus (see p. 267).

Alternate host. Either of the two unlike host-plants of a pest or pathogen that requires both to complete its full life-cycle.

Alternative hosts. Host-plants other than the main host on which an organism (bacterium, fungus, insect, mite etc.) can develop. Weeds and wild plants are often alternative hosts of certain pests and diseases.

Antennae. Paired sensory structures on the heads of insects and other invertebrates. They are often long, thread-like and conspicuous (as in cockroaches, grasshoppers and aphids) but may be small and relatively inconspicuous in many groups of insects.

Ascospore. A spore produced by sexual reproduction in a fungus of the class Ascomycetes.

Basidiospore. A spore produced by sexual reproduction in a fungus of the class Basidiomycetes.

Biological control. Methods developed to control pests by using predators and parasites that feed on them. This may involve introducing and establishing species from other parts of the world. Use of *Phytoseiulus* against glasshouse red spider mites (p. 247) and of *Encarsia* against glasshouse whitefly (p. 158) are examples of the small-scale application of this technique in gardens but many large-scale schemes have been operated overseas to control pests and weeds (see p. 18).

Certified stock. Plant material guaranteed to be of a defined quality, especially of freedom from virus infections.

Chlamydospore. A type of asexually produced fungal spore that is resistant to desiccation or other adverse environmental conditions.

Conidium (pl. conidia). An asexually produced fungal spore.

Cultivar. A variant of a plant species that has arisen (either accidentally or deliberately) in cultivation and is perpetuated by propagation for some desirable horticultural feature (e.g. 'Cox's Orange Pippin' apple). The term 'variety' has a more restricted botanical meaning (see p. 499).

Cocoon. A silken case constructed by an insect larva (e.g. caterpillar) to protect the pupal stage.

Galls. Unusual growths of plant tissues produced as a specific response to attack by various groups of gall-inducing organisms, especially root-knot eelworms (p. 136), gall wasps (p. 229), gall midges (p. 220) and gall mites (p. 252). Galls may develop on roots, stems, buds, leaves, flowers or other parts of plants and they often have a distinctive and elaborate structure.

Honeydew. A sweet excretion produced by aphids, scale insects, mealybugs and other insects that feed on plant sap. It contains sugars derived from sap when the insects feed and it makes infested plants sticky and encourages the growth of sooty moulds (see p. 303).

Host or **host-plant**. A plant on which a pathogen or pest develops. Most pathogens and pests are restricted to a few host-plants on which they can develop successfully.

Host range. The complete range of host-plants on which a pathogen or pest can develop.

Hypha (pl. hyphae). The basic thread-like structure of most fungi.

Immune. Free from infection or infestation; having qualities that prevent the development of a disease or pest.

Infection. The entry of a parasitic organism into a host. A plant may be infected but is not said to be diseased if no visible and damaging symptoms develop.

Infestation. Establishment of pest populations on plants or inside plant tissues.

Larva (pl. larvae). An immature insect. Usually applied to insects that develop through the sequence: egg – larva – pupa – adult (e.g. caterpillars of butterflies, moths and sawflies or maggots of flies). The larva is usually the main feeding stage and is therefore the cause of most damage to plants.

Latent infection. An infection that does not produce obvious symptoms.

Lenticel. One of many pores in the bark of woody plants through which gaseous exchange takes place.

Life-cycle. The succession of stages through which an organism develops, usually between one period of reproductive activity and the next (e.g. egg – larva – pupa – adult in insects or spore – mycelium – fruiting body in fungi).

Mandibles. Horny jaw-like mouth-parts of insects; used to bite and to chew food.

Mould. A vague term applied fairly indiscriminately to the visible mycelial or spore mass of microfungi.

Mouth-parts. The structures that surround the mouth in insects and mites. They are adapted for various functions, especially for biting and chewing solid food or for piercing plant tissues to extract sap.

Mycelium. A mass of fungal hyphae.

Necrotic. Dead and usually dark coloured plant tissues.

Notifiable (pest or disease). Required by law to be reported to officials of the Ministry of Agriculture.

Nymph. The immature stage of those insects that develop directly from nymph to adult without an intermediate pupal stage (e.g. aphids, capsid bugs and related groups).

Oospore. A spore produced by sexual reproduction in a phycomycete fungus of the sub-class Oomycetes.

Ovipositor. The special structure that many female insects have to facilitate egg-laying. This may be relatively simple if eggs are just deposited on plants but in some groups (e.g. sawflies) it consists of a pair of saw-like blades that are used to cut slits in plant tissue before eggs are inserted.

Parasite. An organism living on or in another organism from which it obtains its food and to which it gives no benefit in return.

Parthenogenesis. A type of asexual reproduction in which unfertilised eggs develop normally to produce adults. It occurs quite commonly in aphids, gall wasps and other groups of insects.

Pathogen. A parasitic organism (e.g. fungus, bacterium, virus) that causes disease.

Perithecium (pl. perithecia). A small flask-like body within which asco-spores are produced in some fungi of the class Ascomycetes.

Persistent (of a pesticide). Persisting on plants or in soil for weeks or months after application, during which time it retains some of its pesticidal activity (see p. 28).

Phytotoxic (of a chemical). Injurious to plants. Many pesticides are phytotoxic to some plants (see p. 30).

Predator. An animal that feeds by preying on other animals. The larvae of some insects (e.g. coccinellid beetles, syrphid flies) prey on pests and are therefore beneficial.

Pupa. The non-feeding stage in the metamorphosis of insects that develop through the sequence: egg – larva – pupa – adult. This appears to be a quiescent stage but intense biochemical activity within the pupa re-organises the tissues of the larva to produce the adult, often in only a few days or weeks.

Puparium. A special type of pupa formed during the development of many flies. The general shape is of a cylinder with rounded ends and the fully formed puparium is usually glossy brown or black.

Pycnidium (pl. pycnidia). A small, usually flask-like body, within which conidia are produced in some fungi of the class Deuteromycetes.

Pycniospore. A type of spore produced by a rust fungus (see p. 267).

Race (of a pathogen or pest). A population or strain of a species that can only attack some of the host-plants that are susceptible to the whole species (e.g. races of stem eelworm (see p. 132).

Repellents. Chemicals that are used to repel pests from plants, usually by their unpleasant taste or smell. They are mostly used against birds and mammals.

Resistant (plant). Able to withstand attack by a pest or disease. Many plants are naturally resistant to certain pests and diseases and others have been made resistant through artificial selection and plant breeding.

— (pest). Able to withstand exposure to certain pesticides. Some pests have become highly resistant to pesticides, usually because frequent or continuous treatment of large populations with chemicals selects individuals that have inheritable characters conferring some degree of resistance (see p. 30).

— (spore or sclerotium). Able to survive a period of adverse environmental conditions, such as winter.

Resting (of a spore or sclerotium). Able to become dormant to survive a period of adverse environmental conditions.

Rogueing. The removal and destruction of diseased or infested plants.

Rotation. The alternation, season by season, of different crop plants on a particular area of land to limit the build-up of pests and diseases.

Saprophyte. An organism that feeds on dead organic material.

Sclerotium (pl. sclerotia). A firm, more or less rounded mass of hyphae, often functioning as a resting body (q.v.).

Shot-hole. A symptom of disease in which discrete more or less circular areas of leaf tissue dry and drop out, or a symptom of attack by certain wood-boring beetles on tree trunks and branches (see p. 244).

Siphunculi. Paired tubes near the end of the abdomen of aphids. Shape, size and colour varies appreciably between species. They are the openings of glands that produce waxes and other chemicals that may protect aphids against some predators.

Sooty moulds. Black soot-like growths of fungi that develop on leaves and other parts of plants infested by sap-feeding insects, such as aphids, mealybugs and scale insects. The insects produce a sugary excretion (honeydew) which contaminates plant surfaces and encourages the growth of the sooty moulds (see p. 303).

Spore. A microscopic reproductive structure produced by fungi, bacteria, non-flowering plants (e.g. ferns), and protozoons.

Stag's head. The antler-like appearance of a tree when the upper branches die and are bare of foliage.

Stoma (pl. stomata). A pore in the outer tissues of plants, especially the leaves, through which gaseous exchange takes place.

Swarm spore. A type of spore produced by certain fungi of the class Phycomycetes (and some other organisms) which has limited powers of movement in water films in soil or plant surfaces.

Systemic. Generally distributed within an organism. Used to describe the nature of infection by certain diseases or the mode of action of certain pesticides (see p. 27).

Teliospore. A type of spore produced by a rust fungus (see p. 267).

Tolerant. Able to tolerate attack by a pathogen or pest or action by a chemical. Plants and their cultivars often show different degrees of tolerance.

Urediniospore. A type of spore produced by a rust fungus (see p. 267).

Variety. This term is often used in the same sense as cultivar (q.v.) but should really be restricted to its correct botanical meaning which indicates a variant that has arisen in a plant species growing in the wild and therefore not subjected to selection through cultivation (e.g. *Pinus nigra* var. *maritima*). Many such plant varieties have been taken into cultivation.

Vector. An organism that transports and transmits a pathogen. The most important vectors of plant diseases are aphids and eelworms that transmit viruses but some other pests may also be vectors of certain diseases.

Virulent (of a pathogen). Spreading rapidly and causing serious disease.

Viviparous. Giving birth to live young.

Zone lines. Narrow dark brown or black lines in decayed wood; generally caused by fungi.

General Index

Index of Scientific Names